String-Math 2014

Proceedings of Symposia in

PURE MATHEMATICS

Volume 93

String-Math 2014

String-Math 2014
June 9–13, 2014
University of Alberta
Edmonton, Alberta,
Canada

Vincent Bouchard
Charles Doran
Stefan Méndez-Diez
Callum Quigley
Editors

American Mathematical Society
Providence, Rhode Island

2010 *Mathematics Subject Classification.* Primary 14-XX, 18-XX, 19-XX, 22-XX, 53-XX, 58-XX, 81-XX, 81Txx, 83Exx, 83E30.

Library of Congress Cataloging-in-Publication Data

Names: String-Math (Conference) (2014 : Edmonton, Alta.) — Bouchard, Vincent, 1979- editor.

Title: String-Math 2014 : June 9-13, 2014, University of Alberta, Alberta, Canada / Vincent Bouchard [and three others], editors.

Description: Providence, Rhode Island : American Mathematical Society, [2016] — Series: Proceedings of symposia in pure mathematics ; volume 93 — Includes bibliographical references.

Identifiers: LCCN 2015045551 — ISBN 9781470419929 (alk. paper)

Subjects: LCSH: Geometry, Algebraic–Congresses. — Quantum theory–Mathematics–Congresses. — AMS: Algebraic geometry. msc — Category theory; homological algebra. msc — K-theory. msc — Topological groups, Lie groups. msc — Differential geometry. msc — Global analysis, analysis on manifolds. msc — Quantum theory. msc — Quantum theory – Quantum field theory; related classical field theories – Quantum field theory; related classical field theories. msc — Relativity and gravitational theory – Unified, higher-dimensional and super field theories – Unified, higher-dimensional and super field theories. msc — Relativity and gravitational theory – Unified, higher-dimensional and super field theories – String and superstring theories. msc

Classification: LCC QA564 .S77 2014 — DDC 516.3/5–dc23 LC record available at http://lccn.loc.gov/2015045551

DOI: http://dx.doi.org/10.1090/pspum/093

∞ The paper used in this book is acid-free and falls within the guidelines
established to ensure permanence and durability.
Visit the AMS home page at http://www.ams.org/

10 9 8 7 6 5 4 3 2 1 21 20 19 18 17 16

Contents

Talks from Satellite Events

Preface

The conference 'String-Math 2014' was held June 9–13, 2014 at the University of Alberta. This was the fourth in a series of large meetings exploring the interface of mathematics and string theory. This edition of String-Math is the first to include satellite workshops: 'String-Math Summer School' (June 2–6, 2014 at the University of British Columbia), 'Calabi-Yau Manifolds and their Moduli' (June 14–18, 2014 at the University of Alberta), and 'Quantum Curves and Quantum Knot Invariants' (June 16–20, 2014 at the Banff International Research Station). This volume presents the proceedings of the conference and these three satellite workshops.

For mathematics, string theory has been a source of many significant inspirations, ranging from Seiberg-Witten theory in four-manifolds, to enumerative geometry and Gromov-Witten theory in algebraic geometry, to work on the Jones polynomial in knot theory, to recent progress in the geometric Langlands program and the development of derived algebraic geometry and n-category theory. In the other direction, mathematics has provided physicists with powerful tools, ranging from powerful differential geometric techniques for solving or analyzing key partial differential equations, to toric geometry, to K-theory and derived categories in D-branes, to the analysis of Calabi-Yau manifolds and string compactifications, to modular forms and other arithmetic techniques. The depth, power and novelty of the results obtained in both fields thanks to their interaction is truly mind boggling.

The String-Math series of conferences bring together the leading mathematicians and mathematically-minded physicists working in this interface. They are an excellent vehicle for further promoting such interactions, and for giving attendees greater opportunities to cross cultural boundaries, learn aspects of other fields relevant for their research, and advertise important developments to audiences that might not otherwise hear of them or appreciate their importance. The earlier conferences in this series — String-Math 2011 at UPenn, String-Math 2012 at the Hausdorff Center in Bonn, and String-Math 2013 at the Simons Center for Geometry and Physics in Stony Brook — have helped identify and establish mathematical string theory as a new branch of mathematics, facilitated the entry into the field of young researchers and newcomers, and served to record the state of the art in a rapidly evolving field.

The 'String-Math 2014' conference was organized by Vincent Bouchard (UAlberta), Thomas Creutzig (UAlberta), Emanuel Diaconescu (UAlberta/Rutgers), Charles Doran (UAlberta), David Favero (UAlberta), Terry Gannon (UAlberta), James Lewis (UAlberta), Andreas Malmendier (Colby/UAlberta), Stefan Mendez-Diez (UAlberta), and Callum Quigley (UAlberta). The 'String-Math Summer School' was organized by Jim Bryan (UBC). The 'Calabi-Yau Manifolds and their

Moduli' workshop was organized by Charles Doran (UAlberta), Mark Gross (Cambridge), Shinobu Hosono (Tokyo), Conan Leung (CUHK), James Lewis (UAlberta), and Yongbin Ruan (U Michigan). The 'Quantum Curves and Quantum Knot Invariants' workshop was organized by Vincent Bouchard (UAlberta), Mikhail Khovanov (Columbia), Motohico Mulase (UC Davis), Alexei Oblomkov (UMass), Marko Stovsic (IST), and Piotr Sulkowski (Caltech). These meetings covered a wide array of topics at the interface of mathematics and high energy physics, including:

- Donaldson-Thomas invariants
- Stable quasimaps and Gromov-Witten invariants
- Mirror symmetry, quantization and topological recursion
- Quantization and knot theory
- Mathematical string phenomenology
- Heterotic mirror symmetry
- New and exotic supersymmetric field theories
- Localization and gauge theory
- Gauge theory and Khovanov homology
- Gauge theory angle at integrability
- Conformal field theory
- Homological mirror symmetry
- Gauged linear sigma models
- Categorical constructions of topological field theories
- Non-perturbative dualities, F-theory
- Wall-crossing formulas
- Geometric Langlands
- Arithmetic of strings
- A-twisted Landau-Ginzburg models
- Topological T duality
- String topology
- Elliptic cohomology
- Perturbative amplitudes in gauge theory
- Superstring scattering amplitudes
- Noncommutative geometry

Altogether, the String-Math 2014 conference brought together approximately 130 mathematicians and physicists. There were 25 invited plenary talks given by leaders in both fields. Additionally, there were 22 contributed talks given in parallel sessions. All the talks are available at the conference web site: `https://sites.google.com/a/ualberta.ca/stringmath2014/`.

The math/strings collaboration is clearly here to stay, and we expect this conference series to continue as long the subject remains active and exciting. The venues and years of the first nine conferences of the String-Math series are:

- String-Math 2011, Philadelphia (Penn), June 6–11, 2011
- String-Math 2012, Bonn (Hausdorff center for Mathematics), July 16–21, 2012
- String-Math 2013, Stony Brook (Simons Center for Geometry and Physics), June 17–21, 2013
- String-Math 2014, Edmonton (U of Alberta), June 9–13, 2014
- String-Math 2015, China (Tsinghua Sanya International Mathematics Forum, Sanya, Hainan, China), Dec 31, 2015–Jan 5, 2016

- String-Math 2016, Paris (Institut Poincaré), June 27–July 2, 2016
- String-Math 2017, Hamburg, Germany, July 24–July 29, 2017.
- String-Math 2018, Japan (Tohoku University)
- String-Math 2019, Sweden (Uppsala University)

The conference benefitted from support obtained from the NSF (grant number: NSF DMS 1401390, String Math Conferences 2014), from the Pacific Institute for the Mathematical Sciences through its Collaborative Research Group in Geometry and Physics (2013–2016), from the University of Alberta and from Perimeter Institute for Theoretical Physics.

The editors of String-Math 2014:

Vincent Bouchard
Charles Doran
Stefan Mendez-Diez
Callum Quigley

List of Speakers for String-Math 2014

List of Plenary Speakers for String-Math 2014

Murad Alim
Harvard University

Matt Ballard
University of South Carolina

Christopher E. Beasley
Northeastern University

Tudor Dimofte
Institute for Advanced Study

John Duncan
Case Western Reserve University

Bertrand Eynard
Saclay

Mathias Gaberdiel
ETH Zurich

Davide Gaiotto
Perimeter Institute

Sylvester James Gates
University of Maryland

Jonathan Heckman
Harvard University

Kentaro Hori
Institute for the Physics and Mathematics of the Universe

Sheldon Katz
University of Illinois at Urbana-Champaign

Bumsig Kim
Korea Institute for Advanced Study

Albrecht Klemm
Universität Bonn

Chiu-Chu Melissa Liu
Columbia University

Matilde Marcolli
Caltech

Greg Moore
Rutgers University

David Morrison
University of California, Santa Barbara

Paul Norbury
University of Melbourne

Jonathan Rosenberg
University of Maryland

Yongbin Ruan
University of Michigan

Volker Schomerus
DESY

Christoph Schweigert
Universität Hamburgh

Eric Sharpe
Virginia Tech

Claire Voisin
École Polytechnique

List of Contributed Speakers for String-Math 2014

John Dixon
Tabacon

Richard Eager
Tokyo University

Fabio Ferrari Ruffino
Federal University of São Carlos

Ilmar Gahramanov
Humboldt-Universität zu Berlin

Richard Garavuso
Harish-Chandra Research Institute

Yasuaki Hikida
Rikkyo University

Gerald Hoehn
Kansas State University

Martijn Kool
PIMS/University of British Columbia

Peter Koroteev
Perimeter Institute for Theoretical Physics

Stefan Mendez-Diez
University of Alberta

Hironori Mori
Osaka University

Nicolo Piazzalanga
SISSA/INFN

Helge Ruddat
Johannes Gutenberg-Universität Mainz

Emanuel Scheidegger
Universität Freiburg

Alexander Schmitt
Freie Universität Berlin

Piotr Sułkowski
University of Warsaw

Michel van Garrel
Korea Institute for Advanced Study

Marcel Vonk
University of Amsterdam

Ursula Whitcher
University of Wisconsin-Eau Claire

Simon Wood
Australian National University

Tuna Yildirim
The University of Iowa

Matthew Young
The University of Hong Kong

List of Speakers for the String-Math Summer School

Kevin Costello
Northwestern University

Andy Neitzke
The University of Texas at Austin

Tony Pantev
University of Pennsylvania

Balázs Szendrői
University of Oxford

Eric Zaslow
Northwestern University

List of Speakers for the 'Calabi-Yau Manifolds and their Moduli' Workshop

Xi Chen
University of Alberta

David Favero
University of Alberta

Sara Filippini
University of Zurich

Karl Fredrickson
University of California, Riverside

Sergey Galkin
Moscow State University

Andrew Harder
University of Alberta

Atsushi Kanazawa
University of British Columbia/Harvard University

Tyler Kelly
University of Cambridge

Chiu-Chu Melissa Liu
Columbia University

Andreas Malmendier
Utah State University

Stefan Mendez-Diez
University of Alberta

David Morrison
University of California, Santa Barbara

Howard Nuer
Rutgers University

Andre Perunicic
Queen's University

Mohammad Rahmati
Centro de Investigación en Matemáticas, A.C.

Helge Ruddat
Johannes Gutenberg-Universität Mainz

Alan Thompson
University of Alberta/University of Waterloo

Katrin Wendland
Universität Freiburg

Shing-Tung Yau
Harvard University

Noriko Yui
Queen's University

List of Speakers for the 'Quantum Curves and Quantum Knot Invariants' Workshop

Gaëtan Borot
Max-Planck-Institut für Mathematik

Tudor Dimofte
Institute for Advanced Study

Olivia Dumitrescu
University of Leibniz, Hannover

Pavel Etingof
Massachusetts Institute of Technology

Hiroyuki Fuji
Tsinghua University

Stavros Garoufalidis
Georgia Institute of Technology

Victor Ginzburg
University of Chicago

John Harnad
CRM, Université de Montréal/Concordia University

Lotte Hollands
University of Oxford

Chiu-Chu Melissa Liu
Columbia University

Jun Murakami
Waseda University

Satoshi Nawata
The National Institute for Nuclear Physics and High Energy Physics, Amsterdam

Lenny Ng
Duke University

Paul Norbury
University of Melbourne

Alexei Oblomkov
University of Massachusetts

Anne Schilling
University of California, Davis

Marko Stošić
Instituto Superior Técnico, Portugal

Piotr Sułkowski
University of Warsaw

Katrin Wendland
Universität Freiburg

Proceedings of Symposia in Pure Mathematics
Volume **93**, 2016
http://dx.doi.org/10.1090/pspum/093/01588

All genus mirror symmetry for toric Calabi-Yau 3-orbifolds

Bohan Fang, Chiu-Chu Melissa Liu, and Zhengyu Zong

ABSTRACT. The Remodeling Conjecture proposed by Bouchard-Klemm-Mariño-Pasquetti (BKMP) relates the A-model open and closed topological string amplitudes (open and closed Gromov-Witten invariants) of a symplectic toric Calabi-Yau 3-fold to Eynard-Orantin invariants of its mirror curve. The Remodeling Conjecture can be viewed as a version of all genus open-closed mirror symmetry. After a brief review on mirror symmetry and topological strings on Calabi-Yau geometry, we give a non-technical exposition of our results on the Remodeling Conjecture for toric Calabi-Yau 3-orbifolds.

CONTENTS

1. Mirror Symmetry and Topological Strings on Calabi-Yau Geometry

Mirror symmetry is a string duality relating the A-model on a Calabi-Yau manifold $\mathcal{X}$, defined in terms of the symplectic structure on $\mathcal{X}$, to the B-model on another Calabi-Yau manifold $\check{\mathcal{X}}$ (the "mirror" of $\mathcal{X}$), defined in terms of the complex structure on $\check{\mathcal{X}}$. In particular, mirror symmetry relates A-model topological string theory on $\mathcal{X}$ to B-model topological string theory on $\check{\mathcal{X}}$.

2010 *Mathematics Subject Classification.* 14J33, 14N35, 53D45.
Key words and phrases. Mirror symmetry, toric Calabi-Yau 3-folds, Gromov-Witten invariants, Eynard-Orantin recursion, BKMP conjecture.

1.1. Compact Calabi-Yau Manifolds.

1.1.1. *A-model topological strings on compact symplectic manifolds.* By late 1990's mathematicians had established the foundation of Gromov-Witten (GW) theory, which can be viewed as a mathematical theory of A-model topological closed strings, for any projective manifold [**12**, **13**, **82**], or more generally any compact almost Kähler manifold [**53**, **83**, **105**, **108**], at all genera. (An almost Kähler manifold is a symplectic manifold equipped with an almost complex structure J compatible with the symplectic structure.) GW invariants (A-model topological string amplitudes) are virtual counts of (parametrized) algebraic curves in projective manifolds, or more generally J-holomorphic curves in compact almost Kähler manifolds. In particular, genus-zero GW invariants count rational curves in projective manifolds, or more generally J-holomorphic spheres in compact almost Kähler manifolds.

1.1.2. *B-model topological strings on compact Calabi-Yau manifolds.* The genus-zero B-model topological strings on a Calabi-Yau 3-fold can be described in terms of classical variation of Hodge structures, but a mathematical theory of higher genus B-model on a general compact Calabi-Yau 3-fold was not developed until very recently. In 1993, Bershadsky, Cecotti, Ooguri, and Vafa (BCOV) proposed that the higher genus B-model on a compact Calabi-Yau 3-fold can be constructed from the quantum theory of Kodaira-Spencer gravity [**17**]. In 2011, K. Costello and S. Li initiated a mathematical analysis of the BCOV theory [**85**] based on the effective renormalization method developed by Costello [**34**]. This approach is applicable to compact Calabi-Yau manifolds of any dimension and also yields gravitational descendants.

1.1.3. *Genus-zero mirror symmetry.* In 1991, Candelas, de la Ossa, Green, and Parkes predicted the number of rational curves of arbitrary degree in the quintic 3-fold using mirror symmetry [**24**]. Their stunning prediction motivated the development of GW theory. The genus-zero mirror formula in [**24**] was proved independently by Givental [**54**] and Lian-Liu-Yau [**86**]. Indeed, Givental and Lian-Liu-Yau proved the genus-zero mirror theorem in much more general setting: first for Calabi-Yau and Fano complete intersections in projective spaces [**54**, **86**], and later for semi-Fano complete intersections in projective toric manifolds [**55**, **87**].

1.1.4. *Genus-one mirror symmetry.* In 1993, Bershadsky-Cecotti-Ooguri-Vafa (BCOV) made predictions on genus-one GW invariants of the quintic threefold using mirror symmetry [**17**]. The BCOV genus-one mirror conjecture was open for many years until solved by A. Zinger [**117**] in 2007. To prove the BCOV conjecture, Zinger and his collaborators developed a theory of *reduced* genus-one GW invariants for the quintic 3-fold. Indeed, the theory is defined for a Calabi-Yau hypersurface in a projective space of any dimension, i.e. a degree $r+1$ hypersurface in $\mathbb{P}^r$ for any r. Zinger proved a genus-one mirror formula for a Calabi-Yau hypersurface in a projective space of any dimension [**117**]; the BCOV conjecture corresponds to the $r=4$ case. A. Popa generalized the results in [**117**] to Calabi-Yau complete intersections in projective spaces [**99**].

To define reduced genus-one GW invariants, Vakil-Zinger [**111**] constructed a desingularization of the main component of the moduli space of stable maps to projective spaces using the symplectic approach; later Hu-Li [**66**] provided an algebro-geometric approach to these desingularization results. A key ingredient in Zinger's proof is the Li-Zinger hyperplane theorem of the genus-one GW invariants [**84**], originally proved using analytic methods in symplectic GW theory. Recently,

Chang-Li [**27**] gave an algebro-geometric proof of the Li-Zinger hyperplane property of the genus-one GW invariants of the quintic 3-fold using their previous work on Guffin-Sharpe-Witten invariants [**26**].

1.1.5. *Genus* $g \geq 2$. In 2004, Maulik-Pandharipande [**97**] provided a calculation scheme which determines GW invariants of the quintic 3-fold in all genera and degrees in terms of previously determined GW invariants.

Based on the BCOV theory and geometry of the moduli of Calabi-Yau 3-folds, Yamaguchi-Yau [**113**] showed that the B-model topological string partition functions of the mirror of the quintic 3-fold can be expressed as polynomials of five generators. In 2006, Huang-Klemm-Quackenbush [**67**] conjectured a formula of the generating function F_g of genus-g GW invariants of the quintic 3-fold for $2 \leq g \leq 51$; a key ingredient of their derivation is the work of Yamaguchi-Yau [**113**]. The Huang-Klemm-Quackenbush conjecture has not been verified mathematically yet, even in the $g = 2$ case. Alim-Länge [**6**] showed that the polynomial structure of the topological string partition function found by Yamaguchi-Yau for the quintic 3-fold holds for a general smooth Calabi-Yau 3-fold. This polynomial structure was further explored in a recent work of Alim-Scheidegger-Yau-Zhou [**7**].

A compact Calabi-Yau 1-fold is an elliptic curve. S. Li showed that A-model topological strings (GW theory) on an elliptic curve is equivalent to B-model topological strings (quantum BCOV theory) on the mirror elliptic curve [**85**]. This is the first compact Calabi-Yau example where mirror symmetry is proved at all genera.

Higher genus topological strings on compact Calabi-Yau manifolds and higher genus mirror symmetry for compact Calabi-Yau manifolds are very important, yet extremely difficult problems.

1.2. Toric Calabi-Yau Manifolds/Orbifolds. A nonsingular n-dimensional complex algebraic variety is toric if it contains the algebraic torus $\mathbb{T} = (\mathbb{C}^*)^n$ as a Zariski open dense subset, and the action of $\mathbb{T}$ on itself extends to an algebraic $\mathbb{T}$-action on the algebraic variety. Toric Calabi-Yau manifolds are always noncompact. GW theory and mirror symmetry for toric Calabi-Yau manifolds are much better understood than those for compact Calabi-Yau manifolds.

1.2.1. *A-model topological strings on toric manifolds/orbifolds.* By virtual localization (Graber-Pandharipande [**60**]), all genus full descendant GW invariants of projective toric manifolds can be expressed in terms of Hodge integrals, which are intersection numbers on Deligne-Mumford moduli spaces of stable curves [**37**]. (The terminology "virtual localization" was introduced in [**60**] and the term "Hodge integral" was introduced in [**47**] precisely to study the virtual localization formula in [**60**].) By Mumford's Grothendieck-Riemann-Roch computations [**93**], Hodge integrals can be reduced to descendant integrals on Deligne-Mumford moduli spaces of stable curves, whose values are determined by the Witten's conjecture [**112**]. (The Witten's conjecture was first proved by Kontsevich [**70**].) This gives an algorithm to compute all genus GW invariants (A-model topological closed string amplitudes) of all projective toric manifolds. Moreover, one can use localization to *define* (equivariant) GW invariants of *noncompact* toric manifolds; these invariants can also be determined by the algorithm described above.

1.2.2. *A-model topological strings on toric Calabi-Yau 3-folds.* In 2003, Aganagic-Klemm-Mariño-Vafa [**4**] proposed the topological vertex, an algorithm of computing all genus open and closed GW invariants of an arbitrary smooth toric Calabi-Yau 3-fold $\mathcal{X}$; this algorithm is much more efficient than the algorithm described in Section 1.2.1. The open GW invariants in this context are virtual counts of holomorphic maps from bordered Riemann surfaces to $\mathcal{X}$ relative to a class of Lagrangian submanifolds introduced by Aganagic-Vafa in [**10**]; these open GW invariants are defined and computed by localization with respect to the action of a compact 2-torus which preserves the symplectic and holomorphic structures on $\mathcal{X}$ and also preserves Aganagic-Vafa Lagrangian branes [**38, 73, 81**]. The topological vertex algorithm is proved in the 1-leg case in [**88, 98**] and in the 2-leg case in [**89**]; the full 3-leg case is a consequence of Maulik-Oblomkov-Okounkov-Pandharipande's proof of the Gromov-Witten/Donaldson-Thomas correspondence [**94, 95**] for all smooth toric 3-folds [**96**].

Gromov-Witten (GW) and Donaldson-Thomas (DT) theories can be defined for 3-dimensional orbifolds (smooth DM stacks). Bryan-Cadman-Young [**23**] (resp. Ross [**101**]) developed a vertex formalism for DT (resp. GW) theory of toric Calabi-Yau 3-orbifolds. The GW/DT correspondence has been proved for toric Calabi-Yau 3-orbifolds with transverse A_n-singularities, first in the effective 1-leg case [**118**] and the gerby 1-leg case [**103**], later in the 2-leg case [**104**], and finally in the full 3-leg case [**102**]. The precise statement of the GW/DT correspondence for general toric Calabi-Yau 3-orbifolds is not known, even conjecturally.

1.2.3. *The Remodeling Conjecture.* Let $(\mathcal{X},\omega)$ be a symplectic toric Calabi-Yau 3-manifold/orbifold, where ω is the symplectic form. By Hori-Vafa [**64**], the mirror of $(\mathcal{X},\omega)$ is a non-compact Calabi-Yau 3-fold $(\check{\mathcal{X}},\Omega)$, where

$$\check{\mathcal{X}} = \{(u,v,X,Y) : u,v \in \mathbb{C}, X,Y \in \mathbb{C}^*, uv = H(X,Y)\}$$

is a hypersurface in $\mathbb{C}^2 \times (\mathbb{C}^*)^2$, and

$$\Omega = \mathrm{Res}_{\check{\mathcal{X}}}\Big(\frac{1}{uv - H(X,Y)} du \wedge dv \wedge \frac{dX}{X} \wedge \frac{dY}{Y}\Big)$$

is a holomorphic 3-form on $\check{\mathcal{X}}$. The Bouchard-Klemm-Mariño-Pasquetti (BKMP) Remodeling Conjecture [**20, 21**] says that all genus B-model topological strings on $(\check{\mathcal{X}},\Omega)$ can be reduced to Eynard-Orantin invariants [**44**] of the mirror curve (Σ_q,Φ), where

$$\Sigma_q = \{(X,Y) \in (\mathbb{C}^*)^2 : H(X,Y,q) = 0\} \subset (\mathbb{C}^*)^2, \quad \Phi = \log Y \frac{dX}{X}.$$

Using mirror symmetry, BKMP relates the Eynard-Orantin invariant $\omega_{g,n}$ of the mirror curve to a generating function $F_{g,n}^{\mathcal{X},\mathcal{L}}$ of open GW invariants (A-model topological open string amplitudes) counting holomorphic maps from bordered Riemann surfaces with g handles and n holes to $\mathcal{X}$ with boundaries in an Aganagic-Vafa Lagrangian brane $\mathcal{L}$. The BKMP Remodeling Conjecture was proved for $\mathbb{C}^3$ at all genus g independently by L. Chen [**29**] and J. Zhou [**115**] in the $n > 0$ case (open string sector), and by Bouchard-Catuneanu-Marchal-Sułkowski [**19**] in the $n = 0$ case (closed string sector). In 2012, Eynard-Orantin provided a proof of the BKMP Remodeling conjecture for all symplectic smooth toric Calabi-Yau 3-folds [**46**].

In the rest of this paper, we will give a non-technical exposition of our results on the Remodeling Conjecture for toric Calabi-Yau 3-orbifolds [**50, 51**], which is a version of all genus open-closed mirror symmetry.

2. A-model Geometry and Topology

2.1. An example. We start with a simple example. Let $\mathbb{C}^*$ act on $\mathbb{C}^4$ by

$$t\cdot(z_1,z_2,z_3,z_4)=(tz_1,tz_2,tz_3,t^{-3}z_4)$$

where $t\in\mathbb{C}^*$ and $(z_1,z_2,z_3,z_4)\in\mathbb{C}^4$. This $\mathbb{C}^*$-action restricts to a Hamiltonian $U(1)$-action on the Kähler manifold $(\mathbb{C}^4,\omega_0=\sqrt{-1}\sum_{i=1}^4 dz_i\wedge d\bar{z}_i)$ with the moment map

$$\tilde{\mu}:\mathbb{C}^4\to\mathbb{R},\quad (z_1,z_2,z_3,z_4)\mapsto|z_1|^2+|z_2|^2+|z_3|^2-3|z_4|^2.$$

$$\tilde{\mu}^{-1}(r)/U(1)=\begin{cases}((\mathbb{C}^3-\{0\})\times\mathbb{C})/\mathbb{C}^*=\mathcal{O}_{\mathbb{P}^2}(-3), & r>0,\\ (\mathbb{C}^3\times\mathbb{C}^*)/\mathbb{C}^*=\mathbb{C}^3/\mathbb{Z}_3, & r<0,\end{cases}$$

The total space $\mathcal{O}_{\mathbb{P}^2}(-3)$ is a smooth toric Calabi-Yau 3-fold. The quotient $\mathbb{C}^3/\mathbb{Z}_3$ is a simplicial toric Calabi-Yau 3-fold. The stacky quotient $[\mathbb{C}^3/\mathbb{Z}_3]$ is a toric Calabi-Yau 3-orbifold.

2.2. Toric Calabi-Yau 3-orbifolds. A Calabi-Yau 3-fold X is toric if it contains the algebraic torus $\mathbb{T}=(\mathbb{C}^*)^3$ as a Zariski dense open subset, and the action of $\mathbb{T}$ on itself extends to X. All Calabi-Yau 3-folds are non-compact. There is a rank 2 subtorus $\mathbb{T}'\subset\mathbb{T}$ which acts trivially on the canonical line bundle of X. We call $\mathbb{T}'$ the Calabi-Yau torus. Then $\mathbb{T}\cong\mathbb{T}'\times\mathbb{C}^*$. Let $\mathbb{T}'_{\mathbb{R}}\cong U(1)^2$ be the maximal compact subgroup of $\mathbb{T}'$.

Let $M'=\mathrm{Hom}(\mathbb{T}',\mathbb{C}^*)\cong\mathbb{Z}^2$ and $N'=\mathrm{Hom}(\mathbb{C}^*,\mathbb{T}')$ be the character lattice and the cocharacter lattice of $\mathbb{T}'$, respectively. Then M' and N' are dual lattices. Let X_Σ be a toric Calabi-Yau 3-fold defined by a simplicial fan $\Sigma\subset N'_{\mathbb{R}}\times\mathbb{R}$, where $N'_{\mathbb{R}}:=N'\otimes_{\mathbb{Z}}\mathbb{R}\cong\mathbb{R}^2$ can be identified with the Lie algebra of $\mathbb{T}'_{\mathbb{R}}$. Then X_Σ has at most quotient singularities. We assume that X_Σ is semi-projective, i.e., X_Σ contains at least one $\mathbb{T}$ fixed point, and X_Σ is projective over its affinization $X_0:=\mathrm{Spec}H^0(X_\Sigma,\mathcal{O}_{X_\Sigma})$. Then the support of the fan Σ is a strongly convex rational polyhedral cone $\sigma_0\subset N'_{\mathbb{R}}\times\mathbb{R}\cong\mathbb{R}^3$, and X_0 is the affine toric variety defined by the 3-dimensional cone σ_0. There exists a convex polytope $P\subset N'_{\mathbb{R}}\cong\mathbb{R}^2$ with vertices in the lattices $N'\cong\mathbb{Z}^2$, such that σ_0 is the cone over $P\times\{1\}\subset N'_{\mathbb{R}}\times\mathbb{R}$, i.e. $\sigma_0=\{(tx,ty,t):(x,y)\in P,t\in[0,\infty)\}$. The fan Σ determines a triangulation of P: the 1-dimensional, 2-dimensional, and 3-dimensional cones in Σ are in one-to-one correspondence with the vertices, edges, and faces of the triangulation of P, respectively. This triangulation of P is known as the toric diagram or the dual graph of the simplicial toric Calabi-Yau 3-fold X_Σ; some examples are shown in Figure 1 below. (In Figure 1, $\mathbb{C}^2/\mathbb{Z}_3$ is the A_2 surface singularity, and A_2 is the minimal toric resolution of $\mathbb{C}^2/\mathbb{Z}_3$.)

Let $\Sigma(d)$ be the set of d-dimensional cones in Σ, and let $p=|\Sigma(1)|-3$. Then X_Σ is a GIT quotient

$$X_\Sigma=\mathbb{C}^{3+p}\,/\!\!/\,G_\Sigma=(\mathbb{C}^{3+p}-Z_\Sigma)/G_\Sigma$$

where G_Σ is a p-dimensional subgroup of $(\mathbb{C}^*)^{3+p}$ and Z_Σ is a Zariski closed subset of $\mathbb{C}^{3+p}$ determined by the fan Σ. If X_Σ is a smooth toric Calabi-Yau 3-fold then $G_\Sigma\cong(\mathbb{C}^*)^p$ and G_Σ acts freely on $\mathbb{C}^{3+p}-Z_\Sigma$. In general we have $(G_\Sigma)_0\cong(\mathbb{C}^*)^p$, where $(G_\Sigma)_0$ is the connected component of the identity, and the stabilizers of the G_Σ-action on $\mathbb{C}^{3+p}-Z_\Sigma$ are at most finite and generically trivial. The stacky quotient

$$\mathcal{X}=[(\mathbb{C}^{3+p}-Z_\Sigma)/G_\Sigma]$$

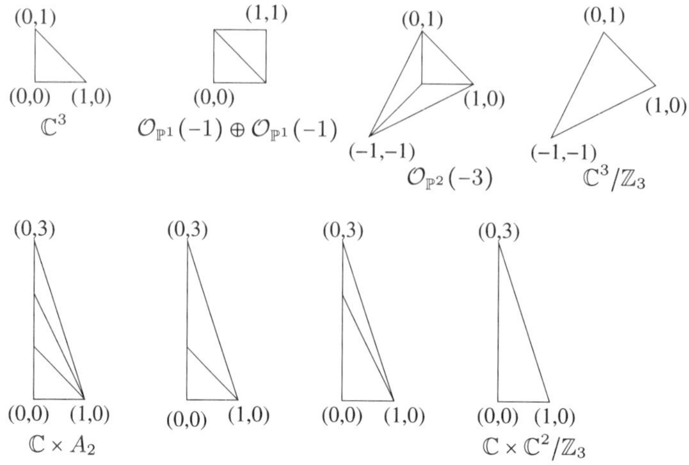

FIGURE 1

is a toric Calabi-Yau 3-orbifold; it is a toric Deligne-Mumford stack in the sense of Borisov-Chen-Smith [**18**].

2.3. Toric crepant resolution and extended Kähler classes. Given a semi-projective simplicial toric Calabi-Yau 3-fold X_Σ which is not smooth, there exists a subdivision Σ' of Σ, such that

$$X_{\Sigma'} = (\mathbb{C}^{3+p+s} - Z_{\Sigma'})/G_{\Sigma'} \longrightarrow X_\Sigma = \big((\mathbb{C}^{3+p} - Z_\Sigma)\times(\mathbb{C}^*)^s\big)/G_{\Sigma'}.$$

is a crepant toric resolution, where $X_{\Sigma'}$ is a smooth toric Calabi-Yau 3-fold, $s = |\Sigma'(1)| - |\Sigma(1)|$, and $G_{\Sigma'}\cong(\mathbb{C}^*)^{p+s}$. $X_{\Sigma'}$ and X_Σ are GIT quotients of the same $G_{\Sigma'}$-action on $\mathbb{C}^{3+p+s}$ with respect to different stability conditions.

Let $K_{\Sigma'}\cong U(1)^{p+s}$ be the maximal compact subgroup of $G_{\Sigma'}\cong(\mathbb{C}^*)^{p+s}$. The $G_{\Sigma'}$-action on $\mathbb{C}^{3+p+s}$ restricts to a Hamiltonian $K_{\Sigma'}$-action on the Kähler manifold $(\mathbb{C}^{3+p+s}, \omega_0 = \sqrt{-1}\sum_{i=1}^{3+p+s} dz_i\wedge d\bar{z}_i)$, with moment map $\widetilde{\mu}:\mathbb{C}^{3+p+s}\to\mathbb{R}^{p+s}$. There exist two (open) cones C and C' in $\mathbb{R}^{p+s}$ such that

$$\widetilde{\mu}^{-1}(\vec{r})/K_{\Sigma'} = \begin{cases}(\mathbb{C}^{3+p+s} - Z_{\Sigma'})/G_{\Sigma'} = X_{\Sigma'}, & \vec{r}\in C',\\ \big((\mathbb{C}^{3+p} - Z_\Sigma)\times(\mathbb{C}^*)^s\big)/G_{\Sigma'} = (\mathbb{C}^{3+p} - Z_\Sigma)/G_\Sigma = X_\Sigma, & \vec{r}\in C\end{cases}$$

$C'\subset\mathbb{R}^{p+s} = H^2(X_{\Sigma'};\mathbb{R})$ is the Kähler cone of $X_{\Sigma'}$ and $C\subset\mathbb{R}^{p+s}$ is the extended Kähler cone of X_Σ.

EXAMPLE 1. $X_{\Sigma'} = \mathcal{O}_{\mathbb{P}^2}(-3)$ is a toric crepant resolution of $X_\Sigma = \mathbb{C}^3/\mathbb{Z}^3$. We have $H^2(X_{\Sigma'};\mathbb{R}) = \mathbb{R}$, $C' = (0,\infty)\subset\mathbb{R}$ is the Kähler cone of $\mathcal{O}_{\mathbb{P}^2}(-3)$, and $C = (-\infty,0)\subset\mathbb{R}$ is the extended Käher cone of $\mathbb{C}^3/\mathbb{Z}_3$.

The parameter $\vec{r}\in C$ determines a Kähler form $\omega(\vec{r})$ on the toric Calabi-Yau 3-orbifold $\mathcal{X} = [(\mathbb{C}^{3+p} - Z_\Sigma)/G_\Sigma]$. The $p+s$ parameters $\vec{r} = (r_1,\ldots,r_{p+s})$ are extended Kähler parameters of $\mathcal{X}$, where $r_1,\ldots,r_p$ are Kähler parameters of $\mathcal{X}$. The A-model closed string flat coordinates are complexified extended Kähler parameters

$$\tau_a = -r_a + \sqrt{-1}\theta_a,\quad a = 1,\ldots,p+s.$$

2.4. Toric Graphs. The action of the Calabi-Yau torus $\mathbb{T}'$ on $\mathcal{X}$ restricts to a Hamiltonian $\mathbb{T}'_{\mathbb{R}}$-action on the Kähler orbifold $(\mathcal{X},\omega(\vec{r}))$, with moment map $\mu' : \mathcal{X} \to M'_{\mathbb{R}} = \mathbb{R}^2$. The 1-skeleton $\mathcal{X}^1$ of the toric Calabi-Yau 3-fold $\mathcal{X}$ is the union of 0-dimensional and 1-dimensional orbits of the $\mathbb{T}$-action on $\mathcal{X}$. $\Gamma := \mu'(\mathcal{X}^1) \subset \mathbb{R}^2$ is a planar trivalent graph, which is known as the toric graph of the symplectic toric Calabi-Yau 3-orbifold $(\mathcal{X},\omega(\vec{r}))$. Some examples are shown in Figure 2.

Note that the toric diagram in Figure 1 depends only on the complex structure on $\mathcal{X}$, where as the toric graph in Figure 2 depends also on the symplectic structure of $\mathcal{X}$: for example, when $\mathcal{X} = \mathcal{O}_{\mathbb{P}^1}(-1) \oplus \mathcal{O}_{\mathbb{P}^1}(-1)$, the length of the unique compact edge in Γ is proportional to the symplectic area of the zero section $\mathbb{P}^1$. The toric graphs in Figure 2 are dual graphs to the toric diagrams presented in Figure 1.

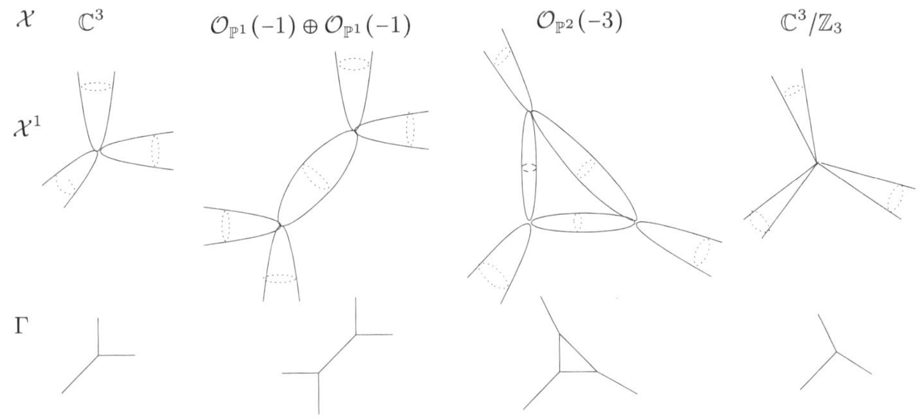

FIGURE 2

2.5. Aganagic-Vafa Lagrangian branes. An Aganagic-Vafa Lagrangian brane in a toric Calabi-Yau 3-orbifold $\mathcal{X}$ is a Lagrangian sub-orbifold of the form

$$\mathcal{L} = [\widetilde{L}/K_{\Sigma'}] \subset \mathcal{X} = [\widetilde{\mu}^{-1}(\vec{r})/K_{\Sigma'}]$$

where

$$\widetilde{L} = \{(z_1,\dots,z_{3+p+s}) \in \widetilde{\mu}^{-1}(\vec{r}) : \sum_{i=1}^{3+p+s} \hat{l}_i^1 |z_i|^2 = c_1, \sum_{i=1}^{3+p+s} \hat{l}_i^2 |z_i|^2 = c_2, \\ \arg(z_1\cdots z_{3+p+s}) = c_3\},$$

c_1, c_2, c_3 are constants, and

$$\sum_{i=1}^{3+p+s} \hat{l}_i^\alpha = 0, \quad \alpha = 1,2.$$

The compact 2-torus $\mathbb{T}'_{\mathbb{R}} \cong U(1)^2$ acts on $\mathcal{L}$, and $\mu'(\mathcal{L})$ is a point on the toric graph $\Gamma = \mu'(\mathcal{X}^1)$ which is not a vertex. $\mathcal{L}$ intersects a unique 1-dimensional $\mathbb{T}$ orbit $\mathfrak{l} \subset \mathcal{X}$. We have $\mathfrak{l} \cong \mathbb{C}^* \times \mathcal{B}\mathbb{Z}_m$ for some positive integer m. When $m = 1$, $\mathcal{L} \cong S^1 \times \mathbb{C}$ is smooth; when $m > 1$, $\mathcal{L}$ is smooth away from $\mathcal{L} \cap \mathfrak{l} \cong S^1 \times \mathcal{B}\mathbb{Z}_m$.

EXAMPLE 2. $\mathcal{X} = \mathbb{C}^3$

$$\mathcal{L} = \{(z_1,z_2,z_3) \in \mathbb{C}^3 : |z_1|^2 - |z_2|^2 = 1, |z_2|^2 - |z_3|^2 = 0, \arg(z_1z_2z_3) = 0\} \cong S^1 \times \mathbb{C}$$

is a Harvey-Lawson special Lagrangian submanifold [**63**].

2.6. Chen-Ruan orbifold cohomology. Let $U = \mathbb{C}^{3+p} - Z_\Sigma$, so that $\mathcal{X} = [U/G_\Sigma]$. Given $v \in G_\Sigma$, let $U^v = \{z \in U : v \cdot z = z\}$. The inertia stack of $\mathcal{X}$ is

$$\mathcal{I}\mathcal{X} = \bigcup_{v \in \mathrm{Box}(\Sigma)} \mathcal{X}_v$$

where $\mathrm{Box}(\Sigma) = \{v \in G_\Sigma : U^v \neq \varnothing\}$ and $\mathcal{X}_v = [U^v/G_\Sigma]$.

We consider cohomology with $\mathbb{C}$-coefficient. As a graded $\mathbb{C}$-vector space, the Chen-Ruan orbifold cohomology [**31**] of $\mathcal{X}$ is

$$H^*_{\mathrm{CR}}(\mathcal{X};\mathbb{C}) = \bigoplus_{v \in \mathrm{Box}(\Sigma)} H^*(\mathcal{X}_v;\mathbb{C})[2\mathrm{age}(v)], \quad \mathrm{age}(v) \in \{0,1,2\}.$$

EXAMPLE 3. Let $\omega = e^{2\pi\sqrt{-1}/3}$, so that $\mathbb{Z}_3 = \{1, \omega, \omega^2\}$. $\mathbb{Z}_3$ acts on $\mathbb{C}^3$ by $\omega\cdot(z_1,z_2,z_3) = (\omega z_1, \omega z_2, \omega z_3)$ for $(z_1,z_2,z_3) \in \mathbb{C}^3$. The inertia stack of $\mathcal{X} = [\mathbb{C}^3/\mathbb{Z}_3]$ is $\mathcal{I}\mathcal{X} = \mathcal{X}_1 \cup \mathcal{X}_\omega \cup \mathcal{X}_{\omega^2}$, where

$$\mathcal{X}_1 = [\mathbb{C}^3/\mathbb{Z}_3], \quad \mathcal{X}_\omega = \mathcal{X}_{\omega^2} = [\{\text{point}\}/\mathbb{Z}_3] = \mathcal{B}\mathbb{Z}_3$$

The Chen-Ruan orbifold cohomology of $\mathcal{X}$ is

$$H^*_{\mathrm{CR}}(\mathcal{X};\mathbb{C}) = \mathbb{C}\mathbf{1} \oplus \mathbb{C}\mathbf{1}_{\frac{1}{3}} \oplus \mathbb{C}\mathbf{1}_{\frac{2}{3}}$$

where $\deg \mathbf{1} = 0$, $\deg \mathbf{1}_{\frac{1}{3}} = 2$, $\deg \mathbf{1}_{\frac{2}{3}} = 4$. The cohomology of $\mathcal{O}_{\mathbb{P}^2}(-3)$ is

$$H^*(\mathcal{O}_{\mathbb{P}^2}(-3);\mathbb{C}) = H^*(\mathbb{P}^2;\mathbb{C}) = \mathbb{C}1 \oplus \mathbb{C}H \oplus \mathbb{C}H^2,$$

where $H \in H^2(\mathbb{P}^2;\mathbb{C})$ is the hyperplane class.

Let $\mathfrak{g} := |\mathrm{Int}(P) \cap N'|$ be the number of lattice points in $\mathrm{Int}(P)$, the interior of the polytope P, and let $\mathfrak{n} := |\partial P \cap N'|$ be the number of lattice points on ∂P, the boundary of the polytope P. Then

$$\begin{aligned}
p &= |\Sigma(1)| - 3 = \dim_{\mathbb{C}} H^2(X_\Sigma;\mathbb{C}),\\
p+s &= |\Sigma'(1)| - 3 = |P \cap N'| - 3 = \dim_{\mathbb{C}} H^2(X_{\Sigma'};\mathbb{C}) = \dim_{\mathbb{C}} H^2_{\mathrm{CR}}(\mathcal{X};\mathbb{C})\\
&= \mathfrak{g} + \mathfrak{n} - 3,\\
\mathfrak{g} &= |\mathrm{Int}(P) \cap N'| = \dim_{\mathbb{C}} H^4(X_{\Sigma'}) = \dim_{\mathbb{C}} H^4_{\mathrm{CR}}(\mathcal{X};\mathbb{C}),\\
\chi &= |\Sigma'(3)| = 2\mathrm{Area}(P) = \dim_{\mathbb{C}} H^*(X_{\Sigma'};\mathbb{C}) = \dim_{\mathbb{C}} H^*_{\mathrm{CR}}(\mathcal{X};\mathbb{C})\\
&= 1 + p + s + \mathfrak{g} = 2\mathfrak{g} - 2 + \mathfrak{n}.
\end{aligned}$$

3. A-model Topological Strings

Let $\mathcal{L}$ be an Aganagic-Vafa Lagrangian brane in a toric Calabi-Yau 3-orbifold $\mathcal{X}$. Then $\mathcal{L}$ is homotopic to $S^1 \times \mathcal{B}\mathbb{Z}_m$, so

$$H_1(\mathcal{L};\mathbb{Z}) = \pi_1(\mathcal{L}) = \mathbb{Z} \times \mathbb{Z}_m.$$

Open GW invariants of $(\mathcal{X},\mathcal{L})$ count holomorphic maps

$$u : (\Sigma, x_1, \dots, x_\ell, \partial\Sigma = \coprod_{j=1}^n R_j) \to (\mathcal{X},\mathcal{L})$$

where Σ is a bordered Riemann surface with stacky points $x_i = B\mathbb{Z}_{r_i}$ and $R_j \cong S^1$ are connected components of $\partial\Sigma$. These invariants depend on the following data:

(1) the topological type (g,n) of the coarse moduli of the domain, where g is the genus of Σ and n is the number of connected components of $\partial\Sigma$,
(2) the degree $\beta' = u_*[\Sigma] \in H_2(\mathcal{X},\mathcal{L};\mathbb{Z})$,

(3) the winding numbers $\mu_1,\dots,\mu_n \in \mathbb{Z}$ and the monodromies $k_1,\dots,k_n \in \mathbb{Z}_m$, where $(\mu_j,k_j) = u_*[R_j] \in H_1(\mathcal{L};\mathbb{Z}) = \mathbb{Z}\times\mathbb{Z}_m$,

(4) the framing $f \in \mathbb{Z}$ of $\mathcal{L}$.

We call the pair $(\mathcal{L},f)$ a framed Aganagic-Vafa Lagrangian brane. We write $\vec{\mu} = ((\mu_1,k_1),\dots,(\mu_n,k_n))$. Let $\mathcal{M}_{(g,n),\ell}(\mathcal{X},\mathcal{L}\mid\beta',\vec{\mu})$ be the moduli space parametrizing maps described above, and let $\overline{\mathcal{M}}_{(g,n),\ell}(\mathcal{X},\mathcal{L}\mid\beta',\vec{\mu})$ be the partial compactification: we allow the domain Σ to have nodal singularities, and an orbifold/stacky point on Σ is either a marked point x_j or a node; we require the map u to be stable in the sense that its automorphism group is finite. Evaluation at the i-th marked point x_i gives a map $\mathrm{ev}_i : \overline{\mathcal{M}}_{(g,n),\ell}(\mathcal{X},\mathcal{L}\mid\beta',\vec{\mu}) \to \mathcal{I}\mathcal{X}$.

Given $\gamma_1,\dots,\gamma_\ell \in H^*_{\mathrm{CR},\mathbb{T}'}(\mathcal{X};\mathbb{C})$, we define

$$\langle\gamma_1,\dots,\gamma_\ell\rangle^{\mathcal{X},(\mathcal{L},f)}_{g,\beta',\vec{\mu}} := \int_{[\overline{\mathcal{M}}_{(g,n),\ell}(\mathcal{X},\mathcal{L}|\beta',\vec{\mu})^{\mathbb{T}'_{\mathbb{R}}}]^{\mathrm{vir}}} \frac{\prod_{i=1}^{\ell}\mathrm{ev}_i^*\gamma_i}{e_{\mathbb{T}'_{\mathbb{R}}}(N^{\mathrm{vir}})}\Bigg|_{(\mathbb{T}_f)_{\mathbb{R}}} \in \mathbb{C}v^{\sum_{i=1}^{\ell}\frac{\deg\gamma_i}{2}-1}$$

where v is the generator of $H^2(\mathcal{B}(\mathbb{T}_f)_{\mathbb{R}};\mathbb{Z}) = H^2(\mathcal{B}U(1);\mathbb{Z}) \cong \mathbb{Z}$.

For $\boldsymbol{\tau} = \sum_{a=1}^{p+s}\tau_a e_a \in H^2_{\mathrm{CR}}(\mathcal{X};\mathbb{C})$, we define

$$\begin{aligned} F^{\mathcal{X},(\mathcal{L},f)}_{g,n}(Z_1,\dots,Z_n,\boldsymbol{\tau}) &= \sum_{\beta',\ell\geq 0}\ \sum_{(\mu_j,k_j)\in\mathbb{Z}\times\mathbb{Z}_m} \frac{\langle\boldsymbol{\tau}^\ell\rangle^{\mathcal{X},(\mathcal{L},f)}_{g,\beta',(\mu_1,k_1)\cdots,(\mu_n,k_n)}}{\ell!} \\ &\quad\cdot \otimes_{j=1}^n\left(Z_j^{\mu_j}(-(-1)^{\frac{-k_j}{m}})\mathbf{1}'_{\frac{-k_j}{m}}\right) \in H^*_{\mathrm{CR}}(\mathcal{B}\mathbb{Z}_m;\mathbb{C})^{\otimes n} \end{aligned}$$

where $H^*_{\mathrm{CR}}(\mathcal{B}\mathbb{Z}_m;\mathbb{C}) = \oplus_{k=0}^{m-1}\mathbb{C}\mathbf{1}'_{\frac{k}{m}}$.

4. B-model Geometry and Topology

4.1. The mirror curve and its compactification. We use the notation in Section 2.6. The convex polytope $P \subset N'_{\mathbb{R}} \cong \mathbb{R}^2$ defines a polarized toric surface (S,L), where S is a toric surface, L is an ample line bundle on S, and

$$\chi(S,L) = h^0(S,L) = |P\cap N'| = 3+p+s.$$

The mirror curve of $\mathcal{X}$ is

$$C = \{(X,Y)\in(\mathbb{C}^*)^2 : H(X,Y) = 0\}$$

where

$$H(X,Y) = \sum_{(m,n)\in P\cap N'} a_{m,n}X^mY^n,\quad a_{m,n}\in\mathbb{C}^*.$$

Note that P is the Newton polytope of $H(X,Y)$, so $H(X,Y)\in H^0((\mathbb{C}^*)^2,\mathcal{O}_{(\mathbb{C}^*)^2})$ is the restriction of a section $s\in H^0(S,L)$. The compactified mirror curve is $\overline{C} := s^{-1}(0)\subset S$.

The element $(t_1,t_2,t_3)\in(\mathbb{C}^*)^3$ acts on $H(X,Y)$ by

$$(t_1,t_2,t_3)\cdot H(X,Y) = t_3H(t_1X,t_2Y) = \sum_{(m,n)\in P\cap N'}(t_1^mt_2^nt_3)a_{m,n}X^mY^n.$$

The complex moduli of the mirror curve is $(\mathbb{C}^*)^{|P\cap N'|}/(\mathbb{C}^*)^3 = (\mathbb{C}^*)^{p+s}$. The mirror curve is parametrized by $q = (q_1,\dots,q_{p+s})\in(\mathbb{C}^*)^{p+s}$. For generic q, the mirror curve $C_q \subset (\mathbb{C}^*)^2$ is a Riemann surface of genus $\mathfrak{g}$ with $\mathfrak{n}$ punctures, and the compactified mirror curve $\overline{C}_q \subset S$ is a compact Riemann surface of genus $\mathfrak{g}$. The Euler characteristic

$$\chi(C_q) = 2 - 2\mathfrak{g} - n = -\dim_{\mathbb{C}} H^*_{\mathrm{CR}}(\mathcal{X};\mathbb{C}) = -\chi(X_{\Sigma'}).$$

4.2. Fundamental normalized differential of the second kind. Let $B(p_1, p_2)$ be the fundamental normalized differential of the second kind on the compactified mirror curve $\overline{C}_q$ (see e.g. [**52**]). It is characterized by the following properties.

(1) $B(p_1, p_2)$ is a bilinear symmetric meromorphic differential on $\overline{C}_q \times \overline{C}_q$.
(2) $B(p_1, p_2)$ is holomorphic everywhere except for a double pole along the diagonal. If p_1, p_2 have local coordinates z_1, z_2 in an open neighborhood U of $p \in \overline{C}_q$ then

$$B(p_1, p_2) = \Big(\frac{1}{(z_1 - z_2)^2} + a(z_1, z_2)\Big)dz_1 dz_2$$

where $a(z_1, z_2)$ is holomorphic on $U \times U$ and symmetric in z_1, z_2. (Therefore, $\mathrm{Res}_{p' \to p} B(p, p') f(p') = df(p)$ for any function f which is defined and holomorphic in an open neighborhood of p.)
(3) (normalization) $\int_{p_1 \in A_i} B(p_1, p_2) = 0$, $i = 1 \dots, \mathfrak{g}$, where $A_1, B_1, \dots, A_{\mathfrak{g}}, B_{\mathfrak{g}}$ is a symplectic basis of $H_1(\overline{C}_q;\mathbb{C})$.

4.3. Framing and holomorphic Morse function. The framing $f \in \mathbb{Z}$ determines a subgroup

$$\mathbb{T}_f \cong \mathbb{C}^* \hookrightarrow \mathbb{T}' \cong (\mathbb{C}^*)^2, \quad t \mapsto (t, t^f).$$

This induces a surjective group homomorphisms between the dual tori:

$$(\mathbb{T}')^\vee \cong (\mathbb{C}^*)^2 \longrightarrow (\mathbb{T}_f)^\vee \cong (\mathbb{C}^*), \quad (X, Y) \mapsto XY^f.$$

We choose f such that all the branch points of the ramified cover

$$\hat{X} := XY^f : C_q \longrightarrow \mathbb{C}^*$$

are simple, so that $\hat{X}$ is a holomorphic Morse function on C_q. The Euler characteristic of $\mathbb{C}^*$ is zero, so by the Riemann-Hurwitz formula,

$$|\mathrm{Crit}(\hat{X})| = -\chi(C_q) = 2\mathfrak{g} - 2 + \mathfrak{n} = \dim_{\mathbb{C}} H^*_{\mathrm{CR}}(\mathcal{X};\mathbb{C}).$$

Let $\hat{X} = e^{-\hat{x}}$. Around each ramification point $p_\sigma \in \mathrm{Crit}(\hat{X})$, one can write $\hat{x} = \hat{x}(p_\sigma) + \zeta_\sigma^2$, where ζ_σ is a local coordinate. For any p in the neighborhood of p_σ, we define $\bar{p}$ such that $\zeta_\sigma(\bar{p}) = -\zeta_\sigma(p)$. Finally, let $\Phi := \log Y \dfrac{d\hat{X}}{\hat{X}}$, which is a multi-valued holomorphic 1-form on $(\mathbb{C}^*)^2$.

5. B-model Topological Strings

The invariants $\omega_{g,n}$ of the mirror curve C_q are defined recursively by the Eynard-Orantin topological recursion [**44**].

- (unstable cases) The initial terms are given by

$$\omega_{0,1} = 0, \quad \omega_{0,2} = B(p_1, p_2).$$

- (stable case) When $2g-2+n>0$, $\omega_{g,n}$ is a symmetric meromorphic differential on $(\overline{C}_q)^n$ which is holomorphic on $(\overline{C}_q \setminus \mathrm{Crit}(\hat{X}))^n$. Locally,
$$\omega_{g,n} = A(z_1,\dots,z_n)dz_1\cdots dz_n,$$
where $A(z_1,\dots,z_n)$ is symmetric and meromorphic. It is defined recursively by the following formula:

$$\begin{aligned}\omega_{g,n}(p_1,\dots,p_n) \;:=\; & \sum_{z\in\mathrm{Crit}(\hat{X})} \mathrm{Res}_{p\to z}\frac{\int_{\xi=p}^{\bar{p}} B(p_n,\xi)}{2(\Phi(p)-\Phi(\bar{p}))}\Big(\omega_{g-1,n+1}(p,\bar{p},p_1,\dots,p_{n-1}) \\ & + \sum_{\substack{g_1+g_2=g\\ I\sqcup J=\{1,\dots,n-1\}}} \omega_{g_1,|I|+1}(p,p_I)\omega_{g_2,|J|+1}(\bar{p},p_J)\Big).\end{aligned}$$

For $\ell\in\mathbb{Z}_m=\mathbb{Z}/m\mathbb{Z}$, let
$$\psi_\ell := \frac{1}{m}\sum_{k=0}^{m-1} e^{\frac{2\pi\sqrt{-1}k\ell}{m}}\mathbf{1}'_{\frac{k}{m}}.$$

Then $\{\psi_\ell : \ell=0,1\dots,m-1\}$ is a canonical basis of $H^*_{\mathrm{CR}}(\mathcal{B}\mathbb{Z}_m;\mathbb{C})$.

Recall that $\mathcal{L}$ intersects a unique 1-dimensional orbit $\mathfrak{l}$ of the $\mathbb{T}$-action on $\mathcal{X}$. We assume that the closure $\bar{\mathfrak{l}}$ of $\mathfrak{l}$ in $\mathcal{X}$ is non-compact, so that $\mathcal{L}$ is an "outer" brane. Then the 2-dimensional cone associated to $\bar{\mathfrak{l}}$ corresponds an edge e on the boundary of the polytope P, and $|e\cap N'|=m+1$. Let $D\subset S$ be the torus invariant divisor associated to the edge e. For generic q, the compactified mirror curve $\overline{C}_q$ intersects D transversally at m points $\bar{p}_0,\dots,\bar{p}_{m-1}$. For $\ell\in\{0,1,\dots,m-1\}$, there exists open neighborhoods U_ℓ of $\bar{p}_\ell$ in the compactified mirror curve $\overline{C}_q$ an open neighborhood U of 0 in $\mathbb{P}^1=\mathbb{C}^*\cup\{0,\infty\}$ such that $\hat{X}|_{U_\ell}: U_\ell\to U$ is biholomorphic. Let $\rho_\ell := (\hat{X}|_{U_\ell})^{-1}: U\to U_\ell$. We define B-model topological open string partition functions as follows.

(1) disk invariants
$$\check{F}_{0,1}(q;X) := \sum_{\ell\in\mathbb{Z}_m}\int_0^X \Big(\big(\log Y(\rho_\ell(X')) - \log Y(\bar{p}_\ell)\big)\frac{dX'}{X'}\Big)\psi_\ell$$
which take values in $H^*_{\mathrm{CR}}(\mathcal{B}\mathbb{Z}_m;\mathbb{C})$.

(2) annulus invariants
$$\check{F}_{0,2}(q;X_1,X_2) := \sum_{\ell_1,\ell_2\in\mathbb{Z}_m}\int_0^{X_1}\int_0^{X_2}\Big((\rho_{\ell_1}\times\rho_{\ell_2})^*\omega_{0,2} - \frac{dX_1'dX_2'}{(X_1'-X_2')^2}\Big)\psi_{\ell_1}\otimes\psi_{\ell_2}$$
which take values in $H^*_{\mathrm{CR}}(\mathcal{B}\mathbb{Z}_m;\mathbb{C})^{\otimes 2}$.

(3) $2g-2+n>0$
$$\check{F}_{g,n}(q;X_1,\dots,X_n) := \sum_{\ell_1,\cdots,\ell_n\in\mathbb{Z}_m}\int_0^{X_1}\cdots\int_0^{X_n}(\rho_{\ell_1}\times\cdots\times\rho_{\ell_n})^*\omega_{g,n}\psi_{\ell_1}\otimes\cdots\otimes\psi_{\ell_n}$$
which take values in $H^*_{\mathrm{CR}}(\mathcal{B}\mathbb{Z}_m;\mathbb{C})^{\otimes n}$.

Each of the m^n components of $\check{F}_{g,n}(q;X_1,\dots,X_n)$ is a power series in $q_1,\dots,q_{p+s}$, $X_1,\dots,X_n$ which converges in an open neighborhood of the origin.

6. All Genus Open-Closed Mirror Symmetry

In this section, $(\mathcal{L}, f)$ is an outer Aganagic-Vafa Lagrangian brane in $\mathcal{X}$, so that the closure of $\mathfrak{l} = \mathbb{C}^* \times \mathcal{B}\mathbb{Z}_m$ contains a unique $\mathbb{T}$ fixed point. Let G be the stabilizer of this fixed point. Then G is a finite abelian group which contains $\mathbb{Z}_m$ as a subgroup. When $\mathcal{X}$ is smooth, we have $m = 1$ and G is trivial.

6.1. Genus-zero open-closed mirror symmetry: disk invariants. Aganagic-Vafa [**10**] and Aganagic-Klemm-Vafa [**5**] conjectured a mirror formula for open GW invariants counting holomorphic disks in a symplectic smooth toric Calabi-Yau 3-folds $\mathcal{X}$ bounded by an (inner or outer) Aganagic-Vafa Lagrangian brane $\mathcal{L}$ in $\mathcal{X}$. This conjecture was proved by Graber-Zaslow for an inner brane at zero framing in $\mathcal{O}_{\mathbb{P}^2}(-3)$ [**61**], by J. Zhou for an outer brane at arbitrary framing in the resolved conifold $\mathcal{O}_{\mathbb{P}^1}(-1) \oplus \mathcal{O}_{\mathbb{P}^1}(-1)$ [**116**], by A. Brini [**22**] for an outer brane at zero framing in $\mathcal{O}_{\mathbb{P}^2}(-3)$, and finally by the first and second authors for an inner or outer brane at arbitrary framing in any symplectic smooth toric Calabi-Yau 3-folds [**48**]. Later, the first two authors and H.-H. Tseng extended the results in [**48**] to the orbifold case [**49**]. We state the main theorem of [**49**] when $(\mathcal{L}, f)$ is a framed outer brane in a symplectic toric Calabi-Yau 3-orbifolds $\mathcal{X}$ as follows.

THEOREM 4.
$$\check{F}_{0,1}(q; X) = |G| F_{0,1}^{\mathcal{X},(\mathcal{L},f)}(\boldsymbol{\tau}; Z)$$
where (q, X) *and* $(\boldsymbol{\tau}, Z)$ *are related by the open-closed mirror map:*
$$\tau_a = \frac{1}{2\pi\sqrt{-1}} \int_{A_a} \Phi = \begin{cases} \log q_a + h_a(q), & a = 1, \dots, p \\ q_a(1 + h_a(q)), & a = p+1, \dots, p+s \end{cases}$$
$$\log Z = \log X + h_0(q)$$
$h_0(q), h_1(q), \dots, h_{p+s}(q)$ *explicit power series in* q *convergent in a neighborhood of the origin,* $h_a(0) = 0$.

We now give a brief outline of the proof of Theorem 4 in [**49**]. By localization, the A-model disk potential $F_{0,1}^{\mathcal{X},(\mathcal{L},f)}$ can be expressed in terms of the J-function, which is related to the hypergeometric functions by the genus-zero mirror theorem for smooth toric DM stacks [**33**]. The B-model disk potential satisfies extended Picard-Fuchs equations, whose solutions are hypergeometric functions.

6.2. All genus open-closed mirror symmetry: the Remodeling Conjecture.

CONJECTURE 5 (Bouchard-Klemm-Mariño-Pasqetti [**20, 21**]).
$$\check{F}_{g,n}(q; X_1, \dots, X_n) = (-1)^{g-1+n} |G|^n F_{g,n}^{\mathcal{X},(\mathcal{L},f)}(\boldsymbol{\tau}; Z_1, \dots, Z_n)$$
where (q, X_j) *and* $(\boldsymbol{\tau}, Z_j)$ *are related by the open-closed mirror map:*
$$\tau_a = \frac{1}{2\pi\sqrt{-1}} \int_{A_a} \Phi = \begin{cases} \log q_a + h_a(q), & a = 1 \dots, p \\ q_a(1 + h_a(q)), & a = p+1, \dots, p+s \end{cases}$$
$$\log Z_j = \log X_j + h_0(q)$$
$h_0(q), h_1(q), \dots, h_{p+s}(q)$ *explicit power series in* q *convergent in a neighborhood of the origin.*

Indeed, the above statement is more general than the original conjecture in [**20**, **21**], where the case $m = 1$ is considered.

Conjecture 5 was proved when $\mathcal{X} = \mathbb{C}^3$ independently by L. Chen [**29**] and J. Zhou [**115**]. In 2012, Eynard-Orantin provided a proof of the BKMP Remodeling conjecture for all symplectic smooth toric Calabi-Yau 3-folds [**46**]. In the orbifold case, the authors prove Conjecture 5 first for affine toric Calabi-Yau 3-orbifolds [**50**] and later for all semi-projective toric Calabi-Yau 3-orbifolds [**51**].

We now give a brief outline of the proof of Conjecture 5 in [**51**]. Givental proved a quantization formula for total descendant potential of equivariant GW theory of GKM manifolds [**56**–**58**]. (See also the book by Lee-Pandharipande [**76**].) The third author generalized this formula to GKM orbifolds [**119**]. The quantization formula is equivalent to a graph sum formula of the total descendant potential, which implies a graph sum formula

$$F_{g,n}^{\mathcal{X},(\mathcal{L},f)} = \sum_{\vec{\Gamma}\in G_{g,n}} \frac{w_A(\vec{\Gamma})}{|\mathrm{Aut}(\vec{\Gamma})|},$$

where $G_{g,n}$ is a certain set of decorated stable graphs. Dunin-Barkowski, Orantin, Shadrin, Spitz [**43**] provide a graph sum formula for the unique solution $\{\omega_{g,n}\}$ to the Eynard-Orantin topological recursion. We expand this graph sum formula at punctures $\{\bar{p}_\ell : \ell \in \mathbb{Z}_m\}$ and obtain a graph sum formula

$$\check{F}_{g,n} = \sum_{\vec{\Gamma}\in G_{g,n}} \frac{w_B(\vec{\Gamma})}{|\mathrm{Aut}(\vec{\Gamma})|}$$

Finally, we use the genus-zero mirror theorem for smooth toric DM stacks [**33**] to prove

$$w_B(\vec{\Gamma}) = (-1)^{g-1+n}|G|^n w_A(\vec{\Gamma})$$

for all decorated graphs $\vec{\Gamma}$.

References

[1] D. Abramovich, B. Fantechi, "Orbifold techniques in degeneration formulas," `arXiv:1103.5132`.

[2] D. Abramovich, T. Graber, and A. Vistoli, *Algebraic orbifold quantum products*, Orbifolds in mathematics and physics (Madison, WI, 2001), Contemp. Math., vol. 310, Amer. Math. Soc., Providence, RI, 2002, pp. 1–24, DOI 10.1090/conm/310/05397. MR1950940 (2004c:14104)

[3] D. Abramovich, T. Graber, and A. Vistoli, *Gromov-Witten theory of Deligne-Mumford stacks*, Amer. J. Math. **130** (2008), no. 5, 1337–1398, DOI 10.1353/ajm.0.0017. MR2450211 (2009k:14108)

[4] M. Aganagic, A. Klemm, M. Mariño, and C. Vafa, *The topological vertex*, Comm. Math. Phys. **254** (2005), no. 2, 425–478, DOI 10.1007/s00220-004-1162-z. MR2117633 (2006e:81263)

[5] M. Aganagic, A. Klemm, and C. Vafa, *Disk instantons, mirror symmetry and the duality web*, Z. Naturforsch. A **57** (2002), no. 1-2, 1–28, DOI 10.1515/zna-2002-9-1001. MR1906661 (2003f:81183)

[6] M. Alim and J. D. Länge, *Polynomial structure of the (open) topological string partition function*, J. High Energy Phys. **10** (2007), 045, 13, DOI 10.1088/1126-6708/2007/10/045. MR2357933 (2008j:81113)

[7] M. Alim, E. Scheidegger, S.-T. Yau, and J. Zhou, *Special polynomial rings, quasi modular forms and duality of topological strings*, Adv. Theor. Math. Phys. **18** (2014), no. 2, 401–467. MR3273318

[8] M. Aganagic, M. Mariño, and C. Vafa, *All loop topological string amplitudes from Chern-Simons theory*, Comm. Math. Phys. **247** (2004), no. 2, 467–512, DOI 10.1007/s00220-004-1067-x. MR2063269 (2005g:81284)

[9] M. F. Atiyah and R. Bott, *The moment map and equivariant cohomology*, Topology **23** (1984), no. 1, 1–28, DOI 10.1016/0040-9383(84)90021-1. MR721448 (85e:58041)

[10] M. Aganagic, C. Vafa, "Mirror symmetry, D-branes and counting holomorphic discs," `arXiv:hep-th/0012041`.

[11] V. V. Batyrev and D. A. Cox, *On the Hodge structure of projective hypersurfaces in toric varieties*, Duke Math. J. **75** (1994), no. 2, 293–338, DOI 10.1215/S0012-7094-94-07509-1. MR1290195 (95j:14072)

[12] K. Behrend, *Gromov-Witten invariants in algebraic geometry*, Invent. Math. **127** (1997), no. 3, 601–617, DOI 10.1007/s002220050132. MR1431140 (98i:14015)

[13] K. Behrend and B. Fantechi, *The intrinsic normal cone*, Invent. Math. **128** (1997), no. 1, 45–88, DOI 10.1007/s002220050136. MR1437495 (98e:14022)

[14] K. Behrend and Yu. Manin, *Stacks of stable maps and Gromov-Witten invariants*, Duke Math. J. **85** (1996), no. 1, 1–60, DOI 10.1215/S0012-7094-96-08501-4. MR1412436 (98i:14014)

[15] M. Bershadsky and A. Radul, *Conformal field theories with additional Z_N symmetry*, Internat. J. Modern Phys. A **2** (1987), no. 1, 165–178, DOI 10.1142/S0217751X87000053. MR880789 (88e:81112)

[16] M. Bershadsky, S. Cecotti, H. Ooguri, and C. Vafa, *Holomorphic anomalies in topological field theories*, Nuclear Phys. B **405** (1993), no. 2-3, 279–304, DOI 10.1016/0550-3213(93)90548-4. MR1240687 (94j:81254)

[17] M. Bershadsky, S. Cecotti, H. Ooguri, and C. Vafa, *Kodaira-Spencer theory of gravity and exact results for quantum string amplitudes*, Comm. Math. Phys. **165** (1994), no. 2, 311–427. MR1301851 (95f:32029)

[18] L. A. Borisov, L. Chen, and G. G. Smith, *The orbifold Chow ring of toric Deligne-Mumford stacks*, J. Amer. Math. Soc. **18** (2005), no. 1, 193–215 (electronic), DOI 10.1090/S0894-0347-04-00471-0. MR2114820 (2006a:14091)

[19] V. Bouchard, A. Catuneanu, O. Marchal, and P. Sułkowski, *The remodeling conjecture and the Faber-Pandharipande formula*, Lett. Math. Phys. **103** (2013), no. 1, 59–77, DOI 10.1007/s11005-012-0588-z. MR3004817

[20] V. Bouchard, A. Klemm, M. Mariño, and S. Pasquetti, *Remodeling the B-model*, Comm. Math. Phys. **287** (2009), no. 1, 117–178, DOI 10.1007/s00220-008-0620-4. MR2480744 (2010i:81352)

[21] V. Bouchard, A. Klemm, M. Mariño, and S. Pasquetti, *Topological open strings on orbifolds*, Comm. Math. Phys. **296** (2010), no. 3, 589–623, DOI 10.1007/s00220-010-1020-0. MR2628817 (2011d:81263)

[22] A. Brini, *Open topological strings and integrable hierarchies: remodeling the A-model*, Comm. Math. Phys. **312** (2012), no. 3, 735–780, DOI 10.1007/s00220-012-1489-9. MR2925133

[23] J. Bryan, C. Cadman, and B. Young, *The orbifold topological vertex*, Adv. Math. **229** (2012), no. 1, 531–595, DOI 10.1016/j.aim.2011.09.008. MR2854183

[24] P. Candelas, X. C. de la Ossa, P. S. Green, and L. Parkes, *A pair of Calabi-Yau manifolds as an exactly soluble superconformal theory*, Nuclear Phys. B **359** (1991), no. 1, 21–74, DOI 10.1016/0550-3213(91)90292-6. MR1115626 (93b:32029)

[25] S. Cecotti and C. Vafa, *Massive orbifolds*, Modern Phys. Lett. A **7** (1992), no. 19, 1715–1723, DOI 10.1142/S0217732392001415. MR1168619 (93d:81138)

[26] H.-L. Chang and J. Li, *Gromov-Witten invariants of stable maps with fields*, Int. Math. Res. Not. IMRN **18** (2012), 4163–4217, DOI 10.1093/imrn/rnr186. MR2975379

[27] H.-L. Chang and J. Li, *An algebraic proof of the hyperplane property of the genus one GW-invariants of quintics*, J. Differential Geom. **100** (2015), no. 2, 251–299. MR3343833

[28] B. Chen, A.-M. Li, S. Sun and G. Zhao, "Relative orbifold Gromov-Witten theory and degeneration formula," `arXiv:1110.6803`.

[29] L. Chen, "Bouchard-Klemm-Marino-Pasquetti Conjecture for $\mathbb{C}^3$," `arXiv:0910.3739`.

[30] W. Chen and Y. Ruan, *Orbifold Gromov-Witten theory*, Orbifolds in mathematics and physics (Madison, WI, 2001), Contemp. Math., vol. 310, Amer. Math. Soc., Providence, RI, 2002, pp. 25–85, DOI 10.1090/conm/310/05398. MR1950941 (2004k:53145)

[31] W. Chen and Y. Ruan, *A new cohomology theory of orbifold*, Comm. Math. Phys. **248** (2004), no. 1, 1–31, DOI 10.1007/s00220-004-1089-4. MR2104605 (2005j:57036)

[32] T.-M. Chiang, A. Klemm, S.-T. Yau, and E. Zaslow, *Local mirror symmetry: calculations and interpretations*, Adv. Theor. Math. Phys. **3** (1999), no. 3, 495–565. MR1797015 (2002e:14064)

[33] T. Coates, A. Corti, H. Iritani, H.-H. Tseng, "A Mirror Theorem for Toric Stacks," `arXiv:1310.4163`, to appear in Compos. Math.

[34] K. Costello, *Renormalization and effective field theory*, Mathematical Surveys and Monographs, vol. 170, American Mathematical Society, Providence, RI, 2011. MR2778558 (2012f:81177)

[35] K. J. Costello, S. Li, "Quantum BCOV theory on Calabi-Yau manifolds and the higher genus B-model," `arXiv:1201.4501`.

[36] D. A. Cox and S. Katz, *Mirror symmetry and algebraic geometry*, Mathematical Surveys and Monographs, vol. 68, American Mathematical Society, Providence, RI, 1999. MR1677117 (2000d:14048)

[37] P. Deligne and D. Mumford, *The irreducibility of the space of curves of given genus*, Inst. Hautes Études Sci. Publ. Math. **36** (1969), 75–109. MR0262240 (41 #6850)

[38] D.-E. Diaconescu and B. Florea, *Localization and gluing of topological amplitudes*, Comm. Math. Phys. **257** (2005), no. 1, 119–149, DOI 10.1007/s00220-005-1323-8. MR2163571 (2006h:14064)

[39] L. Dixon, D. Friedan, E. Martinec, and S. Shenker, *The conformal field theory of orbifolds*, Nuclear Phys. B **282** (1987), no. 1, 13–73, DOI 10.1016/0550-3213(87)90676-6. MR869937 (88f:81120)

[40] L. Dixon, J. A. Harvey, C. Vafa, and E. Witten, *Strings on orbifolds*, Nuclear Phys. B **261** (1985), no. 4, 678–686, DOI 10.1016/0550-3213(85)90593-0. MR818423 (87k:81104a)

[41] L. Dixon, J. Harvey, C. Vafa, and E. Witten, *Strings on orbifolds. II*, Nuclear Phys. B **274** (1986), no. 2, 285–314, DOI 10.1016/0550-3213(86)90287-7. MR851703 (87k:81104b)

[42] S. K. Donaldson and R. P. Thomas, *Gauge theory in higher dimensions*, The geometric universe (Oxford, 1996), Oxford Univ. Press, Oxford, 1998, pp. 31–47. MR1634503 (2000a:57085)

[43] P. Dunin-Barkowski, N. Orantin, S. Shadrin, and L. Spitz, *Identification of the Givental formula with the spectral curve topological recursion procedure*, Comm. Math. Phys. **328** (2014), no. 2, 669–700, DOI 10.1007/s00220-014-1887-2. MR3199996

[44] B. Eynard and N. Orantin, *Invariants of algebraic curves and topological expansion*, Commun. Number Theory Phys. **1** (2007), no. 2, 347–452, DOI 10.4310/CNTP.2007.v1.n2.a4. MR2346575 (2008m:14049)

[45] B. Eynard, *Invariants of spectral curves and intersection theory of moduli spaces of complex curves*, Commun. Number Theory Phys. **8** (2014), no. 3, 541–588, DOI 10.4310/CNTP.2014.v8.n3.a4. MR3282995

[46] B. Eynard and N. Orantin, *Computation of open Gromov-Witten invariants for toric Calabi-Yau 3-folds by topological recursion, a proof of the BKMP conjecture*, Comm. Math. Phys. **337** (2015), no. 2, 483–567, DOI 10.1007/s00220-015-2361-5. MR3339157

[47] C. Faber and R. Pandharipande, *Hodge integrals and Gromov-Witten theory*, Invent. Math. **139** (2000), no. 1, 173–199, DOI 10.1007/s002229900028. MR1728879 (2000m:14057)

[48] B. Fang and C.-C. M. Liu, *Open Gromov-Witten invariants of toric Calabi-Yau 3-folds*, Comm. Math. Phys. **323** (2013), no. 1, 285–328, DOI 10.1007/s00220-013-1771-5. MR3085667

[49] B. Fang, C.-C. M. Liu, H.-H. Tseng, "Open-closed Gromov-Witten invariants of 3-dimensional Calabi-Yau smooth toric DM stacks," `arXiv:1212.6073`.

[50] B. Fang, C.-C. M. Liu, Z. Zong, "All genus open-closed mirror symmetry for affine toric Calabi-Yau 3-orbifolds," `arXiv:1310.4818`.

[51] B. Fang, C.-C. M. Liu, Z. Zong, "On the remodeling conjecture for toric Calabi-Yau 3-orbifolds," preprint 2015.

[52] J. D. Fay, *Theta functions on Riemann surfaces*, Lecture Notes in Mathematics, Vol. 352, Springer-Verlag, Berlin-New York, 1973. MR0335789 (49 #569)

[53] K. Fukaya and K. Ono, *Arnold conjecture and Gromov-Witten invariant*, Topology **38** (1999), no. 5, 933–1048, DOI 10.1016/S0040-9383(98)00042-1. MR1688434 (2000j:53116)

[54] A. B. Givental, *Equivariant Gromov-Witten invariants*, Internat. Math. Res. Notices **13** (1996), 613–663, DOI 10.1155/S1073792896000414. MR1408320 (97e:14015)
[55] A. Givental, *A mirror theorem for toric complete intersections*, Topological field theory, primitive forms and related topics (Kyoto, 1996), Progr. Math., vol. 160, Birkhäuser Boston, Boston, MA, 1998, pp. 141–175. MR1653024 (2000a:14063)
[56] A. B. Givental, *Semisimple Frobenius structures at higher genus*, Internat. Math. Res. Notices **23** (2001), 1265–1286, DOI 10.1155/S1073792801000605. MR1866444 (2003b:53092)
[57] A. B. Givental, *Gromov-Witten invariants and quantization of quadratic Hamiltonians* (English, with English and Russian summaries), Mosc. Math. J. **1** (2001), no. 4, 551–568, 645. Dedicated to the memory of I. G. Petrovskii on the occasion of his 100th anniversary. MR1901075 (2003j:53138)
[58] A. B. Givental, *Symplectic geometry of Frobenius structures*, Frobenius manifolds, Aspects Math., E36, Friedr. Vieweg, Wiesbaden, 2004, pp. 91–112. MR2115767 (2005m:53172)
[59] M. Gross, L. Katzarkov, H. Ruddat, "Towards Mirror Symmetry for Varieties of General Type," `arXiv:1202.4042`.
[60] T. Graber and R. Pandharipande, *Localization of virtual classes*, Invent. Math. **135** (1999), no. 2, 487–518, DOI 10.1007/s002220050293. MR1666787 (2000h:14005)
[61] T. Graber and E. Zaslow, *Open-string Gromov-Witten invariants: calculations and a mirror "theorem"*, Orbifolds in mathematics and physics (Madison, WI, 2001), Contemp. Math., vol. 310, Amer. Math. Soc., Providence, RI, 2002, pp. 107–121, DOI 10.1090/conm/310/05400. MR1950943 (2004b:53149)
[62] S. Hamidi and C. Vafa, *Interactions on orbifolds*, Nuclear Phys. B **279** (1987), no. 3-4, 465–513, DOI 10.1016/0550-3213(87)90006-X. MR867241 (88c:81085)
[63] R. Harvey and H. B. Lawson Jr., *Calibrated geometries*, Acta Math. **148** (1982), 47–157, DOI 10.1007/BF02392726. MR666108 (85i:53058)
[64] K. Hori, C. Vafa, "Mirror symmetry," `arXiv:hep-th/0002222`.
[65] K. Hori, A. Iqbal, C. Vafa, "D-branes and mirror symmetry," `arXiv:hep-th/0005247`.
[66] Y. Hu and J. Li, *Genus-one stable maps, local equations, and Vakil-Zinger's desingularization*, Math. Ann. **348** (2010), no. 4, 929–963, DOI 10.1007/s00208-010-0504-8. MR2721647 (2011i:14022)
[67] M.-x. Huang, A. Klemm, and S. Quackenbush, *Topological string theory on compact Calabi-Yau: modularity and boundary conditions*, Homological mirror symmetry, Lecture Notes in Phys., vol. 757, Springer, Berlin, 2009, pp. 45–102. MR2596635 (2011c:81242)
[68] H. Iritani, *An integral structure in quantum cohomology and mirror symmetry for toric orbifolds*, Adv. Math. **222** (2009), no. 3, 1016–1079, DOI 10.1016/j.aim.2009.05.016. MR2553377 (2010j:53182)
[69] T. J. Jarvis and T. Kimura, *Orbifold quantum cohomology of the classifying space of a finite group*, Orbifolds in mathematics and physics (Madison, WI, 2001), Contemp. Math., vol. 310, Amer. Math. Soc., Providence, RI, 2002, pp. 123–134, DOI 10.1090/conm/310/05401. MR1950944 (2004a:14056)
[70] M. Kontsevich, *Intersection theory on the moduli space of curves and the matrix Airy function*, Comm. Math. Phys. **147** (1992), no. 1, 1–23. MR1171758 (93e:32027)
[71] M. Kontsevich, *Enumeration of rational curves via torus actions*, The moduli space of curves (Texel Island, 1994), Progr. Math., vol. 129, Birkhäuser Boston, Boston, MA, 1995, pp. 335–368. MR1363062 (97d:14077)
[72] J. Kaczorowski and A. Perelli, *A uniform version of Stirling's formula*. part 1, Funct. Approx. Comment. Math. **45** (2011), no. part 1, 89–96, DOI 10.7169/facm/1317045234. MR2865415 (2012h:33002)
[73] S. Katz and C.-C. M. Liu, *Enumerative geometry of stable maps with Lagrangian boundary conditions and multiple covers of the disc*, Adv. Theor. Math. Phys. **5** (2001), no. 1, 1–49. MR1894336 (2003e:14047)
[74] A. Klemm and E. Zaslow, *Local mirror symmetry at higher genus*, Winter School on Mirror Symmetry, Vector Bundles and Lagrangian Submanifolds (Cambridge, MA, 1999), AMS/IP Stud. Adv. Math., vol. 23, Amer. Math. Soc., Providence, RI, 2001, pp. 183–207. MR1876069 (2003a:14082)
[75] V. G. Knizhnik, *Analytic fields on Riemann surfaces. II*, Comm. Math. Phys. **112** (1987), no. 4, 567–590. MR910579 (89b:81199)

[76] Y.P. Lee and R. Pandharipande, *Frobenius manifolds, Gromov-Witten theory and Virasoro constraints.*

[77] A.-M. Li and Y. Ruan, *Symplectic surgery and Gromov-Witten invariants of Calabi-Yau 3-folds*, Invent. Math. **145** (2001), no. 1, 151–218, DOI 10.1007/s002220100146. MR1839289 (2002g:53158)

[78] J. Li, *Stable morphisms to singular schemes and relative stable morphisms*, J. Differential Geom. **57** (2001), no. 3, 509–578. MR1882667 (2003d:14066)

[79] J. Li, "A degeneration formula of Gromov-Witten invariants," J. Diff. Geom. **60** (2002), 199-293.

[80] J. Li, *Recent progress in GW-invariants of Calabi-Yau threefolds*, Current developments in mathematics, 2007, Int. Press, Somerville, MA, 2009, pp. 77–99. MR2532996 (2010i:14102)

[81] J. Li, C.-C. M. Liu, K. Liu, and J. Zhou, *A mathematical theory of the topological vertex*, Geom. Topol. **13** (2009), no. 1, 527–621, DOI 10.2140/gt.2009.13.527. MR2469524 (2009i:14082)

[82] J. Li and G. Tian, *Virtual moduli cycles and Gromov-Witten invariants of algebraic varieties*, J. Amer. Math. Soc. **11** (1998), no. 1, 119–174, DOI 10.1090/S0894-0347-98-00250-1. MR1467172 (99d:14011)

[83] J. Li and G. Tian, *Virtual moduli cycles and Gromov-Witten invariants of general symplectic manifolds*, Topics in symplectic 4-manifolds (Irvine, CA, 1996), First Int. Press Lect. Ser., I, Int. Press, Cambridge, MA, 1998, pp. 47–83. MR1635695 (2000d:53137)

[84] J. Li and A. Zinger, *On the genus-one Gromov-Witten invariants of complete intersections*, J. Differential Geom. **82** (2009), no. 3, 641–690. MR2534991 (2011e:14103)

[85] S. Li, "BCOV theory on the elliptic curve and higher genus mirror symmetry," `arXiv:1112.4063`.

[86] B. H. Lian, K. Liu, and S.-T. Yau, *Mirror principle. I*, Asian J. Math. **1** (1997), no. 4, 729–763. MR1621573 (99e:14062)

[87] B. H. Lian, K. Liu, and S.-T. Yau, *Mirror principle. III*, Asian J. Math. **3** (1999), no. 4, 771–800. MR1797578 (2002g:14080)

[88] C.-C. M. Liu, K. Liu, and J. Zhou, *A proof of a conjecture of Mariño-Vafa on Hodge integrals*, J. Differential Geom. **65** (2003), no. 2, 289–340. MR2058264 (2005m:14104)

[89] C.-C. M. Liu, K. Liu, and J. Zhou, *A formula of two-partition Hodge integrals*, J. Amer. Math. Soc. **20** (2007), no. 1, 149–184, DOI 10.1090/S0894-0347-06-00541-8. MR2257399 (2007e:14091)

[90] C.-C. M. Liu, *Moduli of J-holomorphic curves with Lagrangian boundary conditions*, ProQuest LLC, Ann Arbor, MI, 2002. Thesis (Ph.D.)–Harvard University. MR2703393

[91] C.-C. M. Liu, *Localization in Gromov-Witten theory and orbifold Gromov-Witten theory*, Handbook of moduli. Vol. II, Adv. Lect. Math. (ALM), vol. 25, Int. Press, Somerville, MA, 2013, pp. 353–425. MR3184181

[92] M. Mariño, *Open string amplitudes and large order behavior in topological string theory*, J. High Energy Phys. **3** (2008), 060, 34, DOI 10.1088/1126-6708/2008/03/060. MR2391060 (2009i:81149)

[93] D. Mumford, *Towards an enumerative geometry of the moduli space of curves*, Arithmetic and geometry, Vol. II, Progr. Math., vol. 36, Birkhäuser Boston, Boston, MA, 1983, pp. 271–328. MR717614 (85j:14046)

[94] D. Maulik, N. Nekrasov, A. Okounkov, and R. Pandharipande, *Gromov-Witten theory and Donaldson-Thomas theory. I*, Compos. Math. **142** (2006), no. 5, 1263–1285, DOI 10.1112/S0010437X06002302. MR2264664 (2007i:14061)

[95] D. Maulik, N. Nekrasov, A. Okounkov, and R. Pandharipande, *Gromov-Witten theory and Donaldson-Thomas theory. II*, Compos. Math. **142** (2006), no. 5, 1286–1304, DOI 10.1112/S0010437X06002314. MR2264665 (2007i:14062)

[96] D. Maulik, A. Oblomkov, A. Okounkov, and R. Pandharipande, *Gromov-Witten/Donaldson-Thomas correspondence for toric 3-folds*, Invent. Math. **186** (2011), no. 2, 435–479, DOI 10.1007/s00222-011-0322-y. MR2845622 (2012h:14140)

[97] D. Maulik and R. Pandharipande, *A topological view of Gromov-Witten theory*, Topology **45** (2006), no. 5, 887–918, DOI 10.1016/j.top.2006.06.002. MR2248516 (2007e:14092)

[98] A. Okounkov and R. Pandharipande, *Hodge integrals and invariants of the unknot*, Geom. Topol. **8** (2004), 675–699, DOI 10.2140/gt.2004.8.675. MR2057777 (2006c:14041)

[99] A. Popa, *The genus one Gromov-Witten invariants of Calabi-Yau complete intersections*, Trans. Amer. Math. Soc. **365** (2013), no. 3, 1149–1181, DOI 10.1090/S0002-9947-2012-05550-4. MR3003261

[100] S.-S. Roan, *On Calabi-Yau orbifolds in weighted projective spaces*, Internat. J. Math. **1** (1990), no. 2, 211–232, DOI 10.1142/S0129167X90000137. MR1060636 (91m:14064)

[101] D. Ross, *Localization and gluing of orbifold amplitudes: the Gromov-Witten orbifold vertex*, Trans. Amer. Math. Soc. **366** (2014), no. 3, 1587–1620, DOI 10.1090/S0002-9947-2013-05835-7. MR3145743

[102] D. Ross, *On the Gromov–Witten/Donaldson–Thomas Correspondence and Ruan's Conjecture for Calabi–Yau 3-Orbifolds*, Comm. Math. Phys. **340** (2015), no. 2, 851–864, DOI 10.1007/s00220-015-2438-1. MR3397033

[103] D. Ross and Z. Zong, *The gerby Gopakumar-Mariño-Vafa formula*, Geom. Topol. **17** (2013), no. 5, 2935–2976. MR3190303

[104] D. Ross and Z. Zong, *Cyclic Hodge integrals and loop Schur functions*, Adv. Math. **285** (2015), 1448–1486, DOI 10.1016/j.aim.2015.08.023. MR3406532

[105] Y. Ruan, *Virtual neighborhoods and pseudo-holomorphic curves*, Proceedings of 6th Gökova Geometry-Topology Conference, Turkish J. Math. **23** (1999), no. 1, 161–231. MR1701645 (2002b:53138)

[106] Y. Ruan and G. Tian, *A mathematical theory of quantum cohomology*, J. Differential Geom. **42** (1995), no. 2, 259–367. MR1366548 (96m:58033)

[107] Y. Ruan and G. Tian, *Higher genus symplectic invariants and sigma models coupled with gravity*, Invent. Math. **130** (1997), no. 3, 455–516, DOI 10.1007/s002220050192. MR1483992 (99d:58030)

[108] B. Siebert, "Gromov-Witten invariants of general symplectic manifolds," `arXiv:dg-ga/9608005`.

[109] R. P. Thomas, *A holomorphic Casson invariant for Calabi-Yau 3-folds, and bundles on $K3$ fibrations*, J. Differential Geom. **54** (2000), no. 2, 367–438. MR1818182 (2002b:14049)

[110] H.-H. Tseng, *Orbifold quantum Riemann-Roch, Lefschetz and Serre*, Geom. Topol. **14** (2010), no. 1, 1–81, DOI 10.2140/gt.2010.14.1. MR2578300 (2011c:14147)

[111] R. Vakil and A. Zinger, *A desingularization of the main component of the moduli space of genus-one stable maps into* $\mathbb{P}^n$, Geom. Topol. **12** (2008), no. 1, 1–95, DOI 10.2140/gt.2008.12.1. MR2377245 (2009b:14023)

[112] E. Witten, *Two-dimensional gravity and intersection theory on moduli space*, Surveys in differential geometry (Cambridge, MA, 1990), Lehigh Univ., Bethlehem, PA, 1991, pp. 243–310. MR1144529 (93e:32028)

[113] S. Yamaguchi and S.-T. Yau, *Topological string partition functions as polynomials*, J. High Energy Phys. **7** (2004), 047, 20 pp. (electronic), DOI 10.1088/1126-6708/2004/07/047. MR2095047 (2005k:14122)

[114] E. Zaslow, *Topological orbifold models and quantum cohomology rings*, Comm. Math. Phys. **156** (1993), no. 2, 301–331. MR1233848 (94i:32045)

[115] J. Zhou, "Local Mirror Symmetry for One-Legged Topological Vertex," `arXiv:0910.4320`; "Local Mirror Symmetry for the Topological Vertex," `arXiv:0911.2343`.

[116] J. Zhou, "Open string invariants and mirror curve of the resolved conifold," `arXiv:1001.0447`.

[117] A. Zinger, *The reduced genus 1 Gromov-Witten invariants of Calabi-Yau hypersurfaces*, J. Amer. Math. Soc. **22** (2009), no. 3, 691–737, DOI 10.1090/S0894-0347-08-00625-5. MR2505298 (2010c:14066)

[118] Z. Zong, *Generalized Mariño-Vafa formula and local Gromov-Witten theory of orbi-curves*, J. Differential Geom. **100** (2015), no. 1, 161–190. MR3326577

[119] Z. Zong, *Equivariant Gromov-Witten Theory of GKM Orbifolds*, ProQuest LLC, Ann Arbor, MI, 2015. Thesis (Ph.D.)–Columbia University. MR3312922

Bohan Fang, Beijing International Center for Mathematical Research, Peking University, 5 Yiheyuan Road, Beijing 100871, China
E-mail address: bohanfang@gmail.com

Department of Mathematics, Columbia University, 2990 Broadway, New York, New York 10027
E-mail address: ccliu@math.columbia.edu

Yau Mathematical Sciences Center, Tsinghua University, Jin Chun Yuan West Building, Tsinghua University, Haidian District, Beijing 100084, China
E-mail address: zyzong@math.tsinghua.edu.cn

Proceedings of Symposia in Pure Mathematics
Volume **93**, 2016
http://dx.doi.org/10.1090/pspum/093/01554

Symmetries and defects in three-dimensional topological field theory

Jürgen Fuchs and Christoph Schweigert

Abstract. Boundary conditions and defects of any codimension are natural parts of any quantum field theory. Surface defects in three-dimensional topological field theories of Turaev-Reshetikhin type have applications to two-dimensional conformal field theories, in solid state physics and in quantum computing. We explain an obstruction to the existence of surface defects that takes values in a Witt group. We then turn to surface defects in Dijkgraaf-Witten theories and their construction in terms of relative bundles; this allows one to exhibit Brauer-Picard groups as symmetry groups of three-dimensional topological field theories.

1. Introduction

In this contribution we discuss the relation between two important structures in quantum field theory: symmetries on the one hand, and defects of various codimensions on the other. While it is hardly necessary to emphasize the importance of symmetries, defects in quantum field theories or, in a different formulation, quantum field theories on stratified spaces, and their relation to symmetries are a more recent topic of interest.

Our exposition is organized as follows. We first discuss a few general aspects of topological codimension-1 defects, both in two-dimensional and in three-dimensional quantum field theories. A special emphasis is put on the relation between *invertible* topological defects and symmetries. Afterwards we consider defects for a particular class of three-dimensional topological field theories: Dijkgraaf-Witten theories. An advantage of these models is that they admit a mathematically precise formulation as gauge theories, see [**DW90**, **Fr95**, **Mo13**] and references therein. Apart from providing insights in a language that is close to standard field theoretical

---

2010 *Mathematics Subject Classification.* Primary: 81T45, Secondary: 57R56.

Key words and phrases. Topological field theory, tensor categories, topological defects, Brauer-Picard group.

The first author was supported by VR under project no. 621-2013-4207.

The second author was partially supported by the Collaborative Research Centre 676 "Particles, Strings and the Early Universe - the Structure of Matter and Space-Time", by the RTG 1670 "Mathematics inspired by String theory and Quantum Field Theory" and by the DFG Priority Programme 1388 "Representation Theory".

The authors are grateful to the Erwin-Schrödinger-Institute (ESI) for the hospitality during the programs "Modern trends in topological field theory" and "Topological Phases of Quantum Matter" while part of this work was done.

formulations, this allows one to uncover interesting relations between categories of (relative) bundles and recent results in representation theory.

2. Topological defects in quantum field theories

It has been recently realized in several different contexts that boundary conditions and defects of various codimensions constitute important parts of the structure of a quantum field theory. (Alternatively, one might say that it is instructive to consider the quantum field theory not only on smooth manifolds, but also on stratified spaces.) Codimension-1 defects, also called interfaces, separate regions that can support different quantum field theories. Such defects arise naturally in applications, ranging from condensed matter systems, where they appear as interfaces between different phases of matter, to domain walls in cosmology. The aspect we wish to emphasize in this note is the fact that such defects provide a lot of structural insight into quantum field theories, including in particular their symmetry structures.

Similarly as in the case of boundary conditions, there are various different types of defects one can consider. In particular, one can impose various conservation conditions of physical quantities on defects. An interesting subclass of codimension-1 defects are *topological* defects. They are characterized by the property that the values of correlators for configurations with defects do not change when the location of a defect is only slightly changed, that is, changed by a homotopy without crossing any field insertions or other defects. Examples of such defects have been known for a long time (see e.g. [**DW82**]). Consider, for instance, the two-dimensional Ising model, defined by a $\mathbb{Z}_2$-valued variable on the vertices of a two-dimensional CW-complex. Select a line that crosses bonds transversally, and change the coupling on each bond that is crossed by the line from ferromagnetic to anti-ferromagnetic. In the continuum limit this provides a topological defect line in the critical Ising model.

2.1. Symmetries from invertible topological defects.

A particularly important subclass of topological defects are the *invertible* topological defects. Owing to their topological nature, two topological defects can be brought to coincidence, leading to a fusion product of defects. The precise mathematical formulation of the relevant monoidal structure depends on the dimension in which the quantum field theory is defined. For the moment, let us consider two-dimensional quantum field theories. All statements made in the sequel have the status of theorems [**FFRS04**, **FFRS07**] in the case of two-dimensional rational conformal field theories; the general picture should be much more widely applicable, though. For a two-dimensional rational conformal field theory, the topological line defects form a monoidal category, with morphisms provided by field insertions which can change the type of defect. In particular, there is an "invisible defect", which is a monoidal unit $\mathbf{1}$ in the category of defects.

An *invertible* defect D is characterized by the fact that there exists another defect $D^\vee$ with the property that the fusion of D and $D^\vee$ gives the invisible defect,

$$D \otimes D^\vee \cong \mathbf{1} \cong D^\vee \otimes D \,.$$

(Thus in particular for every theory the invisible defect is invertible.) This behaviour leads to the relation

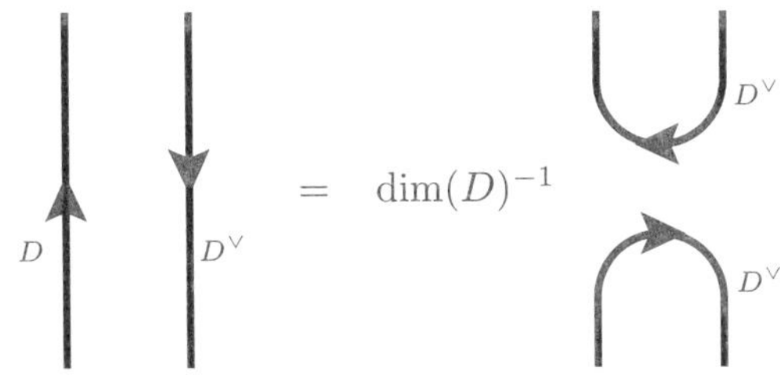

where $\dim(D)$ is the (quantum) dimension of the defect, which for an invertible defect can take the values $+1$ (in unitary theories this is the only possibility) or -1. This relation is to be understood as an identity of correlators when applied locally in any configuration of fields and defect lines. With this relation, it is immediate to deduce a connection with symmetries: one has equalities of the form

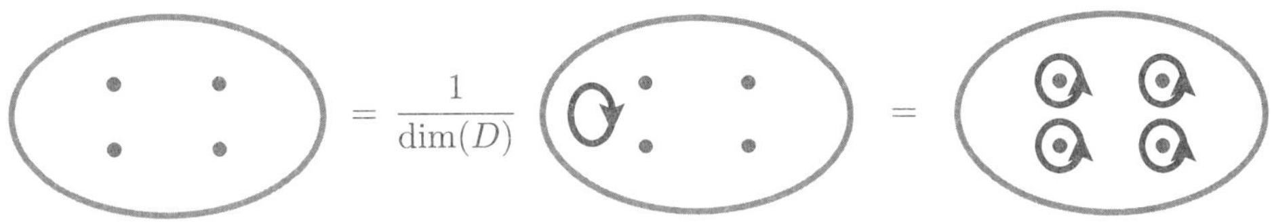

between correlators of different bulk fields.

Let us discuss this issue explicitly for the case of the two-dimensional critical Ising model. There are then three primary fields, commonly referred to as the identity field 1, the spin field σ and the energy field ϵ. Their conformal weights are 0, $\frac{1}{16}$ and $\frac{1}{2}$, respectively. Indecomposable topological line defects of the critical Ising model turn out to be in bijection with these fields: in particular, 1 corresponds to the invisible defect and ϵ to an invertible defect. (It is in fact the latter defect that amounts to changing the couplings from ferromagnetic to antiferromagnetic.) The action of this specific invertible defect on bulk fields is as follows: it leaves the bulk fields corresponding to 1 and to ϵ invariant and changes the sign of σ. Inside RCFT correlators for the critical Ising model we thus have the equalities

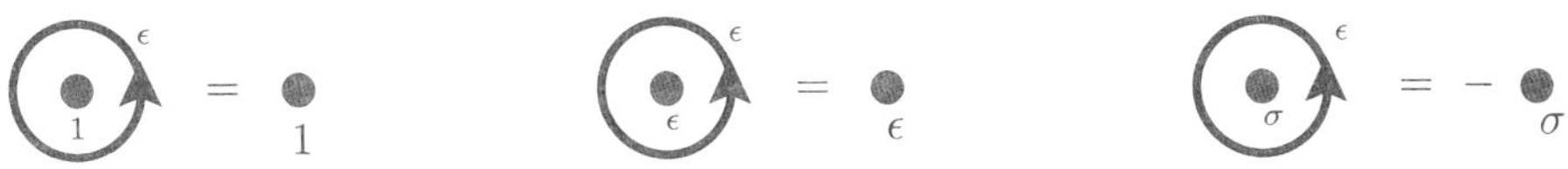

As can be seen from these pictures, there is a natural action of invertible topological line defects on field insertions: wrapping such a defect around a bulk field insertion yields another bulk field.

A further advantage of having a realization of symmetries in terms of invertible topological defects is seen by studying what happens when the defect is moved to a boundary with boundary condition M: this process yields another boundary

condition M', according to

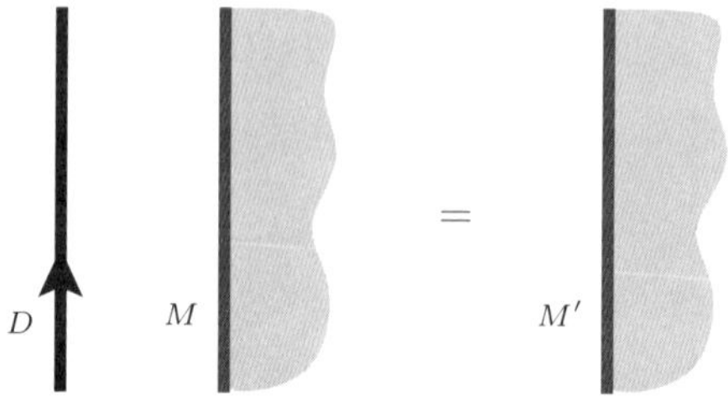

In a similar manner, by wrapping around various structures in an obvious way, the group of invertible topological defects acts as a symmetry group on *all* data of the field theory, including field insertions of bulk, boundary and disorder fields, boundary conditions and types of defects. In the framework of the TFT construction of correlators of two-dimensional rational conformal field theories [**SFR06**], isomorphism classes of invertible topological defects can be explicitly classified [**FFRS04, FFRS07**]. In the example of the critical Ising model one finds a symmetry group $\mathbb{Z}_2$, consisting of the invisible defect and the ϵ-defect described above, while e.g. for the critical three-state Potts model one obtains a non-abelian symmetry group S_3.

For the present purposes, it is important to realize that topological codimension-1 defects can also exist in more general classes of quantum field theories. Moreover, in the general case geometric considerations suggest a natural action of such defects on field theoretic structures, like boundary conditions and defects of various codimensions, as well. In short: defects can wrap around field theoretic structures.

2.2. T-dualities and Kramers-Wannier dualities from topological line defects. Before turning to higher-dimensional field theories, we point out that topological line defects are indeed of much wider use: they can also implement Kramers-Wannier dualities and T-dualities. If a general topological defect wraps around a bulk field, the following situation is created:

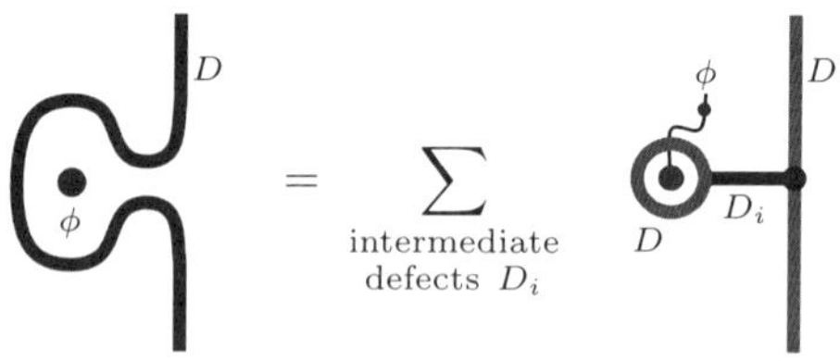

In this way a bulk field ϕ is turned by the defect D into a disorder field. To obtain an order-disorder duality, one also needs the opposite process, turning disorder fields into ordinary local bulk fields. It can be shown [**FFRS04, FFRS07**] that to this end the dual defect $D^\vee$ must be used, and that in this case one turns the disorder field back into a bulk field if and only if the fused defect $D \otimes D^\vee$ is a direct sum of invertible defects. This condition can be examined in concrete models. It is in particular satisfied for the defect corresponding to the spin field σ in the critical Ising model, thanks to the well known fusion rule $\sigma \otimes \sigma \cong 1 \oplus \epsilon$; this defect indeed produces the action of the Kramers-Wannier duality in the critical model.

Again we have a natural action on correlators:

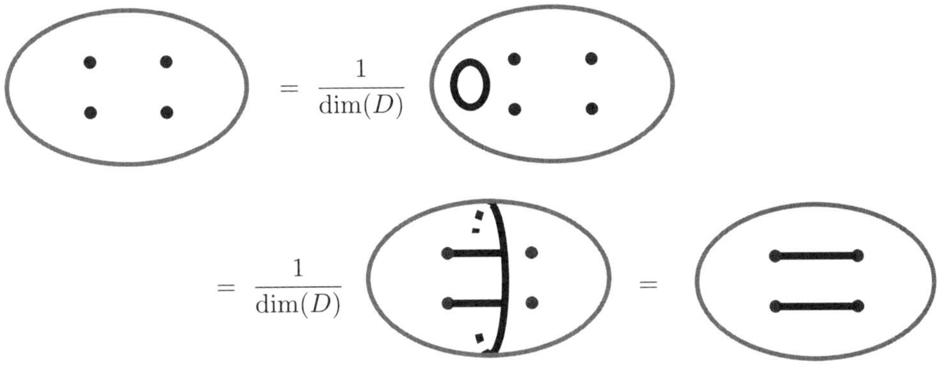

which shows how correlators involving only bulk fields are related to correlators involving disorder fields. Specifically, through the action of the σ-defect the correlator of four spin fields on a sphere can be seen to be equal to the correlator of four disorder fields, according to

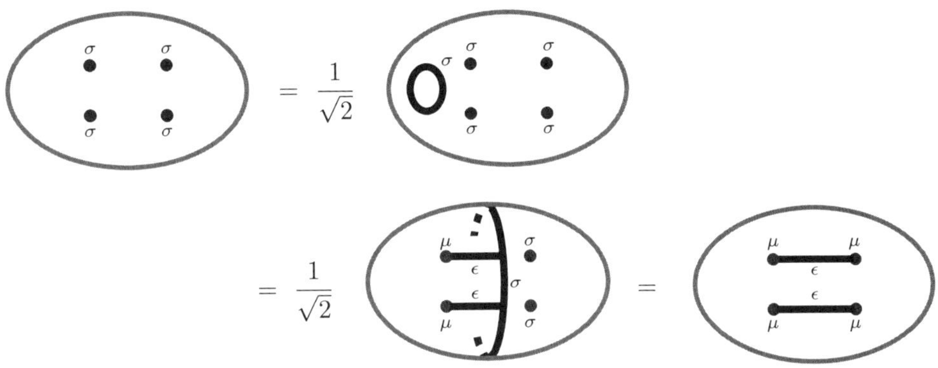

and the correlator of two spin fields on a torus can be expressed as

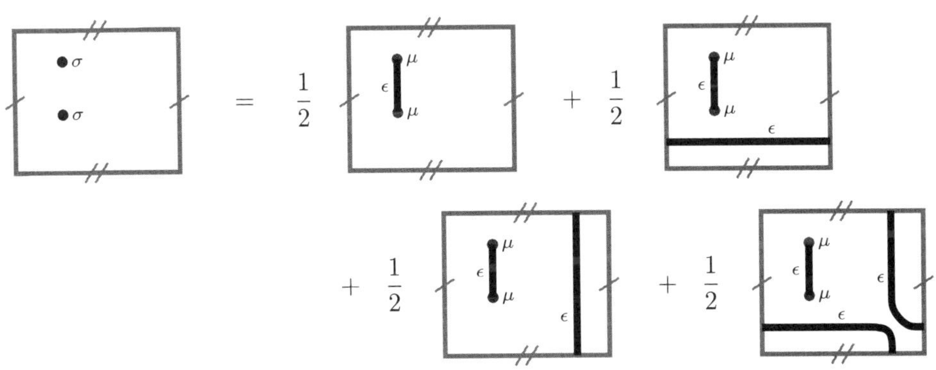

T-duality of compactified free bosons can be understood as a special type of order-disorder duality, see Section 5.4 of [**FGRS07**]. Indeed, the chiral data of two T-dual full conformal field theories can be described by the same theory, namely the one describing a $\mathbb{Z}_2$-orbifold of the free boson theory. The line defect implementing the duality is then constructed from twist fields of this orbifold theory.

2.3. Relative field theories. Before turning to defects in the specific class of three-dimensional Dijkgraaf-Witten theories, we briefly explain two applications of defects. The first is in the context of *relative field theories*. By definition, a $(d-1)$-dimensional relative field theory is a field theory on the boundary of a d-dimensional field theory or, more generally, on a codimension-1 defect. There are interesting cases in which the latter is a topological field theory; we mention two particular situations:

(1) The case that the d-dimensional topological field theory is even an invertible theory. This provides a geometric setup for describing anomalous theories; see [**FT14**].

(2) The case $d=3$, with the three-dimensional topological field theory being of Reshetikhin-Turaev type and built from a modular tensor category that is the representation category of a vertex algebra. In this case the relative theory on a topological surface defect is a local rational conformal field theory whose underlying chiral theory is the one built from the vertex algebra. This provides a geometric interpretation [**KS11**] of the TFT construction [**SFR06**] of RCFT correlators.

2.4. Quantum codes. Another application of surface defects in three-dimensional topological field theories are quantum codes. It is well known that the space of qubits of the toric code on a surface Σ can be described by a topological field theory of Turaev-Viro type. Indeed, the vector space assigned by the topological field theory to the surface Σ contains the information on a vector space together with an action of mapping class groups which is used to construct quantum gates. Two problems are apparent: First, the surfaces Σ of concrete samples typically have a low genus, and thus, according to the Verlinde formula, the TFT state space space $\mathrm{tft}(\Sigma)$ has low dimension. Moreover, the representation of the mapping class group might be too small to allow for universal gates. To circumvent these problems the idea has been put forward [**BW10**, **BQ12**, **BJQ13**] to consider samples with a bilayer (or multi-layer) system and twist defects that create branch cuts. They effectively lead to systems defined on surfaces of higher genus and thus to higher-dimensional quantum codes with richer representations of mapping class groups. Such systems have been analyzed mathematically [**FS14**] by exploiting results on permutation equivariant categories.

3. Defects and boundary conditions in three-dimensional topological field theories

We now turn our attention to a particularly accessible subclass of three-dimensional topological field theories, Dijkgraaf-Witten theories. These are topological field theories of Turaev-Viro type, and they can be constructed explicitly in a gauge-theoretic setting with finite gauge group. We treat them as 3-2-1-extended topological field theories.

3.1. Constructing extended Dijkgraaf-Witten theories from G-bundles. As a first input datum we select a finite group G. The space of field configurations on a closed oriented compact smooth three-manifold M is then taken to be the groupoid $\mathrm{Bun}_G(M)$ of G-bundles on M. By considering formally a path integral with vanishing action, we obtain a partition function that evaluates to

$$\mathrm{tft}_G(M) = \int_{\mathrm{Bun}_G(M)} \mathrm{D}A\ \mathrm{e}^0 = \big|\mathrm{Bun}_G(M)\big|\,.$$

Here we use the fact that the groupoid $\mathrm{Bun}_G(M)$ is essentially finite, so that there is a well-defined counting measure, its groupoid cardinality. The groupoid cardinality of an (essentially) finite groupoid Γ is

$$\big|\Gamma\big| := \sum_{\gamma\in\pi_0(\Gamma)} \frac{1}{|\mathrm{Aut}_\Gamma(\gamma)|}\,,$$

where the summation is over the set $\pi_0(\Gamma)$ of isomorphism classes of objects of Γ, each of which is counted with the inverse of the cardinality of its automorphism group.

Trivially, the number $\mathrm{tft}_G(M) = |\mathrm{Bun}_G(M)|$ defines an invariant of the three-manifold M. It is non-trivial, though, that this invariant is local in the following sense. Let us cut the closed three-manifold into two pieces, consisting of three-manifolds with boundaries that are closed oriented surfaces. Suppose that the topological field theory can associate meaningful quantities to such manifolds as well, which would allow us to reduce the problem of computing a three-manifold invariant to computing invariants of simpler manifolds. To see what the topological field theory tft_G should associate to a closed oriented two-manifold Σ, we take a three-manifold M with boundary Σ. For any G-bundle P on Σ, consider the groupoid $\mathrm{Bun}_G(M,P)$ of G-bundles on M that restrict to the G-bundle P on the boundary ∂M. Then $\mathrm{Bun}_G(M,P)$ is an essentially finite groupoid: its groupoid cardinality provides a function

$$\begin{aligned}\Psi_M:\quad \mathrm{Bun}_G(\Sigma) &\longrightarrow \mathbb{C}\\ P &\longmapsto |\mathrm{Bun}_G(M,P)|\end{aligned}$$

that depends only on the isomorphism class of the bundle P. To the surface Σ itself, we should associate the recipient for all these functions: the vector space $\mathbb{C}[\pi_0(\mathrm{Bun}_G(\Sigma))]$ of complex-valued functions on the set $\pi_0(\mathrm{Bun}_G(\Sigma))$ of isomorphism classes of G-bundles on Σ. We thus recover the well known feature that a three-dimensional topological field theory assigns vector spaces to surfaces. In the case of Dijkgraaf-Witten theories, these vector spaces are obtained by linearization of (isomorphism classes of objects of) categories of bundles.

Closed two-manifolds can, in turn, be decomposed, e.g. in a pair-of-pants decomposition, by cutting them along circles. Hence let S be a closed oriented one-manifold and ask what an extended topological field theory should associate to S. Following the same type of analysis as above, we choose a surface Σ with boundary S. Fixing a G-bundle P on the boundary S, we assign to it a vector space,

$$\Psi_\Sigma:\quad P \longmapsto \mathbb{C}[\pi_0(\mathrm{Bun}_G(\Sigma,P))]\,.$$

The so obtained map $\Psi_\Sigma\colon \mathrm{Bun}_G(S) \to \mathrm{Vect}$ is a vector bundle on the space of field configurations on S. Thus the extended Dijkgraaf-Witten theory based on G should associate to S the $\mathbb{C}$-linear category of vector bundles over the space of field configurations. Denoting, for an essentially finite groupoid Γ, the functor category from Γ to a category $\mathcal{C}$ by $[\Gamma,\mathcal{C}]$, this is the functor category

$$\mathrm{tft}_G(S) = [\mathrm{Bun}_G(S), \mathrm{Vect}]\,.$$

We thus learn that a 3-2-1-extended topological field theory assigns to a one-manifold a $\mathbb{C}$-linear category. The category $\mathrm{tft}_G(S^1)$ can be computed explicitly: the category $\mathrm{Bun}_G(S^1)$ is equivalent to the action groupoid $G//G$ for the adjoint action of G on itself. Indeed, after fixing a point $p \in S^1$, a G-bundle is described by its holonomy $g \in G$, on which gauge transformations act by the adjoint action. Thus $\mathrm{tft}_G(S^1) \simeq [G//G, \mathrm{Vect}]$, and this functor category is equivalent to the category of G-graded vector spaces $V = \bigoplus_{g\in G} V_g$ with a G-action such that $g(V_h) \subset V_{ghg^{-1}}$. Put differently, we deal with G-equivariant vector bundles on the group G.

This category is actually nothing but the *Drinfeld center* of the fusion category G-Vect of G-graded vector spaces. For later reference, we recall the definition of the Drinfeld center $\mathcal{Z}(\mathcal{A})$ of a monoidal category $\mathcal{A}$: an object of $\mathcal{Z}(\mathcal{A})$ is a pair $(U, c_{U,-})$ consisting of an object U of $\mathcal{A}$ and a functorial family of isomorphisms $c_{U,X}: U \otimes X \to X \otimes U$, called a half-braiding, subject to some further coherence properties; a morphism from $(U, c_{U,-})$ to $(V, c_{V,-})$ is a morphism $f: U \to V$ in $\mathcal{A}$ such that $(\mathrm{id}_- \otimes f) \circ c_{U,-} = c_{V,-} \circ (f \otimes \mathrm{id}_-)$. The half-braidings endow the Drinfeld center $\mathcal{Z}(\mathcal{A})$ with the structure of a *braided* monoidal category. Dropping the half-braiding gives a forgetful functor

$$\begin{array}{rcl} \mathcal{Z}(\mathcal{A}) & \to & \mathcal{A} \\ (U, c_{U-}) & \mapsto & U \end{array}$$

which has a natural monoidal structure. If a category $\mathcal{C}$ is already braided, then the braiding provides a braided monoidal functor $\mathcal{C} \to \mathcal{Z}(\mathcal{C})$. The opposite braiding endows the underlying category with another structure of a braided category, denoted by $\mathcal{C}^{\mathrm{rev}}$, and similarly gives a braided monoidal functor $\mathcal{C}^{\mathrm{rev}} \to \mathcal{Z}(\mathcal{C})$. These two functors combine into a braided monoidal functor

$$F: \quad \mathcal{C} \boxtimes \mathcal{C}^{\mathrm{rev}} \to \mathcal{Z}(\mathcal{C}) \tag{1}$$

from the Deligne product of $\mathcal{C}$ and $\mathcal{C}^{\mathrm{rev}}$ to the center. A braided category is called *non-degenerate* iff this functor F is a braided equivalence. A non-degenerate ribbon category is called a *modular* category.

The preceding considerations for Dijkgraaf-Witten theories lead us naturally to the general definition of 3-2-1-extended topological field theories as a symmetric monoidal 2-functor

$$\mathrm{tft}: \quad \mathrm{cobord}_{3,2,1} \to 2\text{-}\mathrm{Vect}(\mathbb{C})\,.$$

This definition naturally generalizes Atiyah's classical definition of (non-extended) topological field theory. In the rest of this subsection we elaborate on this 2-functor. In the definition, 2-Vect($\mathbb{C}$) is the symmetric monoidal bicategory of finitely semisimple $\mathbb{C}$-linear abelian categories, with the Deligne product as a monoidal structure. The bicategory $\mathrm{cobord}_{,3,2,1}$ is an extended cobordism category whose objects are closed oriented one-manifolds S, whose 1-morphisms $S \to S'$ are two-manifolds Σ with boundary together with a decomposition $\partial\Sigma \xrightarrow{\simeq} S' \sqcup -S$, and whose 2-morphisms are three-manifolds with corners, up to diffeomorphism, again with an appropriate decomposition of the boundary.

According to the definition, the 2-functor tft assigns to the oriented circle $\mathbb{S}^1$ a finitely semisimple $\mathbb{C}$-linear category $\mathcal{C} := \mathrm{tft}(\mathbb{S}^1)$. The trinion (pair of pants), regarded as a cobordism $\mathbb{S}^1 \sqcup \mathbb{S}^1 \to \mathbb{S}^1$ gives a functor $\otimes: \mathcal{C} \boxtimes \mathcal{C} \to \mathcal{C}$. Finally, suitable three-manifolds with corners provide natural isomorphisms that endow this functor with an associativity constraint and the category $\mathcal{C}$ with the structure of a braiding. The constraints on these natural transformations, expressed by pentagon and hexagon diagrams, can be deduced from homotopies between three-manifolds with corners. In this way the category $\mathcal{C} = \text{tft}(\mathbb{S}^1)$ is expected to be endowed with the structure of a modular tensor category (see [**BDSV14**] for some recent progress). The Reshetikhin-Turaev construction can be seen as a converse, constructing a 3-2-1-extended topological field theory from a modular tensor category.

3.2. Topological field theories with boundaries and defects. Incorporating defects and boundaries into a topological field theory amounts to consider an enlarged bicategory $\text{cobord}^{\partial}_{3,2,1}$ of cobordisms. As a warm-up, we recall the situation for two-dimensional topological field theories with boundaries, also known as open/closed topological field theory [**La01**, **AN06**, **MS06**, **LP08**]. Beyond disjoint unions of oriented circles, the category $\text{cobord}^{\partial}_{2,1}$ has intervals $\mathbb{I}_{a_1,a_2}$, with boundary conditions a_1, a_2 attached to the end points, as additional objects. Morphisms are now surfaces with boundaries consisting of segments which are are either "cut-and-paste boundaries" implementing locality or true physical boundaries.

A two-dimensional open/closed topological field theory is a symmetric monoidal functor $\text{tft}\colon \text{cobord}^{\partial}_{2,1} \to \text{Vect}$. It assigns a vector space

$$A_a := \text{tft}(\mathbb{I}_{aa})$$

to an interval with boundary condition a at either end point. For any boundary condition a, a disk with three marked intervals on the boundary provides a cobordism $\mathbb{I}_{aa} \sqcup \mathbb{I}_{aa} \to \mathbb{I}_{aa}$; applying the functor tft thus yields a linear map $A_a \otimes A_a \to A_a$. Further analysis [**MS06**, **LP08**] shows that this map furnishes a (not necessarily commutative) associative product on A_a, and that A_a is even endowed with the structure of a symmetric Frobenius algebra. A similar argument shows that for any other boundary condition b, the cobordism $\mathbb{I}_{aa} \sqcup \mathbb{I}_{ab} \to \mathbb{I}_{ab}$ endows the vector space $M_b := \text{tft}(\mathbb{I}_{ab})$ with a linear map $A_a \otimes M_b \to M_b$ and thereby acquires the structure of an A_a-module.

In this way one arrives at the following structure: boundary conditions form a $\mathbb{C}$-linear category that is equivalent to the category A_a-mod of A_a-modules for any given boundary condition a. To the interval $\mathbb{I}_{b_1,b_2}$ the topological field theory assigns

$$\text{tft}(\mathbb{I}_{b_1,b_2}) = \text{Hom}_{A_a\text{-mod}}(b_1, b_2)\,, \tag{2}$$

the vector space of morphisms of A_a-modules. In particular, for any boundary condition b, we find $\text{tft}(\mathbb{I}_{b,b}) = \text{End}_{A_a\text{-mod}}(b)$, which is an algebra that is Morita equivalent to the algebra A_a. If A_a is semisimple, then the commutative Frobenius algebra $C := \text{tft}(S^1)$ that is assigned to the circle turns out to be the center of the algebra A_a:

$$C = \text{tft}(S^1) = Z(\text{tft}(\mathbb{I}_{aa})) = Z(A_a)\,.$$

For three-dimensional topological field theories, we will find a higher-categorical analogue of these structures. In a complete treatment one should start by giving a precise definition for the extended category $\text{cobord}^{\partial}_{3,2,1}$ of cobordisms; several definitions have been proposed for this enlarged category, see e.g. [**Lu09**, Sect. 4.3]. Here we refrain from presenting a formal definition, and instead concentrate on some of the physical requirements that such a category should satisfy. Again we start from oriented three-manifolds, but this time we also allow them to have oriented boundaries that are actual physical boundaries to which boundary conditions need to be assigned, rather than the cut-boundaries that were already present in Section 3.1, which are instead a device for incorporating some aspects of locality into the theory. We also admit codimension-1 submanifolds, called surface defects. Such a defect separates two three-dimensional regions each of which supports a topological field theory, and the theories associated to the two regions may be different. Labels must be assigned to such defects, too. In this way, we get in particular invariants of three-manifold with embedded surfaces. In the present setting of 3-2-1-extended topological field theories, it is natural to go one step further and allow for codimension-2 defects as well. In particular, there will be generalized Wilson lines separating two surface defects from each other, and likewise boundary Wilson lines that are confined to a boundary component.

In a next step one imposes locality by cutting such a three-manifold to produce three-manifolds with cut-boundaries. It is natural to impose the condition that the cut-surface intersects all structure in the three-manifold, e.g. surface defects and boundaries, transversally. Boundary manifolds now carry a decoration consisting of lines labeled by surface defects and boundaries labeled by boundary conditions. Imposing further locality, as in Section 3.1, requires to cut such decorated surfaces transversally. This yields new types of labeled one-manifolds, which are to be taken as the objects in the bicategory $\mathrm{cobord}^{\partial}_{3,2,1}$. Let us describe these objects in detail: There are oriented intervals and circles, with finitely many marked points in the interior that divide them into subintervals. Each subinterval is to be marked with a topological field theory. Points adjacent to two subintervals are to be marked by a surface defect, boundary points by a boundary condition. The possible boundary conditions and types of surface defects have to be determined by a field-theoretic analysis. For the case of Dijkgraaf-Witten theories we present this analysis in the next subsection.

3.3. A model independent analysis of defects and boundaries. We now turn to a general analysis of boundary conditions and defects in three-dimensional topological field theories of Reshetikhin-Turaev type. The labels for the bulk Wilson lines in such a theory are supplied by some modular tensor category $\mathcal{C}$. The boundary conditions of the theory form a bicategory: its objects are boundary conditions, its 1-morphisms are labels for boundary Wilson lines, and its 2-morphisms label point-like insertions on boundary Wilson lines. Our goal is to describe this bicategory in terms related to the modular tensor category $\mathcal{C}$.

We want the boundary Wilson lines for a given boundary condition to be topological as well. As a consequence there is a fusion of boundary Wilson lines and we have invariance under two-dimensional isotopies. This leads to the requirement that for any boundary condition a the boundary Wilson lines and their insertions form a $\mathbb{C}$-linear, finitely semisimple, rigid monoidal category $\mathcal{W}_a$. The category $\mathcal{W}_a$ is thus a fusion category; since boundary Wilson lines are confined to the two-dimensional boundary, $\mathcal{W}_a$ is, however, not endowed with the structure of a braiding.

To derive constraints on the category $\mathcal{W}_a$, we postulate that there should exist a process of moving a bulk Wilson line to the boundary, and that this should be as smooth as possible. (In other words, we exclude possible boundary conditions for which such a smooth process does not exist.) This provides a functor $F_a : \mathcal{C} \to \mathcal{W}_a$. Next we impose the condition that the following two processes are equivalent: fusing two bulk Wilson lines and then moving them into the boundary, and moving the two bulk Wilson lines individually to the boundary and afterwards fusing them in

the boundary; schematically,

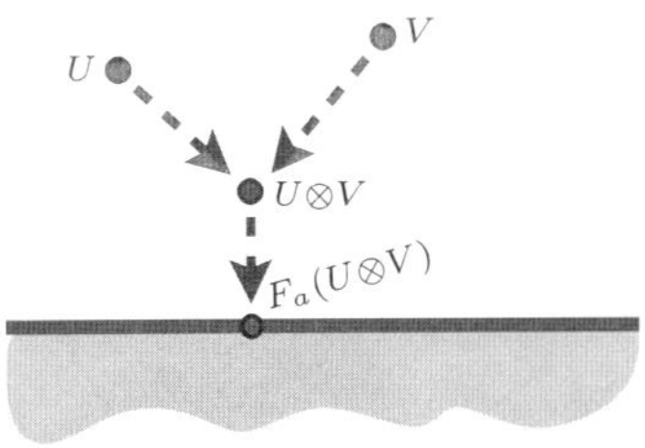

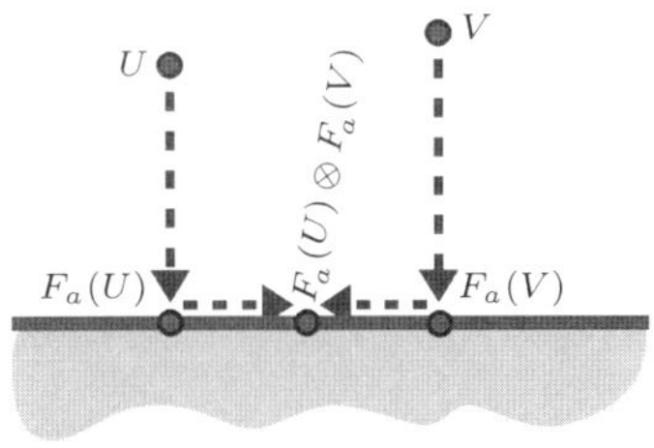

In formulas, what we postulate is the existence of coherent isomorphisms

$$F_a(U \otimes_{\mathcal{C}} V) \cong F_a(U) \otimes_{\mathcal{W}_a} F_a(V)$$

for all $U, V \in \mathcal{C}$. This means that the functor F_a has the structure of a *monoidal* functor.

Similarly, the following two processes should be equivalent:

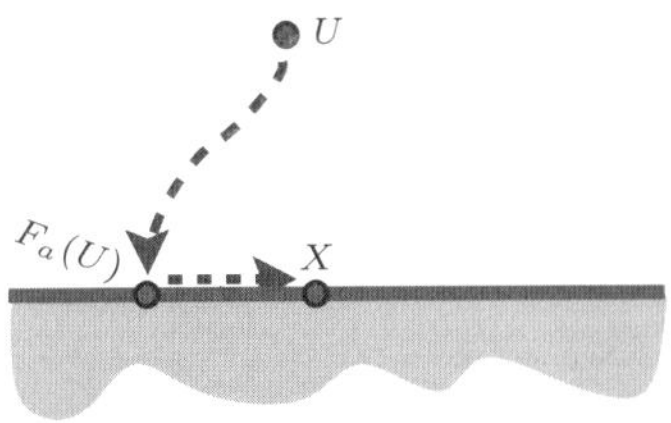

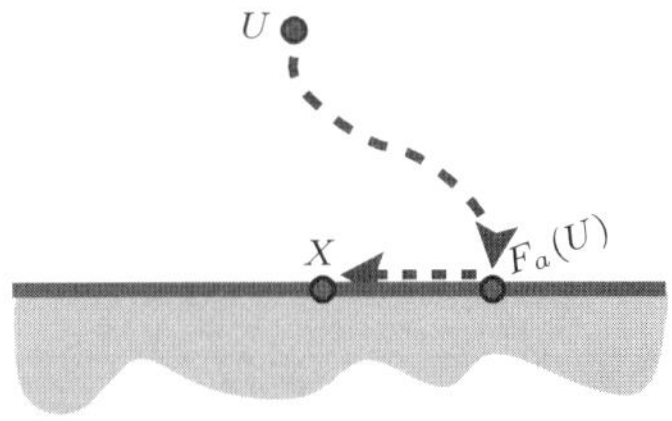

Here a bulk Wilson line labeled by an object U of $\mathcal{C}$ is moved to either side of a boundary Wilson line labeled by an object X of $\mathcal{W}_a$. Accordingly we postulate the existence of coherent isomorphisms

$$F_a(U) \otimes_{\mathcal{W}_a} X \cong X \otimes_{\mathcal{W}_a} F_a(U) .$$

This constitutes an additional structure on a monoidal functor from a braided category to a fusion category, called the structure of a *central* functor. It can be alternatively described as the choice of a lift $\widetilde{F}_a$ of F_a to the center of $\mathcal{W}_a$, i.e. as the existence of a braided monoidal functor $\widetilde{F}$ such that the diagram

$$\begin{array}{ccc} & & \mathcal{Z}(\mathcal{W}_a) \\ & \overset{\widetilde{F}_a}{\nearrow} & \downarrow \\ \mathcal{C} & \xrightarrow{F_a} & \mathcal{W}_a \end{array}$$

commutes, with the vertical arrow the forgetful functor described in Section 3.1. It is now an evident naturality requirement to ask that $\widetilde{F}$ is an *equivalence* of braided categories: the only reason for the existence of a consistent prescription for moving a specific boundary Wilson line past any arbitrary boundary Wilson line should be that this boundary Wilson line actually comes from a Wilson line in the bulk via the process described by the functor F_a.

We thus learn in particular that in order to admit boundary conditions of the type just introduced, a modular tensor category $\mathcal{C}$ must be a Drinfeld center. This constitutes a new phenomenon in three dimensions, for which there is no counterpart in two dimensions: while any commutative algebra (arising, as explained in Section 3.2, as the value of a two-dimensional topological field theory on a circle) is the center of some algebra (namely, of itself), it is *not* true that any braided monoidal fusion category (arising as the value of a three-dimensional topological field theory on a circle) is the Drinfeld center of a fusion category.

In this context, the following notions are relevant [**DMNO13**, **DNO13**]: Two modular tensor categories $\mathcal{C}_1$ and $\mathcal{C}_2$ are said to be *Witt equivalent* iff there exist fusion categories $\mathcal{A}_1$ and $\mathcal{A}_2$ and a braided equivalence

$$\mathcal{C}_1 \boxtimes \mathcal{Z}(\mathcal{A}_1) \xrightarrow{\simeq} \mathcal{C}_2 \boxtimes \mathcal{Z}(\mathcal{A}_2) .$$

The Deligne product induces on the set of equivalence classes a natural group structure; this group is called the Witt group of modular tensor categories. It contains, as a subgroup, the Witt group of finite abelian groups A that are endowed with a quadratic form $q\colon A \to \mathbb{C}^\times$. (Such a form determines a modular tensor category, a fact that is relevant for abelian Chern-Simons theories.) The obstruction to the existence of boundary conditions for a three-dimensional topological field theory based on a modular tensor category $\mathcal{C}$ can thus be restated as the condition that the class of $\mathcal{C}$ in the Witt group must vanish, $[\mathcal{C}] = 0$.

Suppose now we have already found one boundary condition a, i.e. one Witt-trivialization $\widetilde{F}_a\colon \mathcal{C} \xrightarrow{\simeq} \mathcal{Z}(\mathcal{W}_a)$. If b is any other boundary condition, denote by $\mathcal{W}_{a,b}$ the category of boundary Wilson lines separating a piece of boundary in boundary condition a from a piece in boundary condition b. Fusion of Wilson lines inside the piece in boundary condition a with a Wilson line separating a and b then yields a functor

$$\mathcal{W}_a \times \mathcal{W}_{a,b} \longrightarrow \mathcal{W}_{a,b} .$$

Hereby the category $\mathcal{W}_{a,b}$ acquires the structure of a module category over the monoidal category $\mathcal{W}_a$. This amounts to a categorification of the situation in two-dimensional topological field theory analyzed in Section 3.2. The bicategory of module categories over $\mathcal{W}_a$ is therefore a natural candidate for the bicategory of boundary conditions.

In much the same way as a left module M over a ring R is a right module over the ring $\mathrm{End}_R(M)$, the $\mathcal{W}_a$-module category $\mathcal{W}_{a,b}$ is also a right module category over the monoidal category $\mathcal{E}nd_{\mathcal{W}_a}(\mathcal{W}_{a,b})$ of module endofunctors of $\mathcal{W}_{a,b}$. But at the same time it is also a right module category over the fusion category $\mathcal{W}_b$ of Wilson lines inside the boundary with boundary condition b. Naturality thus suggests an equivalence $\mathcal{W}_b \simeq \mathcal{E}nd_{\mathcal{W}_a}(\mathcal{W}_{a,b})$. This can only hold true if the fusion category $\mathcal{E}nd_{\mathcal{W}_a}(\mathcal{W}_{a,b})$ is a Witt trivialization of $\mathcal{C}$ as well. Now indeed it is known [**Sc01**] that for any module category there is a canonical braided equivalence

$$\mathcal{Z}(\mathcal{W}_a) \simeq \mathcal{Z}(\mathcal{E}nd_{\mathcal{W}_a}(\mathcal{W}_{a,b})) ,$$

so that we do have a canonical Witt trivialization for $\mathcal{E}nd_{\mathcal{W}_a}(\mathcal{W}_{a,b})$ as well.

More generally, given two boundary conditions described by $\mathcal{W}_a$-module category $\mathcal{M}_b$ and $\mathcal{M}_c$, the category of boundary Wilson lines is given by the functor category $\mathcal{F}un_{\mathcal{W}_a}(\mathcal{M}_b, \mathcal{M}_c)$ of module functors. This can be regarded as a categorification of Equation (2). At this point a word of warning is in order: via pullback along the forgetful functor

$$\mathcal{C} \xrightarrow{\simeq} \mathcal{Z}(\mathcal{W}_a) \longrightarrow \mathcal{W}_a ,$$

any $\mathcal{W}_a$-module category has a natural structure of a $\mathcal{C}$-module category; however, not every $\mathcal{C}$-module category describes a boundary condition for the Reshetikhin-Turaev theory based on $\mathcal{C}$. As an illustration consider Kitaev's toric code, which is the Dijkgraaf-Witten theory based on $\mathbb{Z}_2$. In that case there are 6 inequivalent indecomposable $\mathcal{C}$-module categories over the Drinfeld center $\mathcal{C} = \mathcal{Z}(\mathrm{Vect}(\mathbb{Z}_2))$ of the category $\mathrm{Vect}(\mathbb{Z}_2)$ of $\mathbb{Z}_2$-graded complex vector spaces, but only 2 indecomposable module categories over the fusion category $\mathrm{Vect}(\mathbb{Z}_2)$. These two $\mathrm{Vect}(\mathbb{Z}_2)$-module categories correspond to the two known [**BK01**] elementary boundary conditions for the toric code.

The results obtained above are easily extended to surface defects that separate three-dimensional topological field theories of Reshetikhin-Turaev type based on two modular tensor categories $\mathcal{C}_1$ and $\mathcal{C}_2$. Again, one starts with a fusion category $\mathcal{W}_d$ of defect Wilson lines that are confined to the defect surface. We now get two monoidal functors

$$\mathcal{C}_1 \to \mathcal{W}_d \qquad \text{and} \qquad \mathcal{C}_2^{\mathrm{rev}} \to \mathcal{W}_d$$

which combine into a single functor

$$F_d : \quad \mathcal{C}_1 \boxtimes \mathcal{C}_2^{\mathrm{rev}} \to \mathcal{W}_d \,.$$

Again this functor F_d has the structure of a central functor, yielding a functor

$$\widetilde{F}_d : \quad \mathcal{C}_1 \boxtimes \mathcal{C}_2^{\mathrm{rev}} \xrightarrow{\simeq} \mathcal{Z}(\mathcal{W}_d)$$

that is again argued to be, by naturality, a braided equivalence. This leads to the following obstruction: surface defects of the type studied here only exist if the classes of the two modular tensor categories in the Witt group are equal, $[\mathcal{C}_1] = [\mathcal{C}_2]$.

Our discussion is, deliberately, not at the highest possible conceptual level. In analogy with the fact that Morita equivalent algebras can be naturally discussed in a bicategorical setting with bimodules as 1-morphisms, in the present situation there is an underlying 3-categorical structure. Also, we did not discuss fusion of topological surface defects. Indeed, surface defects form a monoidal bicategory. The monoidal unit is called the *transparent surface defect*, which separates a modular tensor category $\mathcal{C}$ from itself. The underlying Witt trivialization is based on $\mathcal{C}$, seen as a fusion category, and the braided equivalence

$$\mathcal{C} \boxtimes \mathcal{C}^{\mathrm{rev}} \xrightarrow{\simeq} \mathcal{Z}(\mathcal{C})$$

which follows by Equation (1) from the fact that $\mathcal{C}$ is modular. Ordinary Wilson lines are then recovered as defect Wilson lines separating the transparent defect from itself.

3.4. Defects and boundaries in Dijkgraaf-Witten theories. We now describe an explicit gauge-theoretic construction [**FSV14**] of boundary conditions and surface defects for Dijkgraaf-Witten theories. A key idea is to keep the two-step procedure outlined in Section 3.1: first we assign to manifolds of various dimension groupoids of (generalized) bundles, and then we linearize those categories.

Our ansatz is inspired by the notion of a *relative bundle*: Given a morphism $j : Y \to X$ of smooth manifolds and a group homomorphism $\iota : H \to G$, the category of relative bundles has as objects triples consisting of a G-bundle $P_G \to X$, an H-bundle $P_H \to Y$, and an isomorphism

$$\alpha : \quad \mathrm{Ind}_H^G(P_H) \xrightarrow{\cong} j^* P_G$$

of G-bundles on Y. A morphism is a pair consisting of a morphism $P_G \xrightarrow{\varphi_G} P'_G$ of G-bundles on X and a morphism $P_H \xrightarrow{\varphi_H} P'_H$ on Y of H-bundles, obeying the obvious compatibility constraint

$$\begin{array}{ccc} \mathrm{Ind}_H^G(P_H) & \xrightarrow{\alpha} & j^*P_G \\ {\scriptstyle \mathrm{Ind}_H^G(\varphi_H)}\downarrow & & \downarrow{\scriptstyle j^*\varphi_G} \\ \mathrm{Ind}_H^G(P'_H) & \xrightarrow{\alpha'} & j^*P'_G \end{array}$$

with the isomorphisms α and α'.

In our construction, relative bundles are the appropriate tool for describing boundaries. In the case of a surface defect $\Sigma \hookrightarrow M$ the situation is somewhat more subtle. Suppose that Σ divides the three-manifold M into two disjoint connected components $M_\pm$. We consider the situation that Dijkgraaf-Witten theories for two, possibly different, finite groups $G_\pm$ are associated to the components M_+ and M_-. Denote by $\overline{M}_\pm := M_\pm \cup \Sigma$ the closure of $M_\pm$ in M, and by $j_\pm : \Sigma \hookrightarrow \overline{M}_\pm$ the corresponding embeddings. We then consider, for a given group homomorphism $H \to G_+ \times G_-$ as field configurations the groupoid whose objects consist of two $G_\pm$-bundles $P_\pm \to M_\pm$, an H-bundle $P_H \to \Sigma$ and an isomorphism

$$\beta : \quad \mathrm{Ind}_H^G(P_H) \xrightarrow{\cong} j_+^*P_+ \times j_-^*P_-$$

of $G_+ \times G_-$-bundles on Σ.

We have to add yet one further datum to Dijkgraaf-Witten theories. Namely, typically, the linearization of groupoids is twisted by a 2-cocycle on the relevant groupoid of generalized bundles. This requires the choice of additional data on the finite groups involved. We first discuss the situation without boundaries and without surface defects, for a Dijkgraaf-Witten theory based on a finite group G. Then the additional datum is a three-cocycle $\omega \in Z^3(G, \mathbb{C}^\times)$ on the group. A more geometrically inclined reader is invited to think about this cocycle as a three-cocycle on the stack Bun_G of G-bundles, and thus as a (Chern-Simons) 2-gerbe on Bun_G. Such a three-cocycle leads to a holonomy on closed three-manifolds and thus furnishes a topological Lagrangian, yielding a three-dimensional topological field theory $\mathrm{tft}_{G,\omega}$. To determine the category $\mathrm{tft}_{G,\omega}(\mathbb{S}^1)$, we need a 2-cocycle $\tau(\omega)$ on the groupoid $\mathrm{Bun}_G(\mathbb{S}^1) \simeq G//G$, the action groupoid for the adjoint action of G on itself. The transgressed 2-cocycle $\tau(\omega)$ can be obtained [**Wi08**] from the group cocycle ω by evaluating ω on a suitable triangulated three-manifold. This indeed produces the well-known 2-cocycle [**DPR90**] for Dijkgraaf-Witten theories.

We now illustrate the generalization of this prescription to boundaries and defects in an example. Let $\mathbb{I}$ be an interval with a marked point in its interior, corresponding to a surface defect. To each of the two subintervals $\mathbb{I}_{1,2}$ we a assign a finite group $G_{1,2}$ and three-cocycles $\omega_{1,2} \in Z^3(G_{1,2}, \mathbb{C}^\times)$, of which we think of as topological bulk Lagrangians. To the end point adjacent to the first interval, we assign a group homomorphism $\iota_1 \colon H_1 \to G_1$ and a boundary Lagrangian as follows. We think about the three-cocycle ω_1 on G_1 as a Chern-Simons 2-gerbe on Bun_{G_1}. The group homomorphism ι_1 induces a morphism $\mathrm{Bun}_{H_1} \to \mathrm{Bun}_{G_1}$ of 2-gerbes. A similar situation, one categorical dimension lower, is familiar from the study of D-branes, see e.g. [**FNSW10**]: for D-branes, one has a gerbe on the bulk manifold and a morphism of gerbes on the boundary, from the trivial gerbe to the restriction of the bulk gerbe. This is an instance of the general principle that in gauge theories, holonomies of manifolds with boundary, and thus topological actions on such manifolds, can only be defined if a trivialization on the boundary is chosen.

In the situation at hand, what is to be trivialized is the restriction of the Chern-Simons 2-gerbe to Bun_{H_1}. Thus in terms of group cochains, we are looking for a 2-cochain $\theta_1 \in C^2(H_1, \mathbb{C}^\times)$ that represents a morphism from the trivial 2-gerbe on Bun_H to the 2-gerbe described by the restriction of the three-cocycle ω_1 on G_1 to Bun_{H_1}. Accordingly we impose the condition $\mathrm{d}\theta_1 = \iota_1^*(\omega_1)$. We think about θ_1 as a topological boundary Lagrangian. The situation at the other end point is, in complete analogy, described by a group homomorphism $\iota_2\colon H_2 \to G_2$ and a 2-cochain $\theta_2 \in C^2(H_2, \mathbb{C}^\times)$ such that $\mathrm{d}\theta_2 = \iota_2^*(\omega_2)$.

In a similar vein, the situation for the surface defect is a higher-categorical analogue of bibranes [**FSW08**], which describe the target space physics of topological defects in two-dimensional conformal field theory: We now have a group homomorphism $\iota_{12}\colon H_{12} \to G_1 \times G_2$ and a 2-cochain $\theta_{12} \in C^2(H_{12}, \mathbb{C}^\times)$ satisfying $\mathrm{d}\theta_{12} = \tilde\iota_2^*(\omega_2) \cdot \tilde\iota_1^*(\omega_1)^{-1}$, with the group homomorphisms $\tilde\iota_i\colon H_{12} \xrightarrow{\iota_{12}} G_1 \times G_2 \xrightarrow{p_i} G_i$ obtained from ι_{12} and the canonical projections. Indecomposable boundary conditions and defects correspond to group homomorphisms $\iota_{1,2}$, respectively ι_{12}, that are injective, i.e. are subgroup embeddings.

In [**FSV14**], an explicit prescription was given for transgressing these cochains to a 2-cocycle on the relevant category of (generalized) relative bundles. This prescription is a generalization of the one given in [**Wi08**] for Dijkgraaf-Witten theories without boundaries or defects. It makes the categories associated to one-manifolds with defects and boundaries quite explicitly computable.

On the other hand, there are also general results for boundary conditions and defects in three-dimensional topological field theories of Reshetikhin-Turaev type. Suppose the theory is based on a modular tensor category $\mathcal{C}$. The model independent analysis of Section 3.3 reveals that topological boundary conditions only exist if the category $\mathcal{C}$ is braided equivalent to the Drinfeld center of a fusion category $\mathcal{A}$. For Dijkgraaf-Witten theories based on a finite group G and a three-cocycle $\omega \in Z^3(G, \mathbb{C}^\times)$, this condition is satisfied automatically: the category $\mathcal{C} = \mathcal{C}_{G,\omega}$ is the Drinfeld center of the category $\mathrm{Vect}(G)^\omega$ of G-graded vector spaces with associator twisted by the three-cocycle ω.

Recall from the general analysis in Section 3.3 that boundary conditions are in bijection with module categories over the fusion category $\mathcal{A}$. For the fusion category $\mathrm{Vect}(G)^\omega$ relevant for Dijkgraaf-Witten theories, indecomposable module categories are, in turn, known [**Os03**] to be in bijection with subgroups $\iota\colon H \hookrightarrow G$ and 2-cochains $\theta \in C^2(H, \mathbb{C}^\times)$ such that $\mathrm{d}\theta = \iota^*\omega$. This matches exactly the gauge-theoretic description of types of boundary conditions. According to the general results [**FSV13**], the topological field theory should assign to an interval with Dijkgraaf-Witten theory based on (G, ω) and boundaries with boundary conditions (H_1, θ_1) and (H_2, θ_2) the category

$$\mathrm{Fun}_{\mathrm{Vect}(G)^\omega}(\mathcal{M}(H_1, \theta_1), \mathcal{M}(H_2, \theta_2))$$

of module functors from the indecomposable module category $\mathcal{M}(H_1, \theta_1)$ over the fusion category $\mathrm{Vect}(G)^\omega$ to the indecomposable module category $\mathcal{M}(H_2, \theta_2)$. A non-trivial calculation [**FSV14**] shows that this category coincides, as a finitely semisimple abelian category, with the category obtained by linearizing categories of generalized relative bundles.

A similar calculation can be performed for the circle with surface defects as point-like insertions. This requires a higher-categorical version of the notion of a trace, namely a trace 2-functor on bimodule categories with values in $\mathbb{C}$-linear categories. Such a coherent trace can indeed be constructed [**FSS14**, Sect. 3] and gives rise to results that are in accordance with gauge theoretic calculations in Dijkgraaf-Witten theories [**FSS14**, Sect. 5].

3.5. Symmetries from invertible topological surface defects. We now turn to the relation between symmetries of three-dimensional topological field theories of Turaev-Viro type and invertible surface defects. For such a theory, the modular tensor category $\mathcal{C}$ of bulk Wilson lines is the Drinfeld center of a fusion category $\mathcal{A}$, i.e. $\mathcal{C} = \mathcal{Z}(\mathcal{A})$.

According to the paradigm discussed in Section 1, symmetries should correspond to invertible topological surface defects, and their action should be described by wrapping those defects. By the analysis presented in the preceding subsection, invertible surface defects are described by invertible bimodule categories over the fusion category $\mathcal{A}$. These form a monoidal bicategory; restricting to only invertible 1-morphisms and 2-morphisms in this bicategory, one obtains a categorical 2-group, the Brauer-Picard 2-group [**ENOM10**]. Its isomorphism classes form a finite group $\mathrm{BrPic}(\mathcal{A})$; in the sequel work with this group, rather than the underlying 2-group.

Applying an extended topological field theory to a two-manifold with boundary yields a functor. Take this two-manifold to be a cylinder with a surface defect of type D wrapping the non-contractible cycle:

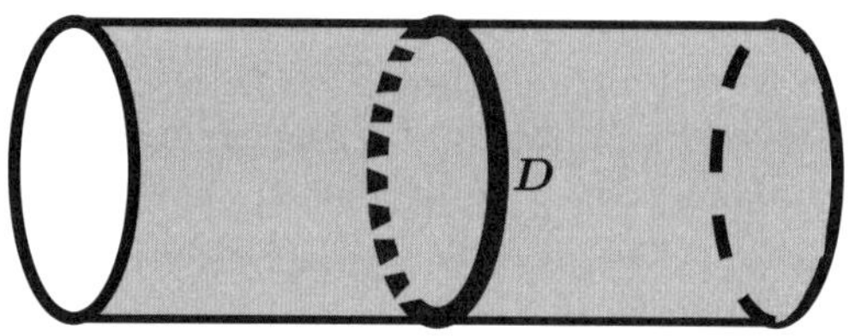

This yields an endofunctor $F_D\colon \mathcal{C} \to \mathcal{C}$. A detailed analysis [**FPSV15**] shows that if the defect D is invertible, then the functor F_D has the structure of a braided monoidal autoequivalence. In the case of Dijkgraaf-Witten theories, the modular tensor category of bulk Wilson lines is $\mathcal{C} = \mathcal{Z}(\mathrm{Vect}(G))$, and an invertible (and thus indecomposable) defect is given by a subgroup $\iota\colon H \hookrightarrow G \times G$ and a 2-cochain $\theta \in C^2(H, \mathbb{C}^\times)$. The functor $F_D \equiv F_{(H,\theta)}$ can then be explicitly obtained by linearizing the span of action groupoids

$$\begin{array}{ccc} & H//H & \\ {}^{\hat\pi_1}\swarrow & & \searrow^{\hat\pi_2} \\ G//G & & G//G \end{array}$$

for adjoint actions which describe the relevant categories of (relative) bundles. Here $\hat\pi_i$ is the functor on groupoids that is induced from the group homomorphism $H \xrightarrow{\iota} G \times G \xrightarrow{p_i} G$. One can then compute a pull-push functor on the linearizations to arrive at the functor

$$F_{(H,\theta)}: \quad \mathcal{Z}(\mathrm{Vect}(G)) = [G//G, \mathrm{Vect}] \longrightarrow [G//G, \mathrm{Vect}]\,,$$

which comes with a natural monoidal structure and which is braided. This construction provides an explicit group homomorphism

$$\mathrm{BrPic}(\mathcal{A}) \longrightarrow \mathrm{brdEq}(\mathcal{Z}(\mathrm{Vect}(G)))$$

from the Brauer-Picard group to the group $\mathrm{brdEq}(\mathcal{Z}(\mathrm{Vect}(G)))$ of braided autoequivalences of the Drinfeld center of $\mathrm{Vect}(G)$. A detailed comparison [**FPSV15**] shows that this group homomorphism is exactly the one considered in [**ENOM10**]; this embeds the construction of that paper naturally into the framework of topological field theories with defects. It is also shown in [**ENOM10**] that the group homomorphism is actually an *iso*morphism. According to the considerations above, this representation theoretic result can be reinterpreted as the statement that symmetries of topological field theories of Turaev-Viro type can already be detected from their action on the category of bulk Wilson lines.

3.6. Symmetries for abelian Dijkgraaf-Witten theories. We now restrict our attention to the class of Dijkgraaf-Witten theories with vanishing three-cocycle and based on a finite abelian group A. In this case the Brauer-Picard group is explicitly known [**ENOM10**]:

$$\mathrm{BrPic}(\mathrm{Vect}(A)) \cong \mathrm{O}_q(A \oplus A^*)\,,$$

where A^* is the character group of A, $q\colon A \oplus A^* \to \mathbb{C}^\times$ is the natural quadratic form that is determined by $q(g,\chi) = \chi(g)$ for $g \in A$ and $\chi \in A^*$, and $\mathrm{O}_q(A \oplus A^*)$ is the subgroup of those automorphisms of the finite group $A \oplus A^*$ that preserve this quadratic form q.

There are three obvious types of symmetries the Dijkgraaf-Witten theory should possess:

(1) *Symmetries of* Bun_A.

As explained in Section 3, we can think about the stack Bun_A as a target space for Dijkgraaf-Witten theories. Accordingly, symmetries of Bun_A, i.e. group automorphisms of the abelian group A, can be expected to be kinematical symmetries of the Dijkgraaf-Witten theory.

Indeed, for $\varphi \in \mathrm{Aut}(A)$ the graph $graph(\varphi) \subset A \oplus A$ is a subgroup; together with the trivial 2-cochain $\theta = 1 \in C^2(graph(\varphi), \mathbb{C}^\times)$ it describes an invertible surface defect. Following the prescription given in Section 3.5, we compute the corresponding braided autoequivalence to be the element

$$\varphi \oplus (\varphi^*)^{-1}: \quad A \oplus A^* \to A \oplus A^*$$

of $\mathrm{O}_q(A \oplus A^*)$.

(2) *Automorphisms of the Chern-Simons 2-gerbe on* Bun_A.

Recall that the three-cocycle $\omega \in C^3(A, \mathbb{C}^\times)$ has the interpretation of a Chern-Simons 2-gerbe on Bun_A. In the case at hand this 2-gerbe is trivial. Nevertheless its automorphisms are not trivial; rather, they are given by 1-gerbes on Bun_A. Their isomorphism classs are given by the cohomology group $H^2(A, \mathbb{C}^\times)$. By transgression [**Wi08**], any cohomology class gives an alternating bihomomorphism $\beta\colon A \times A \to \mathbb{C}^\times$, which in physical terms can be interpreted as a "B-field". (Indeed, for a finite abelian group, transgression provides a group isomorphism between the cohomology group $H^2(A, \mathbb{C}^\times)$ and the group of alternating bihomomorphisms.)

Again we can explicitly identify an invertible topological surface defect: the one given by the diagonal subgroup $A_{\mathrm{diag}} \subset A \oplus A$ together with a representative of a cohomology class in $H^2(A, \mathbb{C}^\times)$, which we identify with the associated alternating bihomomorphism β. The corresponding braided equivalence can be computed to act as

$$\begin{aligned} A \oplus A^* &\longrightarrow A \oplus A^* \\ (g, \chi) &\longmapsto (g, \chi + \beta(g, -)) . \end{aligned}$$

Thus this type of automorphism shifts the character by the contraction of the B-field β with the group element.

(3) *Electric-magnetic dualities.*

It is finally expected that electric-magnetic dualities form a class of symmetries of Dijkgraaf-Witten theories (see e.g. [**BCKA13**] for a discussion of such symmetries for Turaev-Viro theories based on finite-dimensional semisimple Hopf algebras).

To present a simple example, assume that A is a cyclic group and fix a group isomorphism $\delta\colon A \to A^*$. Then there is a natural braided equivalence

$$\begin{aligned} A \oplus A^* &\longrightarrow A \oplus A^* \\ (g, \chi) &\longmapsto (\delta^{-1}(\chi), \delta(g)) . \end{aligned}$$

This can again be described by an invertible surface defect, namely the one given by the diagonal subgroup $A_{\rm diag} \subset A \oplus A$ and the alternating bihomomorphism β defined by

$$\beta(g_1, g_2) := \frac{\delta(g_1)(g_2)}{\delta(g_2)(g_1)} \qquad \text{for} \quad g_1, g_2 \in A .$$

A careful investigation [**FPSV15**] of the structure of the finite group $\mathrm{O}_q(A \oplus A^*)$ shows that in fact the three types of symmetries listed above generate this group, i.e. they already provide *all* symmetries of the theory.

4. Conclusions

Here is a brief summary of crucial insights presented in this contribution:

- Topological defects are important structures in quantum field theories.
- By general principles they describe symmetries and dualities of quantum field theories in such a way that the action of these symmetries and dualities on all different structures of a quantum field theory becomes apparent. This includes the action on field insertions as well as the action on boundary conditions and defects of various codimensions.
- Topological codimension-1 defects and boundaries provide a natural setting for studying relative field theories. This includes in particular anomalous field theories and the TFT construction of RCFT correlators.
- Applications of topological defects virtually concern all fields of physics to which quantum field theory is applied.

There are several classes of quantum field theories in which topological defects can be treated quite explicitly. Besides two-dimensional rational conformal field theories, these include three-dimensional topological field theories of Reshetikhin-Turaev type. We have shown that a natural class of defects and boundary conditions in such theories is obstructed, with an obstruction taking values in the Witt group of non-degenerate braided fusion categories. We have also demonstrated, for the subclass of Dijkgraaf-Witten theories, how defects and boundary conditions can be formulated in a gauge-theoretic setting using (generalizations of) relative bundles.

This provides a natural embedding of structures of categorified representation theory, in particular of the theory of module and bimodule categories over monoidal categories, into the more comprehensive setting of (extended) topological field theory.

References

[AN06] A. Alexeevski and S. Natanzon, *Noncommutative two-dimensional topological field theories and Hurwitz numbers for real algebraic curves*, Selecta Math. (N.S.) **12** (2006), no. 3-4, 307–377, DOI 10.1007/s00029-006-0028-y. MR2305607 (2008a:57032)

[BJQ13] M. Barkeshli, C. M. Jian and X. L. Qi, *Twist defects and projective non-Abelian braiding statistics, Phys. Rev. B* **87** (2013) 045130, cond-mat/1208.4834

[BQ12] M. Barkeshli and X. L. Qi, *Topological nematic states and non-abelian lattice dislocations, Phys. Rev. X* **2** (2012) 031013, cond-mat/1112.3311

[BW10] M. Barkeshli and X. G. Wen, $U(1) \times U(1) \rtimes \mathbb{Z}_2$ *Chern-Simons theory and* $\mathbb{Z}_4$ *parafermion fractional quantum Hall states, Phys. Rev. B* **81** (2010) 045323, cond-mat/0910.2483

[BDSV14] B. Bartlett, C. L. Douglas, C. J. Schommer-Pries and J. Vicary, *Extended 3-dimensional bordism as the theory of modular objects*, math.GT/1411.0945

[BK01] S. B. Bravyi and A. Yu. Kitaev, *Quantum codes on a lattice with boundary, Quantum Computers and Computing* **2** (2001) 43–48, quant-ph/9811052

[BCKA13] O. Buerschaper, M. Christandl, L. Kong, and M. Aguado, *Electric-magnetic duality of lattice systems with topological order*, Nuclear Phys. B **876** (2013), no. 2, 619–636, DOI 10.1016/j.nuclphysb.2013.08.014. MR3143453

[DMNO13] A. Davydov, M. Müger, D. Nikshych, and V. Ostrik, *The Witt group of non-degenerate braided fusion categories*, J. Reine Angew. Math. **677** (2013), 135–177. MR3039775

[DNO13] A. Davydov, D. Nikshych, and V. Ostrik, *On the structure of the Witt group of braided fusion categories*, Selecta Math. (N.S.) **19** (2013), no. 1, 237–269, DOI 10.1007/s00029-012-0093-3. MR3022755

[DPR90] R. Dijkgraaf, V. Pasquier, and P. Roche, *Quasi Hopf algebras, group cohomology and orbifold models*, Nuclear Phys. B Proc. Suppl. **18B** (1990), 60–72 (1991), DOI 10.1016/0920-5632(91)90123-V. Recent advances in field theory (Annecy-le-Vieux, 1990). MR1128130 (92m:81238)

[DW90] R. Dijkgraaf and E. Witten, *Topological gauge theories and group cohomology*, Comm. Math. Phys. **129** (1990), no. 2, 393–429. MR1048699 (91g:81133)

[DW82] K. Drühl and H. Wagner, *Algebraic formulation of duality transformations for abelian lattice models*, Ann. Physics **141** (1982), no. 2, 225–253, DOI 10.1016/0003-4916(82)90286-X. MR673981 (84h:82064)

[ENOM10] P. Etingof, D. Nikshych, and V. Ostrik, *Fusion categories and homotopy theory*, Quantum Topol. **1** (2010), no. 3, 209–273, DOI 10.4171/QT/6. With an appendix by Ehud Meir. MR2677836 (2011h:18007)

[Fr95] D. S. Freed, *Classical Chern-Simons theory. I*, Adv. Math. **113** (1995), no. 2, 237–303, DOI 10.1006/aima.1995.1039. MR1337109 (96h:58019)

[FT14] D. S. Freed and C. Teleman, *Relative quantum field theory*, Comm. Math. Phys. **326** (2014), no. 2, 459–476, DOI 10.1007/s00220-013-1880-1. MR3165462

[FFRS04] J. Fröhlich, J. Fuchs, I. Runkel, and C. Schweigert, *Kramers-Wannier duality from conformal defects*, Phys. Rev. Lett. **93** (2004), no. 7, 070601, 4, DOI 10.1103/PhysRevLett.93.070601. MR2115752 (2005h:81325)

[FFRS07] J. Fröhlich, J. Fuchs, I. Runkel, and C. Schweigert, *Duality and defects in rational conformal field theory*, Nuclear Phys. B **763** (2007), no. 3, 354–430, DOI 10.1016/j.nuclphysb.2006.11.017. MR2299121 (2008d:81208)

[FGRS07] J. Fuchs, M. R. Gaberdiel, I. Runkel, and C. Schweigert, *Topological defects for the free boson CFT*, J. Phys. A **40** (2007), no. 37, 11403–11440, DOI 10.1088/1751-8113/40/37/016. MR2372400 (2009i:81138)

[FNSW10] J. Fuchs, T. Nikolaus, C. Schweigert, and K. Waldorf, *Bundle gerbes and surface holonomy*, European Congress of Mathematics, Eur. Math. Soc., Zürich, 2010, pp. 167–195, DOI 10.4171/077-1/8. MR2648325 (2011i:53022)

[FPSV15] J. Fuchs, J. Priel, C. Schweigert, and A. Valentino, *On the Brauer groups of symmetries of abelian Dijkgraaf-Witten theories*, Comm. Math. Phys. **339** (2015), no. 2, 385–405, DOI 10.1007/s00220-015-2420-y. MR3370609

[FSS14] J. Fuchs, G. Schaumann and C. Schweigert, *A trace for bimodule categoires*, Applied Categ. Structures, in press. DOI 10.1007/s10485-016-9425-3

[FS14] J. Fuchs and C. Schweigert, *A note on permutation twist defects in topological bilayer phases*, Lett. Math. Phys. **104** (2014), no. 11, 1385–1405, DOI 10.1007/s11005-014-0719-9. MR3267664

[FSV13] J. Fuchs, C. Schweigert, and A. Valentino, *Bicategories for boundary conditions and for surface defects in 3-d TFT*, Comm. Math. Phys. **321** (2013), no. 2, 543–575, DOI 10.1007/s00220-013-1723-0. MR3063919

[FSV14] J. Fuchs, C. Schweigert, and A. Valentino, *A geometric approach to boundaries and surface defects in Dijkgraaf-Witten theories*, Comm. Math. Phys. **332** (2014), no. 3, 981–1015, DOI 10.1007/s00220-014-2067-0. MR3262619

[FSW08] J. Fuchs, C. Schweigert, and K. Waldorf, *Bi-branes: target space geometry for world sheet topological defects*, J. Geom. Phys. **58** (2008), no. 5, 576–598, DOI 10.1016/j.geomphys.2007.12.009. MR2419689 (2010b:81134)

[KS11] A. Kapustin and N. Saulina, *Surface operators in 3d topological field theory and 2d rational conformal field theory*, Mathematical foundations of quantum field theory and perturbative string theory, Proc. Sympos. Pure Math., vol. 83, Amer. Math. Soc., Providence, RI, 2011, pp. 175–198, DOI 10.1090/pspum/083/2742429. MR2742429 (2012j:81202)

[LP08] A. D. Lauda and H. Pfeiffer, *Open-closed strings: two-dimensional extended TQFTs and Frobenius algebras*, Topology Appl. **155** (2008), no. 7, 623–666, DOI 10.1016/j.topol.2007.11.005. MR2395583 (2009g:57049)

[La01] C. I. Lazaroiu, *On the structure of open-closed topological field theory in two dimensions*, Nuclear Phys. B **603** (2001), no. 3, 497–530, DOI 10.1016/S0550-3213(01)00135-3. MR1839382 (2002h:81238)

[Lu09] J. Lurie, *On the classification of topological field theories*, Current developments in mathematics, 2008, Int. Press, Somerville, MA, 2009, pp. 129–280. MR2555928 (2010k:57064)

[MS06] G. Moore and G. Segal, *D-branes and K-theory in 2D topological field theory*, hep-th/0609042

[Mo13] J. C. Morton, *Cohomological twisting of 2-linearization and extended TQFT*, *J. Homotopy and Related Structures* August 2013, math.QA/1003.5603

[Os03] V. Ostrik, *Module categories over the Drinfeld double of a finite group*, Int. Math. Res. Not. **27** (2003), 1507–1520, DOI 10.1155/S1073792803205079. MR1976233 (2004h:18005)

[Sc01] P. Schauenburg, *The monoidal center construction and bimodules*, J. Pure Appl. Algebra **158** (2001), no. 2-3, 325–346, DOI 10.1016/S0022-4049(00)00040-2. MR1822847 (2002f:18013)

[SFR06] C. Schweigert, J. Fuchs, and I. Runkel, *Categorification and correlation functions in conformal field theory*, International Congress of Mathematicians. Vol. III, Eur. Math. Soc., Zürich, 2006, pp. 443–458. MR2275690 (2008e:81129)

[Wi08] S. Willerton, *The twisted Drinfeld double of a finite group via gerbes and finite groupoids*, Algebr. Geom. Topol. **8** (2008), no. 3, 1419–1457, DOI 10.2140/agt.2008.8.1419. MR2443249 (2009g:57050)

Teoretisk fysik, Karlstads universitet, Universitetsgatan 21, 651 88 Karlstad, Sweden

E-mail address: juergen.fuchs@kau.se

Algebra und Zahlentheorie, Fachbereich Mathematik, Universität Hamburg, Bundesstrasse 55, 20148 Hamburg, Germany

E-mail address: christoph.schweigert@uni-hamburg.de

Proceedings of Symposia in Pure Mathematics
Volume **93**, 2016
http://dx.doi.org/10.1090/pspum/093/01594

Quantum curves and topological recursion

Paul Norbury

ABSTRACT. This is a survey article describing the relationship between quantum curves and topological recursion. A quantum curve is a Schrödinger operator-like noncommutative analogue of a plane curve which encodes (quantum) enumerative invariants in a new and interesting way. The Schrödinger operator annihilates a wave function which can be constructed using the WKB method, and conjecturally constructed in a rather different way via topological recursion.

CONTENTS

1. Introduction

A *quantum curve* of a plane curve $C=\{(x,y)\in\mathbb{C}^2\,|\,P(x,y)=0\}$ is a Schrödinger-type linear differential equation

$$\widehat{P}(\widehat{x},\widehat{y})\,\psi(p,\hbar)=0 \tag{1.1}$$

where $p\in C$, $\hbar$ is a formal parameter, and $\widehat{P}(\widehat{x},\widehat{y})$ is a differential operator-valued non-commutative quantisation of the plane curve with $\widehat{x}=x\cdot$ and $\widehat{y}=\hbar\frac{d}{dx}$, so in particular

$$[\widehat{x},\widehat{y}]=-\hbar. \tag{1.2}$$

2010 *Mathematics Subject Classification.* Primary 14N10, 05A15, 81S10.
Key words and phrases. Spectral curve, quantum curve, WKB method.

For example, a quantisation of $P(x,y) = y^2 - x$ is the Schrödinger operator $\widehat{P}(\widehat{x},\widehat{y}) = \widehat{y}^2 - \widehat{x} = \hbar^2 \frac{d^2}{dx^2} - x$. The quantum curve can also be a *difference equation* if instead we consider $P(e^x, e^y) = 0$—see Section 1.4.

The equation (1.1) is understood via the WKB method. In other words we require $\psi(p,\hbar)$ to be of the form

$$\psi(p,\hbar) = \exp(\hbar^{-1}S_0(p) + S_1(p) + \hbar S_2(p) + \hbar^2 S_3(p) + \dots) \tag{1.3}$$

and the $S_k(p)$ are calculated recursively via (1.1). A simple consequence of (1.1) is that $S_k(p)$ are meromorphic functions on C, where $S_0(p) = \int^p ydx$ may be multi-valued. The operator $\frac{d}{dx}$ acts on meromorphic functions via composition of the exterior derivative followed by division by dx, and coincides with usual differentiation on an analytic expansion of a meromorphic function with respect to a local coordinate defined by x.

Fundamental question: Can we define $S_k(p)$ directly from the plane curve without using the WKB approximation, and in particular produce a natural choice of $\widehat{P}(\widehat{x},\widehat{y})$?

A conjectural answer to this question is given by

$$S_k(p) = \sum_{2g-1+n=k} \frac{1}{n!} \int^p \int^p \dots \int^p \omega_n^g(p_1,\dots,p_n) \tag{1.4}$$

where $\omega_n^g(p_1,\dots,p_n)$ are multidifferentials for each $g \geq 0$, $n > 0$ recursively defined on the curve C via topological recursion which is described in Section 2. Curves of genus g with n labeled points give a convenient way to encode topological recursion, and often represent an underlying geometric connection. From this viewpoint, (1.4) claims that $S_k(p)$ is related to all punctured curves of Euler characteristic $1-k$. This conjecture is addressed by Gukov and Sułkowski in [**29**] together with the related issue of constructing $\widehat{P}(\widehat{x},\widehat{y})$ algorithmically from the wave function.

The path from the quantum curve to the plane curve is well-defined. It is a little deeper than simple substitution $\widehat{x} \mapsto x$ and $\widehat{y} \mapsto y$ into $\widehat{P}(\widehat{x},\widehat{y})$, since we *deduce* that the differential equation (1.1) is satisfied only on the plane curve $P(x,y) = 0$. This is achieved via the semi-classical limit $\hbar \to 0$, where the differential operator $\widehat{P}(\widehat{x},\widehat{y})$ reduces to a multiplication operator that vanishes precisely on the plane curve. The action of $\hbar\frac{d}{dx}$ on

$$\psi_0(p,\hbar) = \exp(\hbar^{-1}\int^p ydx)$$

is multiplication by y, so

$$\widehat{P}(\widehat{x},\widehat{y})\,\psi(p,\hbar) = \Big[P(x,y) + O(\hbar)\Big]\psi(p,\hbar)$$

and in the $\hbar \to 0$ limit $\widehat{y} = \hbar\frac{d}{dx}$ in $\widehat{P}(\widehat{x},\widehat{y})$ is replaced by its symbol y. Higher order corrections in $\hbar$ are required since $(\hbar\frac{d}{dx})^2 \mapsto y^2 + O(\hbar)$ under its action on $\psi_0(x,\hbar)$.

On the other hand, constructing the quantum curve from the plane curve is not canonical. The main issues lie in the construction of the wave function $\psi(p,\hbar)$ and the ambiguity in ordering the non-commuting operators $\widehat{x}$ and $\widehat{y}$ in $\widehat{P}$. The conjectural formula (1.4) is one attempt to remedy this. Such a wave function is enough to reconstruct the operator $\widehat{P}(\widehat{x},\widehat{y})$. Any $\widehat{P}(\widehat{x},\widehat{y})$ can be expressed as:

$$\widehat{P}(\widehat{x},\widehat{y}) = P(\widehat{x},\widehat{y}) + \hbar P_1(\widehat{x},\widehat{y}) + \hbar^2 P_2(\widehat{x},\widehat{y}) + \dots \tag{1.5}$$

where each $P_k(\widehat{x},\widehat{y})$ is a normal ordered operator valued polynomial—all $\widehat{y}$ terms in a monomial are placed to the right, so $x^m y^n$ is replaced by $\widehat{x}^m\widehat{y}^n$—and has no explicit $\hbar$ dependence. Then the $P_k(\widehat{x},\widehat{y})$ can be reconstructed recursively from the wave function.

The differential operator $\widehat{P}(\widehat{x},\widehat{y})$ generates a principal ideal in the algebra $\mathcal{D}$ of differential operators which act on $\mathbb{C}[x]$. The quotient $\mathcal{D}/\langle\widehat{P}\rangle$ of the algebra $\mathcal{D}$ by the principal ideal $\langle\widehat{P}\rangle = \mathcal{D}\cdot\widehat{P}$ is a $\mathcal{D}$-module which gives a way to study $\widehat{P}(\widehat{x},\widehat{y})$ intrinsically. See [**30**] for a detailed description of this. The wave function $\psi(p,\hbar)$ can be retrieved via the $\mathcal{D}$-module homomorphism it defines:

$$\mathcal{D}/\langle\widehat{P}\rangle \to \mathbb{C}[[x,x^{-1},\hbar,\hbar^{-1}]].$$

1.1. Model enumerative problem. Plane curves and quantum curves naturally arise out of various enumerative problems. A model problem is the enumeration of moments of a given probability measure. Given a measure ρ on $\mathbb{R}$ that is well-behaved, say it is bounded with compact support $K\subset\mathbb{R}$, its Stieltjes transform is analytic in $\overline{\mathbb{C}}-K$,

$$\hat{\rho}(x) = \int_{\mathbb{R}}\frac{\rho(t)dt}{x-t} = \sum_{n\geq 0}\frac{\langle t^n\rangle}{x^{n+1}},\quad \langle t^n\rangle = \int_{\mathbb{R}} t^n\rho(t)dt$$

where the sum is an analytic expansion at $x=\infty$, with coefficients giving the moments of ρ. The function $\hat{\rho}(x)$ extends to a Riemann surface which is a cover of the x-plane. The Riemann surface is equipped with two functions x and $y=\hat{\rho}(x)$ and hence naturally maps to $\mathbb{C}^2$. A toy example of this setup is as follows.

EXAMPLE 1.1. Consider the discrete measure

$$\rho(t) = \sum_{i=1}^{N}\delta_{\lambda_i}$$

with Stieltjes transform

$$\hat{\rho}(x) = \sum_{i=1}^{N}\frac{1}{x-\lambda_i} = \operatorname{tr}(x-M)^{-1}$$

for M a matrix with eigenvalues $\{\lambda_1,\dots,\lambda_N\}$ conveniently chosen to have resolvent $\hat{\rho}(x)$. The Stieltjes transform is a holomorphic function on a plane curve which we call its *spectral curve*

$$\left\{y-\sum_{i=1}^{N}\frac{1}{x-\lambda_i}=0\right\}. \tag{1.6}$$

The function

$$\psi(x) = \det(x-M) = \prod_{i=1}^{N}(x-\lambda_i) = \exp\int^x y\,dx$$

satisfies the first order differential equation

$$\left(\frac{d}{dx}-\sum_{i=1}^{N}\frac{1}{x-\lambda_i}\right)\psi(x)=0 \tag{1.7}$$

which is the *quantum curve*, a non-commutative analogue of the spectral curve since (x, y) is replaced in the equation of the curve (1.6) by operators ($\widehat{x} = x\cdot, \widehat{y} = \frac{d}{dx}$) to produce the differential operator in (1.7).

If we introduce $\hbar$ into the wave function via $\psi(x) = \exp \hbar^{-1} \int^x y dx$ and $\widehat{y} = \hbar \frac{d}{dx}$, then it satisfies the quantum curve equation $(\hat{y} - \sum_{i=1}^N \frac{1}{x-\lambda_i})\psi(x) = 0$ exactly but the perturbative parameter $\hbar$ is not needed here.

REMARK 1.2. The example above is a special case of the more general construction of normal ordered first order quantum curves associated to plane curves linear in y, i.e. $P(x, y) = p(x) + q(x)y$. In these cases the first order approximation of the quantum curve gives the entire quantum curve.

REMARK 1.3. The elementary relation of this example to the spectrum of a matrix anticipates nicely the connection with matrix models which is mentioned briefly in Section 3. Its relation to the spectrum of a matrix means it also has an elementary relation to symmetric polynomials. The Stieltjes transform and wave function are generating functions for power sum symmetric polynomials p_k, respectively elementary symmetric polynomials σ_k:

$$\hat{\rho}(x) = \sum_{k=0}^{\infty} \frac{p_k(\lambda_1, \ldots, \lambda_N)}{x^{k+1}}, \quad \psi(x) = \sum_{k=0}^{\infty} x^N \cdot (-1)^k \frac{\sigma_k(\lambda_1, \ldots, \lambda_N)}{x^k}$$

where the second sum is of course finite. The quantum curve (1.7) gives $\frac{d}{dx}\psi(x) = \hat{\rho}(x)\psi(x)$ which is exactly Newton's identities relating p_k and σ_k.

1.2. WKB method. The quantum curve defines a triangular system in $\frac{d}{dx}S_k(p)$ known as the WKB method. The function $\frac{d}{dx}S_k(p)$ first appears in the system as

$$\hbar^k \frac{\partial P}{\partial y}(x, y) \frac{d}{dx} S_k(p) + \ldots$$

so in particular it is a meromorphic function on the curve $P(x, y) = 0$ with poles at the zeros of dx. Here we present an example in order to go through the WKB method explicitly.

EXAMPLE 1.4. Consider the measure

$$\rho(t) = \frac{1}{2\pi}\sqrt{4 - t^2} \cdot \chi_{[-2,2]}$$

which can arise as the limit $\lim_{N\to\infty} \frac{1}{N} \sum_{i=1}^N \delta_{\lambda_i}$ of a normalised version of Example 1.1.

Its Stieltjes transform extends to a meromorphic function on a rational curve realised as a double cover of $\overline{\mathbb{C}}$ (= the $x-$plane) branched over $x = \pm 2$:

$$\hat{\rho}(x) = \frac{1}{2\pi} \int_{-2}^{2} \frac{\sqrt{4 - t^2}}{x - t} dt = \sum_{n \geq 0} \frac{C_n}{x^{2n+1}} = y, \quad x = y + \frac{1}{y}$$

where the odd moments vanish and $C_n = \frac{1}{n+1}\binom{2n}{n}$ is the nth Catalan number. The series in x is a local statement—it is an analytic expansion of the global meromorphic function $y = \hat{\rho}(x)$ in the local coordinate x on a branch of $x = \infty$. (It does not factor through $x : C \to \mathbb{C}$.) Together x and y define a plane curve

$$C = \{(x, y) \mid P(x, y) = y^2 - xy + 1 = 0\}.$$

The wave function $\psi_0(p) = \exp\int^p ydx = \exp\int^y y(1-\frac{1}{y^2})dy = \frac{1}{y}\exp\frac{1}{2}y^2$ satisfies $\frac{d}{dx}\psi_0(p) = y\cdot\psi_0(p)$ hence on the spectral curve C

$$\left(\frac{d^2}{dx^2} - x\frac{d}{dx} + 1\right)\psi_0(p) = (y^2 - xy + 1 + \frac{d}{dx}y(p))\cdot\psi_0(p) = \frac{d}{dx}y(p)\cdot\psi_0(p) \neq 0$$

since $\frac{d}{dx}y(p) = \frac{y^2}{1-y^2} \neq 0$. We see that when the linear operator is not first order its failure to be as simple as Example 1.1 is due to higher derivatives.

Introduce a perturbative variable $\hbar$ via $\widehat{y} = \hbar\frac{d}{dx}$ and put $\psi_0(p) = \exp\hbar^{-1}\int^p ydx$ then

$$\left(\hbar^2\frac{d^2}{dx^2} - x\hbar\frac{d}{dx} + 1\right)\psi_0(x) = \hbar\frac{y^2}{1-y^2}\psi_0(x) = O(\hbar)$$

which is zero up to order $\hbar$. Following (1.3) we can remove all higher terms in $\hbar$ by replacing ψ_0 with $\psi(p,\hbar) = \exp\left(\hbar^{-1}\sum_{k\geq 0}\hbar^k S_k(p)\right)$ where $\frac{d}{dx}S_0(p) = y$ and $S_k(p)$ are chosen so that $P(\widehat{x},\widehat{y})\psi(p,\hbar) = 0$ for $\widehat{x} = x\cdot$ and $\widehat{y} = \hbar\frac{d}{dx}$ up to all orders of $\hbar$. Concretely,

$$\begin{aligned}
0 &= \psi^{-1}\left(\hbar^2\frac{d^2}{dx^2} - x\hbar\frac{d}{dx} + 1\right)\psi \\
&= \left[\frac{d}{dx}S_0(p)^2 - x\frac{d}{dx}S_0(p) + 1\right] \\
&\quad + \hbar\left[\left(\frac{d}{dx}\right)^2 S_0(p) + (2\frac{d}{dx}S_0(p) - x)\frac{d}{dx}S_1(p)\right] \\
&\quad + \hbar^2\left[\left(\frac{d}{dx}\right)^2 S_1(p) + \frac{d}{dx}S_1(p)^2 + (2\frac{d}{dx}S_0(p) - x)\frac{d}{dx}S_2(p)\right] \\
&\quad + \hbar^3\left[\left(\frac{d}{dx}\right)^2 S_2(p) + 2\frac{d}{dx}S_1(p)\frac{d}{dx}S_2(p) + (2\frac{d}{dx}S_0(p) - x)\frac{d}{dx}S_3(p)\right] \\
&\quad + O(\hbar^4).
\end{aligned}$$

The system is triangular so one can recursively solve for $\frac{d}{dx}S_k(p)$. For example

$$\frac{d}{dx}S_1(p) = -\frac{\left(\frac{d}{dx}\right)^2 S_0(p)}{2\frac{d}{dx}S_0(p) - x} = -\frac{y^3}{(y^2-1)^2}. \tag{1.8}$$

Furthermore, $\frac{d}{dx}S_k(p)$ is rational in $\frac{d}{dx}S_m(p)$ for $m < k$ and $\frac{d}{dx}S_0(p)$ only appears in the denominator $2\frac{d}{dx}S_0(p) - x = \frac{y^2-1}{y}$. Since $\frac{d}{dx}S_1(p)$ is a meromorphic function on C with poles only at $y = \pm 1$ then $\frac{d}{dx}S_k(p)$ is a meromorphic function on C with poles only at $y = \pm 1$ for $k > 0$.

For general curves $S_1(p)$ is discussed in [**29**]. For rational curves one obtains:

$$S_1(p) = -\frac{1}{2}\log\frac{dx}{dz} \tag{1.9}$$

where z is a global rational parameter on the curve C. For example, when $x = z + \frac{1}{z}$, $S_1(z) = -\frac{1}{2}\log(1 - \frac{1}{z^2})$ and hence $\frac{d}{dx}S_1(z) = \frac{1}{x'(z)}\frac{d}{dz}S_1(z) = \frac{-z}{(z^2-1)^2}$ which agrees with (1.8) for $y = 1/z$.

More generally, for any spectral curve C and local parameter z on C, (1.9) gives a first approximation to $S_1(p)$ and involves further terms.

1.3. Relations between quantum curves and topological recursion. There are compelling reasons for conjecturing the relation between quantum curves and topological recursion given by (1.4). There is the close relationship between the disk and annulus invariants which are essentially the input data for topological recursion and the quantum curve. Also, many examples have been verified, which we will discuss in Section 3. We discuss below two more properties of the S_k shared by topological recursion and the quantum curve. The first is the invariance of $S_k(p)$ under a class of isomorphisms between plane curves, where $S_k(p)$ is constructed using either the WKB method or topological recursion. The second is the local behaviour near poles of topological recursion and the quantum curve which exploits the nice fact that the algebra of operators is in some sense commutative near poles.

1.3.1. *Invariance of S_k under isomorphisms.* Consider the following isomorphism between plane curves

$$(x,y) \mapsto (x, y+g'(x)) \tag{1.10}$$

for any polynomial $g(x)$ where $g'(x) = \frac{d}{dx}g(x)$. So their defining polynomials $P(x,y)=0$ and $Q(x,y)=0$ are related by $Q(x,y)=P(x,y-g'(x))$. Now

$$\widehat{P}(\widehat{x},\widehat{y})\psi(p,\hbar)=0 \quad \Rightarrow \quad \widehat{Q}(\widehat{x},\widehat{y})\exp(\hbar^{-1}g(x))\psi(p,\hbar)=0$$

for $P(x(p),y(p))=0$ and $Q(x(p),y(p)+g'(x(p)))=0$ where $\widehat{Q}(\widehat{x},\widehat{y})$ has to be defined carefully as follows: replace each appearance of $\widehat{y}$ in $\widehat{P}(\widehat{x},\widehat{y})$ with the operator $\widehat{y}-g'(x)$ and in particular do not normal order.

The isomorphism (1.10) preserves the underlying curve, not its embedding, together with the function x defined on the curve. The change in wave function for curves related by such an isomorphism only affects the $\hbar^{-1}$ term in the exponent of $\psi(p,\hbar)$ and all $S_k(p)$ for $p>0$ are unchanged under the isomorphism. So we see that $S_k(p)$ for $k>0$ are in some sense intrinsic to the underlying curve equipped with the functions x and y.

Moreover, the ω_n^g, which are defined in Section 2 are unchanged under isomorphisms of type (1.10), lending weight to the conjecture (1.4).

1.3.2. *Local factorisation.* A fundamental plane curve, known as the Airy curve, is $y^2-x=0$. Its quantum curve does indeed satisfy (1.4) [**5**, **42**]. It gives a local model for any curve with dx having simple zeros. We will see below that the operator $\widehat{P}(\widehat{x},\widehat{y})$ has a factor of $(y-y_0)^2-\lambda(x-x_0)=0$ which annihilates part of the wave function corresponding to the wave function of the Airy curve. This uses the fact that the algebra of operators is in some sense commutative near a zero of dx.

The WKB method shows that for $k>1$ the function $S_k(x)$ has a pole of order $3k-3$ at any zero a of dx and $S_1(x)$ has a logarithmic singularity there. We define the largest order term of the $\hbar$ expansion $\hbar^{-1}S_0(p)+S_1(p)+\hbar S_2(p)+\hbar^2S_3(p)+\dots$ at a to be a $\hbar$ expansion with coefficients the largest order terms of each $S_k(p)$.

Fix a zero a of dx and consider only the largest order terms at a in the exponent of ψ. The action of the operator $\widehat{P}(\widehat{x},\widehat{y})$ on highest order terms is rather simple since all operators commute! They behave like differential operators with constant coefficients. For example, in a local coordinate z

$$\begin{aligned}\hbar\frac{d}{dx}x\frac{1}{(z-z(a))^m} &= \hbar\frac{1}{(z-z(a))^m}+\hbar x\frac{d}{dx}(z-z(a))^{-m}\\ &= \hbar x\frac{d}{dx}(z-z(a))^{-m} + \text{ lower order terms}\end{aligned}$$

since $\frac{d}{dx}(z-z(a))^{-m}=O((z-z(a))^{-m-2}$. This does not contradict $[x,\hbar\frac{d}{dx}]=-\hbar$. Multiplication by any function analytic in x, such as $\hbar$, acts next to $\frac{d}{dx}$ like zero on the highest order parts since an analytic function never increases the order of a pole whereas $\frac{d}{dx}$ always increases the order of a pole by 2.

The polynomial factorises as:

$$P(\alpha,y)=\lambda\prod_{k=1}^{n}(y-\lambda_k(x))$$

where $\lambda_k(x)$ are locally defined analytic functions. Exactly two of the $\lambda_k(x)$ coincide when evaluated at a zero a of dx—we may assume $\lambda_1(a)=\lambda_2(a)$. With respect to a local coordinate s near a defined by $x=a+s^2$, we have $y=y(0)+y'(0)s+O(s^2)$. Equivalently, $\lambda_1(x(s))=y(0)+y'(0)s+O(s^2)$ and $\lambda_2(x(s))=y(0)-y'(0)s+O(s^2)$. Hence $(y-\lambda_1(x))(y-\lambda_2(x))=(y-y(0))^2-y'(0)^2s^2+O(s^3)$ and by commutativity near a it can be brought forward and must annihilate the highest order part. The conclusion is that the highest order part of the wave function is given by a rescaled Airy wave function.

The poles of the invariants ω_n^g occur at the zeros of dx and the highest order part of ω_n^g near a pole is given by $y'(0)^{2-2g-n}\omega_n^{g\ \mathrm{Airy}}$. Hence the highest order part of the wave function using (1.4) is $\psi^{\ \mathrm{Airy}}(p,\frac{\hbar}{y'(0)})$ which agrees with the behaviour of the WKB method.

1.4. Why are quantum curves useful? One application of quantum curves is to *predict* topological recursion. The proofs are often easier for quantum curves than for topological recursion. Proofs of both are often equivalent to recursions between enumerative invariants, and the former involves coarser, hence simpler, invariants. A quantum curve assembles enumerative information in an Euler characteristic expansion using a single variable. Whereas topological recursion produces several variable invariants with genus expansions. Furthermore, the nonlinear behaviour of topological recursion which arises out of connectedness assumptions can be simplified to linear behaviour since a wave function satisfies a linear differential equation.

Two examples of this predictive behaviour are as follows. Quantum curves for a family of enumerative examples, so-called *hypermaps*, were proven in [**11**] and there it was conjectured that topological recursion applied to the associated plane curves enumerated hypermaps. The conjecture was verified in low genus numerical calculations. This conjecture was later proven in [**17**]. Currently there is an outstanding conjecture regarding spin Hurwitz numbers. It is known by numerical verification and the quantum curve was proven in [**35**].

Another application of the quantum curve should be to enable one to drop technical assumptions on topological recursion. Topological recursion does not apply to any plane curve. It requires the zeros of dx to be simple. Since the quantum curve requires no such assumptions one would expect to be able to makes sense of topological recursion without the technical assumptions. One such construction is given in [**6**]. This would be useful for curves where x is defined via the quotient of a group action on the curve.

Before we describe the next application we will describe variations on the basic setup of plane curves. This paper is mainly concerned with curves in $\mathbb{C}^2$. However, in some applications spectral curves that are not in $\mathbb{C}^2$ arise. For instance, in knot theory the A-polynomial of a knot (described below) is a curve

$$C \subset \mathbb{C}^* \times \mathbb{C}^*$$

with coordinates $(x,y) = (e^u, e^v)$. If we put $\widehat{u} = u\cdot$ and $\widehat{v} = \hbar\frac{d}{du}$ so $[\widehat{u},\widehat{v}] = -\hbar$, then in (x,y) coordinates we have $\widehat{x} = e^{\widehat{u}} = x\cdot$ and $\widehat{y} = e^{\widehat{v}} = e^{\hbar x\frac{d}{dx}}$ so that

$$\widehat{x}\widehat{y} = q^{-1}\widehat{y}\widehat{x}, \quad q = e^{\hbar}$$

and $\widehat{y}f(x) = e^{\hbar x\frac{d}{dx}}f(x) = f(qx)$. Topological recursion—defined in Section 2—is sometimes modified depending on whether one works in (u,v) or (x,y) variables.

We also consider the case of $\mathbb{C} \times \mathbb{C}^*$ with coordinates $(u,v) = (x, e^y)$, for example the spectral curve of Gromov-Witten invaraints of $\mathbb{P}^1$ treated in Section 3, where the commutator relations are $[\widehat{x},\widehat{y}] = -\hbar\widehat{y}$.

One of the most famous applications of the quantum curve is its relation to the volume conjecture in 3-manifold topology. Conjecturally the behaviour of the quantum curve of the A-polynomial generalises the volume conjecture. Given a knot $K \subset S^3$, its A-polynomial $A_K(m,\ell) = 0$ is defined via its $SL(2,\mathbb{C})$ representation variety $\mathcal{R}(G) = \{\rho \mid \rho : G \to SL(2,\mathbb{C})\}/\!\sim$. The boundary $T^2 = \partial(S^3 - K)$ induces a restriction map

$$\mathcal{R}_G(S^3 - K) \to \mathcal{R}_G(T^2) \cong \mathbb{C}^* \times \mathbb{C}^* \ni \begin{pmatrix} m & 0 \\ 0 & m^{-1} \end{pmatrix}, \begin{pmatrix} \ell & 0 \\ 0 & \ell^{-1} \end{pmatrix}$$

with image a Lagrangian curve defined by $A_K(m,\ell) = 0$. Here $\log \ell = v = S_0'(u)$ is known as the Neumann-Zagier potential [**38**].

The coloured Jones polynomial [**37**] $J_N(K;q) \in \mathbb{Z}[q,q^{-1}]$ has asymptotics as $N \to \infty$ with $N\hbar$ fixed and $q = e^{-\hbar}$ given by

$$\log J_N(K;e^{-\hbar}) \underset{\substack{N \to \infty \\ \hbar \to 0 \\ N\hbar = u + 2\pi i}}{\sim} \hbar^{-1}S_0(u) + \sum_{k=0}^{\infty} \hbar^k S_{k+1}(u).$$

CONJECTURE 1 (Kashaev [**32**], Murakami-Murakami [**37**], Gukov [**28**]). *For the curve defined by $A_K(m,\ell) = 0$, put $(m,\ell) = (e^u, e^v)$ and define $S_0(u)$ via* (1.1) *and* (1.3). *Then*

$$S_0(u) = \textit{ volume of (incomplete) hyperbolic manifold } S^3 - K.$$

CONJECTURE 2 (Dikjgraaf-Fuji-Manabe [**13**]). *Consider the quantisation $\hat{A}_K$ that comes out of topological recursion applied to $A_K(m,\ell) = 0$. In other words, use the $S_k(u)$ calculated via topological recursion and* (1.4) *to construct the wave function which can be used to produce $\hat{A}_K$. Then*

$$\hat{A}_K J_N(K;e^{\hbar}) = 0.$$

Borot and Eynard showed in [**3**] that to get the asymptotics of the Jones polynomial one must supplement the topological recursion with non-perturbative corrections.

Dimofte [**9**] has a beautiful approach to the quantum curve of the A-polynomial. He shows that the character variety of an ideal hyperbolic tetrahedron gives rise to the curve $\{y+x^{-1}-1=0\} \subset \mathbb{C}^* \times \mathbb{C}^*$ with quantum curve $(\widehat{y}+\widehat{x}^{-1}-1)\psi(x,\hbar) = 0$.

As described above, $\hat{y}f(x) = f(qx)$ for $q = e^{\hbar}$ so the quantum curve equation can be written

$$\psi(qx, \hbar) = (1 - x^{-1})\psi(x, \hbar)$$

which is satisfied by a quantum dilogarithm function—a building block of link invariants [**31**]. A hyperbolic three-manifold can be built by gluing hyperbolic tetrahedra and Dimofte studies how such gluing affects the quantum curve in order to build up the quantum curve of the A-polynomial.

Acknowledgements. The author benefited from conversations with numerous people, and in particular would like to thank Gaëtan Borot, Vincent Bouchard, Norman Do, Petya Dunin-Barkowsky, Bertrand Eynard, Peter Forrester, John Harnad, Motohico Mulase, Sergey Shadrin, and participants of String-Math 2014 and the Banff workshop *Quantum Curves and Quantum Knot Invariants.* The author would also like to thank the referee for many useful comments.

2. Topological recursion

Topological recursion as developed by Chekhov, Eynard and Orantin [**7, 23**] arose out of loop equations satisfied by matrix models. It takes as input a *spectral curve* (C, B, x, y) consisting of a compact Riemann surface C, a bidifferential B on C, and meromorphic functions $x, y : C \to \mathbb{C}$. A technical requirement is that the zeros of $\mathrm{d}x$ are simple and disjoint from the zeros of $\mathrm{d}y$ [**23**]. In most cases of interest the bidifferential B is taken to be the fundamental normalised differential of the second kind on C, defined for instance in [**26**]. In the case when C is rational with global rational parameter z, $B = \frac{\mathrm{d}z_1 \otimes \mathrm{d}z_2}{(z_1 - z_2)^2}$ is the Cauchy kernel.

The output of topological recursion is a collection of multidifferentials $\omega_n^g(p_1, \dots, p_n)$ for integers $g \geq 0$ and $n \geq 1$, on C — in other words, a tensor product of meromorphic 1-forms on the product C^n, where $p_i \in C$. When $2g-2+n > 0$, $\omega_n^g(p_1, \dots, p_n)$ is defined recursively in terms of local information around the poles of $\omega_{n'}^{g'}(p_1, \dots, p_{n'})$ for $2g' + 2 - n' < 2g - 2 + n$.

The invariants $\omega_n^g(p_1, \dots, p_n)$ are defined recursively from simpler $\omega_{n'}^{g'}$ for $2g' - 2 + n' < 2g - 2 + n$. The recursion can be represented pictorially via different ways of decomposing a genus g surface with n labeled boundary components into a pair of pants containing the first boundary component and simpler surfaces.

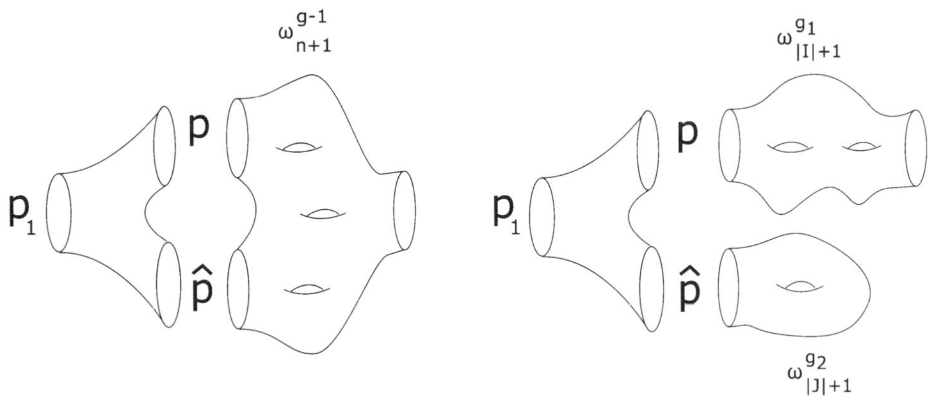

For $2g-2+n>0$ and $S=\{2,\dots,n\}$, define

(2.1)
$$\omega^g_n(p_1,\boldsymbol{p}_S)=\sum_\alpha\operatorname*{Res}_{p=\alpha}K(p_1,p)\left[\omega^{g-1}_{n+1}(p,\hat{p},\boldsymbol{p}_S)+\sum^{\circ}_{\substack{g_1+g_2=g\\ I\sqcup J=S}}\omega^{g_1}_{|I|+1}(p,\boldsymbol{p}_I)\,\omega^{g_2}_{|J|+1}(\hat{p},\boldsymbol{p}_J)\right]$$

where the outer summation is over the zeros α of $\mathrm{d}x$ and the $\circ$ over the inner summation means that we exclude terms that involve ω^0_1. The point $\hat{p}\in C$ is defined to be the unique point $\hat{p}\neq p$ close to α such that $x(\hat{p})=x(p)$. It is unique since each zero α of $\mathrm{d}x$ is assumed to be simple, and we need only consider $p\in C$ close to α. We see that the recursive definition of $\omega^g_n(p_1,\dots,p_n)$ uses only local information around zeros of $\mathrm{d}x$. The recursion takes as input the base cases

$$\omega^0_1=-y(z)\,\mathrm{d}x(z)\qquad\text{and}\qquad\omega^0_2=B(z_1,z_2)$$

that are generally called *unstable terms.* The kernel K is defined by the following formula

$$K(p_1,p)=\frac{-\int_{\hat{p}}^{p}\omega^0_2(p_1,p')}{2[y(p)-y(\hat{p})]\,\mathrm{d}x(p)}$$

which is well-defined in the vicinity of each zero of $\mathrm{d}x$. Note that the quotient of a differential by the differential $\mathrm{d}x(p)$ is a meromorphic function. For $2g-2+n>0$, the multidifferential ω^g_n is symmetric, with poles only at the zeros of $\mathrm{d}x$ and vanishing residues.

The poles of the invariants ω^g_n occur at the zeros of dx and are of order at most $6g-6+4n$ there. One property of ω^g_n that we needed in Section 1.3.2 and proven in [**23**] is that the highest order part of ω^g_n near a pole is given by $y'(0)^{2-2g-n}\omega^{g\text{ Airy}}_n$. Here $\omega^{g\text{ Airy}}_n$ is the invariant obtained from the curve $y^2-x=0$ which is a local model near zeros of dx for any plane curve that is the input of topological recursion.

Now that we have defined topological recursion precisely we need to qualify the $k=1$ part of (1.4) which uses a regularised version of (1.4).

$$S_1(p)=\frac{1}{2!}\int^p\int^p\left[\omega^0_2(p_1,p_2)-\frac{dx_1dx_2}{(x_1-x_2)^2}\right]\tag{2.2}$$

where $\omega^0_2(p_1,p_2)=B$ is part of the input data for the spectral curve.

REMARK 2.1. Heuristically $\omega^g_n(p_1,p_2,\dots,p_n)$ gives the coefficients in the large N expansion of expected values of multiresolvents in a matrix model $\langle\mathrm{Tr}\frac{1}{x(p_1)-A}\cdots\mathrm{Tr}\frac{1}{x(p_n)-A}\rangle_c$, where N is the size of the matrix and g indexes the order in the $1/N$ expansion. The subscript c means cumulant, or the connected part in a graphical expansion. Topological recursion follows from the loop equations satisfied by the resolvents.

For $2g-2+n>0$, the invariants ω^g_n of spectral curves satisfy the following *string equations* for $m=0,1$ [**23**].

$$\sum_\alpha\operatorname*{Res}_{z=\alpha}x^my\omega^g_{n+1}(z,z_S)=-\sum_{j=1}^n dz_j\frac{\partial}{\partial z_j}\left(\frac{x^m(z_j)\omega^g_n(z_S)}{dx(z_j)}\right)\tag{2.3}$$

They also satisfy the dilaton equation [**23**]

$$\sum_\alpha\operatorname*{Res}_{z=\alpha}\Phi(z)\,\omega^g_{n+1}(z,z_1,\dots,z_n)=(2-2g-n)\,\omega^g_n(z_1,\dots,z_n),\tag{2.4}$$

where the summation is over the zeros α of $\mathrm{d}x$ and $\Phi(z) = \int^z y\,\mathrm{d}x(z')$ is an arbitrary antiderivative. The dilaton equation enables the definition of the so-called *symplectic invariants*

$$F_g = \sum_\alpha \operatorname*{Res}_{z=\alpha} \Phi(z)\,\omega_1^g(z).$$

We will see in Section 2.1.1 the importance of the string equations for the quantum curve.

REMARK 2.2. Topological recursion generalises to *local* curves in which C is an open subset of a compact Riemann surface. This is because the recursive definition of $\omega_n^g(p_1,\ldots,p_n)$ uses only local information around zeros of dx. Whereas, quantum curves are global in nature, essentially requiring the full algebraic structure of the curve. Since the formula (1.4) still makes sense for local curves, it suggests that one might be able to make sense of a local quantum curve. This would be important because it was proven in [**18**] that any semi-simple cohomological field theory can be encoded via topological recursion applied to a local curve. Thus it would raise the question of what role quantum curves can play in cohomological field theories.

2.1. Choice of primitive. In (1.4) the expression

$$S_k(p) = \sum_{2g-1+n=k} \int^p \int^p \cdots \int^p \omega_n^g$$

is ambiguous since integration is not uniquely defined. It needs to be interpreted as follows. A *primitive* $F_n^g(p_1,\ldots,p_n)$ of $\omega_n^g(p_1,\ldots,p_n)$ is a meromorphic function on the spectral curve C that satisfies:

$$d_1 \ldots d_n F_n^g(p_1,\ldots,p_n) = \omega_n^g(p_1,\ldots,p_n)$$

where d_i is the exterior derivative in the ith coordinate p_i. For example, choose $p_i, q_i \in C$, $i = 1,\ldots,n$, then one possible primitive of ω_n^g is given by $\int_{q_1}^{p_1}\int_{q_2}^{p_2}\cdots\int_{q_n}^{p_n}\omega_n^g$ which is a function of $p_1,\ldots,p_n$ with q_i fixed. With such a choice of primitive

$$S_k(p) = \sum_{2g-1+n=k} F_n^g(p,p,\ldots,p)$$

for $p \in C$. The sum is finite and the only poles of F_n^g and its specialisation are at the zeros of dx, hence $S_k(p)$ is a meromorphic function on the spectral curve.

The choice of primitive $F_n^g(p_1,\ldots,p_n)$ is rather important. For a function of a single variable one can retrieve the function from any primitive—simply differentiate—so the choice of primitive is not so important. This remains true for a function of severable variables, however it is no longer true for a *specialisation* of a primitive at a single variable. Consider a function of severable variables $f(z_1,\ldots,z_n)$ and any primitive $F(z_1,\ldots,z_n)$, so $\frac{\partial}{\partial z_1}\cdots\frac{\partial}{\partial z_n}F(z_1,\ldots,z_n) = f(z_1,\ldots,z_n)$. Then there is no way to retrieve the function $f(z_1,\ldots,z_n)$ from its specialisation $F(z,\ldots,z)$.

EXAMPLE 2.3. Consider $f(z_1,z_2) = 3z_1^2 + 2z_2$ and $g(z_1,z_2) = 4z_1z_2 + 2z_2$ and choose respective primitives $F(z_1,z_2) = z_1^3z_2 + z_1z_2^2$ and $G(z_1,z_2) = z_1^2z_2^2 + z_1z_2^2$. From the specialisation $F(z,z) = z^4 + z^3 = G(z,z)$ we cannot uniquely determine $f(z_1,z_2)$.

Usually a primitive is obtained by $S_k(p) = \sum_{2g-1+n=k} \int_{p_0}^{p} \int_{p_0}^{p} \cdots \int_{p_0}^{p} \omega_n^g$, i.e. integration from a common point $p_0 \in C$. For rational curves there is another natural primitive. On a rational curve ω_n^g is a sum of its principal parts. Each principal part has its own natural primitive—take the principal part of any primitive. Note that the specialisation $F_n^g(p,..,p)$ of a primitive of $\omega_n^g(p_1,..,p_n)$ obtained in this way from principal parts is in general not of the form $\displaystyle\int_{p'}^{p}\int_{p'}^{p}\cdots\int_{p'}^{p}\omega_n^g(p_1,..,p_n)$.

2.1.1. *A family of choices of primitives.* The string equation (2.3) allows us to generate a t-parametrised family of quantum curves via the action of the operator $e^{t\hbar\frac{d}{dx}}$ on wave functions.

PROPOSITION 2.4. *A quantum curve satisfying* (1.4) *lives in a family of quantum curves. Concretely, if $\psi(p,\hbar)$ satisfies* (1.4), *then the family of wave functions*

$$\psi(p,\hbar,t) = e^{t\hbar\frac{d}{dx}}\psi(p,\hbar) \tag{2.5}$$

also satisfy (1.4) *(up to unstable terms) for different choices of primitive.*

PROOF. The action of $e^{t\hbar\frac{d}{dx}}$ on $S_k(p)$ is given by

$$S_k(p,t) = \sum_{m=0}^{k} t^m \left(\frac{d}{dx}\right)^m S_{k-m}(p)$$

so in particular (2.5) is well-defined.

The new wave function also satisfies a wave equation, with the same semi-classical limit, simply by conjugation:

$$\widehat{P}(\widehat{x},\widehat{y})\psi(p,\hbar) = 0 \Rightarrow e^{t\hbar\frac{d}{dx}}\widehat{P}(\widehat{x},\widehat{y})e^{-t\hbar\frac{d}{dx}}\cdot\psi(p,\hbar,t) = 0.$$

What is a little deeper is that $\psi(p,\hbar,t)$ also satisfies (1.4) up to unstable terms. The unstable terms are not important since one can conjugate by any discrepancy.

Define the linear functional L_p on meromorphic functions on C by

$$L_p f = \sum_{a_i} \operatorname*{Res}_{p=a_i} dy(p) f(p)$$

where the sum is over all points satisfying $dx(a_i) = 0$. Then

$$\frac{d}{dx}F_n^g(p,p,\ldots,p) = \sum_{j=1}^{n}\frac{d}{dx(p_j)}F_n^g(p_1,\ldots,p_n)\bigg|_{p_1=p_2=\cdots=p} = L_pF_{n+1}^g(p,p_1,\ldots,p_n) \tag{2.6}$$

where the first equality is the chain rule and the second equality uses the string equation (2.3) with $m=0$. We can iterate (2.6) to make sense of higher derivatives $(\frac{d}{dx})^m F_n^g$.

The idea is to adjust F_{n+1}^g by terms of the type in the right hand side of (2.6) to again get a symmetric function such as $F_{n+1}^g + t\sum_{j=1}^{n+1} L_{p_j}F_{n+1}^g(p_1,\ldots,p_{n+1})$. Define the function symmetric in p_i

$$\mathcal{F}_n^g(t,p_1,\ldots,p_n) = \sum_{m=0}^{k}\sum_{\{i_1,\ldots,i_m\}} t^m L_{p_{i_1}}\cdots L_{p_{i_m}}F_n^g(p_1,\ldots,p_n) \tag{2.7}$$

where the second sum is over all cardinality m subsets of $\{1,\ldots,n\}$. Notice that

$$d_{p_1}\cdots d_{p_n}\mathcal{F}_n^g(t,p_1,\ldots,p_n) = d_{p_1}\cdots d_{p_n}F_n^g(p_1,\ldots,p_n) = \omega_n^g(p_1,\ldots,p_n)$$

since we have adjusted only by summands independent of at least one variable p_j and hence annihilated by $d_{p_1} \cdots d_{p_n}$.

Hence we see that $S_k(p,t)$ satisfies (1.4) for the choice of primitive $\mathcal{F}^g_n(t,p_1,\ldots,p_n)$ and the proposition is proven. □

The proposition is interesting both for the extra structure it brings to the wave functions—essentially a relationship with the string equation—and to emphasise the fact that there there is a choice of primitive. We see that (1.4) still leaves some ambiguity in the construction of the quantum curve.

REMARK 2.5. The main tool in the proof of Proposition 2.4 is the string equation (2.3) that comes out of topological recursion. In Section 3 we apply this idea to the quantum curve associated to Gromov-Witten invariants of the sphere and we find that the string equation applied there coincides with the string equation that comes out of Gromov-Witten invariants. Furthermore, we see that the t-parametrised family $\mathcal{F}^g_n(t,p_1,\ldots,p_n)$ used there has enumerative meaning—it is related to insertions of the so called puncture operator, and gives a relation to the Toda equations.

3. Enumerative examples

Many examples of the conjectural relation between quantum curves and topological recursion have been proven in the literature. The quantum curve was shown to satisfy (1.4) for simple Hurwitz numbers [**36**, **43**] and simple Hurwitz numbers of an arbitrary base curve [**34**]; monotone Hurwitz numbers [**10**]; Belyi maps [**36**]; bipartite Belyi maps [**12**, **33**]; hypermaps [**11**]; Gromov-Witten invariants with target X for $X = \{\text{pt}\}$ [**42**]; $X = \mathbb{P}^1$ [**16**]; $X =$ the topological vertex and the resolved conifold [**43**]; and spectral curves coming from matrix models [**4**].

There are essentially two types of proofs. Firstly there are the proofs that use the $\hbar$ expansion of $\log\psi(p,\hbar)$ and stay closer to the conjecture, essentially giving a reason for the conjecture. Secondly there are the proofs that use an expansion of $\psi(p,\hbar)$ in x and exploit the fact that the Euler characteristic is a coarser invariant than genus, and hence the proofs may be simpler. It is rather natural to use an x expansion when implementing the quantum curve on a computer. One might also describe the $\hbar$ expansion proof as a proof concentrated around the zeros of dx, while the second type of proof considers expansions at regular values of x. The quantum curve for simple Hurwitz numbers and Belyi maps are examples that have been proven using both approaches—the first approach in [**36**] and the second approach in [**42**], respectively [**11**].

In the remainder of this section we describe two specific examples of quantum curves. In the first of these—the quantum curve of the Gromov-Witten invariants of $\mathbb{P}^1$—we consider a family of quantum curves parametrised by $t \in \mathbb{C}$ with semi-classical limit independent of t. They correspond to different choices of primitives in (1.4). In the second of these—Belyi maps—we present two proofs in order to contrast the two approaches described above used in most proofs.

3.1. Gromov-Witten invariants of $\mathbb{P}^1$.

In the following example the wave function is given by a specialisation of the partition function of Gromov-Witten invariants of $\mathbb{P}^1$. We consider a family of quantum curves parametrised by $t \in \mathbb{C}$ with semi-classical limit independent of t. The $t = 1/2$ case appeared in [**16**]. The t-dependence is rather useful, making a connection with the Toda lattice.

Let $\overline{\mathcal{M}}_{g,n}(\mathbb{P}^1,d)$ denote the moduli space of stable maps of degree d from an n-pointed genus g curve to $\mathbb{P}^1$. The descendant Gromov-Witten invariants of $\mathbb{P}^1$ are defined by

$$\langle\prod_{i=1}^{n}\tau_{b_i}(\alpha_i)\rangle^d_{g.n} := \int_{[\overline{\mathcal{M}}_{g,n}(\mathbb{P}^1,d)]^{vir}}\prod_{i=1}^{n}\psi_i^{b_i}ev_i^*(\alpha_i), \tag{3.1}$$

where $[\overline{\mathcal{M}}_{g,n}(\mathbb{P}^1,d)]^{vir}$ is the virtual fundamental class of the moduli space, of degree given by its virtual dimension $2g-2+n+2d$,

$$ev_i:\overline{\mathcal{M}}_{g,n}(\mathbb{P}^1,d)\longrightarrow\mathbb{P}^1$$

is a natural morphism defined by evaluating a stable map at the i-th marked point of the source curve, $\alpha_i\in H^*(\mathbb{P}^1,\mathbb{Q})$ is a cohomology class of the target $\mathbb{P}^1$, and ψ_i is the tautological cotangent class in $H^2(\overline{\mathcal{M}}_{g,n}(\mathbb{P}^1,d),\mathbb{Q})$. We denote by 1 the generator of $H^0(\mathbb{P}^1,\mathbb{Q})$, and by $\omega\in H^2(\mathbb{P}^1,\mathbb{Q})$ the Poincaré dual to the point class. We call $\tau_k(\omega)$ *stationary* classes since the pull-back $ev_i^*(\omega)\subset\mathcal{M}_{g,n}(\mathbb{P}^1,d)$ restricts to stable maps f with $f(p_i)=x_i$ for a given stationary point $x_i\in\mathbb{P}^1$.

The free energy of the Gromov-Witten invariants of $\mathbb{P}^1$ is defined by

$$F_g=\sum_d q^d\left\langle\exp\left\{\sum_{i\geq 0}^{\infty}\tau_i(\omega)t_i+\tau_i(1)s_i\right\}\right\rangle^g_d$$

and

$$F=\sum_{g\geq 0}F_g=\frac{1}{2}s_0^2t_0+\frac{1}{6}s_0^3t_1+\dots-\frac{1}{24}t_0-\frac{1}{24}s_0t_1+\dots$$
$$+q(1+t_0+\frac{1}{2}t_0^2+\dots+s_0t_1+\frac{1}{2}s_0^2t_2+\dots)+\dots\quad .$$

The partition function is defined to be

$$Z(t_0,t_1,\dots,s_0,s_1,\dots,q)=\exp F.$$

3.1.1. *Quantum curve.* A specialisation of the partition function gives rise to a wave function and quantum curve associated to Gromov-Witten invariants of $\mathbb{P}^1$. We will describe its semi-classical limit below. Define

$$\psi(x,\hbar,q,t)=Z\Big(t_i=i!\left(\frac{\hbar}{x}\right)^{i+1},q=\frac{q}{\hbar^2},s_0=t,s_i=0,i>0\Big)$$
$$=\exp\left\{\frac{q}{\hbar^2}+\frac{-\frac{\hbar}{24}+\frac{q}{\hbar}+\frac{1}{2}t^2\hbar}{x}+\frac{..}{x^2}+\dots\right\}$$

Note that we have switched off all the non-stationary insertions except the class $\tau_0(1)$ which is known as the *puncture operator.* Geometrically we have placed the target stationary points at a single point $p\in\mathbb{P}^1$. The coefficient of x^k counts all covers with local *virtual* degree k over p defined to be the total ramification plus number of preimage points over p. This local virtual degree can differ from the actual local degree. For example, the descendant $\tau_1(\omega)$ can be realised on the space of degree 1 stable maps $\overline{\mathcal{M}}_{1,1}(\mathbb{P}^1,1)$ hence appears as $\langle\tau_1(\omega)\rangle x^{-2}$ in ψ. The exponent -2 of x, which corresponds to local virtual degree 2, reflects that fact that $\tau_1(\omega)$ should arise from a locally 2 to 1 map.

The wave function of the quantum curve is obtained from ψ by modifying the unstable terms:

$$\psi_1(x,\hbar,q,t)=\psi(x,\hbar,q,t)\cdot x^{-t}\exp(\frac{1}{\hbar}(x\ln x-x)). \tag{3.2}$$

THEOREM 1. [16]

$$[e^{\hbar\frac{d}{dx}}+qe^{-\hbar\frac{d}{dx}}-x+(t-\frac{1}{2})\hbar]\psi_1=0.$$

REMARK 3.1. The proof in [16] considers the case $q=1$, $t=1/2$ but it is not difficult to derive the statement here from that special case.

REMARK 3.2. The operator in Theorem 1 is a simple case of the Lax operator for the Toda lattice appearing in Dubrovin-Zhang and Takasaki-Takebe:

$$e^{\hbar\frac{d}{dx}}+v(x)+e^{u(x)}e^{-\hbar\frac{d}{dx}}.$$

It appears in Aganagic-Dijkgraaf-Klemm-Mariño [2] as

$$H=e^{\hbar\frac{d}{dx}}+x+e^{-\hbar\frac{d}{dx}}$$

and the wave function ψ is said to describe the insertion of a D-brane at a fixed x.

The semi-classical limit gives rise to the following spectral curve. Put $\psi_1=\exp(\frac{1}{\hbar}S_0+S_1+\hbar S_2+\dots)$.

$$\begin{aligned}0&=\lim_{\hbar\to0}\exp^{-\frac{1}{\hbar}S_0}\left[e^{\hbar\frac{d}{dx}}+qe^{-\hbar\frac{d}{dx}}-x+(t-\frac{1}{2})\hbar\right]\exp(\frac{1}{\hbar}S_0+S_1+\hbar S_2+\dots)\\&=\lim_{\hbar\to0}[e^{\frac{S_0(x+\hbar)-S_0(x)}{\hbar}}e^{\hbar\frac{d}{dx}}+qe^{\frac{S_0(x-\hbar)-S_0(x)}{\hbar}}e^{-\hbar\frac{d}{dx}}-x+(t-\tfrac{1}{2})\hbar]\exp(S_1+\dots)\\&=[e^{S_0'(x)}+qe^{-S_0'(x)}-x]\exp(S_1).\end{aligned}$$

Hence $e^{S_0'(x)}+qe^{-S_0'(x)}-x=0$ and for $z=e^{S_0'(x)}$ this defines the spectral curve

$$C=\{x=z+\frac{q}{z},\quad y=\ln z\} \tag{3.3}$$

which agrees with the mirror Landau-Ginzburg model [19]. Although it is not algebraic, one can still apply topological recursion to C to produce the stationary Gromov-Witten invariants of $\mathbb{P}^1$.

THEOREM 2 ([18],[40]). *(Analytic expansions around a branch of $\{x_i=\infty\}$ of) the invariants $\omega_n^g(C)$ of the curve C defined in (3.3) are generating functions for the stationary Gromov-Witten invariants of $\mathbb{P}^1$:*

$$\omega_n^g=\sum_{\mathbf{b}}\left\langle\prod_{i=1}^n\tau_{b_i}(\omega)\right\rangle_d^g\cdot\prod_{i=1}^n(b_i+1)!x_i^{-b_i-2}dx_i-\delta_{g0}\delta_{n1}\ln x_1dx_1+\delta_{g0}\delta_{n2}\frac{dx_1dx_2}{(x_1-x_2)^2}.$$

REMARK 3.3. The non-stationary Gromov-Witten invariants of $\mathbb{P}^1$ are also contained inside the $\omega_n^g(C)$ in the form of ancestor invariants. See [18] for details.

Indeed, Theorems 1 and 2 confirm the conjectural form (1.4) that $S_k(p)=F_n^g(p,p,\dots,p)$ for $F_n^g(p_1,\dots,p_n)$ a primitive of ω_n^g of the spectral curve (3.3).

3.1.2. *Relation to Toda equation.* It is useful to replace the derivatives in x with derivatives in the parameter t using the wave operator

$$D := \hbar\frac{d}{dx} + \frac{\partial}{\partial t}$$

and the equation

$$D\psi_1 = 0$$

which follows from the string equation (satisfied quite generally by Gromov-Witten invariants) restricted to the specialisation of the partition function. The quantum curve becomes

$$\psi(t-1) + \frac{q}{x^2}\psi(t+1) - \Big(1 - (t - \frac{1}{2})\frac{\hbar}{x}\Big)\psi(t) = 0 \tag{3.4}$$

where $\psi(t)$ is defined via (3.2) and we suppress all arguments except for t, so for example we write $\psi(t-1)$ for $\psi(x, \hbar, q, t-1)$.

The partition function satisfies the Toda lattice equation

$$\frac{Z(s_0-1)Z(s_0+1)}{Z(s_0)^2} = \frac{1}{q}\frac{\partial^2}{\partial t_0^2}\log Z(s_0)$$

which was conjectured by Eguchi-Yang [**20**] and proven by Okounkov-Pandharipande [**41**] for $s_0 = n \in \mathbb{Z}$ and Dubrovin-Zhang [**14**] for $s_0 \in \mathbb{R}$. Using the divisor equation

$$\frac{\partial F}{\partial t_0} = \frac{1}{2}s_0^2 + q\frac{\partial F}{\partial q}$$

one can restrict the Toda lattice equation to the specialisation to get:

$$\frac{\psi(t-1)\psi(t+1)}{\psi(t)^2} = \hbar^2\frac{d}{dq}\left(q\frac{d}{dq}\log\psi\right).$$

The Toda lattice equation does not uniquely determine the partition function Z, nor does its specialisation determine the wave function ψ. The Toda lattice equation does determine $\psi(x, q, \hbar, t)$ from the initial degree 0 terms $\psi(x, q = 0, \hbar, t)$.

Write (3.4) as $\mathcal{L}\psi = 0$, so

$$\mathcal{L} = e^{-\hbar\frac{\partial}{\partial t}} + \frac{q}{x^2}e^{\hbar\frac{\partial}{\partial t}} - (1 - (t - \frac{1}{2})\frac{\hbar}{x}).$$

Then

$$(D - (t-1)\frac{\hbar}{x}) \circ \mathcal{L} = \mathcal{L} \circ (D - t\frac{\hbar}{x})$$

and ψ is characterised as the solution of (3.4) that is also an eigenfunction of D:

$$D\psi(x, \hbar, q, t) = t\frac{\hbar}{x}\psi(x, \hbar, q, t).$$

Similarly, D is compatible with the Toda lattice equation so it admits an eigenfunction solution if the degree 0 part is an eigenfunction of $\mathcal{L}$. Note that there are solutions of Toda that are not solutions of (3.4) and vice versa.

The *degree zero* Gromov-Witten invariants contrast nicely the difference between the Toda lattice equation and the quantum curve. Put $\psi = \sum_{d\geq 0} q^d\psi_d = \psi_0 + q\psi_1 + \ldots$

Toda equation$|_{q=0}$: $\dfrac{\psi_0(t+1)\psi_0(t-1)}{\psi_0(t)^2} = \hbar^2\psi_1(t)$ has nothing to say about ψ_0.

Quantum curve$|_{q=0}$: $\psi_0(t-1) = (1-(t-\frac{1}{2})\frac{\hbar}{x})\psi_0(t) = 0$ uniquely determines $\psi_0(x,\hbar,t)$ (which is also an eigenfunction of D). An exact formula [**25**, **41**] is given by $\psi_0 = \exp F_0$ where

$$F_0 = \phi(x-\hbar t) + \frac{1}{\hbar}(x-\hbar t)\ln\left(1 - t\frac{\hbar}{x}\right) + t,$$

$$\phi(x) = \sum_{g=1}^{\infty}(1-2^{1-2g})\frac{\zeta(1-2g)}{2g-1}\left(\frac{\hbar}{x}\right)^{2g-1}.$$

Finally, the quantum curve implies rather nice rational behaviour of the wave function. Applying the quantum curve equation (3.4) to

$$\psi = \sum_{d\geq 0} q^d\psi_d = \psi_0\left(1 + q\frac{\psi_1}{\psi_0} + q^2\frac{\psi_2}{\psi_0} + \dots\right),$$

one finds that $\frac{\psi_d}{\psi_0}$ is *rational* in x and with d simple poles. For example,

$$\frac{\psi_1}{\psi_0}(x,\hbar,t) = 1 + \frac{1}{\frac{x}{\hbar} - t - \frac{1}{2}},\quad \frac{\psi_2}{\psi_0}(x,\hbar,t) = \frac{1}{2} + \frac{\frac{1}{2}}{\frac{x}{\hbar} - t - \frac{1}{2}} + \frac{\frac{1}{2}}{\frac{x}{\hbar} - t - \frac{3}{2}}.$$

Put $w = \frac{x}{\hbar} - t$ and

$$\frac{\psi_d}{\psi_0} = r_d(w) = a_{0,d} + \frac{a_{1,d}}{w - \frac{1}{2}} + \frac{a_{2,d}}{w - \frac{3}{2}} + \dots + \frac{a_{d,d}}{w - d + \frac{1}{2}}$$

then the residues satisfy the linear system:

$$a_{i,d} = a_{i+1,d} + \frac{a_{i-1,d-1}}{i(i-1)} \quad (a_{d+1,d} = 0)$$
$$a_{1,d} = a_{2,d} + r_{d-1}(-\tfrac{1}{2}).$$

3.2. Belyi maps. The fundamental example of the spectral curve $y^2 - xy + 1 = 0$ corresponds to the enumeration of Belyi maps. We present two proofs that its quantum curve satisfies (1.4). The first proof uses the $\hbar$ expansion of (the log of) $\psi(p,\hbar)$ while the second proof uses the expansion of $\psi(p,\hbar)$ around $x = \infty$. The first proof uses topological recursion and thus in a sense explains why the conjecture is true. The second proof uses the underlying enumerative problem and highlights a connection to hypergeometric functions.

Let $\mathcal{B}_{g,n}(\mu_1,\dots,\mu_n)$ be the set of all connected genus g *Belyi* maps—meaning branched covers $\pi : \Sigma \to \mathbb{P}^1$ unramified over $\mathbb{P}^1\backslash\{0,1,\infty\}$—with all points over 1 having ramification 2 and ramification divisor over ∞ given by $\pi^{-1}(\infty) = \mu_1 p_1 + \dots + \mu_n p_n$, where the points over ∞ are labelled $p_1,\dots,p_n$. Two Belyi maps $\pi_1 : \Sigma_1 \to \mathbb{P}^1$ and $\pi_2 : \Sigma_2 \to \mathbb{P}^1$ are isomorphic if there exists a homeomorphism $f : \Sigma_1 \to \Sigma_2$ that covers the identity on $\mathbb{P}^1$ and preserves the labelling over ∞.

DEFINITION 3.4. For any $\boldsymbol{\mu} = (\mu_1,\dots,\mu_n) \in \mathbb{Z}_+^n$, define

$$M_{g,n}(\mu_1,\dots,\mu_n) = \sum_{\pi\in\mathcal{B}_{g,n}(\boldsymbol{\mu})} \frac{1}{|\mathrm{Aut}\ \pi|},$$

where Aut π denotes the automorphism group of the branched cover π.

Define

$$F^g_n(x_1,\dots,x_n)=(-1)^n\sum_{\boldsymbol{\mu}>0}M_{g,n}(\mu_1,\dots,\mu_n)x_1^{-\mu_1}\dots x_n^{-\mu_n}+\delta_{g0}\delta_{n1}\log x. \tag{3.5}$$

The exceptional case of $(g,n)=(0,1)$ includes an extra $\log x$ term (to allow for the 0th Catalan number $C_0=1$ missing from $kM_{0,1}(k)=C_k$) so that

$$\frac{d}{dx}F^0_1(x)=\sum_{k\geq 0}C_kx^{-k-1}=y,\quad x=y+\frac{1}{y}.$$

This is the same as Example 1.4, and gives an analytic expansion of the global meromorphic function y in the local coordinate x on a branch of $x=\infty$, of the Stieltjes transform of a probability measure

$$y=\hat{\rho}(x)=\frac{1}{2\pi}\int_{-2}^{2}\frac{\sqrt{4-t^2}}{x-t}dt.$$

The connection between the two goes deeper. The probability measure is the Wigner semicircle distribution of eigenvalues of Hermitian matrices with the Gaussian potential. Associated to any Belyi map $f\in\mathcal{B}_{g,n}(\mu_1,\dots,\mu_n)$ is a fatgraph given by the pull-back of the unit interval $\Gamma=f^{-1}([0,1])$, and these graphs arise when calculating Hermitian matrix integrals. The *spectral curve* of this matrix model is

$$y^2-xy+1=0. \tag{3.6}$$

It was proven in **[24]** that Tutte's equations for discrete surfaces—the fatgraphs $\Gamma=f^{-1}([0,1])$—correspond to the loop equations for this matrix model:

$$\begin{aligned}x_1W_g(x_1,\mathbf{x}_S)=&W_{g-1}(x_1,x_1,\mathbf{x}_S)+\sum_{\substack{g_1+g_2=g\\ I\sqcup J=S}}W_{g_1}(x_1,\mathbf{x}_I)\,W_{g_2}(x_1,\mathbf{x}_J)\\&+\sum_{j=2}^{n}\frac{\partial}{\partial x_j}\frac{W_g(x_1,\mathbf{x}_{S\setminus\{j\}})-W_g(\mathbf{x}_S)}{x_1-x_j}+\delta_{g,0}\,\delta_{n,1}\end{aligned} \tag{3.7}$$

where W_g is related to the generating function (3.5) via

$$\frac{\partial}{\partial x_1}\cdots\frac{\partial}{\partial x_n}F_{g,n}(x_1,\dots,x_n)=W_g(x_1,\dots,x_n).$$

The solution of the loop equations (3.7) for $(g,n)=(0,1)$ defines the spectral curve (3.6). A consequence is that topological recursion applied to the spectral curve (3.6) yields:

(3.8)

$$\omega^g_n(p_1,\dots,p_n)=W_g(x_1,\dots,x_n)\,\mathrm{d}x_1\otimes\dots\otimes\mathrm{d}x_n-\delta_{g0}\delta_{n1}\frac{dx_1}{x_1}+\delta_{g0}\delta_{n2}\frac{dx_1\otimes dx_2}{(x_1-x_2)^2}$$

for $p_i\in C$, the spectral curve, $x_i=x(p_i)$ and equality denotes an analytic expansion in the local coordinate x on a branch of $x=\infty$. We see that the unstable cases $(g,n)=(0,1)$ and $(0,2)$ require minor adjustments, as in the example of Gromov-Witten invariants of $\mathbb{P}^1$.

Consider the following wave function constructed out of $F^g_n(x_1,\dots,x_n)$ defined in (3.5) for $x_i=x(y_i)=y_i+\frac{1}{y_i}$ and specialised to $y_i=y$.

(3.9)

$$\psi(y,\hbar)=\exp\Big(\hbar^{-1}\sum_{k\geq 0}\hbar^kS_k(y)\Big),\quad S_k(y)=\sum_{2g-1+n=k}\frac{1}{n!}F^g_n(x(y),x(y),\dots,x(y))$$

where $x(y) = y + \frac{1}{y}$. The formulae (3.8) show that $\psi(y, \hbar)$ is the conjectural wave function (1.4) for the quantum curve of the spectral curve (3.6). The following theorem confirms the conjectural form.

THEOREM 3. [**29, 36**]

$$\left(\hbar^2 \frac{d^2}{dx^2} - x\hbar \frac{d}{dx} + 1\right) \psi(y, \hbar) = 0 \tag{3.10}$$

for $y^2 - xy + 1 = 0$ *and* $\psi(y, \hbar)$ *defined in* (3.9).

REMARK 3.5. This theorem was known in the physics literature, see for example [**29**]. A rigorous proof using topological recursion was given in [**36**]. A simpler more direct combinatorial proof was given in [**11**].

PROOF. **Proof 1—expansion in $\hbar^k$** following [**36**].

The unstable cases have been proven, since for $(g, n) = (0, 1)$, $\frac{d}{dx} S_0(z) = z$ is shown above and for $(g, n) = (0, 2)$, it follows from the relation

$$d_1 d_2 F_2^0(x(z_1), x(z_2)) = \frac{dz_1 dz_2}{(z_1 - z_2)^2} - \frac{dx(z_1) dx(z_2)}{(x(z_1) - x(z_2))^2} = \frac{dz_1 dz_2}{(1 - z_1 z_2)^2}$$
$$\Rightarrow S_1(z) = -\frac{1}{2} \log(1 - z^2)$$

which agrees with S_1 calculated via the WKB method in (1.8) with $y = 1/z$.

For $\log \psi(z, \hbar) = \hbar^{-1} \sum_{k \geq 0} \hbar^k S_k(z)$ (3.10) becomes

$$\hbar^2 \left(\left(\frac{d}{dx}\right)^2 \log \psi(z, \hbar) + \left(\frac{d}{dx} \log \psi(z, \hbar)\right)^2 \right) - x\hbar \frac{d}{dx} \log \psi(z, \hbar) + 1 = 0$$

which is equivalent to

$$\sum_{k=0}^{\infty} \left(\frac{d}{dx}\right)^2 S_k(z) \hbar^{k+1} + \left(\sum_{k=0}^{\infty} \frac{d}{dx} S_k(z) \hbar^k\right)^2 - x \sum_{k=0}^{\infty} \frac{d}{dx} S_k(z) \hbar^k + 1 = 0. \tag{3.11}$$

Note that if we set $\hbar = 0$ in (3.11) we get $\left(\frac{d}{dx} S_0(z)\right)^2 - x \frac{d}{dx} S_0(z) + 1 = 0$ which is exactly the first of the loop equations (3.7). The essential idea of the proof is that the coefficients of $\hbar^k$ for $k > 0$ have the quadratic form of the topological recursion.

Integrate (3.7) with respect to $x_2, \ldots, x_n$ (when $n > 1$):

$$\begin{aligned} x_1 \frac{d}{dx_1} F_n^g(x_1, \mathbf{x}_S) = {} & \frac{\partial^2}{\partial u_1 \partial u_2} F_{n+1}^{g-1}(u_1, u_2, \mathbf{x}_S) \Big|_{u_1 = u_2 = x_1} \\ & + \sum_{\substack{g_1 + g_2 = g \\ I \sqcup J = S}} \frac{d}{dx_1} F_{|I|+1}^{g_1}(x_1, \mathbf{x}_I) \frac{d}{dx_1} F_{|J|+1}^{g_2}(x_1, \mathbf{x}_J) \\ & + \sum_{j=2}^{n} \frac{\frac{d}{dx_1} F_{n-1}^g(x_1, \mathbf{x}_{S \setminus \{j\}}) - \frac{d}{dx_j} F_{n-1}^g(\mathbf{x}_S)}{x_1 - x_j}. \end{aligned} \tag{3.12}$$

Each term of (3.12) vanishes at $x_j = \infty$ for any $j = 2, \ldots, n$ which determines the constants of integration.

Specialise (3.12) to $x_i = x$

$$\begin{aligned}(3.13)\quad &\frac{1}{n}x\frac{d}{dx}F^g_n(x,\dots,x)\\ &=\sum_{g_1+g_2=g}\binom{n-1}{|I|}\frac{1}{|I|+1}\frac{d}{dx}F^{g_1}_{|I|+1}(x,\dots,x)\,\frac{1}{|J|+1}\frac{d}{dx}F^{g_2}_{|J|+1}(x,\dots,x)\\ &+\frac{1}{n(n+1)}\frac{d^2}{dx^2}F^{g-1}_{n+1}(x,\dots,x)-\frac{1}{n}\frac{d^2}{du^2}F^{g-1}_{n+1}(u,x,\dots,x)\Big|_{u=x}\\ &+(n-1)\frac{d^2}{du^2}F^g_{n-1}(u,x,\dots,x)\Big|_{u=x}\end{aligned}$$

which uses the elementary relations on symmetric polynomials

$$\frac{d}{dx_1}F^g_n(x_1,x,\dots,x)\Big|_{x_1=x}=\frac{1}{n}\frac{d}{dx}F^g_n(x,\dots,x)$$

and

$$\frac{\partial^2}{\partial u_1\partial u_2}F^{g-1}_{n+1}(u_1,u_2,x,\dots,x)\Big|_{u_i=x}=$$
$$\frac{1}{n(n+1)}\frac{d^2}{dx^2}F^{g-1}_{n+1}(x,\dots,x)-\frac{1}{n}\frac{d^2}{du^2}F^{g-1}_{n+1}(u,x,\dots,x)\Big|_{u=x}$$

and the fact that the limit $x_1 \to x_j$ in the last term defines the derivative. See Appendix A in [**36**].

Recall that $S_k(z) = \sum_{2g-1+n=k}\frac{1}{n!}F^g_n(x(z),x(z),\dots,x(z))$. So we multiply (3.13) by $\frac{1}{(n-1)!}$ and sum over all (g,n) such that $2g-1+n=k$. The second derivatives in u cancel and (3.13) becomes

$$x\frac{d}{dx}S_k(z)=\sum_{i+j=k}\frac{d}{dx}S_i(z)\frac{d}{dx}S_j(z)+\frac{d^2}{dx^2}S_{k-1}(z)$$

which is the coefficient of $\hbar^k$ in (3.11) for $k>0$. Hence the theorem is proven.

Proof 2—expansion in x^{-1} following [**11**]. First express the wave function defined in (3.9) using (3.5) as

$$\psi(p,\hbar)=x^{1/\hbar}\,\overline{\psi}(x,\hbar).$$

We write $x = x(p)$ in the argument because we will work with expansions around $x = \infty$. The main idea is to prove the following exact formula.

$$(3.14)\quad \overline{\psi}(x,\hbar)=1+\sum_{e=1}^{\infty}\frac{(-1)^e\hbar^e}{2^e e!}\hbar^{-1}(\hbar^{-1}-1)(\hbar^{-1}-2)\cdots(\hbar^{-1}-2e+1)x^{-2e}.$$

We will see below that the right hand side of (3.14) is given by

$$x^{-N}(N/2)^{-N/2}H_N(x\sqrt{\frac{N}{2}})$$

where $N = 1/\hbar$ and H_N is the Nth Hermite polynomial.

To prove the wave equation (3.10) we simply apply the differential operator directly to the formula (3.14). Actually, we would like to work with $\overline{\psi}(x,\hbar)$ which has an expansion in x^{-1} with coefficients that are Laurent polynomials in $\hbar$ — in other words, $\overline{\psi}(x,\hbar) \in \mathbb{Q}[\hbar^{\pm 1}][[x^{-1}]]$. Thus we first conjugate (3.10) by $x^{1/\hbar}$ to get

$$\left[\hbar^2 \frac{d^2}{dx^2} + \hbar \left((\frac{2}{x} - x) \frac{d}{dx} - \frac{1}{x^2} \right) + \frac{1}{x^2} \right] \overline{\psi}(x,\hbar) = 0. \tag{3.15}$$

which follows immediately by direct application to the formula (3.14).

The remainder of the proof is devoted to proving (3.14). First, consider the logarithm of the modified wave function.

$$\begin{aligned} \log \overline{\psi}(x,\hbar) &= \sum_{g=0}^{\infty} \sum_{n=1}^{\infty} \frac{\hbar^{2g-2+n}}{n!} F_{g,n}(x,x,\ldots,x) \\ &= \sum_{g=0}^{\infty} \sum_{n=1}^{\infty} \frac{\hbar^{2g-2+n}}{n!} \sum_{\mu_1,\ldots,\mu_n=1}^{\infty} (-1)^n M_{g,n}(\mu_1,\ldots,\mu_n)\, x^{-\sum \mu_i} \\ &= \sum_{v=1}^{\infty} \sum_{e=1}^{\infty} (-1)^{e-v} f(v,e)\, \hbar^{e-v} x^{-2e} \end{aligned}$$

Here, $f(v,e)$ denotes the weighted count of connected dessins with v vertices and e edges. To obtain this last expression, we have used the fact that $e - v = 2g - 2 + n$ and $\mu_1 + \cdots + \mu_n = 2e$ for any dessin. The factor $\frac{1}{n!}$ accounts for the fact that we are now considering dessins with unlabelled faces. The weight of a dessin is the reciprocal of its number of automorphisms.

Let $f^\bullet(v,e)$ denote the weighted count of possibly disconnected dessins with v vertices and e edges. Then

$$\overline{\psi}(x,\hbar) = 1 + \sum_{v=1}^{\infty} \sum_{e=1}^{\infty} (-1)^{e-v} f^\bullet(v,e)\, \hbar^{e-v} x^{-2e}.$$

Now $f^\bullet(v,e)$ is equal to $\frac{1}{(2e)!}$ multiplied by the number of triples $(\sigma_0, \sigma_1, \sigma_2)$ of permutations in the symmetric group S_{2e} such that $\sigma_0\sigma_1\sigma_2 = \mathrm{id}$, σ_0 has v disjoint cycles and σ_1 has cycle type 2^e. Clearly we need to sum only over pairs (σ_0, σ_1). Recall that the Stirling number of the first kind $\left[{n \atop k}\right]$ counts the number of permutations in S_n with k disjoint cycles. Since $(2e-1)!!$ is the number of permutations in S_e of cycle type 2^e we have

$$f^\bullet(v,e) = \frac{1}{(2e)!} \begin{bmatrix} 2e \\ v \end{bmatrix} (2e-1)!! = \frac{1}{2^e e!} \begin{bmatrix} 2e \\ v \end{bmatrix}$$

Note that $\overline{\psi}(x,\hbar) \in \mathbb{Q}[\hbar^{\pm 1}][[x^{-1}]]$ since for fixed e we require $2 \le v \le 2e$ to have $f^\bullet(v,e) \neq 0$.

Now we simply use the fact that the generating function for Stirling numbers of the first kind is given by

$$\sum_{k=1}^{n} \begin{bmatrix} n \\ k \end{bmatrix} x^k = x(x+1)(x+2)\cdots(x+n-1).$$

Use this in the expression for the modified wave function as follows.

$$\begin{aligned}\overline{\psi}(x,\hbar)&=1+\sum_{v=1}^{\infty}\sum_{e=1}^{\infty}\frac{(-1)^{e-v}}{2^e e!}\begin{bmatrix}2e\\ v\end{bmatrix}\hbar^{e-v}x^{-2e}\\ &=1+\sum_{e=1}^{\infty}(-1)^e\hbar^e x^{-2e}\frac{1}{2^e e!}\sum_{v=1}^{\infty}\begin{bmatrix}2e\\ v\end{bmatrix}(-\hbar)^{-v}\\ &=1+\sum_{e=1}^{\infty}\frac{(-1)^e\hbar^e}{2^e e!}\hbar^{-1}(\hbar^{-1}-1)(\hbar^{-1}-2)\cdots(\hbar^{-1}-2e+1)x^{-2e}\end{aligned}$$

which proves (3.14) as required. □

REMARK 3.6. The *Hermite polynomials* are defined by

$$H_N(x):=(-1)^N e^{x^2}\frac{d^N}{dx^N}e^{-x^2}=\sum_{k=0}^{\lfloor\frac{N}{2}\rfloor}(-1)^k\binom{N}{2k}(2k-1)!!2^{N-k}x^{N-2k}\tag{3.16}$$

So $H_N(x)$ is a degree N polynomial in x, for example $H_0(x)=1$, $H_1(x)=2x$ and $H_2(x)=4x^2-2$. They satisfy the Hermite equations

$$\Big[\Big(\frac{d}{dx}\Big)^2-2x\frac{d}{dx}+2N\Big]H_N(x)=0.$$

For any function $f:H_N\to\mathbb{C}$ on $N\times N$ Hermitian matrices, define

$$\langle f\rangle_N:=\frac{1}{Z_N}\int_{H_N}f(A)\exp\Big[-\frac{N}{2}\mathrm{tr}A^2\Big]dA,\quad Z_N=\int_{H_N}\exp\Big[-\frac{N}{2}\mathrm{tr}A^2\Big]dA$$

so in particular $\langle A_{ij}\rangle=0$ and $\langle|A_{ij}|^2\rangle=1/N$.

It is well-known that

$$\langle\det(xI-A)\rangle_N=(N/2)^{-N/2}H_N\Big(x\sqrt{\frac{N}{2}}\Big).\tag{3.17}$$

There are numerous proofs of (3.17). The following proof comes from [**27**].

$$\begin{aligned}\langle\det(xI-A)\rangle_N&=\langle\sum_{P\in S_N}\epsilon(P)\prod_{i=1}^{N}(x\delta_{i,P(i)}-A_{i,P(i)})\rangle_N\\ &=\langle\sum_{P\in S_N^{(2)}}\epsilon(P)\prod_{i=1}^{N}(x\delta_{i,P(i)}-A_{i,P(i)})\rangle_N\\ &=\sum_{k=0}^{\lfloor\frac{N}{2}\rfloor}(-1)^k\binom{N}{2k}(2k-1)!!x^{N-2k}N^{-k}=(N/2)^{-N/2}H_N\Big(x\sqrt{\frac{N}{2}}\Big)\end{aligned}$$

where the second equality uses the fact that the only non-zero contributions to the integral come from permutations with no cycles of length greater than 2, denoted by $S_N^{(2)}\subset S_N$. For the third equality, each $P\in S_N^{(2)}$ is a product of say $N-2k$ fixed points and k 2-cycles which each contribute a factor of x, respectively $1/N$. The factor $(-1)^k$ comes from the parity of P and the factor $\binom{N}{2k}(2k-1)!!$ is the number of ways of choosing $N-2k$ fixed points and k 2-cycles. The final equality uses the formula (3.16) for the scaled Hermite polynomial. One consequence of (3.17) is that $\langle\det(xI - A)\rangle_N$ satisfies the (scaled) Hermite equation since

$$\Big[\Big(\frac{1}{N}\frac{d}{dx}\Big)^2 - x\frac{1}{N}\frac{d}{dx} + 1\Big]H_N(x\sqrt{\frac{N}{2}}) = 0.$$

Moreover, for $\hbar = 1/N$

$$\langle\det(x - A)\rangle_N = \psi(x, \hbar)$$

where the left hand side is considered as a function of N. This is obtained by comparing the exact expression (3.14) for $\overline{\psi}(x, \hbar)$ with that for $\langle\det(I - Ax^{-1})\rangle_N$. This identification also gives a proof of the well-known semi-classical limit:

$$\lim_{N\to\infty}\frac{1}{N}\frac{d}{dx}\log\langle\det(xI - A)\rangle = \frac{1}{2\pi}\int_{-2}^{2}\frac{\sqrt{4-t^2}}{x-t}dt, \quad x \notin [-2, 2].$$

REMARK 3.7. Another proof of Theorem 3 appears in [**1**]. It starts from the fact that (3.5) is related to correlation functions of a matrix model, known as the Kontsevich-Penner matrix model [**8**] which is an integral over $N \times N$ Hermitian matrices with a potential that in some sense generalises the integral representation of Hermite polynomials. The wave function defined in (3.9) arises as the specialisation of the partition function of the Kontsevich-Penner matrix model for 1×1 matrices! The proof that it satisfies the differential equation (3.10) is immediate because the integral is simply one dimensional.

References

[1] J. E. Andersen, L. O. Chekhov, P. Norbury, and R. C. Penner, *Models of discretized moduli spaces, cohomological field theories, and Gaussian means*, J. Geom. Phys. **98** (2015), 312–339, DOI 10.1016/j.geomphys.2015.08.018. MR3414961

[2] M. Aganagic, R. Dijkgraaf, A. Klemm, M. Mariño, and C. Vafa, *Topological strings and integrable hierarchies*, Comm. Math. Phys. **261** (2006), no. 2, 451–516, DOI 10.1007/s00220-005-1448-9. MR2191887 (2006i:81175)

[3] G. Borot and B. Eynard, *All order asymptotics of hyperbolic knot invariants from non-perturbative topological recursion of A-polynomials*, Quantum Topol. **6** (2015), no. 1, 39–138, DOI 10.4171/QT/60. MR3335006

[4] M. Bergére, G. Borot, and B. Eynard, *Rational Differential Systems, Loop Equations, and Application to the qth Reductions of KP*, Ann. Henri Poincaré **16** (2015), no. 12, 2713–2782, DOI 10.1007/s00023-014-0391-8. MR3416869

[5] Bergére, Michel and Eynard, Bertrand *Determinantal formulae and loop equations.* IPHT T09/015. Latex, 35 pages. 2009. hal-00354909v1

[6] V. Bouchard and B. Eynard, *Think globally, compute locally*, J. High Energy Phys. **2** (2013), 143, front matter + 34. MR3046532

[7] L. Chekhov and B. Eynard, *Hermitian matrix model free energy: Feynman graph technique for all genera*, J. High Energy Phys. **3** (2006), 014, 18 pp. (electronic), DOI 10.1088/1126-6708/2006/03/014. MR2222762 (2007k:81144)

[8] L. Chekhov and Yu. Makeenko, *The multicritical Kontsevich-Penner model*, Modern Phys. Lett. A **7** (1992), no. 14, 1223–1236, DOI 10.1142/S0217732392003700. MR1164382 (93i:81176)

[9] T. Dimofte, *Quantum Riemann surfaces in Chern-Simons theory*, Adv. Theor. Math. Phys. **17** (2013), no. 3, 479–599. MR3250765

[10] Do, Norman; Dyer, Alastair and Mathews, Daniel *Topological recursion and a quantum curve for monotone Hurwitz numbers.* arXiv:1408.3992

[11] N. Do and D. Manescu, *Quantum curves for the enumeration of ribbon graphs and hypermaps*, Commun. Number Theory Phys. **8** (2014), no. 4, 677–701, DOI 10.4310/CNTP.2014.v8.n4.a2. MR3318387

[12] Do, Norman and Norbury, Paul *Topological recursion for irregular spectral curves.* arXiv:1412.8334

[13] R. Dijkgraaf, H. Fuji, and M. Manabe, *The volume conjecture, perturbative knot invariants, and recursion relations for topological strings*, Nuclear Phys. B **849** (2011), no. 1, 166–211, DOI 10.1016/j.nuclphysb.2011.03.014. MR2795276 (2012g:81196)

[14] B. Dubrovin and Y. Zhang, *Virasoro symmetries of the extended Toda hierarchy*, Comm. Math. Phys. **250** (2004), no. 1, 161–193, DOI 10.1007/s00220-004-1084-9. MR2092034 (2005k:37169)

[15] O. Dumitrescu and M. Mulase, *Quantum curves for Hitchin fibrations and the Eynard-Orantin theory*, Lett. Math. Phys. **104** (2014), no. 6, 635–671, DOI 10.1007/s11005-014-0679-0. MR3200933

[16] Dunin-Barkowski, Petr; Mulase, Motohico; Norbury, Paul; Popolitov, Alexandr and Shadrin, Sergey. *Quantum spectral curve for the Gromov-Witten theory of the complex projective line* Journal für die reine und angewandte Mathematik. Published Online (2014).

[17] Dunin-Barkowski, Petr; Orantin, Nicolas; Popolitov, Alexandr and Shadrin, Sergey. *Combinatorics of loop equations for branched covers of sphere.* arXiv:1412.1698

[18] P. Dunin-Barkowski, N. Orantin, S. Shadrin, and L. Spitz, *Identification of the Givental formula with the spectral curve topological recursion procedure*, Comm. Math. Phys. **328** (2014), no. 2, 669–700, DOI 10.1007/s00220-014-1887-2. MR3199996

[19] T. Eguchi, K. Hori, and S.-K. Yang, *Topological σ models and large-N matrix integral*, Internat. J. Modern Phys. A **10** (1995), no. 29, 4203–4224, DOI 10.1142/S0217751X95001959. MR1357567 (96k:81231)

[20] T. Eguchi and S.-K. Yang, *The topological* $\mathbf{CP}^1$ *model and the large-N matrix integral*, Modern Phys. Lett. A **9** (1994), no. 31, 2893–2902, DOI 10.1142/S0217732394002732. MR1301737 (95j:81190)

[21] B. Eynard, *Invariants of spectral curves and intersection theory of moduli spaces of complex curves*, Commun. Number Theory Phys. **8** (2014), no. 3, 541–588, DOI 10.4310/CNTP.2014.v8.n3.a4. MR3282995

[22] Eynard, Bertrand *Intersection numbers of spectral curves.* arXiv:1104.0176

[23] Eynard, Bertrand and Orantin, Nicolas *Invariants of algebraic curves and topological expansion.* Communications in Number Theory and Physics **1** (2007), 347–452.

[24] B. Eynard and N. Orantin, *Topological recursion in enumerative geometry and random matrices*, J. Phys. A **42** (2009), no. 29, 293001, 117, DOI 10.1088/1751-8113/42/29/293001. MR2519749 (2011a:14109)

[25] C. Faber and R. Pandharipande, *Hodge integrals and Gromov-Witten theory*, Invent. Math. **139** (2000), no. 1, 173–199, DOI 10.1007/s002229900028. MR1728879 (2000m:14057)

[26] J. D. Fay, *Theta functions on Riemann surfaces*, Lecture Notes in Mathematics, Vol. 352, Springer-Verlag, Berlin-New York, 1973. MR0335789 (49 #569)

[27] P. J. Forrester and A. Gamburd, *Counting formulas associated with some random matrix averages*, J. Combin. Theory Ser. A **113** (2006), no. 6, 934–951, DOI 10.1016/j.jcta.2005.09.001. MR2244124 (2008a:15050)

[28] S. Gukov, *Three-dimensional quantum gravity, Chern-Simons theory, and the A-polynomial*, Comm. Math. Phys. **255** (2005), no. 3, 577–627, DOI 10.1007/s00220-005-1312-y. MR2134725 (2006f:58029)

[29] S. Gukov and P. Sułkowski, *A-polynomial, B-model, and quantization*, J. High Energy Phys. **2** (2012), 070, front matter+56. MR2996110

[30] Hollands, Lotte *Topological strings and quantum curves.* http://arxiv.org/abs/0911.3413

[31] R. M. Kashaev, *A link invariant from quantum dilogarithm*, Modern Phys. Lett. A **10** (1995), no. 19, 1409–1418, DOI 10.1142/S0217732395001526. MR1341338 (96j:81060)

[32] R. M. Kashaev, *The hyperbolic volume of knots from the quantum dilogarithm*, Lett. Math. Phys. **39** (1997), no. 3, 269–275, DOI 10.1023/A:1007364912784. MR1434238 (98b:57012)

[33] M. Kazarian and P. Zograf, *Virasoro constraints and topological recursion for Grothendieck's dessin counting*, Lett. Math. Phys. **105** (2015), no. 8, 1057–1084, DOI 10.1007/s11005-015-0771-0. MR3366120

[34] Liu, Xiaojun; Mulase, Motohico and Sorkin, Adam *Quantum curves for simple Hurwitz numbers of an arbitrary base curve* arXiv:1304.0015

[35] M. Mulase, S. Shadrin, and L. Spitz, *The spectral curve and the Schrödinger equation of double Hurwitz numbers and higher spin structures*, Commun. Number Theory Phys. **7** (2013), no. 1, 125–143, DOI 10.4310/CNTP.2013.v7.n1.a4. MR3108774

[36] Mulase, Motohico and Sułkowski, Piotr. *Spectral curves and the Schrödinger equations for the Eynard–Orantin recursion.* arXiv:1210.3006

[37] H. Murakami and J. Murakami, *The colored Jones polynomials and the simplicial volume of a knot*, Acta Math. **186** (2001), no. 1, 85–104, DOI 10.1007/BF02392716. MR1828373 (2002b:57005)

[38] W. D. Neumann and D. Zagier, *Volumes of hyperbolic three-manifolds*, Topology **24** (1985), no. 3, 307–332, DOI 10.1016/0040-9383(85)90004-7. MR815482 (87j:57008)

[39] P. Norbury, *String and dilaton equations for counting lattice points in the moduli space of curves*, Trans. Amer. Math. Soc. **365** (2013), no. 4, 1687–1709, DOI 10.1090/S0002-9947-2012-05559-0. MR3009643

[40] P. Norbury and N. Scott, *Gromov-Witten invariants of* $\mathbb{P}^1$ *and Eynard-Orantin invariants*, Geom. Topol. **18** (2014), no. 4, 1865–1910, DOI 10.2140/gt.2014.18.1865. MR3268770

[41] A. Okounkov and R. Pandharipande, *Gromov-Witten theory, Hurwitz numbers, and matrix models*, Algebraic geometry—Seattle 2005. Part 1, Proc. Sympos. Pure Math., vol. 80, Amer. Math. Soc., Providence, RI, 2009, pp. 325–414, DOI 10.1090/pspum/080.1/2483941. MR2483941 (2009k:14111)

[42] Zhou, Jian *Intersection numbers on Deligne-Mumford moduli spaces and quantum Airy curve.* arXiv:1206.5896

[43] Zhou, Jian *Quantum Mirror Curves for* $\mathbb{C}^3$ *and the Resolved Conifold.* arXiv:1207.0598

Department of Mathematics and Statistics, University of Melbourne, Australia 3010

E-mail address: pnorbury@ms.unimelb.edu.au

Proceedings of Symposia in Pure Mathematics
Volume **93**, 2016
http://dx.doi.org/10.1090/pspum/093/01555

A few recent developments in 2d (2,2) and (0,2) theories

Eric Sharpe

ABSTRACT. In this note we summarize a few of the many recent developments in two-dimensional quantum field theories. We begin with a review of the current state of quantum sheaf cohomology, a heterotic analogue of quantum cohomology. We then turn to dualities: we outline the current status of (0,2) mirror symmetry, and then outline recent work on two-dimensional gauge dualities. In particular, we describe how many two-dimensional gauge dualities in both (2,2) and (0,2) supersymmetric gauge theories can be understood simply as different presentations of the same infrared (IR) geometry. We then discuss (not necessarily supersymmetric) two-dimensional nonabelian gauge theories in which a subgroup of the gauge group acts trivially on massless matter. We describe how these theories 'decompose' into disjoint unions of other theories indexed by discrete theta angles, a fact which in other contexts has proven to have implications for interpretations of certain gauged linear sigma models (GLSMs) and for Gromov-Witten invariants of stacks. We conclude with a discussion of recent developments in infinitesimal moduli of heterotic compactifications.

1. Introduction

Over the last half dozen years, there has been a tremendous amount of progress in gauged linear sigma models (GLSMs) and perturbative string compactifications. A few examples include, but are not limited to:

- Nonperturbative realizations of geometry in GLSMs [**30**, **41**, **61**],
- Perturbative realizations of Pfaffians [**59**, **60**, **63**, **64**],
- Non-birational GLSM phases, and physical realizations of homological projective duality [**15**, **16**, **30**, **41**, **55**, **61**, **69**–**71**],
- Examples of closed strings on noncommutative resolutions [**3**, **30**, **96**],
- Localization techniques, yielding new Gromov-Witten and elliptic genus computations, the role of Gamma classes, and much more (see *e.g.* [**21**–**23**, **42**, **43**, **65**] for a few references),
- Heterotic strings: nonperturbative corrections, 2d dualitites, and non-Kähler moduli [**2**, **6**, **7**, **32**, **33**, **38**–**40**, **52**–**54**, **66**, **73**–**77**, **80**, **81**, **92**–**94**, **101**, **102**].

This talk will largely, though not exclusively, focus on heterotic strings. We will survey some of the results in two-dimensional (0,2) theories over the last six years

2010 *Mathematics Subject Classification.* Primary 14J81, Secondary 14J33, 14N35, 14M25.
Key words and phrases. Mirror symmetry, quantum sheaf cohomology, gerbes.
The author was partially supported by NSF grants PHY-1068725, PHY-1417410.

or so, describing both new results as well as outlining some older results to help provide background and context.

We begin in section 2 with a brief review of the current state of the art in quantum sheaf cohomology. In section 3 we give a brief status report on (0,2) mirror symmetry. In section 4 we discuss recent progress in two-dimensional gauge dualities in theories with (2,2) and (0,2) supersymmetry. We discuss how a number of Seiberg-like dualities can be understood simply as different presentations of the same IR geometry, and use this to predict additional dualities. In section 5 we turn to a different gauge duality, one that applies to both supersymmetric and non-supersymmetric theories in two dimensions. Specifically, in two-dimensional gauge theories in which a finite subgroup of the gauge group acts trivially on the matter, the theory 'decomposes' into a disjoint union of theories. In nonabelian gauge theories, the various components are labelled by different discrete theta angles. Finally, in section 6 we discuss current progress in infinitesimal moduli in heterotic compactifications, specifically, recent developments in understanding moduli in both Calabi-Yau and also non-Kähler heterotic compactifications.

2. Review of quantum sheaf cohomology

Quantum sheaf cohomology is the heterotic string analogue of quantum cohomology. Whereas ordinary quantum cohomology is defined by a space, quantum sheaf cohomology is defined by a space together with a bundle. Specifically, quantum sheaf cohomology is defined by a complex manifold X together with a holomorphic vector bundle $\mathcal{E} \to X$ (often called the 'gauge bundle'), satisfying the conditions

$$\mathrm{ch}_2(\mathcal{E}) = \mathrm{ch}_2(TX), \quad \det \mathcal{E}^* \cong K_X.$$

Briefly, whereas ordinary quantum cohomology is defined by intersection theory on a moduli space of curves, quantum sheaf cohomology is defined by sheaf cohomology (of sheaves induced by $\mathcal{E}$)) over a moduli space of curves. In the special case that $\mathcal{E} = TX$, the quantum sheaf cohomology ring should match the ordinary quantum cohomology ring. See for example [**38**–**40**, **74**, **77**] for a few recent discussions. We shall give here a brief summary oriented more nearly towards physicists; see for example [**40**] for a longer summary oriented towards mathematicians.

In heterotic string compactifications, quantum sheaf cohomology encodes nonperturbative corrections to charged matter couplings. For example, for a heterotic compactification on a Calabi-Yau three-fold X with gauge bundle given by the tangent bundle (known as the standard embedding, or as the (2,2) locus, as in this case (0,2) supersymmetry is enhanced to (2,2)), the low-energy theory has an E_6 gauge symmetry and matter charged under the $\overline{\mathbf{27}}$, counted by $H^{1,1}(X)$. The nonperturbative corrections to the $\overline{\mathbf{27}}^3$ couplings are encoded in Gromov-Witten invariants [**31**] and computed by the A model topological field theory [**106**].

If we now deform the gauge bundle so that it is no longer the tangent bundle, then the $\overline{\mathbf{27}}^3$ couplings will still receive nonperturbative corrections, but those corrections are no longer computed by Gromov-Witten invariants or the A model. Instead, the nonperturbative corrections are encoded in quantum sheaf cohomology. In this more general context, mathematical Gromov-Witten computational tricks no longer seem to apply, and there is no known analogue of periods or Picard-Fuchs equations. New methods are needed, and a few new techniques have been developed, which will be outlined here.

Before working through details, let us give a simple example. Recall the ordinary quantum cohomology ring of $\mathbb{P}^n$ is given by

$$\mathbb{C}[x]/(x^{n+1} - q).$$

When $q \to 0$, this becomes the classical cohomology ring of $\mathbb{P}^n$, hence the name. Now, to compare, the quantum sheaf cohomology ring of $\mathbb{P}^n \times \mathbb{P}^n$ with bundle $\mathcal{E} \to \mathbb{P}^n \times \mathbb{P}^n$ defined by

$$0 \longrightarrow \mathcal{O} \oplus \mathcal{O} \xrightarrow{*} \mathcal{O}(1,0)^{n+1} \oplus \mathcal{O}(0,1)^{n+1} \longrightarrow \mathcal{E} \longrightarrow 0,$$

where

$$* = \begin{bmatrix} Ax & Bx \\ C\tilde{x} & D\tilde{x} \end{bmatrix}$$

($x, \tilde{x}$ vectors of homogeneous coordinates on the two $\mathbb{P}^n$'s, A, B, C, D a set of four $(n+1) \times (n+1)$ constant matrices encoding a deformation of the tangent bundle), is given by

$$\mathbb{C}[x,y]/\left(\det(Ax + By) - q_1,\ \det(Cx + Dy) - q_2\right).$$

Note that in the special case that $A = D = I$, $B = C = 0$, the bundle $\mathcal{E}$ coincides with the tangent bundle of $\mathbb{P}^n \times \mathbb{P}^n$, and in this case, the quantum sheaf cohomology ring above reduces to

$$\mathbb{C}[x,y]/(x^{n+1} - q_1, y^{n+1} - q_2),$$

which is precisely the ordinary quantum cohomology ring of $\mathbb{P}^n \times \mathbb{P}^n$. This is as expected: as mentioned earlier, when $\mathcal{E} = TX$, quantum sheaf cohomology reduces to ordinary quantum cohomology.

Ordinary quantum cohomology can be understood physically as the ring of local operators, known as the OPE ring, of the A model topological field theory in two dimensions. That topological field theory is obtained by twisting a (2,2) nonlinear sigma model along a vector $U(1)$ symmetry. In a (0,2) nonlinear sigma model, if $\det \mathcal{E}^* \cong K_X$, then there is a nonanomalous $U(1)$ symmetry one can twist along, which reduces to the vector $U(1)$ symmetry on the (2,2) locus. If we twist along that nonanomalous $U(1)$, the result is a pseudo-topological field theory known as the A/2 model. Quantum sheaf cohomology is the OPE ring of the A/2 model. (There is also a pseudo-topological analogue of the B model, known as the B/2 model, but in this lecture we shall focus on the A/2 model.)

To be consistent, the ring products must close into the ring, but this is not *a priori* automatic in these quantum field theories, as in principle the products might generate local operators which are not elements of the (pseudo-)topological field theory. In the case of (2,2) supersymmetry, this closure of the OPE ring was argued in *e.g.* [**72**]. Closure in (0,2) theories is also possible – closure does not require (2,2) supersymmetry, but can be accomplished under weaker conditions. This was studied in detail in [**2**]. For example, for a (0,2) SCFT, one can use a combination of worldsheet conformal invariance and the right-moving $N = 2$ algebra to argue closure of the OPE ring on patches on the moduli space.

The local operators in the A model, the additive part of the OPE ring, are BRST-closed states of the form

$$b_{i_1 \cdots i_p \bar{\imath}_1 \cdots \bar{\imath}_q} \chi^{\bar{\imath}_1} \cdots \chi^{\bar{\imath}_q} \chi^{i_1} \cdots \chi^{i_p},$$

which are identified with closed differential forms representing $H^{p,q}(X)$. The analogous operators in the A/2 model are right-BRST-closed states of the form

$$b_{\overline{\imath}_1\cdots\overline{\imath}_q a_1\cdots a_p}\psi_+^{\overline{\imath}_1}\cdots\psi_+^{\overline{\imath}_q}\lambda_-^{a_1}\cdots\lambda_-^{a_p},$$

which are identified with closed bundle-valued differential forms representing elements of $H^q(X,\wedge^p\mathcal{E}^*)$. On the (2,2) locus, where $\mathcal{E}=TX$, the A/2 model reduces to the A model, which in operators follows from the statement

$$H^q(X,\wedge^p T^*X) \;=\; H^{p,q}(X).$$

At a purely schematic level, we can understand correlation functions as follows. Classically, in the A model, correlation functions are of the form

$$\langle \mathcal{O}_1\cdots\mathcal{O}_n\rangle \;=\; \int_X \omega_1\wedge\cdots\wedge\omega_n \;=\; \int_X (\text{top-form}),$$

where $\omega_i \in H^{p_i,q_i}(X)$. In the A/2 model, classical contributions to correlation functions are of the form

$$\langle \mathcal{O}_1\cdots\mathcal{O}_n\rangle \;=\; \int_X \omega_1\wedge\cdots\wedge\omega_n,$$

where $\omega_i\in H^{q_i}(X,\wedge^{p_i}\mathcal{E}^*)$. Now,

$$\omega_1\wedge\cdots\wedge\omega_n \;\in\; H^{\rm top}(X,\wedge^{\rm top}\mathcal{E}^*) \;=\; H^{\rm top}(X,K_X),$$

using the anomaly constraint $\det\mathcal{E}^*\cong K_X$. Thus, again we have a top-form, and so the correlation function yields a number.

In passing, note that the number one gets above depends upon a particular choice of an isomorphism $\det\mathcal{E}^*\cong K_X$. To uniquely define the A/2 theory, one must pick a particular isomorphism, which is a reflection of properties of the corresponding physical heterotic worldsheet theory. Moreover, as one moves on the moduli space of bundles or complex or Kähler structures, that isomorphism may change, so these correlation functions should be understood as sections of bundles over such moduli spaces. Technically, this is closely related to the realization of the Bagger-Witten line bundle in four-dimensional $N=1$ supergravity [**107**] on the worldsheet [**36, 88**], as the action of the global $U(1)$ in the worldsheet $N=2$ algebra on the spectral flow operator. (The original Bagger-Witten paper [**107**] assumed that the SCFT moduli space was a smooth manifold; see for example [**37, 58**] for modern generalizations to the case of moduli stacks.)

Correlation functions as outlined above define functions on spaces of sheaf cohomology groups. Now, we are interested in the relations amongst products of those sheaf cohomology groups, and those relations emerge as kernels of the (correlation) functions.

Let us consider a concrete example, namely the classical sheaf cohomology of $\mathbb{P}^1\times\mathbb{P}^1$ with bundle $\mathcal{E}$ given by a deformation of the tangent bundle, defined as

$$0 \longrightarrow W^*\otimes\mathcal{O} \stackrel{*}{\longrightarrow} \mathcal{O}(1,0)^2\oplus\mathcal{O}(0,1)^2 \longrightarrow \mathcal{E} \longrightarrow 0, \tag{2.1}$$

where $W\cong\mathbb{C}^2$,

$$* = \left[\begin{array}{cc} Ax & Bx \\ C\tilde{x} & D\tilde{x}\end{array}\right],$$

$x,\tilde{x}$ vectors of homogeneous coordinates on the two $\mathbb{P}^1$'s, and A,B,C,D four 2×2 constant matrices encoding the tangent bundle deformation.

We will focus on operators counted by

$$H^1(\mathcal{E}^*) = H^0(W \otimes \mathcal{O}) = W.$$

An n-point correlation function is then a map

$$\mathrm{Sym}^n H^1(\mathcal{E}^*) \; (= \mathrm{Sym}^n W\,) \longrightarrow H^n\left(\wedge^n \mathcal{E}^*\right).$$

The kernel of this map defines the classical sheaf cohomology ring relations, which we shall compute.

Since $\mathcal{E}$ is rank two, we will consider products of two elements of $H^1(\mathcal{E}^*) = W$, *i.e.* a map

$$H^0\left(\mathrm{Sym}^2 W \otimes \mathcal{O}\right) \longrightarrow H^2(\wedge^2 \mathcal{E}^*).$$

This map is implicitly encoded in the resolution

$$0 \longrightarrow \wedge^2 \mathcal{E}^* \longrightarrow \wedge^2 Z \longrightarrow Z \otimes W \longrightarrow \mathrm{Sym}^2 W \otimes \mathcal{O} \longrightarrow 0, \tag{2.2}$$

determined by the definition (2.1), where

$$Z \equiv \mathcal{O}(-1,0)^2 \oplus \mathcal{O}(0,-1)^2.$$

We break the resolution (2.2) into a pair of short exact sequences:

$$0 \longrightarrow \wedge^2 \mathcal{E}^* \longrightarrow \wedge^2 Z \longrightarrow S_1 \longrightarrow 0, \tag{2.3}$$

$$0 \longrightarrow S_1 \longrightarrow Z \otimes W \longrightarrow \mathrm{Sym}^2 W \otimes \mathcal{O} \longrightarrow 0, \tag{2.4}$$

(which define S_1).

The second sequence (2.4) induces

$$H^0(Z \otimes W) \longrightarrow H^0(\mathrm{Sym}^2 W \otimes \mathcal{O}) \stackrel{\delta}{\longrightarrow} H^1(S_1) \longrightarrow H^1(Z \otimes W).$$

Since Z is a sum of $\mathcal{O}(-1,0)$'s and $\mathcal{O}(0,-1)$'s,

$$H^0(Z \otimes W) = 0 = H^1(Z \otimes W),$$

hence the coboundary map

$$\delta : \; H^0(\mathrm{Sym}^2 W \otimes \mathcal{O}) \stackrel{\sim}{\longrightarrow} H^1(S_1)$$

is an isomorphism.

The first sequence (2.3) induces

$$H^1(\wedge^2 Z) \longrightarrow H^1(S_1) \stackrel{\delta}{\longrightarrow} H^2(\wedge^2 \mathcal{E}^*) \longrightarrow H^2(\wedge^2 Z).$$

The last term vanishes, but $H^1(\wedge^2 Z) \cong \mathbb{C}^2$, hence the coboundary map

$$\delta : \; H^1(S_1) \longrightarrow H^2(\wedge^2 \mathcal{E}^*)$$

has a two-dimensional kernel.

The composition of these two coboundary maps is our designed two-point correlation function

$$H^0(\mathrm{Sym}^2 W \otimes \mathcal{O}) \stackrel{\delta,\sim}{\longrightarrow} H^1(S_1) \stackrel{\delta}{\longrightarrow} H^2(\wedge^2 \mathcal{E}^*).$$

The right δ has a two-dimensional kernel, which one can show is generated by

$$\det(A\psi + B\tilde{\psi}), \quad \det(C\psi + D\tilde{\psi}),$$

where A, B, C, D are four matrices defining the deformation $\mathcal{E}$, and $\psi, \tilde{\psi}$ correspond to elements of a basis for W.

Putting this together, we get that the classical sheaf cohmology ring is

$$\mathbb{C}[\psi, \tilde{\psi}] / \left(\det(A\psi + B\tilde{\psi}), \det(C\psi + D\tilde{\psi})\right).$$

So far we have discussed classical physics. Instanton sectors have the same general form, except that X is replaced by a moduli space M of curves, and $\mathcal{E}$ is replaced by an induced sheaf[1] $\mathcal{F}$ over the moduli space M. Broadly speaking, the moduli space M must be compactified, and $\mathcal{F}$ extended over the compactification divisor. The anomaly conditions

$$\mathrm{ch}_2(\mathcal{E}) \;=\; \mathrm{ch}_2(TX), \quad \det \mathcal{E}^* \;\cong\; K_X$$

imply that

$$\det \mathcal{F}^* \;\cong\; K_M,$$

which is needed for the correlation functions to yield numbers.

Within any one instanton sector, in general terms one can follow the same method just outlined. In the case of the example just outlined, it can be shown that in a sector of instanton degree (a, b), the 'classical' ring in that sector is of the form

$$\mathrm{Sym}^{\bullet} W/(Q^{a+1}, \tilde{Q}^{b+1}),$$

where

$$Q \;=\; \det(A\psi + B\tilde{\psi}), \quad \tilde{Q} \;=\; \det(C\psi + D\tilde{\psi}).$$

Now, OPE's can relate correlation functions in different instanton degrees, and so should map ideals to ideals. To be compatible with the ideals above,

$$\langle \mathcal{O} \rangle_{a,b} \;=\; q^{a'-a}\tilde{q}^{b'-b} \langle \mathcal{O} Q^{a'-a} \tilde{Q}^{b'-b} \rangle_{a',b'}$$

for some constants $q, \tilde{q}$. As a result of the relations above, we can read off the OPE's

$$Q \;=\; q, \quad \tilde{Q} \;=\; \tilde{q},$$

which are the quantum sheaf cohomology relations.

More generally [**38**, **39**, **76**], for any toric variety, and any deformation $\mathcal{E}$ of its tangent bundle defined in the form

$$0 \longrightarrow W^* \otimes \mathcal{O} \stackrel{*}{\longrightarrow} \underbrace{\oplus_i \mathcal{O}(\vec{q}_i)}_{Z^*} \longrightarrow \mathcal{E} \longrightarrow 0,$$

the chiral ring is

$$\prod_{\alpha} \left(\det M_{(\alpha)} \right)^{Q^a_\alpha} \;=\; q_a,$$

where the $M_{(\alpha)}$'s are matrices of chiral operators constructed from the map $*$.

So far we have outlined mathematical computations of quantum sheaf cohomology, but there also exist methods based on gauged linear sigma models (GLSMs):

- Ordinary quantum cohomology is computed from (2,2) GLSMs in [**83**],
- Quantum sheaf cohomology is computed from (0,2) GLSMs in [**75**, **76**].

Briefly, for the (0,2) case, one computes quantum corrections to the effective action of the form

$$L_{\mathrm{eff}} \;=\; \int d\theta^+ \sum_a \Upsilon_a \log \left(\prod_{\alpha} (\det M_{(\alpha)})^{Q^a_\alpha} / q_a \right),$$

[1] If there are vector zero modes ('excess' intersection in the (2,2) case), then this story is more complicated – for example, there is a second induced sheaf, and one must utilize four-fermi terms in the action. For simplicity, for the purposes of this outline, we shall focus on the simpler case of no vector zero modes.

from which one derives the conditions for vacua

$$\prod_{\alpha} \left(\det M_{(\alpha)}\right)^{Q^a_\alpha} = q_a.$$

These are the quantum sheaf cohomology relations, and those derived in [**38, 39**] match these.

The current state of the art in quantum sheaf cohomology are computations on toric varieties. Our goal is to eventually perform these computations on compact Calabi-Yau manifolds, and as an intermediate step, we are currently studying Grassmannians.

Briefly, we need better computational methods. Conventional Gromov-Witten tricks seem to revolve around the idea that the A model is independent of complex structure, which is not necessarily true for the A/2 model. That said, it has been argued [**76**] that the A/2 model is independent of some moduli. Despite attempts to check [**47**], however, this is still not perfectly well-understood.

3. (0,2) mirror symmetry

Let us begin our discussion of dualities with a review of progress on a conjectured generalization of mirror symmetry, known as (0,2) mirror symmetry.

Now, ordinary mirror symmetry, in its most basic form, is a relation between Calabi-Yau manifolds, ultimately because a (2,2) supersymmetric nonlinear sigma model is defined by a manifold. Nonlinear sigma models with (0,2) supersymmetry are defined by a space X together with a holomorphic vector bundle $\mathcal{E} \to X$ satisfying certain consistency conditions discussed earlier, so (0,2) mirror symmetry, in its most basic form, is a statement about complex manifolds together with holomorphic vector bundles.

In this language, a prototypical[2] (0,2) mirror is defined by a space Y with holomorphic vector bundle $\mathcal{F} \to Y$, such that

$$\begin{aligned} \dim X &= \dim Y, \\ \operatorname{rank} \mathcal{E} &= \operatorname{rank} \mathcal{F}, \\ \mathrm{A}/2(X, \mathcal{E}) &= \mathrm{B}/2(Y, \mathcal{F}), \\ H^p(X, \wedge^q \mathcal{E}^*) &= H^p(Y, \wedge^q \mathcal{F}), \\ \text{(moduli)} &= \text{(moduli)}. \end{aligned}$$

In the special case that $\mathcal{E} = TX$, (0,2) mirror symmetry should reduce to ordinary mirror symmetry.

Some of the first significant evidence for (0,2) mirror symmetry was numerical: the authors of [**24**] wrote a computer program to scan a large number of examples and compute pertinent sheaf cohomology groups. The resulting data set was mostly invariant under the exchange of sheaf cohomology groups outlined above, giving a satisfying albeit limited test of the existence of (0,2) mirrors.

In [**25**], the Greene-Plesser orbifold construction [**50**] was extended to (0,2) models. This construction (and its (0,2) generalization) creates mirrors to Fermat-type Calabi-Yau hypersurfaces and complete intersections, as resolutions of certain

[2]Described here is the most basic incarnation of (0,2) mirror symmetry. For example, ordinary mirrors are sometimes given by Landau-Ginzburg models instead of spaces, and there are analogous statements in the (0,2) case. For simplicity, we focus on prototypical incarnations in which both sides of the mirror relation are defined by spaces (and bundles).

orbifolds of the original hypersurface or complete intersection. Because the construction can be understood as utilizing symmetries of what are called 'Gepner models' (see *e.g.* [**48**]), the fact that the SCFT's match is automatic, and so one can build what are necessarily examples of (0,2) mirrors. Unfortunately, this construction does not generate families of mirrors, only isolated examples.

In another more recent development, the Hori-Vafa-Morrison-Plesser-style GLSM duality picture of mirror symmetry [**62, 84**] was repeated for (0,2) theories in [**1**]. Unfortunately, unlike the case of ordinary mirror symmetry, understanding duality in (0,2) GLSMs requires additional input beyond the machinery of [**62, 84**].

More recently, a promising approach was discussed in [**80**], generalizing Batyrev's mirror construction [**18, 19**] to (0,2) models defined by certain special ('reflexively plain') hypersurfaces in toric varieties, with bundles given by deformations of the tangent bundles. The authors of [**80**] are able to make a proposal for a precise mapping of parameters in these cases, *i.e.* to relate families of (0,2) models, which they check by matching singularity structures in moduli spaces.

This represents significant progress, but there is still much to do before (0,2) mirror symmetry is nearly as well understood as ordinary mirror symmetry. Beyond the [**80**] construction, we would still like a more general mirror construction that applies to a broader class of varieties, and bundles beyond just deformations of the tangent bundle. Fully developing (0,2) mirror symmetry will also require further developments in quantum sheaf cohomology.

4. Two-dimensional gauge dualities

Next, we shall give an overview of recent progress in two-dimensional gauge theoretic dualities, in which different-looking gauge theories renormalization-group (RG) flow to the same infrared (IR) fixed point, *i.e.* become isomorphic at low energies and long distances.

Such dualities are of long-standing interest in the physics community, and there has been significant recent interest (see *e.g.* [**21, 44, 45, 59, 63, 67, 68**]). In two dimensions, we will see that such dualities can at least sometimes be understood as different presentations of the same geometry. This not only helps explain why these dualities work, but also implies a procedure to generate further examples (at least for Calabi-Yau and Fano geometries).

A prototypical example of a two-dimensional gauge duality, closely analogous to the central example of four-dimensional Seiberg duality [**90**], was described in [**21**] and relates a pair of theories with (2,2) supersymmetry:

$U(k)$ gauge group n chirals in fundamental, $n > k$ A chirals in antifundamental, $A < n$	$U(n-k)$ gauge group n chirals Φ in fundamental A chirals P in antifundamental nA neutral chirals M superpotential $W = M\Phi P$

The theory on the left RG flows to a nonlinear sigma model on

$$\mathrm{Tot}\left(S^{\oplus A} \longrightarrow G(k,n)\right) \;=\; \left(\mathbb{C}^{kn} \times \mathbb{C}^{kA}\right)//GL(k),$$

where S is the universal subbundle on the Grassmannian $G(k,n)$. The RG flow for the theory on the right is a bit more subtle, but can be analyzed by realizing that the superpotential is realizing a map in the short exact sequence

$$0 \longrightarrow S \stackrel{\Phi}{\longrightarrow} \mathcal{O}^n \longrightarrow Q \longrightarrow 0,$$

which implies that the theory on the right RG flows to a nonlinear sigma model on

$$\mathrm{Tot}\left((Q^*)^{\oplus A} \longrightarrow G(n-k,n) \right) \;=\; \mathrm{Tot}\left(S^{\oplus A} \longrightarrow G(k,n) \right),$$

the same geometry as for the theory on the left. Since the two theories RG flow to nonlinear sigma models on the same geometry, the RG flows of the two theories eventually coincide, and so the two gauge theories are Seiberg dual. In particular, this particular version of Seiberg duality has a purely geometric understanding.

We can apply the ideas above to make predictions for further two-dimensional dualities, at least for Fano and Calabi-Yau geometries. (For other cases, GLSM phases can be decorated with discrete Coulomb vacua [**78, 79**], which complicate the analysis.)

Our next example will be constructed utilizing the fact that the Grassmannian $G(2,4)$ is a quadric hypersurface in $\mathbb{P}^5$. The corresponding duality relates the theories

$U(2)$ gauge theory 4 chirals ϕ_i in fundamental	$U(1)$ gauge theory 6 chirals $z_{ij} = -z_{ji}$, $i,j = 1\cdots 4$, charge $+1$ one chiral P, charge -2 $W = P(z_{12}z_{34} - z_{13}z_{24} + z_{14}z_{23})$

The theory on the left RG flows to a nonlinear sigma model on $G(2,4)$. The theory on the right RG flows to a nonlinear sigma model on the corresponding quadric hypersurface. Since the geometries match, we see that the RG flows converge, and so the theories are Seiberg dual. As a consistency check, the chirals on the right and left are related by

$$z_{ij} \;=\; \epsilon_{\alpha\beta}\phi_i^\alpha \phi_j^\beta .$$

Both theories admit a global $GL(4)$ action, which acts as

$$\phi_i^\alpha \;\mapsto\; V_i^j \phi_j^\alpha, \quad z_{ij} \;\mapsto\; V_i^k V_j^\ell z_{k\ell}.$$

Chiral rings, anomalies, and Higgs moduli spaces match automatically.

This particular example is interesting because it relates abelian and nonabelian gauge theories, which in four dimensions would be difficult at best. In two dimensions, since gauge fields have no dynamics, abelian and nonabelian gauge theories are more closely related than in four dimensions.

In two dimensions, this understanding of Seiberg dualities in terms of matching geometries is not only entertaining but serves a more concrete purpose. In four dimensions, renormalizability heavily constrains possible superpotentials, which means as a practical matter that theories tend to have a number of global symmetries which can be used as guides to help confirm possible Seiberg duals. In two dimensions, by contrast, renormalizability does not constrain superpotentials at all, and generic superpotentials wll break all symmetries. Thus, identifying gauge duals as different presentations of the same geometry allows us to construct duals when standard tricks from four dimensions do not apply.

We can build on the previous example to construct a simple set of (2,2) supersymmetric examples in which global symmetries are broken. Specifically, consider the two theories

$U(2)$ gauge theory	$U(1)$ gauge theory
4 chirals ϕ_i in fundamental	6 chirals $z_{ij} = -z_{ji}$ of charge $+1$
chirals p_a, charge $-d_a$ under $\det U(2)$	one chiral P of charge -2
$W = \sum_a p_a f_a(\epsilon_{\alpha\beta}\phi_i^\alpha \phi_j^\beta)$	chirals P_a of charge $-d_a$
	$W = P(z_{12}z_{34} - z_{13}z_{24} + z_{14}z_{23}) + \sum_a P_a f_a(z_{ij})$

The two theories above RG flow to nonlinear sigma models on the complete intersection

$$G(2,4)[d_1, d_2, \cdots] \;=\; \mathbb{P}^5[2, d_1, d_2, \cdots]$$

and so, as above, are Seiberg dual.

An even more complex-appearing (2,2) gauge duality can be described as follows:

$U(2)$ gauge theory	$U(n-2) \times U(1)$ gauge theory
n chirals in fundamental	n chirals X in fundamental of $U(n-2)$
	n chirals P in antifundamental of $U(n-2)$
	(n choose 2) chirals $z_{ij} = -z_{ji}$, charge $+1$ under $U(1)$
	$W = \operatorname{tr} PAX$

Each of these two theories RG flows to a nonlinear sigma model on $G(2,n)$, using the fact that $G(2,n)$ can be described as the rank 2 locus of an $n \times n$ matrix A over $\mathbb{P}^{\frac{n!}{(n-2)!2!}-1}$, where A is defined by

$$A(z_{ij}) \;=\; \left[\begin{array}{cccc} z_{11} = 0 & z_{12} & z_{13} & \cdots \\ z_{21} = -z_{12} & z_{22} = 0 & z_{23} & \cdots \\ z_{31} = -z_{13} & z_{32} = -z_{23} & z_{33} = 0 & \cdots \\ \cdots & \cdots & \cdots & \cdots \end{array} \right],$$

using the perturbative description of Pfaffians in [**59**, **64**]. Since the RG flows converge, the two gauge theories above are necessarily Seiberg dual.

The same techniques can be extended to two-dimensional theories with (0,2) supersymmetry. Consider for example the two theories

$U(2)$ gauge theory	$U(1)$ gauge theory
4 chirals in fundamental	6 chirals, charge $+1$
1 Fermi in $(-4,-4)$	2 Fermis, charge $-2,-4$
8 Fermis in $(1,1)$	8 Fermis, charge $+1$
1 chiral in $(-2,-2)$	1 chiral, charge -2
2 chirals in $(-3,-3)$	2 chirals, charge -3
plus suitable superpotential	plus suitable superpotential

(Matter supermultiplets in (0,2) supersymmetry come in two types labelled 'chiral' and 'Fermi'. In the left column, $U(2)$ representations are indicated with a non-increasing pair of integers as in [**63**].) These theories will RG flow to the (0,2) nonlinear sigma model on the Calabi-Yau

$$G(2,4)[4] \;=\; \mathbb{P}^5[2,4],$$

with holomorphic vector bundle $\mathcal{E}$ given as

$$0 \longrightarrow \mathcal{E} \longrightarrow \oplus^8 \mathcal{O}(1) \longrightarrow \mathcal{O}(2) \oplus^2 \mathcal{O}(3) \longrightarrow 0.$$

Since the RG flows converge, these two theories are Seiberg dual. As a consistency test, it can be shown that the elliptic genera of these two theories match [**63**], applying recent GLSM-based computational methods described in [**22, 23, 43**].

A different example is provided by 'triality' [**44, 45**]. Here, triples of (0,2) GLSMs are believed to flow to the same IR fixed point. Each GLSM has two different geometric phases; however, unlike previous cases, not all of the geometric phases describe the same geometry. Schematically, we can understand the relationship between the phases as follows [**63**]:

$$
\begin{array}{ccc}
S^A \oplus (Q^*)^{2k+A-n} \to G(k,n) & - - - - - - & (S^*)^A \oplus (Q^*)^n \to G(k, 2k+A-n) \\
\updownarrow \cong & & \\
(Q^*)^A \oplus S^{2k+A-n} \to G(n-k,n) & - - - - - & (Q^*)^n \oplus (S^*)^{2k+A-n} \to G(n-k, A) \\
 & & \updownarrow \cong \\
S^n \oplus (Q^*)^A \to G(A-n+k, 2k+A-n) & - - - & (S^*)^n \oplus (Q^*)^{2k+A-n} \to G(A-n+k, A) \\
\updownarrow \cong & & \\
(Q^*)^n \oplus (S^*)^A \to G(k, 2k+A-n) & - - - - - - & (Q^*)^{2k+A-n} \oplus S^A \to G(k,n).
\end{array}
$$

Each phase also has a bundle summand, either $(\det S)^{\oplus 2}$ or $(\det S^*)^{\oplus 2}$, which we have omitted for brevity. Horizontal dashed lines indicate phase transitions to different geometries; vertical arrows indicate equivalent geometries. The fourth line is physically equivalent to the first: the bottom right corner is equivalent to the upper left, and the bottom left, to the upper right. In writing the diagram above, we have used the fact that in (0,2) theories, dualizing the gauge bundle is an equivalence of the theories:

$$\mathrm{QFT}(X, \mathcal{E} \to X) \cong \mathrm{QFT}(X, \mathcal{E}^* \to X).$$

(See for example [**93**] for a discussion of corner cases of this duality.) A test of triality recently appeared in [**54**].

How do these gauge dualities relate to (0,2) mirrors as discussed in the previous section? As we have seen, gauge dualities often relate different presentations of the same geometry, whereas (0,2) mirrors exchange different geometries. The existence of (0,2) mirrors seems to imply that there ought to exist more 'exotic' gauge dualities, that present different geometries.

So far in this section we have used mathematics to make predictions for physics. In the next section we shall turn that around, and use physics to make predictions for mathematics.

5. Decomposition in two-dimensional nonabelian gauge theories

In a two-dimensional orbifold or gauge theory, if a finite subgroup of the gauge group acts trivially on all massless matter, the theory decomposes as a disjoint union [**57**].

For example, a trivially-acting $\mathbb{Z}_2$ orbifold of a nonlinear sigma model on a space X is equivalent to a nonlinear sigma model on two copies of X:

$$\mathrm{CFT}\left([X/\mathbb{Z}_2]\right) = \mathrm{CFT}\left(X \coprod X\right).$$

In the $\mathbb{Z}_2$ orbifold, since the $\mathbb{Z}_2$ acts trivially on X, there is a dimension zero twist field. Linear combinations of that twist field and the identity operator form projection operators onto the two copies of X.

For another example, consider a D_4 orbifold of a nonlinear sigma model on a space X, where the center $\mathbb{Z}_2 \subset D_4$ acts trivially on X. This orbifold is equivalent to the disjoint union of a pair of $\mathbb{Z}_2 \times \mathbb{Z}_2$ orbifolds, one with and one without discrete torsion:

$$\mathrm{CFT}\left([X/D_4]\right) \;=\; \mathrm{CFT}\left([X/\mathbb{Z}_2 \times \mathbb{Z}_2] \coprod [X/\mathbb{Z}_2 \times \mathbb{Z}_2]_{\mathrm{d.t.}}\right),$$

where $D_4/\mathbb{Z}_2 = \mathbb{Z}_2 \times \mathbb{Z}_2$.

These are examples in physics of what is meant by 'decomposition.'

Decomposition is also a statement about mathematics. Briefly, over the last several years, the following dictionary has been built:

2d Physics	Math
D-brane	Derived category [**91**]
Gauge theory	Stack [**85**–**87**]
Gauge theory with trivially-acting subgroup	Gerbe [**57**, **85**–**87**]
Universality class of renormalization group flow	Categorical equivalence

In particular, decomposition is a statement about the physics of strings propagating on gerbes, detailed in the 'decomposition conjecture' [**57**], which for banded gerbes can be summarized as:

$$\mathrm{CFT}\left(G - \text{gerbe on } X\right) \;=\; \mathrm{CFT}\left(\coprod_{\hat{G}} (X, B)\right),$$

where $\hat{G}$ is the set of irreducible representations of G, and the B field on each component is determined by the image of the characteristic class of the gerbe:

$$H^2(X, Z(G)) \stackrel{Z(G)\mapsto U(1)}{\longrightarrow} H^2(X, U(1)).$$

The decomposition conjecture has been checked in a wide variety of ways, including, for example:

- multiloop orbifold partition functions: partition functions decompose in the desired form,
- quantum cohomology ring relations as derived from GLSMs match the implicit prediction above,
- D-branes, K theory, sheaves on gerbes: the physical decomposition of D-branes matches the mathematical decomposition of K theory and sheaves on gerbes.

Decomposition also has a number of applications, including

- Predictions for Gromov-Witten invariants of gerbes, as checked in *e.g.* [**8**–**11**, **49**, **103**–**105**],
- Understanding certain GLSM phases [**30**, **56**, **59**, **95**], via giving a physical realization of Kuznetsov's homological projective duality [**71**],

and these works serve implicitly as further checks on the decomposition conjecture above.

To understand the decomposition conjecture in orbifolds, one can compare (multi)loop partition functions, state spaces, and D-branes, and they all imply the same result. In gauge theories, there are further subtleties. For example, let us compare the following two theories:

- Ordinary $\mathbb{CP}^n$ model: a $U(1)$ gauge theory with $n+1$ chiral superfields, each of charge $+1$,
- Gerby $\mathbb{CP}^n$ model: a $U(1)$ gauge theory with $n+1$ chiral superfields, each of charge k, $k > 1$.

In order for these two theories to be distinct, the physics of the second must be different from the first – but how can multiplying the charges by a factor change anything? Naively, this is just a convention, and physics should not depend upon conventions.

Perturbatively, multiplying all the charges by a factor does not modify the physics; however, nonperturbatively[3], there can be a difference between these two theories. On a compact worldsheet, to make manifest the distinction, one must specify which bundles the fields couple to, to unambiguously specify the theory. If the chiral fields are sections of a line bundle L in the first theory, then in the second they are sections of a different bundle, $L^{\otimes k}$, and hence have different zero modes, different anomalies, and hence different nonperturbative physics.

On a noncompact worldsheet, one can instead appeal to the periodicity of the θ angle in the two-dimensional gauge theory. The θ angle acts as an electric field, so by building a sufficiently large capacitor, one can excite states of arbitary mass. In particular, we can distinguish the second theory from the first by adding a pair of massive minimally charged fields, which a sufficiently large capacitor can excite. In this fashion, essentially through different periodicities of the θ angle, one can distinguish the two theories.

Now, decomposition has been extensively checked for orbifolds and abelian gauge theories, but tests in nonabelian gauge theories in two dimensions have only appeared more recently [**97**]. Since two-dimensional gauge theories do not have propagating degrees of freedom, an analogous phenomena ought to take place in nonabelian gauge theories with center-invariant matter. Specifically, it was proposed in [**97**] that for G semisimple, a G-gauge theory with center-invariant matter should decompose into a sum of theories with variable discrete theta angles. For example, an $SU(2)$ gauge theory with only adjoints or other center invariant matter should decompose into a pair of $SO(3)$ gauge theories with the same matter but different discrete theta angles, schematically:

$$SU(2) \; = \; SO(3)_+ \; + \; SO(3)_- . \tag{5.1}$$

Before working through this in detail, let us first remind the reader of how discrete theta angles are defined, as they are relatively new [**4, 46**]. Consider a two-dimensional gauge theory, with gauge group $G = \tilde{G}/K$, $\tilde{G}$ compact, semisimple, and simply-connected, K a finite subgroup of the center of $\tilde{G}$. This theory has a degree-two K-valued characteristic class which we will denote w. (For example, in an $SO(3)$ gauge theory, this is the second Stiefel-Whitney class.) For any character

[3]We would like to thank A. Adams, J. Distler, and R. Plesser for explaining the distinction, on both compact and noncompact worldsheets, at an Aspen workshop in 2004.

λ of K, we can add the topological term $\lambda(w)$ to the action. This is the discrete theta angle term, and we see in this fashion that the possible values of the discrete theta angle are classified by characters of K.

For example, let us consider an $SO(3)$ gauge theory. Now,

$$SO(3) = SU(2)/\mathbb{Z}_2,$$

hence as $\mathbb{Z}_2$ has two characters, we see that an $SO(3)$ gauge theory in two dimensions has two discrete theta angles.

Let us check the decomposition conjecture for nonabelian gauge theories in the case of pure $SU(2)$ gauge theory in two dimensions. The partition function for pure (nonsupersymmetric) two-dimensional gauge theories can be found in *e.g.* [**51**, **82**, **89**], from which we derive

$$\begin{aligned} Z(SU(2)) &= \sum_R (\dim R)^{2-2g} \exp(-AC_2(R)), \\ Z(SO(3)_+) &= \sum_R (\dim R)^{2-2g} \exp(-AC_2(R)). \end{aligned}$$

In the expressions above, g is the genus of the two-dimensional surface, A is its area, R a representation, and $C_2(R)$ a Casimir of the representation R. The $SU(2)$ partition function sums over all representations R of $SU(2)$, and the $SO(3)_+$ partition function sums over all representations R of $SO(3)$. (For $SO(3)_+$, the discrete theta angle vanishes, so $SO(3)_+$ is the ordinary $SO(3)$ gauge theory.) The partition function of $SO(3)_-$ was described in [**100**], and has the form

$$Z(SO(3)_-) = \sum_R (\dim R)^{2-2g} \exp(-AC_2(R)),$$

where the sum is now over representations of $SU(2)$ that are not representations of $SO(3)$. Combining these three expressions, it should be clear that

$$Z(SU(2)) = Z(SO(3)_+) + Z(SO(3)_-).$$

More generally, for G gauge theories with G semisimple, K a finite subgroup of the center of G, and matter invariant under K, we can express decomposition schematically as

$$G = \sum_{\lambda \in \hat{K}} (G/K)_\lambda.$$

This can be checked for pure gauge theories using partition functions as above, and can also similarly be checked for correlation functions of Wilson lines in pure gauge theories. In addition, it can also be checked in supersymmetric theories using expressions for partition functions given in [**21**, **42**]. The arguments in this case revolve around details of cocharacter lattices, which for brevity we omit here; see [**97**] for details.

6. Heterotic moduli

It was known historically that for large-radius heterotic nonlinear sigma models on the (2,2) locus, there were three classes of infinitesimal moduli[4]:

- Kähler moduli, counted by $H^1(X, T^*X)$,
- Complex moduli, counted by $H^1(X, TX)$,

[4]Physically, the moduli are indistinguishable from one another; the distinction we list is purely mathematical in origin.

- Bundle moduli, counted by $H^1(X, \mathrm{End}\,\mathcal{E})$,

for a compactification on a space X with gauge bundle $\mathcal{E} = TX$ (the (2,2) locus).

When the gauge bundle $\mathcal{E}$ is different from the tangent bundle TX, the correct counting is more complicated. It was shown in the physics literature in *e.g.* [**5**] that the correct counting is given by

- Kähler moduli, counted by $H^1(X, T^*X)$,
- Compatible complex and bundle moduli, counted by $H^1(Q)$ where Q is defined by the Atiyah sequence

$$0 \longrightarrow \mathrm{End}\,\mathcal{E} \longrightarrow Q \longrightarrow TX \longrightarrow 0. \tag{6.1}$$

The extension class is determined by the curvature of the bundle. Specifically, it is an element of

$$\mathrm{Ext}^1(TX, \mathrm{End}\,\mathcal{E}) \;=\; H^1(T^*X \otimes \mathrm{End}\,\mathcal{E})$$

given by the curvature.

In particular, as $\mathcal{E}$ is required to be a *holomorphic* bundle, the complex and bundle moduli are not independent of one another, and in fact a given bundle may not be compatible with all complex structure moduli, a result that was well-known in mathematics but whose relevance the physics community only recently digested.

At the time, however, this still left unresolved the question of understanding moduli of heterotic non-Kähler compactifications [**99**]. In a non-Kähler compactification, there is no version of Yau's theorem relating metric moduli to complex and Kähler moduli, so in principle, in close-to-large-radius[5] non-Kähler compactifications, the moduli need not have any meaningful connection to Calabi-Yau moduli. (That said, it should also be noted that even in a Calabi-Yau (0,2) compactification, although the space admits a Kähler metric, away from the large-radius limit the metric solving the supergravity equations is necessarily non-Kähler, because the Green-Schwarz condition forces H to be nonzero.)

A partial solution to this problem was discovered in [**81**]. There, it was argued from a worldsheet analysis that for non-Kähler compactifications in a purely formal $\alpha' \to 0$ limit, the infinitesimal moduli are counted by $H^1(S)$, where

$$0 \longrightarrow T^*X \longrightarrow S \longrightarrow Q \longrightarrow 0,$$

where Q is the extension determined by the Atiyah sequence (6.1). The extension above is determined by an element of

$$\mathrm{Ext}^1(TX, T^*X)$$

determined by the H flux, which is assumed to obey $dH = 0$.

As non-Kähler compactifications do not exist in the $\alpha' \to 0$ limit, the solution above was necessarily incomplete. It was improved upon in [**6**, **33**], which gave an overcounting of heterotic moduli valid through first order in α'. On manifolds satisfying the $\partial\overline{\partial}$-lemma, the moduli are overcounted by $H^1(S')$, where

$$0 \longrightarrow T^*X \longrightarrow S' \longrightarrow Q' \longrightarrow 0$$

[5]Non-Kähler heterotic compactifications do not have a large-radius limit. The best one can do is to hope for solutions "close" to large-radius, where geometry is still valid. In this section, we implicitly assume the non-Kähler compactifications being considered are all in that regime, close enough to large radius that geometry is a valid description.

(defined by H satisfying the Green-Schwarz condition), for Q' given by

$$0 \longrightarrow \mathrm{End}\,\mathcal{E} \oplus \mathrm{End}TX \longrightarrow Q' \longrightarrow TX \longrightarrow 0,$$

with the extension defined by the curvatures of the gauge bundle and TX.

The overcounting above is the current state-of-the-art; currently work is in progress to find the correct counting and to extend to higher orders in α'.

So far we have outlined infinitesimal moduli, corresponding to marginal operators on the worldsheet. These can be obstructed by *e.g.* nonperturbative effects, and there is an interesting story behind this. Initially, in the mid-80s, it was observed in [**34**, **35**] that a single worldsheet instanton can generate a superpotential term obstructing deformations off the (2,2) locus, but in the early 90s it was observed that for moduli realizable in GLSMs, the sum of the contributions from different contributing rational curves all cancel out, and so the moduli are unobstructed. This led to a revitalization of interest in (0,2) models, and paved the way for work on F theory, for example. The original GLSM arguments have found alternate presentations[6] in *e.g.* [**17**, **20**]. Current work on the subject, such as [**12**–**14**, **26**–**29**], has focused on understanding non-GLSM moduli, for which nonperturbative corrections to obstructions often do not cancel out.

7. Conclusions

In this note we have given an overview of recent developments in two-dimensional theories, focusing primarily though not exclusively on (0,2) theories. We began in section 2 with a brief review of the current state of the art in quantum sheaf cohomology. In section 3 we gave a brief status report on (0,2) mirror symmetry. In section 4 we discussed recent progress in two-dimensional gauge dualities in theories with (2,2) and (0,2) supersymmetry. We showed how a number of Seiberg-like dualities can be understood simply as different presentations of the same IR geometry, and use this to predict additional dualities. In section 5 we described a different gauge duality, one that applies to both supersymmetric and nonsupersymmetric theories in two dimensions. Specifically, in two-dimensional gauge theories in which a finite subgroup of the gauge group acts trivially on the matter, the theory 'decomposes' into a disjoint union of theories. In nonabelian gauge theories, the various components are labelled by different discrete theta angles. Finally, in section 6 we discussed current progress in infinitesimal moduli in heterotic compactifications, specifically, recent developments in understanding moduli in both Calabi-Yau and also non-Kähler heterotic compactifications.

References

[1] Allan Adams, Anirban Basu, and Savdeep Sethi, $(0,2)$ *duality*, Adv. Theor. Math. Phys. **7** (2003), no. 5, 865–950. MR2045304 (2005b:81167)

[2] Allan Adams, Jacques Distler, and Morten Ernebjerg, *Topological heterotic rings*, Adv. Theor. Math. Phys. **10** (2006), no. 5, 657–682. MR2281544 (2008b:81263)

[6]In our experience, sometimes these papers are mis-quoted as claiming that the spacetime superpotential vanishes in heterotic compactifications. The correct statement is that nonperturbative corrections to gauge singlet moduli interactions arising from moduli realizable in GLSMs cancel out. However, gauge non-singlet interactions can and will receive nonperturbative corrections, and can even be nonzero classically. For example, on the (2,2) locus, in a heterotic compactification on a Calabi-Yau 3-fold, the $\overline{\mathbf{27}}^3$ couplings are nonzero: in addition to the classical contribution described in [**98**], they also receive nonperturbative corrections corresponding to the Gromov-Witten invariants of the Calabi-Yau [**31**, **106**].

[3] Nicolas M. Addington, Edward P. Segal, and Eric R. Sharpe, *D-brane probes, branched double covers, and noncommutative resolutions*, Adv. Theor. Math. Phys. **18** (2014), no. 6, 1369–1436. MR3285612

[4] Ofer Aharony, Nathan Seiberg, and Yuji Tachikawa, *Reading between the lines of four-dimensional gauge theories*, J. High Energy Phys. **8** (2013), 115, front matter+41. MR3106323

[5] Lara B. Anderson, James Gray, Andre Lukas, and Burt Ovrut, *Stabilizing the complex structure in heterotic Calabi-Yau vacua*, J. High Energy Phys. **2** (2011), 088, 16, DOI 10.1007/JHEP02(2011)088. MR2820787 (2012f:81209)

[6] Lara B. Anderson, James Gray, and Eric Sharpe, *Algebroids, Heterotic Moduli Spaces and the Strominger System*, JHEP **1407** (2014), 037.

[7] Lara B. Anderson, Bei Jia, Ryan Manion, Burt Ovrut, and Eric Sharpe, *General aspects of heterotic string compactifications on stacks and gerbes*, (2013).

[8] Elena Andreini, Yunfeng Jiang, and Hsian-Hua Tseng, *On Gromov-Witten theory of root gerbes*, (2008).

[9] Elena Andreini, Yunfeng Jiang, and Hsian-Hua Tseng, *Gromov-Witten theory of product stacks*, (2009).

[10] Elena Andreini, Yunfeng Jiang, and Hsian-Hua Tseng, *Gromov-Witten theory of root gerbes I: structure of genus 0 moduli spaces*, (2009).

[11] Elena Andreini, Yunfeng Jiang, and Hsian-Hua Tseng, *Gromov-Witten theory of banded gerbes over schemes*, (2011).

[12] Paul S. Aspinwall and Benjamin Gaines, *Rational curves and* $(0,2)$*-deformations*, J. Geom. Phys. **88** (2015), 1–15, DOI 10.1016/j.geomphys.2014.09.012. MR3293392

[13] Paul S. Aspinwall, Ilarion V. Melnikov, and M. Ronen Plesser, $(0,2)$ *elephants*, J. High Energy Phys. **1** (2012), 060, 50. MR2949305

[14] Paul S. Aspinwall and M. Ronen Plesser, *Elusive worldsheet instantons in heterotic string compactifications*, String-Math 2011, Proc. Sympos. Pure Math., vol. 85, Amer. Math. Soc., Providence, RI, 2012, pp. 33–51, DOI 10.1090/pspum/085/1373. MR2985325

[15] Matthew Ballard, Dragos Deliu, David Favero, M. Umut Isik, and Ludmil Katzarkov, *Homological Projective Duality via Variation of Geometric Invariant Theory Quotients*, (2013).

[16] Matthew Ballard, David Favero, and Ludmil Katzarkov, *Variation of geometric invariant theory quotients and derived categories*, (2012).

[17] Anirban Basu and Savdeep Sethi, *World-sheet stability of* $(0,2)$ *linear sigma models*, Phys. Rev. D (3) **68** (2003), no. 2, 025003, 8, DOI 10.1103/PhysRevD.68.025003. MR2033807

[18] Victor V. Batyrev, *Dual polyhedra and mirror symmetry for Calabi-Yau hypersurfaces in toric varieties*, J. Algebraic Geom. **3** (1994), no. 3, 493–535. MR1269718 (95c:14046)

[19] Victor V. Batyrev and Lev A. Borisov, *Dual cones and mirror symmetry for generalized Calabi-Yau manifolds*, Mirror symmetry, II, AMS/IP Stud. Adv. Math., vol. 1, Amer. Math. Soc., Providence, RI, 1997, pp. 71–86. MR1416334 (98b:14033)

[20] Chris Beasley and Edward Witten, *Residues and world-sheet instantons*, J. High Energy Phys. **10** (2003), 065, 39 pp. (electronic), DOI 10.1088/1126-6708/2003/10/065. MR2030598 (2005f:81223)

[21] Francesco Benini and Stefano Cremonesi, *Partition functions of* $N = (2,2)$ *gauge theories on* S^2 *and vortices*, (2012).

[22] Francesco Benini, Richard Eager, Kentaro Hori, and Yuji Tachikawa, *Elliptic genera of 2d N=2 gauge theories*, (2013).

[23] Francesco Benini, Richard Eager, Kentaro Hori, and Yuji Tachikawa, *Elliptic genera of two-dimensional* $\mathcal{N} = 2$ *gauge theories with rank-one gauge groups*, Lett. Math. Phys. **104** (2014), no. 4, 465–493, DOI 10.1007/s11005-013-0673-y. MR3177993

[24] Ralph Blumenhagen, Rolf Schimmrigk, and Andreas Wißkirchen, $(0,2)$ *mirror symmetry*, Nuclear Phys. B **486** (1997), no. 3, 598–628, DOI 10.1016/S0550-3213(96)00698-0. MR1436158 (98a:81156)

[25] Ralph Blumenhagen and Savdeep Sethi, *On orbifolds of* $(0,2)$ *models*, Nuclear Phys. B **491** (1997), no. 1-2, 263–278, DOI 10.1016/S0550-3213(97)00095-3. MR1449061 (98f:81294)

[26] Volker Braun, Maximilian Kreuzer, Burt A. Ovrut, and Emanuel Scheidegger, *Worldsheet instantons, torsion curves and non-perturbative superpotentials*, Phys. Lett. B **649** (2007), no. 4, 334–341, DOI 10.1016/j.physletb.2007.03.066. MR2308810 (2008b:81211)

[27] Volker Braun, Maximilian Kreuzer, Burt A. Ovrut, and Emanuel Scheidegger, *Worldsheet instantons and torsion curves*, Advances in string theory, AMS/IP Stud. Adv. Math., vol. 44, Amer. Math. Soc., Providence, RI, 2008, pp. 231–240. MR2494190 (2010e:14008)

[28] Volker Braun, Burt A. Ovrut, Maximilian Kreuzer, and Emanuel Scheidegger, *Worldsheet instantons and torsion curves. A. Direct computation*, J. High Energy Phys. **10** (2007), 022, 58, DOI 10.1088/1126-6708/2007/10/022. MR2357956 (2009i:14080)

[29] Volker Braun, Burt A. Ovrut, Maximilian Kreuzer, and Emanuel Scheidegger, *Worldsheet instantons and torsion curves. B. Mirror symmetry*, J. High Energy Phys. **10** (2007), 023, 53, DOI 10.1088/1126-6708/2007/10/023. MR2357955 (2009i:14079)

[30] Andrei Căldăraru, Jacques Distler, Simeon Hellerman, Tony Pantev, and Eric Sharpe, *Non-birational twisted derived equivalences in abelian GLSMs*, Comm. Math. Phys. **294** (2010), no. 3, 605–645, DOI 10.1007/s00220-009-0974-2. MR2585982 (2011f:81182)

[31] Philip Candelas, Xenia C. de la Ossa, Paul S. Green, and Linda Parkes, *A pair of Calabi-Yau manifolds as an exactly soluble superconformal theory*, Nuclear Phys. B **359** (1991), no. 1, 21–74, DOI 10.1016/0550-3213(91)90292-6. MR1115626 (93b:32029)

[32] Jin Chen, Xiaoyi Cui, Mikhail Shifman, and Arkady Vainshtein, *N=(0,2) deformation of (2, 2) sigma models: Geometric structure, holomorphic anomaly, and exact β functions*, Phys.Rev. **D90** (2014), no. 4, 045014.

[33] Xenia de la Ossa and Eirik E. Svanes, *Holomorphic bundles and the moduli space of $N = 1$ supersymmetric heterotic compactifications*, J. High Energy Phys. **10** (2014), 123, front matter+54. MR3324801

[34] M. Dine, N. Seiberg, X.-G. Wen, and E. Witten, *Nonperturbative effects on the string world sheet*, Nuclear Phys. B **278** (1986), no. 4, 769–789, DOI 10.1016/0550-3213(86)90418-9. MR862903 (88c:81081)

[35] M. Dine, N. Seiberg, X.-G. Wen, and E. Witten, *Nonperturbative effects on the string world sheet. II*, Nuclear Phys. B **289** (1987), no. 2, 319–363, DOI 10.1016/0550-3213(87)90383-X. MR895317 (88k:81195)

[36] Jacques Distler, *Notes on N=2 sigma models*, (1992).

[37] Jacques Distler and Eric Sharpe, *Quantization of Fayet-Iliopoulos Parameters in Supergravity*, Phys.Rev. **D83** (2011), 085010.

[38] Ron Donagi, Josh Guffin, Sheldon Katz, and Eric Sharpe, *Physical aspects of quantum sheaf cohomology for deformations of tangent bundles of toric varieties*, Adv. Theor. Math. Phys. **17** (2013), no. 6, 1255–1301. MR3262522

[39] Ron Donagi, Josh Guffin, Sheldon Katz, and Eric Sharpe, *A mathematical theory of quantum sheaf cohomology*, Asian J. Math. **18** (2014), no. 3, 387–417, DOI 10.4310/AJM.2014.v18.n3.a1. MR3257832

[40] Ron Donagi, Joshua Guffin, Sheldon Katz, and Eric Sharpe, $(0,2)$ *quantum cohomology*, String-Math 2011, Proc. Sympos. Pure Math., vol. 85, Amer. Math. Soc., Providence, RI, 2012, pp. 83–103, DOI 10.1090/pspum/085/1375. MR2985327

[41] Ron Donagi and Eric Sharpe, *GLSMs for partial flag manifolds*, J. Geom. Phys. **58** (2008), no. 12, 1662–1692, DOI 10.1016/j.geomphys.2008.07.010. MR2468445 (2009m:81223)

[42] Nima Doroud, Jaume Gomis, Bruno Le Floch, and Sungjay Lee, *Exact results in $D = 2$ supersymmetric gauge theories*, J. High Energy Phys. **5** (2013), 093, front matter+69. MR3080568

[43] Abhijit Gadde and Sergei Gukov, *2d index and surface operators*, J. High Energy Phys. **3** (2014), 080, front matter+40. MR3190966

[44] Abhijit Gadde, Sergei Gukov, and Pavel Putrov, *(0, 2) trialities*, JHEP **1403** (2014), 076.

[45] Abhijit Gadde, Sergei Gukov, and Pavel Putrov, *Exact Solutions of 2d Supersymmetric Gauge Theories*, (2014).

[46] Davide Gaiotto, Gregory W. Moore, and Andrew Neitzke, *Framed BPS states*, Adv. Theor. Math. Phys. **17** (2013), no. 2, 241–397. MR3250763

[47] Richard S. Garavuso and Eric Sharpe, *Analogues of Mathai-Quillen forms in sheaf cohomology and applications to topological field theory*, (2013).

[48] Doron Gepner, *Exactly solvable string compactifications on manifolds of* $\mathrm{SU}(N)$ *holonomy*, Phys. Lett. B **199** (1987), no. 3, 380–388, DOI 10.1016/0370-2693(87)90938-5. MR929596 (89h:83035)

[49] Amin Gholampour and Hsian-Hua Tseng, *On Donaldson-Thomas invariants of threefold stacks and gerbes*, Proc. Amer. Math. Soc. **141** (2013), no. 1, 191–203, DOI 10.1090/S0002-9939-2012-11346-2. MR2988722

[50] B. R. Greene and M. R. Plesser, *Duality in Calabi-Yau moduli space*, Nuclear Phys. B **338** (1990), no. 1, 15–37, DOI 10.1016/0550-3213(90)90622-K. MR1059831 (91h:32018)

[51] David J. Gross and Washington Taylor IV, *Two-dimensional QCD is a string theory*, Nuclear Phys. B **400** (1993), no. 1-3, 181–208, DOI 10.1016/0550-3213(93)90403-C. MR1227260 (94h:81140)

[52] Josh Guffin, *Quantum sheaf cohomology, a précis*, Mat. Contemp. **41** (2012), 17–26. MR3087571

[53] Josh Guffin and Sheldon Katz, *Deformed quantum cohomology and* $(0,2)$ *mirror symmetry*, J. High Energy Phys. **8** (2010), 109, 27, DOI 10.1007/JHEP08(2010)109. MR2756043 (2012b:81223)

[54] Jirui Guo, Bei Jia, and Eric Sharpe, *Chiral operators in two-dimensional (0,2) theories and a test of triality*, J. High Energy Phys. **6** (2015), 201, front matter+43. MR3370134

[55] Daniel Halpern-Leistner and Ian Shipman, *Autoequivalences of derived categories via geometric invariant theory*, (2013).

[56] James Halverson, Vijay Kumar, and David R. Morrison, *New methods for characterizing phases of 2D supersymmetric gauge theories*, J. High Energy Phys. **9** (2013), 143, front matter+31, DOI 10.1007/JHEP09(2013)143. MR3107702

[57] Simeon Hellerman, André Henriques, Tony Pantev, Eric Sharpe, and Matt Ando, *Cluster decomposition, T-duality, and gerby CFTs*, Adv. Theor. Math. Phys. **11** (2007), no. 5, 751–818. MR2365496 (2008m:81168)

[58] Simeon Hellerman and Eric Sharpe, *Sums over topological sectors and quantization of Fayet-Iliopoulos parameters*, Adv. Theor. Math. Phys. **15** (2011), no. 4, 1141–1199. MR2929685

[59] Kentaro Hori, *Duality in two-dimensional* $(2,2)$ *supersymmetric non-Abelian gauge theories*, J. High Energy Phys. **10** (2013), 2013:121, front matter+74, DOI 10.1007/JHEP10(2013)121. MR3118316

[60] Kentaro Hori and Johanna Knapp, *Linear sigma models with strongly coupled phases - one parameter models*, JHEP **1311** (2013), 070.

[61] Kentaro Hori and David Tong, *Aspects of non-abelian gauge dynamics in two-dimensional* $\mathcal{N} = (2,2)$ *theories*, J. High Energy Phys. **5** (2007), 079, 41 pp. (electronic), DOI 10.1088/1126-6708/2007/05/079. MR2318130 (2009d:81351)

[62] Kentaro Hori and Cumrun Vafa, *Mirror symmetry*, (2000).

[63] Bei Jia, Eric Sharpe, and R. Wu, *Notes on nonabelian (0,2) theories and dualities*, JHEP **1408** (2014), 017.

[64] Hans Jockers, Vijay Kumar, Joshua M. Lapan, David R. Morrison, and Mauricio Romo, *Nonabelian 2D gauge theories for determinantal Calabi-Yau varieties*, J. High Energy Phys. **11** (2012), 166, front matter + 46. MR3036443

[65] Hans Jockers, Vijay Kumar, Joshua M. Lapan, David R. Morrison, and Mauricio Romo, *Two-sphere partition functions and Gromov-Witten invariants*, Comm. Math. Phys. **325** (2014), no. 3, 1139–1170, DOI 10.1007/s00220-013-1874-z. MR3152749

[66] Sheldon Katz and Eric Sharpe, *Notes on certain* $(0,2)$ *correlation functions*, Comm. Math. Phys. **262** (2006), no. 3, 611–644, DOI 10.1007/s00220-005-1443-1. MR2202305 (2006m:81232)

[67] David Kutasov and Jennifer Lin, *(0,2) ADE Models From Four Dimensions*, (2014).

[68] David Kutasov and Jennifer Lin, *(0,2) Dynamics From Four Dimensions*, Phys.Rev. **D89** (2014), 085025.

[69] Alexander Kuznetsov, *Homological projective duality for grassmannians of lines,*.

[70] Alexander Kuznetsov, *Homological projective duality*, Publ. Math. Inst. Hautes Études Sci. **105** (2007), 157–220, DOI 10.1007/s10240-007-0006-8. MR2354207 (2008k:14043)

[71] Alexander Kuznetsov, *Derived categories of quadric fibrations and intersections of quadrics*, Adv. Math. **218** (2008), no. 5, 1340–1369, DOI 10.1016/j.aim.2008.03.007. MR2419925 (2009g:14019)

[72] Wolfgang Lerche, Cumrun Vafa, and Nicholas P. Warner, *Chiral rings in* $N = 2$ *superconformal theories*, Nuclear Phys. B **324** (1989), no. 2, 427–474, DOI 10.1016/0550-3213(89)90474-4. MR1025424 (91d:81132)

[73] Hai Lin, Baosen Wu, and Shing-Tung Yau, *Heterotic String Compactification and New Vector Bundles*, (2014).

[74] Jock McOrist, *The revival of* $(0,2)$ *sigma models*, Internat. J. Modern Phys. A **26** (2011), no. 1, 1–41, DOI 10.1142/S0217751X11051366. MR2763382 (2012e:81160)

[75] Jock McOrist and Ilarion V. Melnikov, *Half-twisted correlators from the Coulomb branch*, J. High Energy Phys. **4** (2008), 071, 19, DOI 10.1088/1126-6708/2008/04/071. MR2425232 (2009h:81272)

[76] Jock McOrist and Ilarion V. Melnikov, *Summing the instantons in half-twisted linear sigma models*, J. High Energy Phys. **2** (2009), 026, 61, DOI 10.1088/1126-6708/2009/02/026. MR2486403 (2011b:81228)

[77] Ilarion Melnikov, Savdeep Sethi, and Eric Sharpe, *Recent developments in* $(0,2)$ *mirror symmetry*, SIGMA Symmetry Integrability Geom. Methods Appl. **8** (2012), Paper 068, 28, DOI 10.3842/SIGMA.2012.068. MR2988026

[78] Ilarion V. Melnikov and M. Ronen Plesser, *The Coulomb branch in gauged linear sigma models*, J. High Energy Phys. **6** (2005), 013, 33, DOI 10.1088/1126-6708/2005/06/013. MR2158560 (2007a:81135)

[79] Ilarion V. Melnikov and M. Ronen Plesser, *A-model correlators from the Coulomb branch*, J. High Energy Phys. (2006), no. 2, 044, 21. MR2219460 (2007d:81209)

[80] Ilarion V. Melnikov and M. Ronen Plesser, *A* $(0,2)$ *mirror map*, J. High Energy Phys. **2** (2011), 001, 15, DOI 10.1007/JHEP02(2011)001. MR2820839

[81] Ilarion V. Melnikov and Eric Sharpe, *On marginal deformations of* $(0,2)$ *non-linear sigma models*, Phys. Lett. B **705** (2011), no. 5, 529–534, DOI 10.1016/j.physletb.2011.10.055. MR2860523

[82] Alexander A. Migdal, *Recursion Equations in Gauge Theories*, Sov.Phys.JETP **42** (1975), 413.

[83] David R. Morrison and M. Ronen Plesser, *Summing the instantons: quantum cohomology and mirror symmetry in toric varieties*, Nuclear Phys. B **440** (1995), no. 1-2, 279–354, DOI 10.1016/0550-3213(95)00061-V. MR1336089 (96f:32036)

[84] David R. Morrison and M. Ronen Plesser, *Towards mirror symmetry as duality for two-dimensional abelian gauge theories*, Strings '95 (Los Angeles, CA, 1995), World Sci. Publ., River Edge, NJ, 1996, pp. 374–387. MR1660724 (99m:81224)

[85] Tony Pantev and Eric Sharpe, *Notes on gauging noneffective group actions*, (2005).

[86] Tony Pantev and Eric Sharpe, *GLSMs for gerbes (and other toric stacks)*, Adv. Theor. Math. Phys. **10** (2006), no. 1, 77–121. MR2222223 (2007a:81121)

[87] Tony Pantev and Eric Sharpe, *String compactifications on Calabi-Yau stacks*, Nuclear Phys. B **733** (2006), no. 3, 233–296, DOI 10.1016/j.nuclphysb.2005.10.035. MR2192505 (2006h:81243)

[88] Vipul Periwal and Andrew Strominger, *Kähler geometry of the space of* $N = 2$ *superconformal field theories*, Phys. Lett. B **235** (1990), no. 3-4, 261–267, DOI 10.1016/0370-2693(90)91961-A. MR1037713 (91g:32028)

[89] B. Ye. Rusakov, *Loop averages and partition functions in* $\mathrm{U}(N)$ *gauge theory on two-dimensional manifolds*, Modern Phys. Lett. A **5** (1990), no. 9, 693–703, DOI 10.1142/S0217732390000780. MR1051372 (91m:81152)

[90] N. Seiberg, *Electric-magnetic duality in supersymmetric non-abelian gauge theories*, Nuclear Phys. B **435** (1995), no. 1-2, 129–146, DOI 10.1016/0550-3213(94)00023-8. MR1314365 (96b:81117)

[91] Eric Sharpe, *D-branes, derived categories, and Grothendieck groups*, Nuclear Phys. B **561** (1999), no. 3, 433–450, DOI 10.1016/S0550-3213(99)00535-0. MR1726156 (2001k:14024)

[92] Eric Sharpe, *Notes on correlation functions in* $(0,2)$ *theories*, Snowbird lectures on string geometry, Contemp. Math., vol. 401, Amer. Math. Soc., Providence, RI, 2006, pp. 93–104, DOI 10.1090/conm/401/07556. MR2222532 (2007e:81100)

[93] Eric Sharpe, *Notes on certain other* $(0,2)$ *correlation functions*, Adv. Theor. Math. Phys. **13** (2009), no. 1, 33–70. MR2471852 (2010a:81217)

[94] Eric Sharpe, *An introduction to quantum sheaf cohomology* (English, with English and French summaries), Ann. Inst. Fourier (Grenoble) **61** (2011), no. 7, 2985–3005. MR3112514

[95] Eric Sharpe, *A few Ricci-flat stacks as phases of exotic GLSM's*, Phys.Lett. **B726** (2013), 390–395.

[96] Eric Sharpe, *Predictions for Gromov-Witten invariants of noncommutative resolutions*, J. Geom. Phys. **74** (2013), 256–265, DOI 10.1016/j.geomphys.2013.08.012. MR3118585

[97] Eric Sharpe, *Decomposition in diverse dimensions*, Phys.Rev. **D90** (2014), 025030.

[98] Andrew Strominger, *Yukawa Couplings in Superstring Compactification*, Phys.Rev.Lett. **55** (1985), 2547.

[99] Andrew Strominger, *Superstrings with torsion*, Nuclear Phys. B **274** (1986), no. 2, 253–284, DOI 10.1016/0550-3213(86)90286-5. MR851702 (87m:81177)

[100] Yuji Tachikawa, *On the 6d origin of discrete additional data of 4d gauge theories*, JHEP **1405** (2014), 020.

[101] Meng-Chwan Tan, *Two-dimensional twisted sigma models and the theory of chiral differential operators*, Adv. Theor. Math. Phys. **10** (2006), no. 6, 759–851. MR2302273 (2009h:81267)

[102] Meng-Chwan Tan, *Two-dimensional twisted sigma models, the mirror chiral de Rham complex, and twisted generalised mirror symmetry*, J. High Energy Phys. **7** (2007), 013, 80 pp. (electronic), DOI 10.1088/1126-6708/2007/07/013. MR2326774 (2009g:81184)

[103] Xiang Tang and Hsian-Hua Tseng, *Duality theorems for étale gerbes on orbifolds*, Adv. Math. **250** (2014), 496–569, DOI 10.1016/j.aim.2013.10.002. MR3122175

[104] Hsian-Hua Tseng, *On degree-0 elliptic orbifold Gromov-Witten invariants*, Int. Math. Res. Not. IMRN **11** (2011), 2444–2468, DOI 10.1093/imrn/rnq158. MR2806585 (2012h:14143)

[105] Hsian-Hua Tseng, *Notes on orbifold Gromov-Witten theory*, Fifth International Congress of Chinese Mathematicians. Part 1, 2, AMS/IP Stud. Adv. Math., 51, pt. 1, vol. 2, Amer. Math. Soc., Providence, RI, 2012, pp. 327–340. MR2908078

[106] Edward Witten, *Mirror manifolds and topological field theory*, Essays on mirror manifolds, Int. Press, Hong Kong, 1992, pp. 120–158. MR1191422 (94c:81194)

[107] Edward Witten and Jonathan Bagger, *Quantization of Newton's constant in certain supergravity theories*, Phys. Lett. B **115** (1982), no. 3, 202–206, DOI 10.1016/0370-2693(82)90644-X. MR669354 (83m:81107)

DEPARTMENT OF PHYSICS, VIRGINIA TECH
Current address: Department of Physics, Virginia Tech
E-mail address: `ersharpe@vt.edu`

Proceedings of Symposia in Pure Mathematics
Volume **93**, 2016
http://dx.doi.org/10.1090/pspum/093/01587

Codimension two defects and the Springer correspondence

Aswin Balasubramanian

ABSTRACT. One can associate an invariant to a large class of regular codimension two defects of the six dimensional $(0, 2)$ SCFT $\mathcal{X}[j]$ using the classical Springer correspondence. Such an association allows a simple description of S-duality of associated Gaiotto-Witten boundary conditions in $\mathcal{N} = 4$ SYM for arbitrary gauge group G and by extension, a determination of certain local aspects of class $\mathcal{S}$ constructions. I point out that the problem of *classifying* the corresponding boundary conditions in $\mathcal{N} = 4$ SYM is intimately tied to possible symmetry breaking patterns in the bulk theory. Using the Springer correspondence and the representation theory of Weyl groups, I construct a pair of functors between the class of boundary conditions in the theory in the phase with broken gauge symmetry and those in the phase with unbroken gauge symmetry.

CONTENTS

1. Introduction

In recent years, the study and possible classification of defect operators in supersymmetric quantum field theories of various dimensions has been pursued with some vigor. Two major themes in this study have been connections between the study of defect operators and various ideas in geometric representation theory and the close relationship between defects in a d dimensional (S)QFT and various l (for $l < d$) dimensional (S)QFTs. In this note, I will discuss a particular instance where both of these themes play a prominent role. This involves a class of regular codimension two defects of the six dimensional $(0, 2)$ SCFT that we will henceforth

2010 *Mathematics Subject Classification.* Primary 20G05, 20G45, 22E57.
Key words and phrases. Supersymmetric quantum field theories, geometric representation theory.
UTTG-27-14.
The author's work was made possible by the hospitality of the Theory Group, UT Austin. The author would also like to thank the Mathematical Sciences Research Institute, Berkeley for hospitality during the *Introductory Workshop on Geometric Representation Theory.*

denote by $\mathcal{X}[j]$ for every Lie algebra j of type A, D, E. On the one hand, the problem of classifying these defects turns out to be connected intimately to the classical Springer correspondence and closely related aspects of the representation theory of Weyl groups. And on the other hand, the existence of these defects in the six dimensional theory is tied to the existence of certain lower dimensional SCFTs with eight supercharges. One set of examples of the three dimensional SCFTs in question are the Gaiotto-Witten theories $T^{\rho}[G]$ and their mirror duals $T_{\rho^\vee}[G]$. These theories are interesting objects by themselves and also serve as important ingredients in constructions of other four and three dimensional theories. Specifically, in four dimensions, a subset of the theories dubbed 'class S' [**1, 2**] are of this type.

In what follows, these Gaiotto-Witten theories will play an important role. In section 2, their specific role in the study of boundary conditions in four dimensional $\mathcal{N} = 4$ SYM and the relationship to codimension-two defects of $\mathcal{X}[j]$ is recalled. In section 3, the close connection of these topics to the Springer Correspondence is recalled. In section, 4, the connection is exploited to provide a classification scheme that relates the existence of the Gaiotto-Witten theories to the possibility of specific symmetry breaking patters in four dimensional $\mathcal{N} = 4$ SYM theory. A pair of functors between the classes of boundary conditions of the theory with gauge group $G^\vee$ and the theory with gauge group $L^\vee$ (for $L^\vee$ a suitable subgroup of $G^\vee$) play an important role in this classification scheme. While this note is largely a review of the work done in [**3**], the narrower focus of this note allows for an elaboration of some subtler aspects like, for example, the difference between the two finite groups usually denoted by $A(O)$ and $\overline{A}(O)$ and how the difference between the two can be understood in terms of the Springer invariant.

2. Boundary conditions for $\mathcal{N} = 4$ SYM

The existence of the regular codimension two defects of the six dimensional $\mathcal{X}[j]$ theory can be related to the existence of certain 1/2 BPS boundary conditions of $\mathcal{N} = 4$ SYM in four dimensions. This can be inferred by considering the following setup (see Fig 1) . In six dimensions, formulate theory $\mathcal{X}[j]$ on a spacetime of the form $\mathbb{R}^{1,2} \times \mathbb{S}^1 \times H$, where H is a two dimensional disc with a cigar metric. One can reduce to four dimensions in two different ways. First, we reduce along the circle fiber $\tilde{S}^1$ of the cigar H to get five dimensional $\mathcal{N} = 4$ SYM with gauge group G (a compact group) together with a boundary condition that is specified by a Nahm pole. If a twist by Dynkin diagram automorphism of j is included in the $\tilde{S}^1$ reduction, then G is the compact group corresponding to the 'folded' Dynkin diagram. This allows for the appearance of non-simply laced G. If the twist is not included, G is the compact group corresponding to complex Lie algebra j and is thus simply laced. Reducing further on the circle S^1, one gets four dimensional $\mathcal{N} = 4$ SYM with gauge group G formulated on a four manifold M_4 that is topologically $\mathbb{R}^{1,2} \times \mathbb{R}^+$ together with a boundary condition that is labeled by the same Nahm pole ρ. If the two dimensional reductions are done in the opposite order, one gets $G^\vee$ $\mathcal{N} = 4$ theory with a boundary condition that involves coupling to a boundary three dimensional $\mathcal{N} = 4$ theory called $T_{\rho^\vee}[G]$. Let us pick a co-ordinate y for $\mathbb{R}^+$ direction such that $y = 0$ forms the boundary of M_4. The presence of a Nahm pole boundary condition of type ρ for the four dimensional $\mathcal{N} = 4$ theory with gauge group G implies (by definition), the following equations for three out

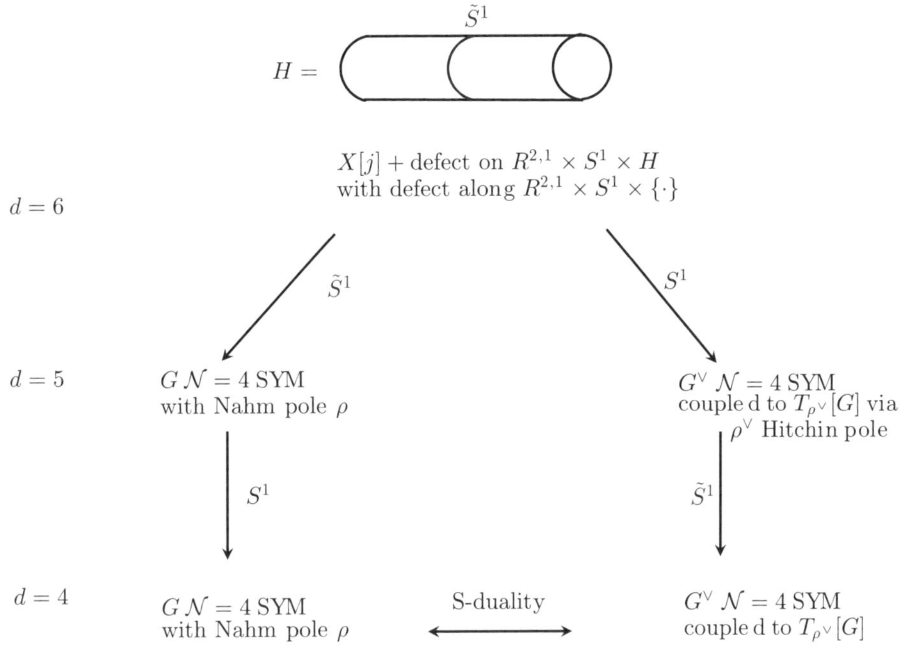

FIGURE 1. The connection between codimension two defects of $\mathcal{X}[j]$ and boundary conditions for 4d $\mathcal{N} = 4$ SYM. The Nahm (ρ) and Hitchin ($\rho^\vee$) labels associated to a regular codimension two defect are highlighted.

of the six scalars in the theory,

$$
\begin{aligned}
\frac{dX^i}{dy} &= \epsilon^{ijk}[X^j, X^k] \\
X^i &= \frac{\rho(\tau^i)}{y},
\end{aligned}
\tag{1}
$$

where ρ denotes an embedding of $\mathfrak{sl}_2$ into the complex Lie algebra $\mathfrak{g}$ and τ^i are the standard generators of $\mathfrak{sl}_2$. They obey $[\tau^1, \tau^3] = 2\tau^2, [\tau^2, \tau^1] = \tau^1, [\tau^2, \tau^3] = -\tau^3$. Here, the three scalars have been arranged into a vector X and they transform in the three dimensional representation of the $SO(3)_X$ part of the R symmetry group. There is a corresponding set of equations that are obeyed by the gauge fields in the theory. When the $\mathfrak{sl}_2$ embedding is trivial, the gauge field obey the usual Dirichlet boundary conditions. Hence, the most general Nahm pole boundary conditions can be viewed as generalized Dirichlet boundary conditions.

In **[4]**, Gaiotto-Witten constructed a vast class of 1/2 BPS conformal boundary conditions in four dimensional $\mathcal{N} = 4$ SYM theory formulated on $\mathbb{R}^{1,2} \times \mathbb{R}^+$ of which the Nahm pole boundary conditions form a small but very interesting subset. Their classification was by associating a triple $(\mathcal{O}^\rho, H, \mathcal{B})$ to every boundary condition, where $\mathcal{O}^\rho$ is a nilpotent orbit in the Lie algebra $\mathfrak{g}$ (by the Jacobson-Morozov theorem, this is equivalent to a choice of an embedding $\rho : \mathfrak{sl}_2 \to \mathfrak{g}$) and H is a subgroup of the centralizer $Z_{\mathfrak{g}}(\rho)$ of the associated $\mathfrak{sl}_2$ triple and $\mathcal{B}$ is a three dimensional SCFT with eight supercharges.

The translation from the indexing set of triples to the boundary condition is as follows. First, one imposes Nahm pole boundary conditions for a triplet of the scalars (labeled X^i) as in Eqns. 1. The second datum implies that one considers a decomposition of the Lie algebra into $\mathfrak{h} \oplus \mathfrak{h}^\perp$ and Higgs the theory down to H symmetry at the boundary. The third datum in the triple involves choosing a a three dimensional $\mathcal{N} = 4$ SCFT $\mathcal{B}$ with H global symmetry and gauging the global symmetry and coupling the theory at the boundary to the four dimensional theory.

In [**5**], they further studied the action of S-duality on this vast class of boundary conditions. One of the important instances of this duality of boundary conditions is the case of a pure Nahm pole boundary condition of type ρ. The S-dual of this boundary condition happens to be a boundary condition in the $G^\vee$ theory that corresponds to coupling to the boundary theory $T_{\rho^\vee}$. In terms of the indexing set of triple, this can be stated as,

$$(\mathcal{O}^\rho, Id, \varnothing) \leftrightarrow_{S-dual} (\mathcal{O}^0, G^\vee, T_{\rho^\vee}[G]). \tag{2}$$

In order the make precise the idea of 'coupling at the boundary', one needs to recall some facts about $\mathcal{N} = 4$ SCFTs in three dimensions. In general, these theories admit at least two quantum moduli spaces of vacua called the Higgs branch and the Coulomb branch. The $\mathcal{N} = 4$ supersymmetry of the theory ensures that both of these are hyper-Kähler spaces. In theories with Lagrangian descriptions, the Higgs can be described as a hyper-Kähler quotient and the metric is not 'quantum corrected'. The Coulomb branch, on the other hand *is* changed by quantum corrections and is, generically, much harder to understand. For theories without Lagrangian descriptions, one is, *a priori*, at a loss to describe to either of the branches. The $\mathcal{N} = 4$ theories in four dimensional admit a duality called *three dimensional mirror symmetry* that, among other things, exchanges the Higgs and Coulomb branches. For the theories denoted by $T_{\rho^\vee}[G]$, the Coulomb branches are specific strata inside the nilpotent cone of $\mathfrak{g}$ (denoted by $\mathcal{N}_\mathfrak{g}$) called Slodowy slices and the Higgs branches are certain strata inside the nilpotent cone of $\mathfrak{g}^\vee$ (denoted by $\mathcal{N}_{\mathfrak{g}^\vee}$) corresponding to nilpotent orbit closures. So, the coupling at the boundary for the $G^\vee$ theory is to be understood to proceed via the gauging of the global symmetry on the Higgs branch of the $T_{\rho^\vee}[G]$ theory. This is an effective IR description of what it means to 'couple to a boundary theory'. On the G side, the space of solutions to the Nahm equations can be non-trivial and if this is the case, the four dimensional theory is be interpreted to have a moduli space of vacua given by this space of solutions. Not coincidentally, these solutions are exactly the Slodowy slices that arise as Coulomb branches of the $T_{\rho^\vee}[G]$ theories. The fact that the same moduli spaces have different realizations on the G and $G^\vee$ is the justification for the S-duality map in Eqn. 2.

Now, note that on the G side of the duality, the classical gauge fields are G valued. So, if one were to start with the data describing the boundary condition on the G side, it seems reasonable that one is able to directly conclude something about the part of the vacuum moduli space of $T_{\rho^\vee}[G]$ that is a stratum inside $\mathcal{N}_\mathfrak{g}$. This, however, constitutes 'classical' data for the mirror dual of $T_{\rho^\vee}[G]$ and not for $T_{\rho^\vee}[G]$ itself. This mirror dual is usually denoted by $T^\rho[G]$. For this reason, S-duality of boundary conditions in the four dimensional $\mathcal{N} = 4$ theory is inextricably connected with mirror symmetry for three dimensional $\mathcal{N} = 4$ SCFTs.

In order to identify precisely which $T_{\rho^\vee}[G]$ appear as duals of a given Nahm pole boundary conditions, Gaiotto-Witten proposed using the dimension of the moduli space as an invariant. It can be shown that for $G = SU(N)$, this is an adequate invariant and matching just the dimensions of moduli spaces on both sides leads to a complete specification of the duality map. However, for arbitrary G, this is inadequate. This motivates the search for a finer invariant of the moduli spaces. In the following section, one such invariant is discussed.

3. Springer correspondence and the Springer invariant

The classical Springer correspondence is a construction that relates the geometric world of nilpotent orbits and Slodowy slices to the algebraic world of irreducible representations of the Weyl group. What follows is an extremely short review of a rich subject. The reader is referred to [**3**] for a longer discussion. The starting point in the construction is the existence of the *Springer resolution* $\mu : \tilde{\mathcal{N}} \to \mathcal{N}$. This is a simultaneous resolution of the singularities of $\mathcal{N}$. From a physical perspective, the *existence* of such a resolution of the vacuum moduli spaces is related to the existence of Fayet-Iliopoulos and Mass parameters in the $\mathcal{N} = 4$ theory. Typically, turning on real values for either the FI or mass parameters leads to a resolution of the Higgs branch of a given theory $T_{\rho^\vee}[G]$ or Higgs branch of its mirror $T^\rho[G]$.

The existence of such a map μ ties together topological properties of the resolved space $\mathcal{N}$ and the singular space $\mathcal{N}$ in fairly intricate ways. An example of such a relationship is the decomposition theorem. The Springer correspondence can be viewed as a small but powerful part of the results obtained by applying the decomposition theorem to the particular setting of the Springer resolution (for a review, see [**6**]). To be more specific, let e be a representative of a nilpotent orbit $\mathcal{O}$. The Springer correspondence is a statement about how the top dimensional cohomology of the Springer fiber $\mathcal{B}_e(= \mu^{-1}(e))$ decomposes as a module for the Weyl group. The correspondence can be encapsulated in an injective map,

$$Sp : Irr(W) \to \tilde{\mathcal{O}}, \tag{3}$$

where $Irr(W)$ is the set of all irreducible representation of the Weyl group associated to $\mathfrak{g}$ and $\tilde{\mathcal{O}}$ is the set of all pairs $(\mathcal{O}, \chi)$, where $\mathcal{O}$ is a nilpotent orbit and χ is an irreducible representation of the component group $A(\mathcal{O})$ associated to the nilpotent orbit $\mathcal{O}$. The component group is defined to be the quotient $C_G(e)/C^0_G(e)$, where $C_G(e)$ is the centralizer of the nilpotent element in the group and $C^0_G(e)$ is its connected component. It is often useful to consider the inverse of this map and denote it as $Sp^{-1} : \tilde{\mathcal{O}} \to Irr(W)$. Some care is required in using the inverse since the map Sp is guaranteed only to be injective. In what follows, we will be able to use the inverse without worrying about this subtlety. Now, since both nilpotent orbits and Slodowy slices are resolved under the Springer map, the constructions like the Springer correspondence apply to both kinds of strata. Physically, this means one can separately attach a Higgs branch Springer invariant (denoted by r) and a Coulomb branch Springer invariant (denoted by $\bar{r}$) for theories like $T_{\rho^\vee}[G]$. Outside of case $G = SU(N)$, there is an asymmetry between the properties of the two possible Springer invariants. This asymmetry singles out one of them as being the more useful Springer invariant. In this note, the Coulomb branch Springer invariant of $T_{\rho^\vee}[G]$ is the one that plays this role. [1]

[1] In the work [**3**], all statements were made using the $T^\rho[G]$ theories and hence the same invariant was called the Higgs branch Springer invariant.

4. Classification via Symmetry breaking

In this section, we will see that the invariant constructed in Sec 3 is strictly smarter than the dimension of the associated vacuum moduli space. In particular, it allows for an elegant description of which exact strata pair up to form the Higgs and Coulomb branches of $T_{\rho^\vee}[G]$. Further, it allows a re-organization of the classification of the associated boundary conditions in the four dimensional theory. The seemingly ad-hoc nature (from a physical standpoint) of the association of the invariant is compensated by the incredibly simple and powerful statements that one can make using this invariant.

As discussed in the earlier section, coupling to $T_{\rho^\vee}[G]$ in the boundary is via a nilpotent representative $\rho^\vee$ of the nilpotent orbit that is identified as the Higgs branch of the theory. Now, let us think of the coupling to the boundary theory under possible symmetry breaking in the bulk. Note that this is a *different* situation from the one where we consider symmetry breaking only at the boundary. The symmetric breaking is triggered by a vev m for the bulk scalar and unbroken gauge group is the centralizer $Z_{\mathfrak{g}^\vee}(m)$ of the semi-simple element $m \in \mathfrak{g}^\vee$. Let the connected part of the unbroken gauge group be $L^\vee$. Now, a necessary condition for a coupling via $\rho^\vee$ to make sense is that there should be enough unbroken gauge symmetry at the boundary so that $\rho^\vee$ is still a nilpotent representative of some nilpotent orbit in the Lie algebra $l^\vee$. Clearly, for any non-zero nilpotent element $\rho^\vee$, making $l^\vee$ too small would violate this requirement. The most that one can do is demand that $\rho^\vee$ correspond to a 'a very big orbit' in $l^\vee$. Loosely speaking, it is so big that it can't fit into a smaller subalgebra. A mathematically precise way of enforcing this notion is to demand that $\rho^\vee$ is a distinguished orbit in $l^\vee$.

Those familiar with the structure theory of nilpotent orbits will immediately recognize the close similarity between what has just been said and Bala-Carter classification of nilpotent orbits. There is, however, one extremely important difference. In the work of Bala-Carter, the problem of classifying nilpotent orbits is mapped to the problem of classifying distinguished nilpotent orbits in *Levi* subalgebras[2]. Note that in our discussion of symmetry breaking, no mention of $l^\vee$ being related to Levi subalgebra was mentioned. The relationship that one can hope for is that the Lie algebra of the connected component of the unbroken gauge group ($l^\vee$, in current notation) occurs as the semi-simple part of a Levi subalgebra. Such a relationship indeed exists for groups of type A. But, outside of groups of type A, the above statement is *not true* in general. For groups outside of type A, only a proper subset of the centralizers of semi-simple elements are related to Levi subalgebras. This is part of what makes this extremely interesting. A study of distinguished nilpotent orbits in these more general centralizers was done by Sommers in [7] and this extends the work of Bala-Carter. It is this extended form of the Bala-Carter classification, one that can be termed Bala-Carter-Sommers classification, that is relevant for the above discussion treating the classification of defects in conjunction with symmetry breaking patterns. For a given theory $T_{\rho^\vee}[G]$, let $e^\vee$ be the nilpotent representative of the orbit that represents the Higgs branch and let $(\mathfrak{l}^\vee, e^\vee)$ be an associated Sommers pair.

Now, let us consider a *distinguished* symmetry breaking in the bulk of the $G^\vee$ theory. This means the following. One picks a vev m for the scalar in the vector

---

[2]Equivalently, one looks at distinguished nilpotent orbits in parabolic subalgebras. Each parabolic subalgebra has a canonical Levi decomposition whose 'Levi part' is called a Levi subalgebra.

multiplet such that the RG flow triggered by the vev m lands in the theory with a unbroken gauge symmetry whose connected component is the group $L^\vee$. In other words, the RG flow is picked so that in the UV, one starts with a random nilpotent orbit in g and in the IR, one ends up recovering the Sommers pair $(l^\vee, e^\vee)$. The fact that the Sommers pair is a characterization of the Higgs branch of $T_{\rho^\vee}[G]$ is obvious from the way it has been constructed. It will, in fact, turn out to identify it uniquely after we impose a subtle equivalence relation on Sommers pairs (this is discussed below). But, a natural question is how does one relate the Coulomb branch of $T_{\rho^\vee}[G]$ to their corresponding Sommers pair of the Higgs branch. It turns out that the following matching conditions for the Springer invariants specifies this relationship completely,

(4)
$$\begin{aligned} s &= \epsilon_{\mathfrak{l}} \times Sp^{-1}[\mathfrak{l}^\vee] \\ Sp^{-1}[\mathfrak{g}, \mathcal{O}_N] &= j_{W[\mathfrak{l}^\vee]}^{W[\mathfrak{g}^\vee]}(s), \end{aligned}$$

where $\mathcal{O}_N$ is the nilpotent orbit that enters the description of the Nahm boundary condition (and is called the Nahm datum in Fig 1) and $\mathcal{O}_H$ is the nilpotent orbit that corresponds to the Higgs branch of $T_{\rho^\vee}[G]$ and is called the Hitchin datum in Fig 1. The invariant that was previously dubbed the Coulomb branch Springer invariant denoted by $\bar{r}$ is the same as $Sp^{-1}[\mathfrak{g}, \mathcal{O}_N]$ in the matching condition above. For the class of boundary conditions being considered, it is sufficient to restrict to the case where $\mathcal{O}_H$ is a special nilpotent orbit. The operation denoted by j is a functor between the irreducible representations of the Weyl group $W[\mathfrak{l}^\vee]$ and that of the Weyl group $W[\mathfrak{g}^\vee]$. It is a truncated version of the usual induction functor and the nature of its action is explained in greater detail in **[3]**.

Now, for a given nilpotent representative $e^\vee$, there could be different choices (non conjugate) for $\mathfrak{l}^\vee$ such that $(\mathfrak{l}^\vee, e^\vee)$ is a Sommers pair. What does this indicate? This indicates that one is dealing with two different theories $T^\rho[G]$ and $T^{\rho'}[G]$ whose associated Coulomb branches correspond to the Sommers pairs $(\mathfrak{l}^\vee, e^\vee)$ and $(\mathfrak{l}^{\vee'}, e^\vee)$. The number of such distinct Sommers pairs associated to a given special nilpotent orbit is indicative of the size of its associated $\overline{A}(O)$ group.

One of the esoteric features of this setup is the appearance of the group $\overline{A}(O)$.[3] In several cases, it is the same as the group $A(O)$ that we encountered before, but it is in general a smaller group. This group is a quotient of the component group $A(O)$. It was originally encountered by Lusztig in his work **[9]** and is thus sometimes called "Lusztig's quotient". In terms of the matching conditions for the Springer invariant, the appearance of the smaller group can be understood in the following way. For some $G^\vee$, there exist different $L^\vee$ (not conjugate to each other) for which the output of j-induction (invoked as in Eqn. 4) happens to be the same irrep of $W[\mathfrak{g}^\vee] \cong W[\mathfrak{g}]$. Now, define an equivalence relation on Sommers pairs for a particular nilpotent orbit such that they are identified if the output of j-induction is the same. The Sommers pairs, together with equivalence relation now give a characterization of the group $\overline{A}(O)$. As an example of the cases where $\overline{A}(O) \neq A(O)$, we list here two such instances for $G^\vee = E_8$.

[3]To paint a more complete picture, one actually needs to involve more data about conjugacy classes/subgroups in $\overline{A}(O)$. For this, we refer to **[3,8]**

$\mathcal{O}_H$	$A(O)$	$\overline{A}(O)$
$E_7(a_4)$	S_2	1
$E_8(b_6)$	S_3	S_2

4.1. A pair of functors. In the above discussion, we have really encountered two functors between classes of boundary conditions. The first is distinguished symmetry breaking in the presence of a boundary condition. Let us called this the functor s. The functor s goes from theory with gauge symmetry $G^\vee$ to the theory with gauge symmetry $L^\vee$. The second is the functor that goes between boundary conditions in theory with gauge symmetry $L^\vee$ and the theory with gauge symmetry $G^\vee$. Let us call this functor i. These functors can be viewed as being analogous to the restrict and induce functors (which form an adjoint pair) between the categories of representation of a finite group $\mathcal{F}$ and a subgroup $\mathcal{H} \subset \mathcal{F}$. In terms of the associated Springer invariants, the functors encountered above are directly the Res and j induction functors in the study of Weyl group representations. Note however that one is dealing here with the class of boundary conditions in a four dimensional quantum field theory. This is morally modeled by an associated 3-category. So, a complete understanding of the pair of functors (s, i) should be developed in this setting.

4.2. Mirror symmetry and Symplectic duality. We have seen that S-duality of boundary conditions in four dimensional $\mathcal{N} = 4$ has intimate connections with mirror symmetry for certain associated three dimensional SCFTS. From a mathematical perspective, it turns out that this 3d mirror symmetry has close connections with the idea of symplectic duality in the sense of BLPW **[10]**. According to BLPW, two conical symplectic resolutions $\mu : M \to M_0$ and $\mu' : M' \to M_0'$ are defined to be symplectic duals of each other if certain associated categories are related to each other via Koszul duality. From a physical perspective, the two resolved spaced M and M' are resolved Higgs branches of two different 3d SCFTs which are mirror duals of each other. The case of $T^\rho[G]$ and $T_{\rho^\vee}[G]$ is a particular instance of such a mirror pair and the relevant Higgs branches are, respectively, Slodowy slices and nilpotent orbit closures. The relevant symplectic resolution is the Springer resolution discussed in section 3. This relationship has several consequences, one of which is a relationship between pieces of the cohomology of the resolved spaces M and M' (discussed in **[10, 11]**). It would be interesting to understand the precise relationship between the cohomological consequences of symplectic duality and the matching condition that occurs in section 4.

The relationship at the level of categories between the mathematical definition of symplectic duality and 3d mirror symmetry has been studied recently for many 3d $\mathcal{N} = 4$ theories **[12]**. The braid group actions that play an important role in the duality between categories yield Weyl group actions at the level of cohomology upon suitable 'decategorification'.

Acknowledgements

The author would like to thank P. Achar, D. Ben Zvi, J. Distler, A. Neitzke, N. Proudfoot for discussions.

References

[1] D. Gaiotto, $N = 2$ *dualities*, J. High Energy Phys. **8** (2012), 034, front matter + 57. MR3006961

[2] D. Gaiotto, G. W. Moore, and A. Neitzke, *Wall-crossing, Hitchin systems, and the WKB approximation*, Adv. Math. **234** (2013), 239–403, DOI 10.1016/j.aim.2012.09.027. MR3003931
[3] A. Balasubramanian, *Describing codimension two defects*, *JHEP* **1407** (2014) 095, [arXiv:1404.3737].
[4] D. Gaiotto and E. Witten, *Supersymmetric boundary conditions in $\mathcal{N} = 4$ super Yang-Mills theory*, J. Stat. Phys. **135** (2009), no. 5-6, 789–855, DOI 10.1007/s10955-009-9687-3. MR2548595 (2011a:81238)
[5] D. Gaiotto and E. Witten, *S-duality of boundary conditions in $\mathcal{N} = 4$ super Yang-Mills theory*, Adv. Theor. Math. Phys. **13** (2009), no. 3, 721–896. MR2610576 (2011j:81322)
[6] M. A. A. de Cataldo and L. Migliorini, *The decomposition theorem, perverse sheaves and the topology of algebraic maps*, Bull. Amer. Math. Soc. (N.S.) **46** (2009), no. 4, 535–633, DOI 10.1090/S0273-0979-09-01260-9. MR2525735 (2011a:14012)
[7] E. Sommers, *A generalization of the Bala-Carter theorem for nilpotent orbits*, Internat. Math. Res. Notices **11** (1998), 539–562, DOI 10.1155/S107379289800035X. MR1631769 (99g:20086)
[8] O. Chacaltana, J. Distler, and Y. Tachikawa, *Nilpotent orbits and codimension-2 defects of 6d $\mathcal{N} = (2,0)$ theories*, Internat. J. Modern Phys. A **28** (2013), no. 3-4, 1340006, 54, DOI 10.1142/S0217751X1340006X. MR3027737
[9] G. Lusztig, *Characters of reductive groups over a finite field*, Annals of Mathematics Studies, vol. 107, Princeton University Press, Princeton, NJ, 1984. MR742472 (86j:20038)
[10] T. Braden, A. Licata, N. Proudfoot, and B. Webster, *Quantizations of conical symplectic resolutions ii: category $\mathcal{O}$ and symplectic duality*, *arXiv preprint arXiv:1407.0964* (2014).
[11] N. Proudfoot and T. Schedler, *Poisson-de rham homology of hypertoric varieties and nilpotent cones*, *arXiv preprint arXiv:1405.0743* (2014).
[12] T. Dimofte, *Symplectic Duality and Knot Homologies, (talk at String-Math 2014).*

THEORY GROUP, DEPARTMENT OF PHYSICS, UNIVERSITY OF TEXAS AT AUSTIN
Current address: Department of Mathematics, University of Hamburg and DESY, Hamburg

Proceedings of Symposia in Pure Mathematics
Volume **93**, 2016
http://dx.doi.org/10.1090/pspum/093/01602

Higher spin AdS_3 holography and superstring theory

Thomas Creutzig, Yasuaki Hikida, and Peter B. Rønne

ABSTRACT. It has been believed for a long time that the tensionless limit of superstring theory can be described by a higher spin gauge theory. Recently, a concrete realization of this idea was proposed via 3d Aharony-Bergman-Jafferis (ABJ) theory with the help of holographic duality. In this note we review our work on finding a similar relation involving 2d coset type models. We start by proposing and examining holographic dualities between 3d higher spin gauge theories with matrix valued fields and the large N limit of 2d coset type models. After that we discuss possible relations to superstring theory with emphasis on the role of the matrix form of the higher spin fields and the extended supersymmetry.

1. Introduction

The gauge theory of higher spin fields can be introduced as a natural extension of the electromagnetic theory with a spin-1 field and gravity theory described by a spin-2 field. Superstring theory includes a large spectrum of massive higher spin states and it is believed that the tensionless limit of superstring theory can be described by a higher spin gauge theory. Moreover, several examples of AdS/CFT dualities involving higher spin gauge theories are known, and while these dualities are much simpler than the full superstring dualities they do share important non-trivial features.

The most famous example of a non-trivial higher spin gauge theory is given by the Vasiliev theory [**Vas03**]. It was proposed that 4d Vasiliev theory is dual to the 3d $O(N)$ vector model [**KP02**] (see also [**SS02**]) and this proposal was confirmed by examining the correlation functions [**GY10**, **GY11**]. The role of higher spin symmetry in the correspondence was clarified in [**MZ13a**, **MZ13b**]. Further, it is possible to extend the Vasiliev theory to include matrix valued fields i.e. to associate Chan-Paton (CP) factors. Recently, it was argued in [**CMSY13**] that the 4d extended Vasiliev theory with CP factors is dual to the 3d Aharony-Bergman-Jafferis (ABJ) theory. Since the ABJ theory is known to be dual to the superstring

2010 *Mathematics Subject Classification.* Primary 81T40, 83E30.

Key words and phrases. Higher spin gravity, AdS/CFT correspondence, conformal and W symmetry.

The work of the first author was supported by NSERC grant number RES0019997.

The work of the second author was supported by JSPS KAKENHI Grant Number 24740170.

The work of the third author was funded by AFR grant 3971664 from Fonds National de la Recherche, Luxembourg, and partial support by the Internal Research Project GEOMQ11 (Martin Schlichenmaier), University of Luxembourg, is also acknowledged.

theory on $AdS_3 \times \mathbb{C}P^3$ [**ABJM08, ABJ08**], the duality in [**CMSY13**] suggests a non-trivial relation (called as ABJ triality) between higher spin theory, ABJ theory and superstring theory. In other words, through the AdS/CFT correspondence, it becomes possible to examine superstring theory in terms of gauge theory with a large amount of higher spin symmetry.

The lower dimensional version of the proposal by [**KP02**] was introduced in [**GG11**] (see [**GG13b**] for a review), and the claim is that the 3d Vasiliev theory in [**PV99**] is dual to a 2d large N minimal model. This proposal is motivated by the enhancement of the asymptotic symmetry in 3d higher spin gauge theory found in [**HR10, CFPT10, CFP11**]. There are several generalizations of this proposal. A truncated version was proposed in [**Ahn11, GV11**]. Moreover, supersymmetric versions were introduced in [**CHR12, CHR13b, BCGG13**] and several supporting checks can be found, e.g., in [**CG13, HP13, HLGPR12, CHR13c, MZ13c**]. It is also possible to extend the 3d Vasiliev theory to include matrix valued fields [**PV99**]. Therefore, it is natural to expect that the AdS/CFT correspondence with the 3d extended Vasiliev theory with CP factors leads to a non-trivial correspondence between higher spin gauge theory and superstring theory as in [**CMSY13**]. We expect to obtain a deeper understanding about such trialities by studying the lower dimensional version since, in general, lower dimensional theories are more tractable than higher dimensional ones. In this note we would like to review our works on this subject in [**CHR13a, CHR14**]. Similar works may be found in [**GG13a, GP14, BCG14, GG14, CPV14**].

The rest of this note is organized as follows; In the next section we review the ABJ triality in [**CMSY13**] and explain why it is important to introduce CP factors to the higher spin fields. In section 3 we propose a duality between 3d higher spin gauge theory with CP factors and a 2d coset type model, and give several supporting arguments for the proposal. In section 4 we slightly generalize the duality to accommodate extended supersymmetry, and discuss the relation to superstring theory by making use of the $\mathcal{N} = 3$ superconformal symmetry of the 2d coset type model. In section 5 we conclude this note.

2. A review of ABJ triality

In order to explain the ideas on the ABJ triality in [**CMSY13**], let us start form the original proposal by Klebanov and Polyakov [**KP02**], where the minimal Vasiliev theory on AdS_4 is dual to the 3d $O(N)$ vector model. The Vasiliev theory includes totally symmetric tensor fields $\varphi_{\mu_1 \ldots \mu_s}$, which transform under the gauge transformation as

$$\varphi_{\mu_1 \ldots \mu_s} \sim \varphi_{\mu_1 \ldots \mu_s} + \partial_{(\mu_1} \xi_{\mu_2 \ldots \mu_s)} \,. \tag{2.1}$$

Here $\xi_{\mu_1 \ldots \mu_{s-1}}$ are gauge parameters and the parenthesis means symmetrization of the indices. In the minimally truncated case, the theory includes a gauge field for each even spin $s = 2, 4, 6, \ldots$. The dual theory is proposed to be 3d $O(N)$ vector model, which consists of N free bosons h_i $(i = 1, 2, \ldots, N)$ in the vector representation of the $O(N)$ global symmetry. We need to take the large N limit to relate to the classical theory of higher spin fields. We also assign an $O(N)$ singlet condition to the operators, which may then be given by bilinears of h_i such as $h_i h^i$.

The operators dual to the higher spin gauge fields are then constructed as

$$J_{\mu_1\ldots\mu_s} = h_i \partial_{(\mu_1} \cdots \partial_{\mu_s)} h^i + \cdots \tag{2.2}$$

by the action of derivatives.

In [**CMSY13**] they extended this duality in order to relate to superstring theory. There are several crucial points such as the introduction of supersymmetry, deformation from the free theory, and so on. Among them, here we would like to focus on the CP factor for the higher spin gauge fields. In the Vasiliev theory, it is not so difficult to extend the theory to include matrix valued fields. Field equations in the Vasiliev theory are given in terms of a non-commutative $*$-product. Now we require the fields to take $M \times M$ matrix values such as $[\varphi_{\mu_1\ldots\mu_s}]^\alpha{}_\beta$ with $\alpha, \beta = 1, 2, \ldots, M$. Due to the replacement, we need to change the $*$-multiplication by including also the multiplication in the matrix algebra. Even after the changes, we can use the same field equations since the $*$-product already has a non-abelian nature.

A version of the 4d extended Vasiliev theory is proposed in [**CMSY13**] to be dual to the ABJ theory, which is a 3d U(N) $\times$ U(M) Chern-Simons matter theory. The theory includes bi-fundamental matter, i.e. fields A_i^α, B_β^j in the bi-fundamental representation of U(N)$\times$U(M). We need to take a large N limit to get the relation to the classical higher spin theory. As in the case without CP factor, let us assign a U(N) invariant condition. Then we can construct higher spin currents in terms of bi-linears of bi-fundamentals as

$$[J_{\mu_1\ldots\mu_2}]^\alpha{}_\beta = A_i^\alpha \partial_{(\mu_1} \cdots \partial_{\mu_s)} B_\beta^i + \cdots . \tag{2.3}$$

In this way we can construct $M \times M$ matrix valued currents dual to $M \times M$ matrix valued higher spin fields. In order to have operators dual to string states, we have to also assign a U(M) invariant condition as well. The U(M) singlets are given by single trace operators in the form of tr ($ABAB \cdots AB$). Since the bi-linear AB is supposed to correspond to a single-particle state in higher spin theory, this implies that a generic string corresponds to a multi-particle state of higher spin fields in the singlet of the U(M) symmetry for the CP factor. This is a lesson we obtained from the ABJ triality in [**CMSY13**], and we shall utilize it in order to construct a lower dimensional analogue.

3. Higher spin AdS$_3$ holography with CP factor

We now try to find an AdS$_3$ version of the ABJ triality. This should relate higher spin theory on AdS$_3$, 2d CFT and superstring theory on AdS$_3$$\times$M$_7$. Here M$_7$ represents a 7d manifold. From the analysis of the ABJ triality, we should extend the 3d Vasiliev theory such that the theory includes $M \times M$ matrix valued fields, which was actually constructed in [**PV99**]. First we propose which model is the 2d CFT dual to the 3d Vasiliev theory with CP factors, and check the proposal. Then we will discuss the relation to superstring theory.

We consider the 3d Vasiliev theory with $M \times M$ matrix valued fields and also with $\mathcal{N} = 2$ supersymmetry. The theory includes massive matter fields along with higher spin gauge fields. The gauge algebra is a supersymmetric higher spin algebra denoted by shs$_M[\lambda]$, and the masses of the matter fields are also parametrized by the same parameter λ. The proposal is that the dual theory is given by the following coset [**CHR13a**]

$$\frac{\mathrm{su}(N+M)_k \oplus \mathrm{so}(2NM)_1}{\mathrm{su}(N)_{k+M} \oplus \mathrm{u}(1)_\kappa} \tag{3.1}$$

with $\kappa = NM(N+M)(N+M+k)$. In order to relate to the classical higher spin theory, we take a large N limit where $N, k \to \infty$ while we keep M finite as well as the 't Hooft parameter

$$\lambda = \frac{N}{N+M+k} . \tag{3.2}$$

This 't Hooft parameter is identified with λ appearing in the dual higher spin theory. Moreover, M is set to be the same as the size of CP factor. For $M = 2$ our duality reduces to the one in [**GG13a**] obtained independently.

There are several results that support our conjecture. First of all, we can see that the proposal is a natural extension of the previously known duality without CP factor. Indeed, the coset (3.1) with $M = 1$ reduces to the coset used in the duality of [**CHR12**], which is an $\mathcal{N} = 2$ supersymmetric extension of the original proposal in [**GG11**]. Moreover, in the limit of large level $k \to \infty$ (or $\lambda \to 0$ in terms of the 't Hooft parameter) the coset can be shown to reduce to a free system with bi-fundamentals. The group manifold $\mathrm{SU}(N+M)$ may be described by an $(N+M) \times (N \times M)$ matrix as

$$\begin{pmatrix} A & B \\ C & D \end{pmatrix} . \tag{3.3}$$

Ignoring the U(1) factor, A corresponds to the gauge factor $\mathrm{SU}(N)$ in the denominator of the coset (3.1), while D represents $\mathrm{SU}(M)$ symmetry which can be shown to decouple in the limit. The other blocks B, C transform as bi-fundamental representations under the $\mathrm{SU}(N) \times \mathrm{SU}(M)$ transformation. The limit of large level k corresponds to the small curvature limit of the coset manifold, and the bi-fundamentals become free boson fields. Therefore, we can apply the arguments for the ABJ triality in section 2. From the bilinears of the bi-fundamentals, we can construct higher spin currents of $M \times M$ matrix form, and we can see that they are dual to the higher spin gauge fields with $\mathrm{U}(M)$ CP factor at the parameter value $\lambda = 0$.

Even with generic M and λ, we have evidence for our conjecture. In [**CHR13a**] we have shown that the one-loop partition function of the higher spin theory can be reproduced by the 't Hooft limit of the coset (3.1), see also [**CV14**]. For the higher spin gauge theory, the one-loop partition function can be written in terms of a one-loop determinant, and the explicit expression may be found in [**CHR13a**] and references therein. For the dual coset (3.1) a state is labeled by $(\Lambda_{N+M}; \Lambda_N)$ in the 't Hooft limit, where Λ_L represents the highest weight for $\mathrm{SU}(L)$. In order to determine the spectrum of the theory, we need to specify how to take pairs of holomorphic and anti-holomorphic parts. Here we take a diagonal modular invariant as

$$\mathcal{H} = \bigoplus_{\Lambda_{N+M}, \Lambda_N} \mathcal{H}_{(\Lambda_{N+M};\Lambda_N)} \otimes \bar{\mathcal{H}}_{(\Lambda_{N+M};\Lambda_N)^*} \tag{3.4}$$

where the charge conjugated states are paired. Using the methods developed in [**GGHR11, CG13**], we can show the match of the partition functions in the 't Hooft limit once we assume the decoupling of so called "light states." We can also show that the symmetry algebra matches for first few spins explicitly.

We may now be able to say that the duality between the 3d Vasiliev theory with U(M) CP factor and the coset (3.1) is more or less concrete. Thus the next question would be how the duality relates to superstring theory. Let us first review the arguments by Gaberdiel and Gopakumar [**GG13a**, **GG14**]. They focused on the case with $M = 2$. In this case the coset model coincides with the Wolf space model, which is known to possess a large $\mathcal{N} = 4$ superconformal symmetry [**SSTVP88**, **VP89**, **ST90**].[1] With the large supersymmetry, we can identify the target space of superstring theory involved as $\mathrm{AdS}_3\times\mathrm{S}^3\times\mathrm{S}^3\times\mathrm{S}^1$. However, now the higher spin theory includes only 2×2 matrix valued fields and it is not obvious how to see the relation to string states since the situation is much different from the one in [**CMSY13**]. Recently, they examined their conjecture more closely when the radius of one of two S^3's becomes very large [**GG14**]. In this case the dual CFT has a small $\mathcal{N} = 4$ superconformal symmetry, and there is a considerable amount of literature on the duality between the 2d CFT and superstring theory.

However, if we want to apply the picture obtained in [**CMSY13**], it is better to keep M generic. Therefore, we utilize the coset (3.1) with generic M. As we saw in section 2, we need to assign a U(M) invariant condition to the CP factor of higher spin fields. Thus, instead of the coset (3.1), we consider the following coset as [**CHR13a**, **CV14**]

$$\frac{\mathrm{su}(N+M)_k \oplus \mathrm{so}(2NM)_1}{\mathrm{su}(N)_{k+M} \oplus \mathrm{su}(M)_{k+N} \oplus \mathrm{u}(1)_\kappa}, \tag{3.5}$$

which is a Kazama-Suzuki model with $\mathcal{N} = 2$ superconformal symmetry [**KS89b**, **KS89a**]. The target space of superstring theory dual to the coset is of the form $\mathrm{AdS}_3\times\mathrm{M}_7$, but the $\mathcal{N} = 2$ superconformal symmetry is not enough to determine M_7. Therefore, we cannot identify which superstring theory is involved in our triality. However, we noticed that the $\mathcal{N} = 2$ supersymmetry of the coset (3.5) is enhanced to $\mathcal{N} = 3$ at a specific value of the level $k = N + M$ [**CHR14**]. With the extended supersymmetry the candidates of M_7 are quite restricted, and it is expected that we can construct an AdS_3 version of the ABJ triality by making use of the critical level coset model.

4. Relations to superstring theory

Let us first look for a 3d Vasiliev theory with $\mathcal{N} = p > 2$ supersymmetry. It is known to be difficult to extend the 3d Vasiliev theory to have extended supersymmetry with generic parameter λ. However for $\lambda = 1/2$ the supersymmetry can be enhanced to a generic $\mathcal{N} = p > 2$ [**PV99**, **HLGPR12**]. At this value of the parameter, the matter become massless and conformally coupled to gravity, and we consistently truncate the field content by half. We consider the case with $\mathcal{N} = 2n+1$ $(n = 0, 1, 2, \ldots)$,[2] whose supersymmetry algebra $\mathrm{so}(2n + 1|2)$ is generated by

$$T_{\alpha\beta} = \{y_\alpha, y_\beta\}, \quad Q^I_\alpha = y_\alpha \otimes \phi^I, \quad M^{IJ} = [\phi^I, \phi^J]. \tag{4.1}$$

Here we have introduced two types of parameters y_α $(\alpha = 1, 2)$ and ϕ^I $(I = 1, 2, \ldots, 2n + 1)$ with the properties

$$[y_\alpha, y_\beta] = 2i\epsilon_{\alpha\beta}, \quad \{\phi^I, \phi^J\} = 2\delta^{IJ}. \tag{4.2}$$

[1] It was already suggested in [**HLGPR12**] to use the Wolf space model for the construction of a higher spin holography with extended supersymmetry.

[2] See [**CPV14**] for the cases with $\mathcal{N} = 2n$.

The twister variables y_α organize fields with higher spin in a neat way. Moreover, ϕ^I generates the Clifford algebra, which can be realized by $2^n \times 2^n$ matrices. The fields now depend on ϕ^I, or in other words they are associated with a U(2^n) CP factor.

The proposal here is that the higher spin theory with extended supersymmetry is dual to the coset (3.1) with $M = 2^{n-1}$ and a specific value of the level $k = N - M$ (after assuming some decoupling of free fermions) [**CHR14**]. We assume the following non-standard Hilbert space as

$$\mathcal{H} = \bigoplus_{\Lambda_{N+M}} \mathcal{H}_{\Lambda_{N+M}} \otimes \bar{\mathcal{H}}_{\Lambda^*_{N+M}}, \quad \mathcal{H}_{\Lambda_{N+M}} = \bigoplus_{\Lambda_N \in \Omega} (\Lambda_{N+M}; \Lambda_N). \tag{4.3}$$

At large N the label Λ_N can be represented by two Young diagrams $(\Lambda^l_N, \Lambda^r_N)$ and Ω means that the sum is taken over $\Lambda^l_N = (\Lambda^r_N)^t$. Here t represents the transpose of the Young diagram. For $n = 0$, the duality reduces to the one proposed in [**BCGG13**]. Developing the techniques used in [**BCGG13**], we can show that the partition function in the large N limit (with the assumption of the decoupling of light states again) reproduces the one from the dual classical higher spin theory. This means that the spectrum matches between the proposed dual theories.

In order to understand the meaning of the Hilbert space in (4.3), we move to another expression by making use of the level-rank duality in [**KS89b, NS97**]. We considered the coset (3.1) with $k = N - M$, but a level-rank dual expression is given by the coset with $k = N + M$. We should remark that the number of decoupled fermions is changed. See [**CHR14**] for more detailed explanation. In the level-rank dual expression, a su($N + M$) factor with the level $k = N + M$ appears in the numerator of the coset (3.1). A crucial point here is that the su$(N + M)_{N+M}$ factor has a realization in terms of free fermions in the adjoint representation of su($N+M$). With this realization, the Hilbert space is generated by the free fermions modulo the factors in the denominator of (3.1) and the repeated fusions of adjoint fermions yield the states in the representations of the form $\Lambda \in \Omega$ as discussed in [**BCGG13**]. Going back to the original form before applying the level-rank duality, the Hilbert space becomes the one in (4.3). The symmetry generators of the critical level coset can be constructed by the free fermions. For examples, spin one currents can be given by bi-linears of fermions, while spin 3/2 currents are written in terms of the product of three fermions. Explicit forms of these currents can be found in [**CHR14**].

In this way we found another type of duality between 3d higher spin theory and 2d coset type model. Let us try to figure out the superstring theory related to these dual models (assuming that it exists). From the lesson obtained in section 2, we have to deal with singlets in the sense of the CP factor of the higher spin fields. Here we set M to be a generic positive integer. Then, it is natural to think of the Kazama-Suzuki model (3.5) with the critical level $k = N + M$. The higher spin theory dual to the Kazama-Suzuki model includes fields of the $2M \times 2M$ matrix form, but with U(M) invariant condition assigned, see also [**CV14**]. One of the main results obtained in [**CHR14**] is that the critical level Kazama-Suzuki model has a $\mathcal{N} = 3$ superconformal symmetry. From the dual conformal symmetry, the target space of superstring theory is fixed to be of the form AdS$_3\times$M$_7$, as mentioned above. For the cases with $\mathcal{N} = 3$ superconformal symmetry, the known explicit examples are M$_7$ =(S$^3\times$S$^3\times$S^1)/$\mathbb{Z}_2$ in [**YIS99**] and M$_7$ =SO(3)/U(1) or

SO(5)/SO(3) in [**AGS00a**]. The BPS spectrum and marginal deformations are studied in [**AGS00b**] for the latter two models, and their result is consistent with those for our coset [**CHR14**]. Therefore, we may conjecture that superstring theory on AdS$_3\times$M$_7$ with M$_7$ =SO(3)/U(1) or SO(5)/SO(3) and our coset are dual to each other. In order to examine whether this conjecture is true or not, we need to investigate our proposed triality in further detail.

5. Conclusion

In this note we have reviewed our works on a lower dimensional analogue of ABJ triality in [**CMSY13**]. Extending the duality by Klebanov and Polyakov in [**KP02**], the authors in [**CMSY13**] proposed a triality between 4d extended Vasiliev theory, superstring theory and the ABJ theory. We have explained why the extension of Vasiliev theory with CP factor is important to see relations to superstring theory. Inspired by the work, we have extended the duality by Gaberdiel and Gopakumar in [**GG11**] such that the 3d extended Vasiliev theory with U(M) CP factor in [**PV99**] is involved. Our conjecture in [**CHR13a**] is that the dual theory is given by the coset model in (3.1). We gave several supporting arguments for the duality, for instance, by showing the match of one-loop partition functions. In order to see relations to superstring theory, we extend the duality to have more supersymmetry. In [**CHR14**] we proposed a duality between 3d Vasiliev theory with extended supersymmetry and the coset (3.1) at a critical level. Based on the duality we proposed that the Kazama-Suzuki model (3.5) with the critical level $k = N + M$ is dual to a superstring theory with the help of $\mathcal{N} = 3$ superconformal symmetry of the critical level model.

We have worked on a lower dimensional version since it is expected to allow to study the triality in more detail than the original ABJ triality. Indeed the 2d coset type models in (3.1) and (3.5) can be solved exactly, in principle. Moreover, the gauge sector of 3d Vasiliev theory is topological and dynamical degrees of freedom exist only in the matter sector. However, at least for our case, the supersymmetry is not so large to fix the dual superstring theory uniquely. We are currently working to make our conjecture more concrete, and we would like to report on our findings in the near future. Recently, we started to understand the nature of marginal deformations of the 2d coset models. Higher spin gauge theory should correspond to the tensionless limit of superstring theory, so we need to deform the coset model to compare with superstring theory at a typical point of the moduli space. The higher spin symmetry is generically broken by the marginal deformation of the critical level Kazama-Suzuki model and the mass of higher spin fields generated through the breaking of higher spin symmetry is computed [**HR15**, **CH15**].

References

[ABJ08] Ofer Aharony, Oren Bergman, and Daniel Louis Jafferis, *Fractional M2-branes*, J. High Energy Phys. **11** (2008), 043, 27, DOI 10.1088/1126-6708/2008/11/043. MR2460990 (2010c:81188)

[ABJM08] Ofer Aharony, Oren Bergman, Daniel Louis Jafferis, and Juan Maldacena, $\mathcal{N} = 6$ *superconformal Chern-Simons-matter theories, M2-branes and their gravity duals*, J. High Energy Phys. **10** (2008), 091, 38, DOI 10.1088/1126-6708/2008/10/091. MR2452954 (2010c:81228)

[AGS00a] Riccardo Argurio, Amit Giveon, and Assaf Shomer, *Superstring theory on* AdS$_3 \times$ G/H *and boundary* $\mathcal{N} = 3$ *superconformal symmetry*, J. High Energy Phys. **4** (2000), Paper 10, 22, DOI 10.1088/1126-6708/2000/04/010. MR1761100 (2002e:81153)

[AGS00b] Riccardo Argurio, Amit Giveon, and Assaf Shomer, *The spectrum of $\mathcal{N} = 3$ string theory on $AdS_3 \times G/H$*, JHEP **0012** (2000), 025.

[Ahn11] Changhyun Ahn, *The large N 't Hooft limit of coset minimal models*, J. High Energy Phys. **10** (2011), 125, 23, DOI 10.1007/JHEP10(2011)125. MR2909683

[BCG14] Matteo Beccaria, Constantin Candu, and Matthias R. Gaberdiel, *The large $\mathcal{N} = 4$ superconformal $\mathcal{W}_\infty$ algebra*, J. High Energy Phys. **6** (2014), 117, front matter+37. MR3234309

[BCGG13] Matteo Beccaria, Constantin Candu, Matthias R. Gaberdiel, and Michael Groher, *$\mathcal{N} = 1$ extension of minimal model holography*, J. High Energy Phys. **7** (2013), 174, front matter+28. MR3106125

[CFP11] A. Campoleoni, S. Fredenhagen, and S. Pfenninger, *Asymptotic $\mathcal{W}$-symmetries in three-dimensional higher-spin gauge theories*, J. High Energy Phys. **9** (2011), 113, 57, DOI 10.1007/JHEP09(2011)113. MR2889842

[CFPT10] A. Campoleoni, S. Fredenhagen, S. Pfenninger, and S. Theisen, *Asymptotic symmetries of three-dimensional gravity coupled to higher-spin fields*, J. High Energy Phys. **11** (2010), 007, 36, DOI 10.1007/JHEP11(2010)007. MR2795130 (2012c:83038)

[CG13] Constantin Candu and Matthias R. Gaberdiel, *Supersymmetric holography on* AdS_3, J. High Energy Phys. **9** (2013), 071, front matter+30, DOI 10.1007/JHEP09(2013)071. MR3102214

[CH15] Thomas Creutzig and Yasuaki Hikida, *Higgs phenomenon for higher spin fields on AdS_3*, arXiv:1506.04465.

[CHR12] Thomas Creutzig, Yasuaki Hikida, and Peter B. Rønne, *Higher spin* AdS_3 *supergravity and its dual CFT*, J. High Energy Phys. **2** (2012), 109, front matter+33. MR2996096

[CHR13a] Thomas Creutzig, Yasuaki Hikida, and Peter B. Rønne, *Extended higher spin holography and Grassmannian models*, JHEP **1311** (2013), 038.

[CHR13b] Thomas Creutzig, Yasuaki Hikida, and Peter B. Rønne, *$\mathcal{N} = 1$ supersymmetric higher spin holography on* AdS_3, J. High Energy Phys. **2** (2013), 019, front matter + 28. MR3046595

[CHR13c] Thomas Creutzig, Yasuaki Hikida, and Peter B. Rønne, *Three point functions in higher spin* AdS_3 *supergravity*, J. High Energy Phys. **1** (2013), 171, front matter + 47. MR3045478

[CHR14] Thomas Creutzig, Yasuaki Hikida, and Peter B. Rønne, *Higher spin AdS_3 holography with extended supersymmetry*, JHEP **1410** (2014), 163.

[CMSY13] Chi-Ming Chang, Shiraz Minwalla, Tarun Sharma, and Xi Yin, *ABJ triality: from higher spin fields to strings*, J. Phys. A **46** (2013), no. 21, 214009, 97, DOI 10.1088/1751-8113/46/21/214009. MR3062135

[CPV14] Constantin Candu, Cheng Peng, and Carl Vollenweider, *Extended supersymmetry in AdS_3 higher spin theories*, JHEP **1412** (2014), 113.

[CV14] Constantin Candu and Carl Vollenweider, *On the coset duals of extended higher spin theories*, JHEP **1404** (2014), 145.

[GG11] Matthias R. Gaberdiel and Rajesh Gopakumar, *An AdS_3 dual for minimal model CFTs*, Phys.Rev. **D83** (2011), 066007.

[GG13a] Matthias R. Gaberdiel and Rajesh Gopakumar, *Large $\mathcal{N} = 4$ holography*, JHEP **1309** (2013), 036.

[GG13b] Matthias R. Gaberdiel and Rajesh Gopakumar, *Minimal model holography*, J. Phys. A **46** (2013), no. 21, 214002, 47, DOI 10.1088/1751-8113/46/21/214002. MR3062128

[GG14] Matthias R. Gaberdiel and Rajesh Gopakumar, *Higher spins & strings*, J. High Energy Phys. **11** (2014), 044, front matter+42. MR3290804

[GGHR11] Matthias R. Gaberdiel, Rajesh Gopakumar, Thomas Hartman, and Suvrat Raju, *Partition functions of holographic minimal models*, J. High Energy Phys. **8** (2011), 077, 47, DOI 10.1007/JHEP08(2011)077. MR2876054

[GP14] Matthias R. Gaberdiel and Cheng Peng, *The symmetry of large $\mathcal{N} = 4$ holography*, JHEP **1405** (2014), 152.

[GV11] Matthias R. Gaberdiel and Carl Vollenweider, *Minimal model holography for* $\mathrm{SO}(2N)$, J. High Energy Phys. **8** (2011), 104, 20, DOI 10.1007/JHEP08(2011)104. MR2876039

[GY10] Simone Giombi and Xi Yin, *Higher spin gauge theory and holography: the three-point functions*, J. High Energy Phys. **9** (2010), 115, 80, DOI 10.1007/JHEP09(2010)115. MR2776932 (2012a:81177)

[GY11] Simone Giombi and Xi Yin, *Higher spins in AdS and twistorial holography*, J. High Energy Phys. **4** (2011), 086, 20, DOI 10.1007/JHEP04(2011)086. MR2833232 (2012g:81133)

[HLGPR12] Marc Henneaux, Gustavo Lucena Gómez, Jaesung Park, and Soo-Jong Rey, *Super-W_∞ asymptotic symmetry of higher-spin* AdS$_3$ *supergravity*, J. High Energy Phys. **6** (2012), 037, front matter+47. MR3006882

[HP13] Kentaro Hanaki and Cheng Peng, *Symmetries of holographic super-minimal models*, J. High Energy Phys. **8** (2013), 030, front matter+29. MR3106408

[HR10] Marc Henneaux and Soo-Jong Rey, *Nonlinear W_∞ as asymptotic symmetry of three-dimensional higher spin AdS gravity*, J. High Energy Phys. **12** (2010), 007, 20, DOI 10.1007/JHEP12(2010)007. MR2818488 (2012g:81138)

[HR15] Yasuaki Hikida and Peter B. Rønne, *Marginal deformations and the Higgs phenomenon in higher spin AdS$_3$ holography*, arXiv:1503.03870 [hep-th].

[KP02] I. R. Klebanov and A. M. Polyakov, *AdS dual of the critical* O(N) *vector model*, Phys. Lett. B **550** (2002), no. 3-4, 213–219, DOI 10.1016/S0370-2693(02)02980-5. MR1948547 (2003k:81137)

[KS89a] Yoichi Kazama and Hisao Suzuki, *Characterization of $\mathcal{N} = 2$ superconformal models generated by the coset space method*, Phys. Lett. B **216** (1989), no. 1-2, 112–116, DOI 10.1016/0370-2693(89)91378-6. MR978153 (90k:81215)

[KS89b] Yoichi Kazama and Hisao Suzuki, *New $\mathcal{N} = 2$ superconformal field theories and superstring compactification*, Nuclear Phys. B **321** (1989), no. 1, 232–268, DOI 10.1016/0550-3213(89)90250-2. MR1003762 (90g:81249)

[MZ13a] Juan Maldacena and Alexander Zhiboedov, *Constraining conformal field theories with a higher spin symmetry*, J. Phys. A **46** (2013), no. 21, 214011, 42, DOI 10.1088/1751-8113/46/21/214011. MR3062137

[MZ13b] Juan Maldacena and Alexander Zhiboedov, *Constraining conformal field theories with a slightly broken higher spin symmetry*, Classical Quantum Gravity **30** (2013), no. 10, 104003, 37, DOI 10.1088/0264-9381/30/10/104003. MR3055086

[MZ13c] Heidar Moradi and Konstantinos Zoubos, *Three-point functions in $\mathcal{N} = 2$ higher-spin holography*, J. High Energy Phys. **4** (2013), 018, front matter + 46. MR3065963

[NS97] Stephen G. Naculich and Howard J. Schnitzer, *Superconformal coset equivalence from level-rank duality*, Nuclear Phys. B **505** (1997), no. 3, 727–748, DOI 10.1016/S0550-3213(97)00579-8. MR1490787 (99g:81071)

[PV99] S. F. Prokushkin and M. A. Vasiliev, *Higher-spin gauge interactions for massive matter fields in 3D AdS space-time*, Nuclear Phys. B **545** (1999), no. 1-3, 385–433, DOI 10.1016/S0550-3213(98)00839-6. MR1697868 (2000d:81101)

[SS02] E. Sezgin and P. Sundell, *Massless higher spins and holography*, Nuclear Phys. B **644** (2002), no. 1-2, 303–370, DOI 10.1016/S0550-3213(02)00739-3. MR1935590 (2003h:81226)

[SSTVP88] Ph. Spindel, A. Sevrin, W. Troost, and A. Van Proeyen, *Extended supersymmetric σ-models on group manifolds. I. The complex structures*, Nuclear Phys. B **308** (1988), no. 2-3, 662–698, DOI 10.1016/0550-3213(88)90582-2. MR967938 (89m:53138)

[ST90] Alexander Sevrin and Georgios Theodoridis, *$\mathcal{N} = 4$ superconformal coset theories*, Nuclear Phys. B **332** (1990), no. 2, 380–390, DOI 10.1016/0550-3213(90)90100-R. MR1046356 (91j:81085)

[Vas03] M. A. Vasiliev, *Nonlinear equations for symmetric massless higher spin fields in* (A)dS$_d$, Phys. Lett. B **567** (2003), no. 1-2, 139–151, DOI 10.1016/S0370-2693(03)00872-4. MR2004217 (2004i:81175)

[VP89] Antoine Van Proeyen, *Realizations of $\mathcal{N} = 4$ superconformal algebras on Wolf spaces*, Class.Quant.Grav. **6** (1989), 1501.

[YIS99] Satoshi Yamaguchi, Yukitaka Ishimoto, and Katsuyuki Sugiyama, AdS$_3$/CFT$_2$ *correspondence and space-time $\mathcal{N} = 3$ superconformal algebra*, J. High Energy Phys. **2** (1999), Paper 26, 25 pp. (electronic), DOI 10.1088/1126-6708/1999/02/026. MR1694600 (2000g:81074)

Department of Mathematical and Statistical Sciences, 632 CAB, University of Alberta, Edmonton, Alberta T6G 2G1, Canada
E-mail address: creutzig@ualberta.ca

Department of Physics, Rikkyo University, 3-34-1 Nishi-Ikebukuro, Toshima, Tokyo 171-8501, Japan
E-mail address: hikida@rikkyo.ac.jp

University of Luxembourg, Mathematics Research Unit, FSTC, Campus Kirchberg, 6, rue Coudenhove-Kalergi, L-1359 Luxembourg-Kirchberg, Luxembourg
E-mail address: peter.roenne@uni.lu

Proceedings of Symposia in Pure Mathematics
Volume **93**, 2016
http://dx.doi.org/10.1090/pspum/093/01559

Humbert surfaces and the moduli of lattice polarized K3 surfaces

Charles F. Doran, Andrew Harder, Hossein Movasati, and Ursula Whitcher

ABSTRACT. In this article we introduce a collection of partial differential equations in the moduli of lattice polarized K3 surfaces whose algebraic solutions are the loci of K3 surfaces with lattice polarizations of higher rank. In the special case of rank 17 polarization such loci encode the well-known Humbert surfaces. The differential equations treated in the present article are directly derived from the Gauss-Manin connection of families of lattice polarized K3 surfaces. We also introduce some techniques to calculate the Gauss-Manin connection with the presence of isolated singularities.

1. Introduction

Elliptic curves have two different types of analogue in the realm of surfaces: abelian surfaces and K3 surfaces. An elliptic curve has a unique nonvanishing holomorphic one-form (up to an overall scaling), and Abelian surfaces and K3 surfaces have a unique nonvanishing holomorphic two-form. Unlike elliptic curves or abelian surfaces, K3 surfaces are simply connected.

The second cohomology $H^2(X,\mathbb{Z})$ of a K3 surface X is equipped with a lattice structure by means of the cup product. This lattice is isomorphic to

$$H \oplus H \oplus H \oplus E_8 \oplus E_8.$$

Here, H is the standard hyperbolic lattice, and E_8 is the unique even, unimodular, and negative definite lattice of rank 8.

By the Lefschetz $(1,1)$ theorem, the elements of the lattice $\mathrm{NS}(X) := H^2(X,\mathbb{Z}) \cap H^{1,1}$ are Poincaré dual to the Néron-Severi group of divisors in X. For an algebraic K3 surface, the signature of $\mathrm{NS}(X)$ must be $(1,a)$ for some $a \leq 19$; the *Picard rank* of the surface is $1+a$. Let L be an even non-degenerate lattice of signature $(1, 19-m)$, $m \geq 0$. A lattice polarization on the K3 surface X is given by a primitive embedding

$$i : L \hookrightarrow \mathrm{NS(X)}.$$

whose image contains a pseudo-ample class, that is, a numerically effective class with positive self-intersection. We denote an L-polarized K3 surface by (X,i), or simply by X if the choice of polarization is clear. One may define lattice-polarized abelian surfaces in a similar fashion.

2010 *Mathematics Subject Classification.* Primary 14J28, 14J15, 14G35.
The authors thank Jacob Lewis for many interesting conversations.

Let M be the rank-18 lattice $H \oplus E_8 \oplus E_8$, and let N be the lattice $H \oplus E_8 \oplus E_7$. In [**CD07**] and [**CD12**], Clingher and the first author investigated a correspondence between abelian surfaces and K3 surfaces polarized by M or N. The correspondence is clear for Hodge-theoretic reasons, and can be realized in an explicit geometric fashion; M-polarized K3 surfaces are in one-to-one correspondence with products of elliptic curves, while N-polarized K3 surfaces correspond to arbitrary abelian surfaces.

We study families of M- and N-polarized K3 surfaces. We say that a family $\mathcal{X} \to T$ of surfaces over an quasi-projective variety T is polarized by a lattice L if its fibers X_t are smooth and equipped with an L-polarization which varies continuously with t. We further assume that for a general fiber the image of the polarization is precisely $\mathrm{NS}(X)$. The locus of points in T such that the lattice polarization i is not surjective is a (typically infinite) union of irreducible varieties of codimension one. In this article we characterize such a locus using partial differential equations in T. Our first example is the case $X = E_1 \times E_2$, where E_i, $i = 1, 2$ are two elliptic curves. In this case a third order differential equation in the j-invariants of E_i, $i = 1, 2$ is tangent to modular curves; for treatments of this problem using the corresponding K3 surfaces, see [**CDLW09**] and §3.

The main results of our work are as follows. In Theorem 2.3, we characterize the failure of surjectivity for N-polarized K3 surfaces. We compute the Picard-Fuchs system of partial differential equations for N-polarized K3 surfaces. In the N-polarized case, surjectivity of the polarization fails on regions of the moduli space corresponding to Humbert surfaces. We illustrate the corresponding simplification of the Picard-Fuchs system for the Humbert surface $\mathcal{H}_5$. We then compute the Gauss-Manin connection, a full system of differential equations annihilating the periods. In order to do so, we develop techniques for computing the Gauss-Manin connection in the presence of isolated singularities.

The text is organized in the following way. In §2, we describe the relationship between a lattice polarization and the number of Picard-Fuchs equations making up the Gauss-Manin connection. As an example, we treat the classical case of a product of two elliptic curves in §3. In § 4, we recall the properties of the families of M- and N-polarized K3 surfaces studied in [**CD07**], [**CDLW09**], and [**CD12**]; the surfaces are realized as singular hypersurfaces in projective space. We review the Griffiths-Dwork method for computing Picard-Fuchs equations in § 5. The method fails to compute the full Gauss-Manin connection for the singular hypersurface realization of N-polarized K3 surfaces. However, we are able to extract Picard-Fuchs equations in different coordinate systems. As an example, we describe the simplification of the Picard-Fuchs system on a Humbert surface. In § 6, we describe algorithms for calculating the full Gauss-Manin connection of the M- and N-polarized K3 surface families. This system of differential equations which annihilates the period ω is canonical, in the sense that any other differential operator which annihilates ω is in the left ideal generated by those described in § 6.10. In particular, this is the case for the Picard-Fuchs equations computed by the Griffiths-Dwork method in § 5. The techniques of § 6 are derived from [**Mov11**]. The main technical innovation is the calculation of a Gauss-Manin connection with the presence of isolated singularities. The result of the computations discussed in this section can be obtained from the third author's webpage (see [**Mov**]).

2. Lattice polarizations and the Gauss-Manin connection

Let $\mathcal{X} \to T$ be a family of algebraic surfaces polarized by a lattice L of signature $(1, b-m-3)$, where b is the second Betti number of the surfaces and T is an affine variety. Throughout the text, we work with the function ring R of T. For simplicity, we may take R to be any localization of the polynomial ring $\mathbb{Q}[a, b, c, \cdots]$. We denote by k the field of fractions of R; we may consider a family $\mathcal{X} \to T$ as a family of L-polarized surfaces over k. For an analytic space X and $x \in X$, (X, x) denotes a small neighborhood of x in X and $\mathcal{O}_{(X,x)}$ is the ring of germs of holomorphic functions in a neighborhood of x in X.

The relative algebraic de Rham cohomology $H^2_{\mathrm{dR}}(\mathcal{X}/T)$ (see [**Gro66**]) is a free R-module of finite rank. It carries the *Gauss-Manin connection* $\nabla : H^2_{\mathrm{dR}}(\mathcal{X}/T) \to \Omega^1_T \otimes_{\mathsf{R}} H^2_{\mathrm{dR}}(\mathcal{X}/T)$, where Ω^1_T is the R-module of differential forms in R (cf. [**KO68**]). The elements in the image of the polarization $i : L \to H^2_{\mathrm{dR}}(\mathcal{X}/T)$ are constant sections of the connection, that is $\nabla(\alpha) = 0$ for any α in the image of i, so we have the induced connection

$$\nabla : H^2_{\mathrm{dR}}(\mathcal{X}/T)_i \to \Omega^1_T \otimes_{\mathsf{R}} H^2_{\mathrm{dR}}(\mathcal{X}/T)_i$$

which we denote again by ∇. Here $H^2_{\mathrm{dR}}(\mathcal{X}/T)_i = H^2_{\mathrm{dR}}(\mathcal{X}/T)/i(L)$. For any algebraic vector field v in T, we have $\nabla_v : H^2_{\mathrm{dR}}(X)_i \to H^2_{\mathrm{dR}}(X)_i$, so we can talk about the iteration $\nabla^i_v = \nabla_v \circ \nabla_v \circ \cdots \circ \nabla_v$, i-times.

In the topological context, the above algebraic connection can be viewed in the following way. The elements of $H^2_{\mathrm{dR}}(\mathcal{X}/T)_i$ are global sections of the cohomology bundle $\mathcal{H} := \bigcup_{t \in T} H^2(X_t, \mathbb{C})$, and the constant sections of ∇ are given by $\mathbb{C}$-linear combinations of sections with values in $\bigcup_{t \in T} H^2(X_t, \mathbb{Z})$. In this way, we can talk about ∇_v for any local analytic vector field v defined on some open set U in T. The connection ∇_v acts on holomorphic sections of $\mathcal{H}$ over U. We sometimes take a local holomorphic map $t : (\mathbb{C}^n, 0) \to T$ with $t_0 := t(0)$. In this case we denote by $(u_1, u_2, \dots, u_n)$ a coordinate system in $(\mathbb{C}^n, 0)$ and by $\frac{\partial}{\partial u_i}$ the corresponding vector fields. The notation $\nabla_{\frac{\partial}{\partial u_i}}$ refers to the pull-back of the connection ∇ to $(\mathbb{C}^n, 0)$ and then its composition with the vector field $\frac{\partial}{\partial u_i}$. If the image of t is a subset of a subvariety S of T, then, by abuse of notation, we say that the local vector fields $\frac{\partial}{\partial u_i}$ are tangent to S around t_0.

We wish to use ∇ to characterize loci of T where the polarization $i : L \to \mathrm{NS}(X_t)$ is not surjective. Let ω be a meromorphic differential 2-form on $\mathcal{X}$ which restricted to X_t, $t \in T$, is holomorphic everywhere. The restriction gives us an element in $H^2_{\mathrm{dR}}(X)_i$ which we denote again by ω.

THEOREM 2.1. *Let $S \subset T$ be an algebraic subset of codimension one in T. If for all $t_0 \in S$ the polarization $i : L \to \mathrm{NS}(X_{t_0})$ is not surjective, then for all $t_0 \in S$ and any local vector field $\frac{\partial}{\partial u}$ tangent to S in a neighborhood of t_0, the $\mathbb{C}$-vector space generated by*

$$\nabla^i_{\frac{\partial}{\partial u}} \omega, \ i = 0, 1, 2, \dots,$$

in $H^2_{\mathrm{dR}}(X_{t_0})_i$ has dimension strictly less than $m+2$.

PROOF. Let Γ be the Poincaré dual of the image of L in $H^2_{\mathrm{dR}}(X_{t_0})$. Then Γ is a primitive sublattice of $H_2(X_{t_0}, \mathbb{Z})$. We define

$$H_2(X_{t_0}, \mathbb{Z})_i := H_2(X_{t_0}, \mathbb{Z})/\Gamma.$$

The hypothesis of the theorem implies that there is a continuous family of cycles $0 \neq \delta_s \in H_2(X_s, \mathbb{Z})_i$, $s \in (S, t_0)$ such that the integral $\int_{\delta_s} \omega$ is identically zero. If $\frac{\partial}{\partial u}$ is a local vector field tangent to (S, t_0) then we have

$$0 = \frac{\partial^i}{\partial u^i} \int_{\delta_{t(u)}} \omega = \int_{\delta_{t(u)}} \nabla^i_{\frac{\partial}{\partial u}} \omega, \ i = 0, 1, \ldots$$

Thus, the integration of all $\nabla^i_{\frac{\partial}{\partial u}} \omega$, $i = 0, 1, 2, \ldots$ over $\delta_{t(u)}$ is identically zero, so they cannot generate the whole $\mathbb{C}$-vector space $H^2_{\mathrm{dR}}(X_{t(u)})_i$. Note that $t(0) = t_0$. □

2.1. The differential rank-jump property. It is natural to ask whether the converse of Theorem 2.1 holds: can we use the dimension of the vector space $\nabla^i_{\frac{\partial}{\partial u}} \omega$ to detect whether a lattice polarization is surjective?

DEFINITION 2.2. Let $\mathcal{X} \to T$ be a family of algebraic surfaces polarized by a lattice L of signature $(1, b - m - 3)$, where b is the second Betti number of the surfaces. Suppose $\mathcal{X}$ has the property that if $S \subset T$ is an algebraic subset of codimension one such that for all $t_0 \in S$ and any local vector field $\frac{\partial}{\partial u}$ tangent to S in a neighborhood of t_0, the $\mathbb{C}$-vector space generated by

$$\nabla^i_{\frac{\partial}{\partial u}} \omega, \ i = 0, 1, 2, \ldots,$$

in $H^2_{\mathrm{dR}}(X_{t_0})_i$ has dimension strictly less than $m + 2$, then the polarization $i : L \to \mathrm{NS}(X_{t_0})$ is not surjective for all $t_0 \in S$. Then we say $\mathcal{X}$ has the *differential rank-jump* property.

THEOREM 2.3. *Let $\mathcal{X} \to T$ be a family of rank 17 N-polarized K3 surfaces which has open image in the moduli space of N-polarized K3 surfaces. Then $\mathcal{X}$ has the differential rank-jump property.*

We prove Theorem 2.3 in §4.2. We now describe conditions that will guarantee a family $\mathcal{X} \to T$ has the differential rank-jump property. Let us take a codimension one irreducible subvariety S of T and assume that for any local vector field $\frac{\partial}{\partial u}$ tangent to (S, t_0), the $\mathbb{C}$-vector space generated by

$$\nabla^i_{\frac{\partial}{\partial u}} \omega, \ i = 0, 1, 2, \ldots,$$

in $H^2_{\mathrm{dR}}(X_{t_0})_i$ has dimension strictly less than $m + 2$. By the same argument as in the proof of Corollary 2.6, for any collection of local vector fields $\frac{\partial}{\partial u_i}$, $i = 1, 2, \cdots, n$ all tangent to (S, t_0), the $\mathbb{C}$-vector space generated by the forms described in Equation 2.2 in $H^2_{\mathrm{dR}}(X_{t_0})_i$ is also of dimension strictly less than $m + 2$.

Let $n = \dim_{\mathbb{C}} T$ and $(u_1, u_2, \ldots, u_{n-1})$ be a coordinate system around a smooth point t of S. Let us choose a basis $\tilde{\Omega} := [\omega_1, \omega_2, \ldots, \omega_r]^{\mathrm{t}}$, $r < m + 2$, $\omega_1 = \omega$ for the $\mathcal{O}_{(S,t_0)}$-module generated by the forms given in Equation 2.2 and write the Gauss-Manin connection in this basis:

$$\nabla \tilde{\Omega} = A \otimes_{\mathcal{O}_{(S,t)}} \tilde{\Omega},$$

where A is a matrix with entries which are differential 1-forms in (S, t_0). A fundamental system of solutions for this system is given by the integration of $\tilde{\Omega}$ over n linearly independent continuous family of cycles with $\mathbb{C}$ coefficients. These are $\mathbb{C}$-linear combination of continuous family of cycles in $H_2(X_t, \mathbb{Z})_i$. It follows that the $\mathbb{C}$-vector space spanned by the periods $\int_{\delta_s} \omega$, $\delta_s \in H_2(X_s, \mathbb{Z})_i$, $s \in (S, t_0)$ has dimension strictly less than $m+2$ and so there are constants $a_i \in \mathbb{C}$, $i = 1, \ldots, m+2$ such that

$$\int_{\sum a_i\delta_{i,s}} \omega = \sum a_i \int_{\delta_{i,s}} \omega = 0, \quad s \in (S, t_0).$$

where $\delta_{i,s} \in H_2(X_s, \mathbb{Z})_i$, $i = 1, 2, \ldots, m+2$, $s \in (S, t_0)$ is a continuous family of cycles which form a basis for the $\mathbb{Z}$-module $H_2(X_s, \mathbb{Z})_i$. Let $\delta_s := \sum a_i \delta_{i,s}$.

DEFINITION 2.4. Let $\mathcal{X} \to T$ be a proper smooth family of algebraic K3 surfaces or abelian surfaces. We say that the family is *perfect* if the following condition is satisfied: For a continuous family of cycles $\delta_s \in H_2(X_s, \mathbb{C})$, $s \in (T, t_0)$, if the locus of parameters s such that $\int_{\delta_s} \omega = 0$ is a part of a codimension one algebraic set, then δ_s up to multiplication by a constant is in $H_2(X_s, \mathbb{Z})$.

Perfect families have the differential rank-jump property.

2.2. The period domain and perfect families. Let $V_{\mathbb{Z}}$ be a free $\mathbb{Z}$-module of rank $m+2$, and let $\psi_{\mathbb{Z}} : V_{\mathbb{Z}} \times V_{\mathbb{Z}} \to \mathbb{Z}$ be a non-degenerate (not necessarily unimodular) symmetric bilinear form on $V_{\mathbb{Z}}$. The *period domain* determined by $\psi_{\mathbb{Z}}$ is

$$D := \mathbb{P}(\{\omega \in V_{\mathbb{C}} \mid \psi_{\mathbb{C}}(\omega, \omega) = 0, \ \psi_{\mathbb{C}}(\omega, \bar{\omega}) > 0\})$$

The group $\Gamma_{\mathbb{Z}} := \mathrm{Aut}(V_{\mathbb{Z}}, \psi_{\mathbb{Z}})$ acts from the left on D. The quotient $\Gamma\backslash D$ is the moduli of polarized Hodge structures of type $1, m, 1$ with the polarization $\psi_{\mathbb{Z}}$.

Let (X, i) be an L-polarized surface. Because L is non-degenerate, $i(L) \oplus i(L)^{\perp}$ has finite index in $H^2(X, \mathbb{Z})$. The restriction of the cup product to $i(L)^{\perp}$ determines a non-degenerate lattice of signature $(2, m)$. Let $\psi_{\mathbb{Z}}$ be the associated bilinear form. The holomorphic 2-form $\omega \in H^2_{\mathrm{dR}}(X) = H^2(X, \mathbb{C})$ lies in $i(L)^{\perp} \otimes \mathbb{C}$. Therefore, we have a *period map*

$$p : \mathcal{M} \to \Gamma^0_{\mathbb{Z}}\backslash D$$

where $\mathcal{M}$ is the coarse moduli space of pseudo-ample L-polarized surfaces. Here $\Gamma^0_{\mathbb{Z}}$ is the finite index subgroup of $\Gamma_{\mathbb{Z}}$ which acts trivially on the discriminant group of $V_{\mathbb{Z}}$. For K3 surfaces with pseudo-ample polarizations, the Torelli problem is true and p is a biholomorphism of analytic spaces (see [**Dol96**]).

We need explicit affine coordinates on D. Let $\delta_1, \delta_2, \ldots, \delta_{m+2}$ be a basis of the $\mathbb{Z}$-module $V_{\mathbb{Z}}$, and let $\Psi_0 = [\psi_{\mathbb{Z}}(\delta_i, \delta_j)]$. For $\omega \in V_{\mathbb{C}}$, let

$$\omega = \sum_{i=1}^{m+2} x_i \delta_i, \ x_i \in \mathbb{C}.$$

We have

$$\psi_{\mathbb{C}}(\omega, \omega) = x\Psi_0 x^{\mathrm{t}}, \ \psi_{\mathbb{C}}(\omega, \bar{\omega}) = x\Psi_0 \bar{x}^{\mathrm{t}} \text{ where } x = [x_1, x_2, \ldots, x_{m+2}],$$

and so

$$D = \{[x] \in \mathbb{P}^{m+1} \mid x\Psi_0 x^{\mathrm{t}} = 0, \ x\Psi_0 \bar{x}^{\mathrm{t}} > 0 \}.$$

If we view $[x] \in \mathbb{P}^{m+1}$ as a $(m+2) \times 1$ matrix, then the group

$$\Gamma_{\mathbb{Z}} := \{A \in \mathrm{GL}(m+2, \mathbb{Z}) \mid A^{\mathrm{t}} \Psi_0 A = \Psi_0\}$$

acts on D from the left by matrix multiplication.

We say L-polarized surfaces have a *perfect moduli space* if the following condition is satisfied:

CONDITION 2.5. *Let* $c = [c_1 : c_2 : \ldots : c_{m+2}] \in \mathbb{P}^{m+1}_{\mathbb{C}}$. *The set* $\{x \in D \mid \sum_{i=1}^{m+2} c_i x_i = 0\}$ *induces an analytic subvariety in* $\Gamma_{\mathbb{Z}} \backslash D$ *if and only if* $c \in \mathbb{P}^{m+1}_{\mathbb{Q}}$.

Families with perfect moduli spaces satisfy the differential rank-drop property. We prove that N-polarized K3 surfaces of rank 17 have perfect moduli spaces in §4.3.

2.3. Rank jumps and partial differential equations. Let us choose a basis ω_i, $i = 1, 2, \ldots, m+2$, $\omega_1 = \omega$ for the R-module $H^2_{\mathrm{dR}}(\mathcal{X}/T)_i$. The Gauss-Manin connection matrix A with entries in Ω^1_T is determined uniquely by the equality

$$\nabla(\Omega) = A \otimes_{\mathsf{R}} \Omega,\ \Omega := [\omega_1, \omega_2, \ldots, \omega_{m+2}]^{\mathsf{t}}.$$

(See § 5, § 6, and [**Mov11**] for techniques to compute A.) From now on we assume that T is of dimension m. Let us consider a local holomorphic map $t : (\mathbb{C}^{m-1}, 0) \to (T, t_0)$. We denote by $(u_1, u_2, u_3, \ldots, u_{m-1})$ the coordinate system in $(\mathbb{C}^{m-1}, 0)$ and by $\frac{\partial}{\partial u_i}$, $i = 1, 2, \ldots, m-1$ the corresponding local vector fields. For simplicity we write

$$\alpha_{u_i} := \nabla_{\frac{\partial}{\partial u_i}} \alpha,\ \alpha \in H^2_{\mathrm{dR}}(\mathcal{X}/T)_i,$$

$$f_{u_i} := \frac{\partial f}{\partial u_i},\ f \in \mathsf{k}.$$

For any pair of words x, y in the variables u_i, let $R^{x,y}$ be the $(m+2) \times (m+2)$ matrix satisfying

$$\begin{pmatrix} \omega \\ \omega_{u_1} \\ \vdots \\ \omega_{u_{m-1}} \\ \omega_y \\ \omega_x \end{pmatrix} = R^{x,y} \begin{pmatrix} \omega_1 \\ \omega_2 \\ \vdots \\ \omega_{m+2} \end{pmatrix}.$$

Let us analyze the matrix $R^{x,y}$ in more detail. We take a collection x_i, $i = 1, 2, \ldots, n$, of n regular functions on T forming a quasi-affine coordinate system on T, that is, the map $(x_1, x_2, \cdots, x_n) : T \to \mathbb{C}^n$ is an embedding of T as a quasi-affine subvariety of $\mathbb{C}^n$. Any regular function on T can be written as a polynomial in $x_1, \ldots, x_n$ and so the R-module Ω^1_T is generated by dx_i, $i = 1, 2, 3, \ldots, n$. This implies that the Gauss-Manin connection matrix A can be written as

$$A = \sum_{i=1}^{m} A_i dx_i,\ A_i \in \mathrm{Mat}(m+2, \mathsf{R}).$$

Since the Gauss-Manin connection is integrable we have $dA = A \wedge A$ and so

$$\frac{\partial A_i}{\partial x_j} + A_i A_j = \frac{\partial A_j}{\partial x_i} + A_j A_i.$$

Let

$$
\begin{aligned}
A^{u_1} &:= \sum_{i=1}^{n} A_i x_{i,u_1} \\
A^{u_1u_2} &:= A^{u_1}A^{u_2} + (A^{u_1})_{u_2} \\
&= (\sum_{i=1}^{n} A_i x_{i,u_1})(\sum_{i=1}^{n} A_i x_{i,u_2}) + \sum_{i=1}^{n}((A_i)_{u_2}x_{i,u_1} + A_i x_{i,u_1u_2}) \\
&= \sum_{i=1}^{n}\sum_{j=1}^{n}(A_iA_j + \frac{\partial A_i}{\partial x_j})x_{i,u_1}x_{j,u_2} + \sum_{i=1}^{n} A_i x_{i,u_1u_2} \\
A^{u_1u_2u_3} &:= A^{u_1u_2}A^{u_3} + (A^{u_1u_2})_{u_3} = \cdots
\end{aligned}
$$

and so on. These are uniquely determined by the equalities $\Omega_{u_1} = A^{u_1}\Omega$, $\Omega_{u_1u_2} = A^{u_1u_2}\Omega$ and $\Omega_{u_1u_2u_3} = A^{u_1u_2u_3}\Omega$. Now, the first row of $R^{x,y}$ is just $[1, 0, 0, \ldots, 0]$ and the i-th row, $2 \leq i \leq m$, of $R^{x,y}$ is the first row of the matrix A^{u_i}. The $(m+1)$-th and $(m+2)$-th row of $R^{x,y}$ are respectively the first row of A^y and A^x.

COROLLARY 2.6. *Let $\mathcal{X} \to T$ be a family of L-polarized algebraic surfaces with the differential rank-jump property, and let $S \subset T$ be an algebraic subset of codimension one in T. Then for all $t_0 \in S$ the polarization $i : L \to \mathrm{NS}(X_{t_0})$ is not surjective if and only if for x and $t : (\mathbb{C}^{m-1}, 0) \to (S, t_0)$ as above we have the following collection of partial differential equations:*

$$
\det(R^{x,y}) = 0, \ \forall x \textit{ of length 2, } y \textit{ of length 2 or 3} \tag{2.1}
$$

For the proof of the reverse direction of the above corollary, we need the following lemma.

LEMMA 2.7. *Let $t : (\mathbb{C}^m, 0) \to (T, t_0)$ be a coordinates system around a point $t_0 \in T$. The differential forms ω, ω_{u_i}, $i = 1, 2, \ldots, m$ are linearly independent for a generic point in the image of t.*

PROOF. If the assertion is not true, then we have a meromorphic vector field V in $(\mathbb{C}^m, 0)$ such that $\nabla_V\omega = \omega$. Let $\gamma(s) \in (\mathbb{C}^m, 0)$, $s \in (\mathbb{C}, 0)$ be a solution of V. Integrating $\nabla_V\omega = \omega$ over topological two-cycles of $X_{\gamma(s)}$, we conclude that the periods $\int_{\delta_{\gamma(s)}} \omega$ are all of the form $c_\delta e^s$, where c_δ is a constant depending only on δ. We conclude that $\frac{\omega}{e^s}$ has constant periods which is in contradiction with the local Torelli problem for K3 surfaces and the fact that t is a coordinates system. □

PROOF OF COROLLARY 2.6. Let us first prove the forward direction. Note that if $\frac{\partial}{\partial u_i}$, $i = 1, 2, \ldots, k$, is a collection of local vector fields tangent to S then the $\mathbb{C}$-vector space generated by

$$
\omega_x, \quad \text{where } x \text{ is any word in } u_i \ i = 1, 2, \ldots, k \tag{2.2}
$$

in $H^2_{\mathrm{dR}}(X_{t(u)})_i$ is also of dimension strictly less than $m+2$. This follows by taking the vector field $\sum_{i=1}^{k} a_i \frac{\partial}{\partial u_i}$, where a_i's are unknown constants, and applying Theorem 2.1. Take $k = m-1$. We conclude that $\omega, \omega_{u_1}, \cdots, \omega_{u_{m-1}}, \omega_y, \omega_x$ are linearly independent in $H^2_{\mathrm{dR}}(X_{t(u)})_i$, whereas $\omega_1, \omega_2, \ldots, \omega_{m+2}$ form a basis for $H^2_{\mathrm{dR}}(X_{t(u)})_i$. It follows that the determinant of the matrix $R^{x,y}$ is identically zero.

Now, let us prove the reverse direction. Let V be the vector space generated by ω, ω_i, $i = 1, 2, \ldots, m-1$. By Lemma 2.7 V is of dimension m. If all the second derivatives $\omega_{u_i u_j}$ are in V then by further derivations of the corresponding equalities we conclude that V is closed under all derivations and so for all $t_0 \in S$ and any local vector field $\frac{\partial}{\partial u}$ tangent to S in a neighborhood of t_0, the $\mathbb{C}$-vector space generated by $\nabla^i_{\frac{\partial}{\partial u}}\omega$, $i = 0, 1, 2, \ldots$, in $H^2_{\rm dR}(X_{t_0})_i$ has dimension strictly less than $m+2$. Using the converse of Theorem 2.1, the proof is finished. In a similar way, if at least one of the second derivatives, say $\omega_{u_1u_1}$ is not in V, then by our hypothesis (Equation 2.1), all other second derivatives and third derivatives are in the vector space generated by V and $\omega_{u_1u_1}$. Taking further derivatives of the corresponding equalities, we obtain the hypothesis of the converse of Theorem 2.1. □

2.4. A remark on the number of partial differential equations in Corollary 2.6. Assume that $\mathcal{X} \to T$ has the differential rank-jump property and $\omega_{u_1u_1}$ is linearly independent with ω, ω_{u_i}, $i = 1, 2, \ldots, m-1$. Then in Corollary 2.6 we can reduce the number of partial differential equations to

$$\det(R^{x,u_1u_1}) = 0,\ x = u_1u_i, 2 \le i \le m-1,\ u_iu_j,\ 2 \le i \le j \le m-1, \tag{2.3}$$

$$\det(R^{x,u_1u_1}) = 0,\ x = u_1u_1u_i,\ 1 \le i \le m-1. \tag{2.4}$$

Note that we have in total $\frac{(m-1)(m+2)}{2} - 1$ partial differential equations. Of these equations, $m-1$ are third-order and the rest are second order. Let us take a coordinate system $(u_1, u_2, \ldots, u_{m-1})$ around t_0 for S such that $u_1 = u$. From the partial differential equations (Equations 2.1 and 2.4) it follows that the $\mathbb{C}$-vector space generated by $\omega, \omega_{u_1}, \cdots, \omega_{u_{m-1}}, \omega_{u_1u_1}, \omega_x$ has dimension less than $m+2$; by our hypothesis, the first $m+1$ differential forms are linear independent. We conclude that ω_x is in the vector space V generated by $\omega, \omega_{u_1}, \cdots, \omega_{u_{m-1}}, \omega_{u_1u_1}$. Therefore, the vector space V is closed under all derivations $\frac{\partial}{\partial u_i}$. In particular, the $\mathbb{C}$-vector space generated by $\nabla^i_{\frac{\partial}{\partial u}}\omega$, $i = 0, 1, 2, \ldots$, in $H^2_{\rm dR}(X_{t_0})_i$ has dimension strictly less than $m+2$.

3. Product of two elliptic curves

As an example, we consider the Gauss-Manin connection of the family of abelian surfaces which are products of elliptic curves. Let $X = E_1 \times E_2$ be a product of two elliptic curves and let the polarization be given by the divisor $E_1 \times \{p_2\} + \{p_1\} \times E_2$. In this case $m = 2$ and we consider each E_k parametrized by the classical j-invariant $j_k \in \mathbb{P}^1$. In Weierstrass coordinates we have:

$$E_k\colon y^2 + xy - x^3 + \frac{36}{j_k - 1728}x + \frac{1}{j_k - 1728} = 0,\ j_k \neq 0, 1728,\ k = 1, 2$$

We consider $\mathbb{P}^1 = \mathbb{C} \cup \{\infty\}$ as the compactification of the moduli of elliptic curves with a cusp ∞. Therefore, $\mathbb{P}^1 \times \mathbb{P}^1$ is the compactification of the moduli of pairs of elliptic curves.

3.1. Gauss-Manin connection. Let f be the defining polynomial of $E = E_k$, $k = 1, 2$. We calculate the Gauss-Manin connection of E in the basis $[\alpha_k, \omega_k]^{\rm t} = [\frac{dx\wedge dy}{df}, \frac{dx\wedge dy}{df}]^{\rm t}$

$$\nabla_{\frac{\partial}{\partial j}}\begin{pmatrix}\alpha_k\\ \omega_k\end{pmatrix} = A \cdot \begin{pmatrix}\alpha_k\\ \omega_k\end{pmatrix}$$

where

$$A = \frac{1}{j(j-1728)} \begin{pmatrix} -432 & -60 \\ -(j-1728) & 432 \end{pmatrix}.$$

Let $A_k,\ k = 1,2$ be two copies of A corresponding to $j_k,\ k = 1,2$. We have

$$H^2_{\mathrm{dR}}(E_1 \times E_2)_i = H^1_{\mathrm{dR}}(E_1) \otimes_{\mathbb{C}} H^1_{\mathrm{dR}}(E_2)$$

for which we choose the basis $[\tilde{\omega}_k]_{k=1,2,3,4} := [\alpha_1 \otimes, \alpha_2, \alpha_1 \otimes \omega_2, \omega_1 \otimes \alpha_2, \omega_1 \otimes \omega_2]^{\mathrm{t}}$. In this basis the Gauss-Manin connection matrix is given by:

$$A = \begin{pmatrix} (A_1)_{11} & 0 & (A)_{12} & 0 \\ 0 & (A_1)_{11} & 0 & (A_1)_{12} \\ (A_1)_{21} & 0 & (A_1)_{22} & 0 \\ 0 & (A_1)_{21} & 0 & (A_1)_{22} \end{pmatrix} dj_1 +$$

$$\begin{pmatrix} (A_2)_{11} & (A_2)_{12} & 0 & 0 \\ (A_2)_{21} & (A_2)_{22} & 0 & 0 \\ 0 & 0 & (A_2)_{21} & (A_2)_{22} \\ 0 & 0 & (A_2)_{21} & (A_2)_{22} \end{pmatrix} dj_2$$

3.2. The box equation. After simplifying the differential equation $\det(R^{uuu}) = 0$, where

$$\begin{pmatrix} \omega \\ \omega_u \\ \omega_{uu} \\ \omega_{uuu} \end{pmatrix} = R^{uuu,uu} \begin{pmatrix} \tilde{\omega}_1 \\ \tilde{\omega}_2 \\ \tilde{\omega}_3 \\ \tilde{\omega}_4 \end{pmatrix}$$

and $\omega = \tilde{\omega}_1 = \alpha_1 \otimes \alpha_2$, we obtain the *box equation*

$$\Box(j_1) = \Box(j_2) \tag{3.1}$$

where

$$\Box(j(u)) = j'(u)^2 \frac{36j(u)^2 - 41j(u) + 32}{144(j(u)-1)^2 j(u)^2} + \frac{1}{2}\{j(u), u\}$$

and

$$\{j(u), u\} = \frac{2j'(u)j'''(u) - 3j''(u)^2}{2j'(u)^2}$$

is the Schwarzian derivative.

The Lefschetz $(1,1)$-theorem implies that the subloci of $\mathbb{P}^1 \times \mathbb{P}^1$ where the polarization is not surjective are given by pairs of isogenous elliptic curves. Let $X_0(d) \subset \mathbb{P}^1 \times \mathbb{P}^1$ be the modular curve of isogenous elliptic curves $f : E_1 \to E_2$ with $\deg(f) = d$. The stronger version of Corollary 2.6 in this case is

PROPOSITION 3.1. *Let S be an algebraic curve in $\mathbb{P}^1 \times \mathbb{P}^1$ and let $(j_1(u), j_2(u)) \in S$ be a local holomorphic parametrization of S. Then*

$$\Box(j_1) = \Box(j_2)$$

if and only if S is a modular curve $X_0(d)$ of degree d isogenous elliptic curves, for some $d \in \mathbb{N}$.

See § 5 and [**CDLW09**] for a discussion of the corresponding special loci in the moduli of M-polarized K3 surfaces.

PROOF. We know that $\det(R^{uu,uuu})=0$ is equivalent to $\Box(j_1)=\Box(j_2)$. In turn, this is equivalent to saying that the Picard-Fuchs equation of ω with respect to the parameter u is of order 3. In this case we have $H_2(E_1\times E_2,\mathbb{Z})_i=H_1(E_1,\mathbb{Z})\otimes H_1(E_2,\mathbb{Z})$ and so there are constants $a_{ij}\in\mathbb{C},\ i,j=1,2$ such that

$$\delta=\sum a_{ij}\delta_{1,i}\otimes\delta_{2,i},\ a_{ij}\in\mathbb{C}$$

and

$$\int_{\sum a_{ij}\delta_{1,i}\otimes\delta_{2,j}}\alpha_1\otimes\alpha_2=\sum a_{ij}\int_{\delta_{1,i}}\alpha_1\int_{\delta_{2,j}}\alpha_2=0.$$

where $\{\delta_{i,1},\delta_{i,2}\},\ i=1,2$ is a basis of $H_1(E_i,\mathbb{Z})$ with $\langle\delta_{i,1},\delta_{i,2}\rangle=1$. Therefore, we need only prove that the universal family of pairs of two elliptic curves is perfect in the sense of Definition 2.4.

Let $\tau_k=\frac{\int_{\delta_2}\omega_k}{\int_{\delta_1}\omega_k},\ k=1,2$. The above equality becomes

$$\tau_2=A(\tau_1),\ A\in\mathrm{GL}(2,\mathbb{C}),\tag{3.2}$$

where $A(\tau_1)$ is the Möbius transformation of τ_1. Now, let assume that the locus described by Equation 3.2 is algebraic in $\mathbb{P}^1\times\mathbb{P}^1$. From this we only use the following: For any fixed $j_1\in\mathbb{P}^1$ there are a finitely many $j_2\in\mathbb{P}^1$ such that $(j_1,j_2)\in X$. This property using periods is:

$$\#\{\ \mathrm{SL}(2,\mathbb{Z})AB\ \mid\ B\in\mathrm{SL}(2,\mathbb{Z})\}<\infty.$$

This implies that A up to multiplication by a constant has rational coefficients. $\Box$

REMARK 3.2. Let C be a curve such that in the decomposition of its Jacobian into simple abelian varieties, there appear two elliptic curves $E_k,\ k=1,2$. We have the canonical map $C\to E_1\times E_2$. It follows from the above arguments that if E_1 and E_2 are not isogenous, then the homology class $[C]\in H_2(E_1\times E_2,\mathbb{Z})$ of the image of C satisfies $[C]=a_1[E_1]+a_2[E_2]$. That is, no contribution comes from $H_1(E_1,\mathbb{Z})\otimes H_1(E_2,\mathbb{Z})$.

REMARK 3.3. Let P be a reduced polynomial in j_1,j_2. Suppose $P=0$ is tangent to Equation 3.1. This property can be written in a purely algebraic way. Consider a solution of the Hamiltonian differential equation

$$\begin{cases} j_1'=\frac{\partial P}{\partial j_2}\\ j_2'=-\frac{\partial P}{\partial j_1}\end{cases}\tag{3.3}$$

passing through a point of $P=0$. The solution is entirely contained in $P=0$. Now, further derivatives of j_i's are polynomials in j_1,j_2 and we can substitute all these in $\Box(j_1)=\Box(j_2)$. We obtain an operator $\hat{\Box}=\Box(j_1)-\Box(j_2)$ from $\mathbb{C}[j_1,j_2]$ to itself. $P=0$ is tangent to the Box equation if and only if

$$P\text{ divides }\hat{\Box}(P).$$

REMARK 3.4. All the curves $X_0(d)$ cross $(\infty,\infty)\in\mathbb{P}^1\times\mathbb{P}^1$, and they are uniquely determined by the box equation. One may conjecture that this data is enough to calculate their equations. Around (∞,∞) we have the local coordinates (q_1,q_2), where

$$q_i=e^{2\pi i\tau_i}.$$

The curve $X_0(d)$ near (∞,∞) is reducible and its irreducible components are given by

$$q_1^{d_1}=q_2^{d_2},\ d_1d_2=d.$$

One may calculate an explicit equation of $X_0(d)$ using the q-expansion of the j-function: $j(q)=\frac{1}{q}+744+\cdots$. The equation of $X_0(d)$ is a polynomial $P_d(j_1,j_2)$ in two variables such $P_d(j(q),j(q^d))=0$.

3.3. The box equation and the Ramanujan differential equation. One may also derive Equation 3.1 using the following Ramanujan ordinary differential equation:

$$\text{(3.4)}\qquad \mathrm{R}:\begin{cases}\dot t_1=t_1^2-\frac{1}{12}t_2\\ \dot t_2=4t_1t_2-6t_3\\ \dot t_3=6t_1t_3-\frac{1}{3}t_2^2\end{cases}\quad \dot t_k=2\pi\sqrt{-1}q\frac{\partial t_k}{\partial q}=\frac{\partial t_k}{\partial \tau}$$

Note that these are not precisely the differential equations as stated by Ramanujan in [**Ram00**], but they agree after scaling solutions as indicated below. They were determined by geometric considerations in [**Mov08**]. These equations are satisfied by the scaled Eisenstein series:

$$\text{(3.5)}\quad t_k=a_kE_{2k}(q):=a_k\left(1+b_k\sum_{n=1}^{\infty}\left(\sum_{d|n}d^{2k-1}\right)q^n\right),\quad k=1,2,3,\ q=e^{2\pi i\tau}$$

and

$$(b_1,b_2,b_3)=(-24,240,-504),\quad (a_1,a_2,a_3)=(2\pi i,12(2\pi)^2,8(2\pi i)^3).$$

Let

$$j=\frac{t_2^3}{t_2^3-27t_3^2}$$

be the j-function. From Equation 3.3 we can calculate j',j'',j''' as rational functions in t_1,t_2,t_3. Thus, there is a polynomial in four variables which annihilate (j,j',j'',j'''). After calculating this we obtain the Schwarzian differential equation:

$$\Box(j(\tau))=S(j)+Q(j)(j')^2=0,$$

where $Q(j)=\frac{36j^2-41j+32}{72(j-1)^2j^2}$ and $S(j)$ is the Schwarzian derivative of j with respect to τ. The Schwarzian derivative satisfies the properties

$$S(f\circ g)=(S(f)\circ g)\cdot(g')^2+S(g).$$

$$S\left(\frac{a\tau+b}{c\tau+d}\right)=0.$$

Therefore if g is a Möbius transformation then $S(f\circ g)=(S(f)\circ g)\cdot(g')^2$ and if f is a Möbius transformation then $S(f\circ g)=S(g)$.

Now, if τ is a function of another parameter t then

$$\Box(j\circ\tau)=S(\tau)$$

and so for $A\in\mathrm{SL}(2,\mathbb{C})$:

$$\Box(j\circ A\circ\tau)=S(A\circ\tau)=S(\tau)$$

4. Lattice-polarized K3 surfaces

In this section we consider three families of lattice-polarized K3 surfaces described by singular hypersurfaces in projective space $\mathbb{P}^3$.

4.1. M-polarized K3 surfaces. In [**CD07**] and [**CDLW09**], the authors studied the surfaces described by the following family of polynomials in $\mathbb{P}^3$:

$$q_{a,b,d} = y^2zw - 4x^3z + 3axzw^2 + bzw^3 - \frac{1}{2}(dz^2w^2 + w^4) \quad d \neq 0 \tag{4.1}$$

After resolving singularities, we obtain a family of M-polarized K3 surfaces $X(a,b,d)$, where M is the rank 18 lattice $H \oplus E_8 \oplus E_8$. In fact, two such K3-surfaces are isomorphic if and only if the corresponding parameters are in the same orbit of the $\mathbb{C}^*$-action:

$$k, (a,b,d) \mapsto (k^2a, k^3b, k^6d), \quad k \in \mathbb{C}^*.$$

The coarse moduli space of such K3 surfaces is the subset of the weighted projective space $\mathbb{P}^{(2,3,6)}$ where $d \neq 0$. In $\mathbb{P}^{(2,3,6)}$ the loci of parameters such that the polarization $\mathrm{M} \to \mathrm{NS}(X_t)$ is not surjective is given by the curves C_n, $n \in \mathbb{N}$, which parametrize K3-surfaces with polarization $M_d := H \oplus E_8 \oplus E_8 \oplus \langle -2d \rangle$. There is a Hodge-theoretic correspondence between pairs of elliptic curves and M-polarized surfaces. Under this correspondence, C_d corresponds to the modular curve $X_0(d)$ (see § 5 and [**CDLW09**]).

4.2. N-polarized family of K3-surfaces. The next step is the N-polarized family of K3-surfaces $X(a,b,c,d)$, where $\mathrm{N} = H \oplus E_8 \oplus E_7$, which is studied in [**CD12**]. These surfaces are realized as the resolution of singularities of the hypersurfaces in $\mathbb{P}^3$ described by the following polynomials:

$$q_{a,b,c,d} = y^2zw - 4x^3z + 3axzw^2 + bzw^3 + cxz^2w - \frac{1}{2}(dz^2w^2 + w^4). \tag{4.2}$$

$$c \neq 0 \text{ or } d \neq 0.$$

Two N-polarized K3-surfaces are isomorphic if and only if the corresponding parameters are in the same orbit of the $\mathbb{C}^*$-action

$$k, (a,b,c,d) \mapsto (k^2a, k^3b, k^5c, k^6d), \ k \in \mathbb{C}^*.$$

The space $\mathbb{P}^{(2,3,5,6)} \backslash \{c = d = 0\}$ is the coarse moduli space of N-polarized K3 surfaces (see [**CD12**]).

Let D and $\Gamma_{\mathbb{Z}}$ be associated to N-polarized K3 surfaces as in § 2.2, and let $\mathbb{H}_2$ be the Siegel upper half plane of genus 2:

$$\mathbb{H}_2 = \left\{ z = \begin{pmatrix} z_1 & z_2 \\ z_2 & z_3 \end{pmatrix} \mid \mathrm{Im}(z_1)\mathrm{Im}(z_3) > \mathrm{Im}(z_2)^2, \ \mathrm{Im}(z_1) > 0 \right\} \tag{4.3}$$

The rank-five lattice $V_{\mathbb{Z}}$ is naturally isomorphic to the orthogonal direct sum $\mathrm{H} \oplus \mathrm{H} \oplus \langle 2 \rangle$. We select an integral basis for $V_{\mathbb{Z}}$ such that the intersection form $\psi_{\mathbb{Z}}$ in this basis is

$$\Psi_0 := [\psi_{\mathbb{Z}}(\delta_i, \delta_j)] = \begin{pmatrix} 0 & 0 & 0 & 0 & 1 \\ 0 & 0 & 0 & 1 & 0 \\ 0 & 0 & -2 & 0 & 0 \\ 0 & 1 & 0 & 0 & 0 \\ 1 & 0 & 0 & 0 & 0 \end{pmatrix}.$$

We have an isomorphism of groups

$$\wedge^2 : \mathrm{Sp}(4,\mathbb{Z})/\pm\mathrm{id} \to \Gamma_{\mathbb{Z}}/\pm\mathrm{id},$$

The images of generators of $\mathrm{Sp}(4,\mathbb{Z})$ under this isomorphism are

$$\begin{pmatrix} 0 & I_2 \\ -I_2 & 0 \end{pmatrix} \mapsto S := \begin{pmatrix} 0 & 0 & 0 & 0 & -1 \\ 0 & 0 & 0 & -1 & 0 \\ 0 & 0 & 1 & 0 & 0 \\ 0 & -1 & 0 & 0 & 0 \\ -1 & 0 & 0 & 0 & 0 \end{pmatrix},$$

(4.4)

$$\begin{pmatrix} I_2 & \tilde{B} \\ 0 & I_2 \end{pmatrix} \mapsto B := \begin{pmatrix} 1 & -b_1 & 2b_2 & -b_3 & b_2^2 - b_1 b_3 \\ 0 & 1 & 0 & 0 & b_3 \\ 0 & 0 & 1 & 0 & b_2 \\ 0 & 0 & 0 & 0 & b_1 \\ 0 & 0 & 0 & 0 & 1 \end{pmatrix}, \text{ where } \tilde{B} = \begin{pmatrix} b_1 & b_2 \\ b_2 & b_3 \end{pmatrix}.$$

and

$$\begin{pmatrix} \tilde{U}^{-t} & 0 \\ 0 & \tilde{U} \end{pmatrix} \mapsto U := \det(\tilde{U}) \begin{pmatrix} 1 & 0 & 0 & 0 & 0 \\ 0 & a^2 & -2ab & b^2 & 0 \\ 0 & -ac & ad+bc & -bd & 0 \\ 0 & c^2 & -2cd & d^2 & 0 \\ 0 & 0 & 0 & 0 & 1 \end{pmatrix}$$

where $\tilde{U} = \begin{pmatrix} a & b \\ c & d \end{pmatrix}$. We have a biholomorphism

$$\mathbb{H}_2 \to D, \quad \begin{pmatrix} z_1 & z_2 \\ z_2 & z_3 \end{pmatrix} \mapsto x = [z_2^2 - z_1 z_3; z_3; z_2; z_1; 1].$$

and so we obtain the isomorphism

$$\mathrm{Sp}(4,\mathbb{Z})\backslash\mathbb{H}_2 \cong \Gamma_{\mathbb{Z}}\backslash D \tag{4.5}$$

between the period domain of principally polarized abelian surfaces and the period domain of N-polarized K3 surfaces as above. For proof, see [**GN97**]. Since on both sides the period map is a biholomorphism, this determine a bijection between the corresponding coarse moduli spaces. On the left hand side of Equation 4.5 we have Humbert surfaces which are given by

$$c_1(z_2^2 - z_1 z_2) + c_2 z_1 + c_3 z_2 + c_4 z_3 + c_5 = 0, c_i \in \mathbb{Z} \tag{4.6}$$

The Humbert surfaces parametrize abelian surfaces where the endomorphism ring $\mathrm{End}(A)$ is isomorphic to an order in a real quadratic field. Under the correspondence described in Equation 4.5, Humbert surfaces parametrize K3 surfaces with a polarization $N \to \mathrm{NS}(X)$of rank 18 and with $H \oplus E_8 \oplus E_7 \subset N$ (see [**EK14**]). In this case showing Condition 2.5 holds is equivalent to showing that the hypersurface given by Equation 4.6 with $c_i \in \mathbb{C}$ induces a hypersurface in $\mathrm{Sp}(4,\mathbb{Z})\backslash\mathbb{H}_2$ if and only if, up to multiplication by a constant, $c_i \in \mathbb{Z}$.

4.3. The differential rank-jump property for N-polarized K3 surfaces. To show that N-polarized K3 surfaces have the differential rank-jump property, it is enough to prove that the moduli of N-polarized K3 surfaces of rank 17 have a perfect moduli space, as characterized in Condition 2.5. Let us assume that $cx = 0$ induces an analytic subvariety of $\Gamma_{\mathbb{Z}}\backslash D$. Using the isomorphism described in Equation 4.5, we consider the Satake compactification

$$\overline{\Gamma_{\mathbb{Z}}\backslash D} \cong \overline{\mathrm{Sp}(4,\mathbb{Z})\backslash\mathbb{H}_2} = (\mathrm{Sp}(4,\mathbb{Z})\backslash\mathbb{H}_2) \cup D_\infty, \qquad D_\infty := (\mathrm{SL}(2,\mathbb{Z}))\backslash\mathbb{H}) \cup \{\infty\}$$

see for instance [**Fre83**]. Using this topology, if we set $z_2 = 0$ and let either $\mathrm{Im}(z_1)$ or $\mathrm{Im}(z_2)$ go to $+\infty$ then the point converges to a point in D_∞ inside $\overline{\mathrm{Sp}(4,\mathbb{Z})\backslash\mathbb{H}_2}$. By our hypothesis, for all $A \in \Gamma_{\mathbb{Z}}$ the set $cAx = 0$ induces an analytic subvariety of $\mathrm{Sp}(4,\mathbb{Z})\backslash\mathbb{H}_2$. Since the codimension of D_∞ inside $\overline{\mathrm{Sp}(4,\mathbb{Z})\backslash\mathbb{H}_2}$ is bigger than one we conclude that it induces an analytic subvariety H_A in the compactification $\overline{\Gamma_{\mathbb{Z}}\backslash D}$ (Hartog's extension theorem, [**Gun90**]). Now, we set $z_2 = 0$ and send $\mathrm{Im}(z_1)$ to $+\infty$. We conclude that

$$H_A \cap D_\infty = \left\{ \frac{(cA)_2}{(cA)_1} \mid A \in \Gamma_{\mathbb{Z}} \right\}$$

where for a vector v, v_i is its i-the coordinate. Now this set has no accumulation points in $\mathbb{H}$ and intersects $\mathbb{R}$ only in rational numbers. The same set with matrices $A = C(B^n)D,\ n \in \mathbb{N}$, where $C, D \in \Gamma_{\mathbb{Z}}$ are arbitrary elements and B is given by Equation 4.4, has no accumulation point in $\mathbb{H}$. We assume that $b_2^2 = b_1 b_2$, let n go to infinity and compute the accumulation set and conclude that it is

$$\left\{ \frac{(cA_2)}{(cA_1)} \mid \quad A = CB_\infty D, \quad C, D \in \Gamma_{\mathbb{Z}} \right\} \subset \mathbb{Q} \tag{4.7}$$

where

$$B_\infty := \begin{pmatrix} 0 & -b_1 & 2b_2 & -b_3 & 0 \\ 0 & 0 & 0 & 0 & b_3 \\ 0 & 0 & 0 & 0 & b_2 \\ 0 & 0 & 0 & 0 & b_1 \\ 0 & 0 & 0 & 0 & 0 \end{pmatrix}.$$

Here A_i denotes the ith column of the matrix A.

For a generic choice of seven matrices $A = CB_\infty D$ as described in Equation 4.7, the $\mathbb{Q}$-vector space generated by seven vectors $A_2 - rA_1$, where $r := \frac{cA_2}{cA_1}$ is of dimension 4, and so the vector c which is orthogonal to it, has rational coordinates (up to multiplication by a constant). For instance, take B_∞ with $b_1 = b_2 = b_3 = 1$, a B matrix with $b_1 = b_2 = 1, b_3 = 0$ and a U matrix with $a = d = 0,\ b = 1,\ c = -1$. The seven matrices

$$A_i A_{i-1} \cdots A_1 B_\infty S, \quad i = 1, 2, \ldots, 7$$

satisfy the desired conditions, where the sequence $A_1, \cdots, A_7$ is given by B, U, S, B, S, B, U. For the computer code of these computations see [**Mov**].

4.4. A lattice-polarized family of K3 surfaces of rank 12. In this section we consider a nine-parameter family of K3 surfaces

$$X(t_i, i = 4, 6, 7, 10, 12, 15, 16, 18, 24)$$

polarized by a lattice of rank 12. The family is obtained by resolving the singularities of the hypersurfaces described by the following polynomials in $\mathbb{P}^3$:

(4.8)

$$q_{t_i} = -4x^3 + y^2w - \frac{1}{2}w^4 + t_4zxw^2 + t_6z^2y^2 + t_7z^2yx + t_{10}z^2xw + t_{12}z^2w^2 + t_{15}z^3y + t_{16}z^3x + t_{18}z^3w + t_{24}z^4$$

The parameters t_i satisfy the condition that two such K3 surfaces are isomorphic if and only if the corresponding parameters are in the same orbit of the $\mathbb{C}^*$-action

$$k, (t_i,\ i \in I) \mapsto (k^{\frac{i}{2}}t_i,\ i \in I), \quad I := \{4, 6, 7, 10, 12, 15, 16, 18, 24\}.$$

See also §6.

For a generic choice of the parameters, the surface given by $q_{t_i} = 0$ has a unique singularity

$$P_1 = [0 : 1 : 0 : 0]$$

After a blow-up in P_1 we obtain the configuration of lines and curves in X illustrated in Figure 1. (We have labeled the lines L and the curves a_i.) In § 6, we will use a

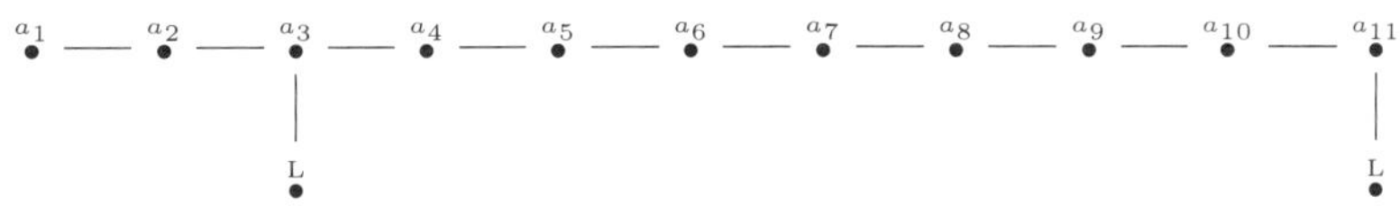

FIGURE 1. A rank 12 lattice

subfamily of the family described by Equation 4.8 in which all t_i's are zero except $t_4 = 3a$.

5. The Griffiths-Dwork technique

5.1. The Griffiths-Dwork technique for smooth hypersurfaces. We want to compute the Picard-Fuchs equations satisfied by the period $\int \omega$ of the holomorphic $(2,0)$-form on our K3 surfaces. Let us recall the Griffiths-Dwork method for a family of smooth hypersurfaces $Y(t)$ in $\mathbb{P}^m$ given by a family of polynomials $q(t)$. (For a more detailed exposition, see [**CK99**] or [**DGJ08**].)

First, we observe that we may write forms in $H^{m-1}(Y(t))$ as residues of meromorphic forms in $H^m(\mathbb{P}^m - Y(t))$:

$$\mathrm{Res}\left(\frac{p\Omega}{q^k}\right) \in H^{m-k,k-1}(Y).$$

Here Ω is the usual holomorphic form on $\mathbb{P}^m$ and p is a homogeneous polynomial satisfying

$$\deg p = k \deg q - (m+1).$$

In particular, $\omega = \mathrm{Res}\left(\frac{\Omega}{q}\right)$. The image of the residue map is the *primitive cohomology*, which consists of the classes in $H^{m-1}(Y(t))$ orthogonal to the hyperplane class.

Next, we differentiate the period with respect to our parameter. Note that we may move the derivative under the integral sign. For any class $\alpha = \operatorname{Res}\left(\frac{p\Omega}{q^k}\right)$ and for any 3-cycle γ in the family of surfaces $Y(t)$ which restricts to each fiber as a 2-cycle γ_t,

$$\begin{aligned}
\frac{d}{dt}\int_{\gamma_t} \alpha &= \frac{d}{dt}\int_{\gamma_t} \operatorname{Res}\left(\frac{p\Omega}{q^k}\right) \\
&= \int_{\gamma_t} \operatorname{Res}\left(\Omega \frac{d}{dt}\frac{p}{q^k}\right).
\end{aligned} \tag{5.1}$$

We then observe that because $H^{m-1}(Y(t))$ is finite-dimensional, if we take enough derivatives, we will obtain a $\mathbb{C}(t)$-linear relationship between $\int \operatorname{Res}(\frac{\Omega}{q})$ and its derivatives. In order to compute this relationship, we use a *reduction of pole order* formula to compare classes of the form $\operatorname{Res}(\frac{r\Omega}{q^{k+1}})$ to classes of the form $\operatorname{Res}(\frac{p\Omega}{q^k})$. Suppose $r = \sum_i A_i \frac{\partial q}{\partial x_i}$, where the A_i are polynomials of the appropriate degree. Then,

$$\frac{\Omega}{q^{k+1}} \sum_i A_i \frac{\partial q}{\partial x_i} = \frac{1}{k}\frac{\Omega}{q^k} \sum_i \frac{\partial A_i}{\partial x_i} + \text{exact terms.} \tag{5.2}$$

We may generalize the Griffiths-Dwork technique to a multi-parameter family $Y(t_1, \dots, t_j)$ by taking partial derivatives with respect to each parameter.

If our hypersurfaces $Y(t)$ are K3 surfaces, then $H^2(Y(t), \mathbb{C})$ is a 22-dimensional vector space, and the primitive cohomology $PH(Y)$ is 21-dimensional. Thus, we are guaranteed a 21st order ordinary differential equation, since there must be a linear relationship between $\int \omega$ and its first 21 derivatives. If $Y(t)$ has high Picard rank, however we expect a much lower order differential equation: the holomorphic $(2,0)$ form ω and its derivatives are orthogonal to $\operatorname{Pic}(Y(t))$, so we should obtain an ordinary differential equation of order $22 - \rho$, where ρ is the generic Picard rank of $Y(t)$.

To implement the Griffiths-Dwork method, we shift our attention from the cohomology ring to a related ring. Let $S = \mathbb{C}(t)[x_0, \dots, x_m]$, and let

$$J(q) = \left\langle \frac{\partial q}{\partial x_0}, \dots, \frac{\partial q}{\partial x_m} \right\rangle$$

be the *Jacobian ideal* of q. Using Equation 5.2, we see that if $r \in J(q)$ and $\deg r = (k+1)\deg q - (m+1)$, we may reduce the pole order of $\operatorname{Res}(\frac{r\Omega}{q^{k+1}})$. Let $R(q) = S/J(q)$ be the *Jacobian ring*. The grading on S induces a grading on $R(q)$, and we have injective maps

$$R(q)_{k \deg q - (m+1)} \hookrightarrow H^{m-k,k-1}(Y) \tag{5.3}$$

for $k = 1, 2, \dots$. The image of these *induced residue maps* is the primitive cohomology $PH(Y(t))$. Thus, we may implement the Griffiths-Dwork method by using a computer algebra system such as [**BCP97**] to work with graded pieces of $R(q)$.

5.2. The Griffiths-Dwork technique for M- and N-polarized surfaces. If the K3 surfaces under examination have ADE singularities, the residue map is still well-defined, since we can extend ω uniquely to the resolution of singularities. The authors of [**CDLW09**] applied the Griffiths-Dwork technique to the singular K3 surfaces $X(a, b, d)$ described in Equation 4.1. The computation yields the following characterization of one-parameter loci in moduli where the map $\mathrm{M} \to NS(X_t)$ fails to be surjective (compare Proposition 3.1):

THEOREM 5.1 ([**CDLW09**]). *A one-parameter family of* M*-polarized K3 surfaces X_t generically has Picard-Fuchs equation of rank* 4. *The following are equivalent:*

- *Each surface X_t is polarized by the enhanced lattice* $\mathrm{M}_n = H \oplus E_8 \oplus E_8 \oplus \langle -2n \rangle$
- *The Picard-Fuchs equation drops to rank* 3
- *The corresponding pairs of elliptic curves* $\mathrm{E}_1(t)$ *and* $\mathrm{E}_2(t)$ *are n-isogenous*
- *The j-invariants of* $\mathrm{E}_1(t)$ *and* $\mathrm{E}_2(t)$ *satisfy* $\square(j_1(t)) = \square(j_2(t))$.

One may attempt to compute the Picard-Fuchs equations for N-polarized K3 surfaces by the Griffiths-Dwork technique, using the realization as singular hypersurfaces given in Equation 4.2. In practice, the computation is sensitive to the choice of parametrization on the moduli space. Recall that an N-polarized K3 surface is given by a point $(a, b, c, d) \in \mathbb{P}^{(2,3,5,6)}$, where c and d are not simultaneously 0. The locus where $c = 0$ yields the M-polarized K3 surfaces. We may work in the affine chart on $\mathbb{P}^{(2,3,5,6)}$ where $c \neq 0$ by applying the Griffiths-Dwork technique to the polynomials $q_{a,b,1,d}$. If we do so, we find that the elements of $R(q_{a,b,1,d})$ corresponding to $\mathrm{Res}\left(\frac{\Omega}{q}\right)$ and its first and second derivatives with respect to the parameters a, b, and d lie in the $\mathbb{C}(a, b, d)$-vector space spanned by the 5 elements

$$1, w^4, w^3x, w^3z, w^2z^2.$$

Thus, we will obtain $10 - 5 = 5$ second-order Picard-Fuchs equations, and a generic one-parameter subfamily specified by choosing $a(t)$, $b(t)$, and $d(t)$ will satisfy a fifth-order ordinary differential equation, as expected.

If we choose a chart on $\mathbb{P}^{(2,3,5,6)}$ where c is not constant, such as the chart where $d = 1$, we obtain a different number of equations. The elements of $R(q_{a,b,c,1})$ corresponding to $\mathrm{Res}\left(\frac{\Omega}{q}\right)$ and its first and second derivatives with respect to the parameters a and b lie in the $\mathbb{C}(a, b, c)$-vector space spanned by the same 5 basis elements, $1, w^4, w^3x, w^3z$, and w^2z^2. The first derivative $\frac{\partial}{\partial c}\mathrm{Res}\left(\frac{\Omega}{q}\right)$ and the mixed second derivatives involving c also lie in this vector space. However, the element of $R(q_{a,b,c,1})$ corresponding to $\frac{\partial^2}{\partial c^2}\mathrm{Res}\left(\frac{\Omega}{q}\right)$ contains a term in a sixth basis element, wz^3. Thus, in this chart we find only 4 second-order Picard-Fuchs equations, and an arbitrary one-parameter subfamily specified by equations $a(t)$, $b(t)$, and $c(t)$ yields a sixth-order ordinary differential equation. A similar result holds if we work with the full polynomials $q_{a,b,c,d}$: the element of $R(q_{a,b,c,d})$ corresponding to $\frac{\partial^2}{\partial c^2}\mathrm{Res}\left(\frac{\Omega}{q}\right)$ is independent of the ring elements corresponding to $\mathrm{Res}\left(\frac{\Omega}{q}\right)$ and the other first and second derivatives.

By using the Griffiths-Dwork technique applied to the polynomials $q_{a,b,c,d}$ or the methods of § 6, one can show that $\int \omega$ and its first derivatives are linearly dependent:

$$4a\frac{\partial}{\partial a}\int \omega + 6b\frac{\partial}{\partial b}\int \omega + 10c\frac{\partial}{\partial c}\int \omega + 12d\frac{\partial}{\partial d}\int \omega + \int \omega = 0. \tag{5.4}$$

Thus, $\frac{\partial^2}{\partial c^2}\mathrm{Res}\left(\frac{\Omega}{q}\right)$ cannot be linearly independent of the other first and second derivatives of $\mathrm{Res}\left(\frac{\Omega}{q}\right)$. The discrepancy demonstrates that the induced residue maps are not injective; this problem arises because our representative polynomials are not smooth.

5.3. Griffiths-Dwork for weighted projective hypersurfaces. In order to fix the problem that we encountered at the end of § 5.2, we will pass from the expression for N-polarized K3 surfaces as singular hypersurfaces in projective space to an expression as generic hypersurfaces in a weighted projective space.

According to Table 1.1 of [**Bel97**], any generic hypersurface in $\mathbb{WP}^3(3,4,10,13)$ has an embedding of the lattice N into its Néron-Severi lattice. Conversely, if X is an N-polarized K3 surface it can be written in the form of Equation 4.2, then X is birational to an anticanonical hypersurface in $\mathbb{WP}^3(3,4,10,13)$ of the form

$$x_0^{10} + bx_0^6x_1^3 + \frac{d}{4}x_0^2x_1^6 + 3ax_0^4x_1^2x_2 - \frac{c}{2}x_1^5x_2 + x_0^2x_1x_2^2 + 2x_0x_1x_2x_3 - 4x_2^3 + x_1x_3^2 = 0 \tag{5.5}$$

with parameters $(a,b,c,d) \in \mathbb{WP}^3(2,3,5,6)$, and where the variables x_0, x_1, x_2 and x_3 have weights $3,4,10$ and 13 respectively. To see how this birational transformation comes about, one restricts the weighted projective family of K3 surfaces to a copy of $(\mathbb{C}^\times)^3 \subseteq \mathbb{WP}^3(3,4,10,13)$, exhibits an elliptic fibration over $\mathbb{CP}^1$ with singular fibers of types II^* and III^*, then matches parameters with the natural fibration of this form on the projective hypersurfaces in Equation 4.2.

In § 4 of [**Dol82**] it is shown that one may apply version of Griffiths residues to compute the orbifold cohomology of the hypersurfaces in Equation 5.5, since a generic member of the family in Equation 5.5 is quasismooth (in other words, its only singularities are inherited from the weighted projective space in which it lives). Since the primitive orbifold Hodge structure contains the transcendental Hodge structure of the minimal resolution of an orbifold K3 surface as a direct summand, the Griffiths-Dwork method will succeed in producing differential relations between the periods of the family of K3 surfaces in Equation 4.2. See § 5.3.2 of [**CK99**] for details of how this technique differs from the Griffiths-Dwork technique for hypersurfaces in projective space as described in § 5.1.

Therefore, if we attempt to compute the Picard-Fuchs equation of the N-polarized family on the chart $d = 1$, then this technique must produce the correct results and the problem encountered at the end of § 5.2 vanishes. However if one does this computation, the result is a family of differential equations with complicated polynomial coefficients. For the sake of obtaining a differential equation that we can write down in a few pages, we will choose the family of K3 surfaces over $\mathbb{C}^3$ obtained by setting $a = 1$ in Equation 5.5. The resulting family of K3 surfaces is a family over $\mathbb{C}^3$ with coordinates b, c and d and according to [**SY89**] the differential ideal annihilating periods of the three parameter family in Equation 5.5 is generated by a system of 5 equations, and each equation is expressed in the form

$$A\frac{\partial^2}{\partial a_i \partial a_j}\int\omega = A^{a_i,a_j}_{d,d}\frac{\partial^2}{\partial d^2}\int\omega + A^{a_i,a_j}_{b}\frac{\partial}{\partial b}\int\omega + A^{a_i,a_j}_{c}\frac{\partial}{\partial c}\int\omega + A^{a_i,a_j}_{d}\frac{\partial}{\partial d}\int\omega + A^{a_i,a_j}_{0}\int\omega$$

for (a_i, a_j) one of the pairs

$$(b,d), (b,c), (b,b), (c,c) \text{ and } (c,d).$$

We refer to these linear differential equations as $\mathcal{D}_{(a_i,a_j)}$

The polynomial A does not depend upon our choice of a_i and a_j. In our situation, we find that

$$A = 1296b^4c - 2340b^2cd - 2592b^2c - 4320bc^2 - 875c^3 - 432bd^2 + 900cd^2 - 2412cd + 1296c.$$

The other coefficients of our Picard-Fuchs equations are given as follows.

5.3.1. *The equation $\mathcal{D}_{(b,d)}$.*

$$\begin{aligned}
A^{b,d}_{d,d} =& \frac{2}{3}(-648b^3cd - 1296b^2c^2 - 2700bc^3 - 625c^4 + 648b^2d^2 + 1080bcd^2 + 648bcd \\
&- 810c^2d + 648d^3 + 1296c^2 - 648d^2) \\
A^{b,d}_{b} =& -\frac{1}{6}c(1296b^2 - 8100bc - 3125c^2 + 180d - 1296) \\
A^{b,d}_{c} =& -\frac{1}{2}(-3060b^2c - 625bc^2 - 432bd + 1050cd - 1260c) \\
A^{b,d}_{d} =& -648b^3c + 432b^2d + 2730bcd + 625c^2d + 648bc + 450c^2 + 648d^2 - 432d \\
A^{b,d}_{0} =& \frac{5}{12}(360bc + 125c^2 + 36d)
\end{aligned}$$

5.3.2. *The equation $\mathcal{D}_{(b,c)}$.*

$$\begin{aligned}
A^{b,c}_{d,d} =& -4(216b^3c^2 + 150b^2c^3 - 108b^3d^2 + 108b^2cd - 45bc^2d - 125c^3d + 108bd^3 \\
&- 216bc^2 - 60c^3 + 108bd^2 + 216cd^2 - 108cd) \\
A^{b,c}_{b} =& -\frac{1}{2}(-900b^3c - 432b^2d + 750bcd - 3420bc - 1225c^2) \\
A^{b,c}_{c} =& -c(216b^3 - 750b^2c - 330bd + 625cd - 216b - 960c) \\
A^{b,c}_{d} =& -6(-72b^3d - 150b^2cd + 108b^2c + 105bc^2 + 48bd^2 + 125cd^2 + 72bd \\
&+ 31cd - 108c) \\
A^{b,c}_{0} =& -\frac{5}{2}(-30b^2c - 6bd + 25cd - 30c)
\end{aligned}$$

5.3.3. *The equation $\mathcal{D}_{(b,b)}$.*

$$
\begin{aligned}
A^{b,b}_{d,d} =& -4(-720b^2c^3 - 250bc^4 - 36b^2cd^2 - 1296bc^2d - 475c^3d + 432bd^3 \\
&+ 180cd^3 - 144c^3 + 36cd^2) \\
A^{b,b}_{b} =& 6c(-432b^3 - 125b^2c + 60bd + 432b + 245c) \\
A^{b,b}_{c} =& -2c(720b^2c + 625bc^2 - 432bd + 750cd - 1152c) \\
A^{b,b}_{d} =& -6(264b^2cd + 250bc^2d - 432bc^2 - 175c^3 + 288bd^2 + 480cd^2 - 408cd) \\
A^{b,b}_{0} =& -5c(36b^2 + 25bc + 30d - 36)
\end{aligned}
$$

5.3.4. *The equation $\mathcal{D}_{(c,c)}$.*

$$
\begin{aligned}
A^{c,c}_{d,d} =& 4c^{-1}(90b^3c^3 + 324b^4d^2 - 9b^2c^2d - 150bc^3d - 648b^2d^3 + 126bc^3 + 25c^4 \\
&- 648b^2d^2 - 1404bcd^2 - 495c^2d^2 + 324d^4 + 117c^2d - 648d^3 + 324d^2) \\
A^{c,c}_{b} =& -18c^{-1}(15b^4c - 36b^3d - 25b^2cd + 6b^2c + 10bc^2 + 6bd^2 + 36bd \\
&+ 42cd - 21c) \\
A^{c,c}_{c} =& -3c^{-1}(432b^4c + 150b^3c^2 - 936b^2cd - 250bc^2d - 864b^2c - 1662bc^2 - 425c^3 \\
&- 72bd^2 + 180cd^2 - 648cd + 432c) \\
A^{c,c}_{d} =& 18c^{-1}(72b^4d - 30b^3cd + 21b^2c^2 - 120b^2d^2 + 50bcd^2 - 144b^2d - 234bcd \\
&- 55c^2d + 108d^3 + 15c^2 - 180d^2 + 72d) \\
A^{c,c}_{0} =& \frac{5}{4}c^{-1}(-36b^3c + 36b^2d + 60bcd + 36bc + 25c^2 + 36d^2 - 36d)
\end{aligned}
$$

5.3.5. *The equation $\mathcal{D}_{(c,d)}$.*

$$
\begin{aligned}
A^{c,d}_{d,d} =& -2(648b^4d - 360b^2c^2 - 125bc^3 - 1188b^2d^2 - 1296b^2d - 2808bcd \\
&- 675c^2d + 540d^3 - 72c^2 - 1188d^2 + 648d) \\
A^{c,d}_{b} =& \frac{3}{2}(-432b^3 - 125b^2c + 60bd + 432b + 245c) \\
A^{c,d}_{c} =& -\frac{1}{2}(720b^2c + 625bc^2 - 432bd + 750cd - 1152c) \\
A^{c,d}_{d} =& -\frac{1}{2}(2592b^4 - 3888b^2d + 750bcd - 5184b^2 - 9936bc - 2275c^2 \\
&+ 3240d^2 - 6048d + 2592) \\
A^{c,d}_{0} =& -\frac{5}{4}(36b^2 + 25bc + 30d - 36)
\end{aligned}
$$

5.3.6. *Comments.* First note that in order to produce the Picard-Fuchs equations in § 5.3 we have chosen a family of K3 surfaces whose period map is dominant onto the moduli space of N-polarized K3 surfaces with degree 2. Other choices of families that we have tried produced differential equations which are too complicated to be written concisely.

Secondly, the construction that we have described is effective for any chart on the moduli space of K3 surfaces, but we have not been able to use this to produce global results on the moduli space of N-polarized K3 surfaces. In § 6, we will use another technique based upon the method of tame polynomials in order to produce an expression for the Gauss-Manin connection on the moduli space of N-polarized

K3 surfaces which encompasses all of the data that we might be able to determine from the Griffiths-Dwork method applied to various charts on the moduli space of N-polarized K3 surfaces, and in particular its restriction to the chart where $a = 1$ should reproduce the equations above. The only disadvantage of the technique in § 6 is that it produces very large equations.

Finally, we would like to point out that, in theory, the technique described above is valid for the moduli space of L-polarized K3 surfaces for any lattice L which appears as the Picard lattice of a generic anticanonical K3 surface in Reid's list of 95 weighted projective threefolds as listed in Table 1.1 of [**Bel97**].

5.4. The Elkies-Kumar parametrizations and Picard-Fuchs equations. Before moving on, we will describe how one may apply the technique of Griffiths-Dwork for weighted projective hypersurfaces along with Equation 5.5 and the work of Elkies and Kumar [**EK14**] to produce Picard-Fuchs equations for any family of K3 surfaces living over a Humbert surface in the moduli space of N-polarized K3 surfaces.

In Theorem 11 of [**Kum08**], Kumar determines an expression for the Shioda-Inose partner for the Jacobian of a given curve of genus 2. We recall that there are invariants I_2, I_4, I_6 and I_{10} which determine a curve of genus 2 up to isomorphism, called Igusa-Clebsch invariants. Kumar writes down a family of elliptically fibered K3 surfaces varying with parameters the Igusa-Clebsch invariants so that the Shioda-Inose partner of such a K3 surface is the Jacobian of the genus 2 curve determined by the Igusa-Clebsch invariants appearing in the equation for the K3 surface. This family is written as

$$y^2 = x^3 - t^3\left(\frac{I_4}{12}t + 1\right)x + t^5\left(\frac{I_{10}}{4}t^2 + \frac{(I_2I_4 - 3I_6)}{108}t + \frac{I_2}{24}\right).$$

In § 6-35 of [**EK14**], Elkies and Kumar determine explicit parametrizations for all rational Humbert surfaces with square-free discriminant less than 100. One can use these parametrizations to provide Picard-Fuchs equations for the corresponding Humbert surfaces in the moduli space of N-polarized K3 surfaces. First, it is understood that the family written by Kumar is exactly the same as the family of K3 surfaces written in [**CD12**], since both are Shioda-Inose partners of Jacobians of genus 2 curves, and thus are N-polarized K3 surfaces. The exact relationship between sets of parameters is given by the map

$$(a, b, c, d) = \left(\frac{I_4}{36}, \frac{3I_6 - I_2I_4}{216}, \frac{I_{10}}{4}, \frac{I_2I_{10}}{96}\right).$$

For any of the Humbert surfaces whose parametrization is determined by Elkies and Kumar, it is then possible to compute the corresponding Picard-Fuchs equation by straightforward application of the Griffiths-Dwork method in § 5.3, since an N-polarized K3 surface of Picard rank 18 expressed as in Equation 5.5 is quasi-smooth if its transcendental lattice has discriminant greater than 4.

As an example, the Humbert surface $\mathcal{H}_5$ is parametrized by

$$u, v \mapsto (a(u,v), b(u,v), c(u,v), d(u,v)) = \left(\frac{u^2}{4}, \frac{u^3}{8} + \frac{v}{2}, v^2, uv^2 + \frac{v^2}{4}\right).$$

Applying the Griffiths-Dwork method to the family of weighted projective hypersurfaces obtained from this parametrization and Equation 5.5 produces the Picard-Fuchs operators

$$
\begin{aligned}
&-4u^2(270u^3+99u^2+9u+125v)\frac{\partial^2}{\partial u^2}+18u^2v(10u+3)(15u+2)\frac{\partial^2}{\partial u\partial v}\\
&+2u(-810u^3-207u^2-9u+250v)\frac{\partial}{\partial u}+54uv(5u-1)(5u+1)\frac{\partial}{\partial v}-375v
\end{aligned}
$$

and

$$
\begin{aligned}
&2uv(270u^3+99u^2+9u+125v)\frac{\partial^2}{\partial v^2}+2u(108u^4+36u^3+3u^2+100uv+10v)\frac{\partial^2}{\partial u\partial v}\\
&+10u(10u+1)\frac{\partial}{\partial u}+2(270u^4+99u^3+9u^2+150uv-5v)\frac{\partial}{\partial v}-5(6u+1)
\end{aligned}
$$

which generate the differential ideal which annihilates the periods of the family of K3 surfaces over the Humbert surface $\mathcal{H}_5$. These Picard-Fuchs equation should agree, after change of variables, with the Picard-Fuchs equations given in § 3 of [**Nag**].

6. Calculating the Gauss-Manin connection

In this section we calculate the Gauss-Manin connection for M- and N-polarized K3 surfaces using *tame polynomials*; the relevant techniques were first introduced in [**Mov11**]. The method of tame polynomials differs from our naive application of the Griffiths-Dwork technique in two ways: we work with affine hypersurfaces rather than hypersurfaces in projective space, and we use a one-parameter deformation to remove computational problems caused by singularities in the representative hypersurfaces.[1]

6.1. Tame polynomials. We follow the notation of [**Mov11**, Chapter 4]. Let $\mathsf{R}=\mathbb{Q}[a_1,\dots,a_n]$, where we view the a_i as arbitrary parameters. Let us consider the homogeneous polynomial of degree 24

$$g := -4x^3+y^2w-\frac{1}{2}w^4$$

in the weighted ring

$$\mathsf{R}[x,y,z],\ \deg(x)=8,\ \deg(y)=9,\ \deg(w)=6 \tag{6.1}$$

The polynomial has an isolated singularity at the origin $0\in\mathbb{C}^3$. A *tame polynomial in* R *with the last homogeneous polynomial* g is a polynomial of the form $f:=g+f_1$, where $f_1\in\mathsf{R}[x,y,w]$ is of degree strictly less than 24.

Our main example is the case $\mathsf{R}=\mathbb{Q}[a,b,c,d]$ and

$$f=y^2w-4x^3+3axw^2+bw^3+cxw-\frac{1}{2}(dw^2+w^4). \tag{6.2}$$

We may obtain f by starting with the family of polynomials $q_{a,b,c,d}$ which describe N-polarized K3 surfaces given in Equation 4.2, and converting to the affine chart where $z=1$. Note, however, that we have changed the weights on w, x, and y.

[1]See `http://w3.impa.br/~hossein/k3surfaces` for explicit equations for all calculations in this section.

REMARK 6.1. $\{g = 0\}$ induces a rational curve L_2 in $\mathbb{P}^{(8,9,6)}$. After a resolution of singularities, L_2 is the same L_2 as in [**CD12**]. We consider the weights

$$\deg(a) = 4,\ deg(b) = 6,\ \deg(c) = 10,\ \deg(d) = 12$$

and in this way f becomes a homogeneous polynomial of degree 24 in 7 variables x, y, w, a, b, c, d. These weights are compatible with the weights of Eisenstein series computed in [**CD12**], Theorem 1.7.

REMARK 6.2. The monomials of degree less than 24 in Equation 6.1 are:

$$y^2, y, yw, yw^2, yx, yxw, 1, w, w^2, w^3, x, xw, xw^2, x^2, x^2w.$$

Thus, the most general tame polynomial f that we can write is g plus a linear combination of the above monomials with coefficients in R. If we are interested in such a tame polynomial up to linear transformations $x \mapsto x + * + *w$, $y \mapsto * + *x + *w$, $w \mapsto w + *$, all $*$ in R, then we may discard the monomials $x^2, x^2w, yw, yxw, yw^2, w^3$ in the definition of f. In this way we obtain an affine version of the equation for the family of rank 12 K3 surfaces introduced in § 4.4:

(6.3)

$$f = -4x^3 + y^2w - \frac{1}{2}w^4 + t_4xw^2 + t_6y^2 + t_7yx + t_{10}xw + t_{12}w^2 + t_{15}y + t_{16}x + t_{18}w + t_{24}$$

6.2. Algebraic De Rham cohomology. The R-module $V_g := \mathsf{R}[x, y, w]/\text{Jacob}(g)$ is free of rank 10. In fact, the set of monomials

$$I := \{xw^3, xw^2, w^3, xy, xw, w^2, y, x, w, 1\} \tag{6.4}$$

form a basis for both V_g and $V_f := \mathsf{R}[x, y, w]/\text{Jacob}(f)$ (see [**Mov11**, Proposition 4.6]). Let

$$\mathbb{U}_0 := \text{Spec}(\mathsf{R}),\ \mathbb{U}_1 = \text{Spec}(\mathsf{R}[x, y, w]).$$

The Brieskorn module or the relative de Rham cohomology of $\mathbb{U}_1/\mathbb{U}_0$ is by definition:

$$\mathsf{H} = H^2_{\text{dR}}(\mathbb{U}_1/\mathbb{U}_0) := \frac{\Omega^3_{\mathbb{U}_1/\mathbb{U}_0}}{f\Omega^3_{\mathbb{U}_1/\mathbb{U}_0} + df \wedge d\Omega^1_{\mathbb{U}_1/\mathbb{U}_0}} \tag{6.5}$$

where $\Omega^i_{\mathbb{U}_1/\mathbb{U}_0}$ is the set of differential i-forms of $\mathsf{R}[x, y, w]$ over R (the differential of the elements of R is zero). It is an R-module in a canonical way. It can be shown that H is a free R-module generated by

$$\alpha dx \wedge dy \wedge dw, \alpha \in I. \tag{6.6}$$

(See [**Mov11**, Theorem 4.1 and Corollary 4.1].)

6.3. Discriminant. Let

$$A_f : V_f \to V_f,\ \ A(P) = P \cdot f.$$

We use the monomial basis of the free R-module V_f defined in Equation 6.4 in order to write A_f as a matrix. Let $\Delta(s)$ be the minimal polynomial of A_f. It is a factor of the characteristic polynomial $\det(A_f - s \cdot I) \in \mathsf{R}[s]$. By definition, the discriminant of f is $\Delta := \Delta(0)$.

6.4. Gauss-Manin connection. In order to introduce the Gauss-Manin connection on H it is more convenient to use the R-module

$$\mathsf{M} \quad := \quad \frac{\Omega^{n+1}_{\mathbb{U}_1/\mathbb{U}_0}[\frac{1}{f}]}{\Omega^{n+1}_{\mathbb{U}_1/\mathbb{U}_0} + d(\Omega^{n}_{\mathbb{U}_1/\mathbb{U}_0}[\frac{1}{f}])}$$

which we call the Gauss-Manin system of f. Here $\Omega^{i}_{\mathbb{U}_1/\mathbb{U}_0}[\frac{1}{f}]$ is the set of polynomials in $\frac{1}{f}$ with coefficients in $\Omega^{i}_{\mathbb{U}_1/\mathbb{U}_0}$. The Gauss-Manin system has a natural filtration given by the pole order along $\{f = 0\}$, namely

$$\mathsf{M}_i := \{[\frac{\omega}{f^i}] \in \mathsf{M} \mid \omega \in \Omega^{n+1}_{\mathbb{U}_1/\mathbb{U}_0}\},$$

$$\mathsf{M}_1 \subset \mathsf{M}_2 \subset \cdots \subset \mathsf{M}_i \subset \cdots \subset \mathsf{M}_\infty := \mathsf{M}.$$

We have a canonical map $\mathsf{H} \to \mathsf{M}$, $\omega \mapsto \frac{\Omega}{f}$. If the discriminant $\Delta \in \mathsf{R}$ of f is not zero then this map is an inclusion and an isomorphism of R-modules $\mathsf{H} \cong \mathsf{M}_1$. The Gauss-Manin connection on M is the map

$$\nabla : \mathsf{M} \to \Omega^1_{\mathbb{U}_0} \otimes_{\mathsf{R}} \mathsf{M}$$

which is obtained by derivation with respect to the elements of R (the derivation of x, y and w is zero). More precisely

$$\nabla(\frac{P\omega}{f^i}) = \frac{d_{\mathsf{R}}P \cdot f - iP \cdot d_{\mathsf{R}}f}{f^{i+1}}\omega, \ \ P \in \mathsf{R}[x, y, w], \tag{6.7}$$

where $\omega = dx \wedge dy \wedge dw$ and $d_{\mathsf{R}} : \mathsf{R}[x] \to \mathsf{R}[x]$ is the differential with respect to elements in R. (Note that Equation 6.7 is the algebraic, affine counterpart of Equation 5.1.)

6.5. Gauss-Manin connection, $\Delta \neq 0$. If the discriminant of f is not zero then the Gauss-Manin connection on M induces a connection on all M_i, $i \in \mathbb{N}$:

$$\nabla : \mathsf{M}_i \to \frac{1}{\Delta}\Omega^1_{\mathbb{U}_0} \otimes_{\mathsf{R}} \mathsf{M}_i.$$

The R-module $\mathsf{H} = \mathsf{M}_1$ is freely generated by the forms described in Equation 6.6, so theoretically we could calculate a 10×10 matrix $\tilde{A}$ with entries in $\Omega_{\mathbb{U}_0}$ such that

$$\nabla\Omega = \frac{1}{\Delta}\tilde{A} \otimes \Omega$$

where Ω is a 10×1 matrix made up of the expressions in Equation 6.6. However, in practice performing this calculation for the tame polynomial described in Equation 6.3 using the algorithms in [**Mov11**] was beyond the power of our computers.

The discriminant of the tame polynomial given in Equation 6.2 with $a, b, c, d \in \mathsf{R}$ is zero. In order to obtain a tame polynomial with non-zero discriminant, we introduce a new parameter s and work with the ring $\tilde{\mathsf{R}} := \mathsf{R}[s]$ and the tame polynomial $\tilde{f} = f - s$. The polynomial $\Delta(s)$ turns out to be the discriminant of the tame polynomial $f - s$ with coefficients in $\mathsf{R}[s]$. We use the algorithms in [**Mov11**] to calculate the Gauss-Manin connection of $\tilde{f}$. This means that we take the 10×1 matrix Ω made up of the expressions in Equation 6.6 and calculate the 10×10 matrices $A(s), B(s), C(s), D(s)$ and $S(s)$ with entries in $\mathsf{R}[s]$ satisfying the equality

$$\nabla\Omega = \frac{1}{\Delta(s)}\tilde{A}(s) \otimes \Omega,$$

$$\tilde{A}(s) = A(s)da + B(s)db + C(s)dc + D(s)dd + S(s)ds^2$$

Here d stands for both a differential and a parameter. Therefore, dd means the differential of $d \in \mathsf{R}$. For our example (6.2) we have $\Delta(s) = s(\Delta + \Delta_1 s + \cdots)$, $\Delta_i \in \mathsf{R}$. It turns out that $A(0) = B(0) = C(0) = D(0) = 0$. Therefore, we obtain the calculation of the Gauss-Manin connection for f.

$$\nabla\Omega = \frac{1}{\Delta}\tilde{A} \otimes \Omega,$$

$$\tilde{A} = A \cdot da + B \cdot db + C \cdot dc + D \cdot dd$$

where A, B, C and D are 10×10 matrices with entries in R. Using these calculations we can check that:

PROPOSITION 6.3. *We have*

(1) *The free* R*-module generated by*

$$\alpha dx \wedge dy \wedge dw, \alpha = xw^3, xw^2, w^3, xw, w^2, \tag{6.8}$$

is invariant under $\Delta \cdot \nabla$.

(2) $\omega = \alpha dx \wedge dy \wedge dw, \alpha = xy, y$ *is a flat section, that is,* $\nabla(\omega) = 0$.

6.6. Gauss-Manin connection, $\Delta = 0$. Let f be a tame polynomial with zero discriminant. For $P \in \ker(A_f)$ we have $fPdx = df \wedge \omega_P$ for some $\omega_P \in \Omega^n_{\mathbb{U}_1/\mathbb{U}_0}$ and so

$$\begin{aligned} \frac{Pdx}{f^i} &= \frac{fPdx}{f^{i+1}} = \frac{df \wedge \omega_P}{f^{i+1}} \\ &= \frac{1}{i}\left(\frac{d\omega_P}{f^i} - d\left(\frac{\omega_P}{f^i}\right)\right) \\ &= \frac{1}{i}\frac{d\omega_P}{f^i} \quad \text{in } \mathsf{M} \end{aligned}$$

We conclude that

$$\frac{Pdx - \frac{1}{i}d\omega_P}{f^i} = 0, \quad \text{in } \mathsf{M}$$

For $i = 1$ we conclude that there are many R-linear relations between the expressions in Equation 6.6 in M. We find

PROPOSITION 6.4. *For arbitrary a, b, c, d, we have the following equalities in* M*:*

$$\alpha dx \wedge dy \wedge dw = 0 \text{ in } \quad \mathsf{M} \tag{6.9}$$

where α is one of the polynomials:

$xy,$

$y,$

$300dxw^3 - 432bdxw^2 - (216a^2d - 17c^2)w^3 - (48ac^2 - 132d^2)xw$

$- (102acd + 24bc^2)w^2 - 6c^3x - 3c^2dw,$

$150xw^3 - 216bxw^2 - 108a^2w^3 + 66dxw - 51acw^2 - 5c^2w,$

$- (1350ac^2 - 1800d^2)xw^3 + (1944abc^2 - 2592bd^2 + 114c^3)xw^2 + (972a^3c^2 - 1296a^2d^2 + 102c^2d)w^3$

$- (882ac^2d + 144bc^3 - 792d^3)xw + (387a^2c^3 - 612acd^2$

$- 144bc^2d)w^2 \quad 6c^3dx + (18ac^4 - 18c^2d^2)w - c^5.$

---

[2]The data of $\Delta(s), A(s), \cdots$ as a text file is around 500KB.

and for $c = 0$ we have the equalities given in Equation 6.9, where α is one of the polynomials:

$$25xw^3 - 36bxw^2 - 18a^2w^3 + 11dxw,$$
$$xy,$$
$$y,$$
$$-75bxw^3 + (108b^2 - 19d)xw^2 + 54a^2bw^3 - 9bdxw + 12a^2dw^2 - 5d^2x,$$
$$-150axw^3 + 216abxw^2 + (108a^3 - 17d)w^3 - 18adxw + 24bdw^2 - 7d^2w,$$
$$-900abxw^3 + (1296ab^2 - 114ad)xw^2 + (648a^3b - 51bd)w^3 - 108abdxw + (72a^3d + 72b^2d - 11d^2)w^2 - 6ad^2x - 9bd^2w - d^3.$$

6.7. M-**polarized K3 surfaces.** In this section we compute the Gauss-Manin connection for M-polarized K3 surfaces by setting $c = 0$ in Equation 6.2. In Proposition 6.4 we found 6 relations between the differential forms described in Equation 6.6. Let $\mathsf{R} = \mathbb{Q}[a, b, d]$. We consider the R-module generated by four elements

$$\alpha dx \wedge dy \wedge dw, \quad \alpha = 1, w, w^2, w^3,$$

of H and write the Gauss-Manin connection on this module. This means that we consider the 4×1 matrix Ω with the above entries and calculate:

$$\nabla\Omega = \frac{1}{\Delta}(A \cdot da + B \cdot db + D \cdot dd) \otimes \Omega$$

where Δ and all entries of A, B, D are explicit polynomials in a, b, d with coefficients in $\mathbb{Q}$ (see [**Mov**]). For instance,

$$\Delta := a(a^6d^6 - 2a^3b^2d^6 - 2a^3d^7 + b^4d^6 - 2b^2d^7 + d^8)$$

We can use this data and calculate the differential equations (22) and (23) of [**CDLW09**]. We can also calculate the Picard-Fuchs equations of $\omega = dx \wedge dy \wedge dw$ when a, b and d depend on a parameter t. As an example, we compute Picard-Fuchs equations for the modular curves $Y_0(n) + n$, where $n = 2, 3$, using the parametrizations given in [**CDLW09**, §3.2].

For $n = 2$, we have:

$$a = (16 + t)(256 + t), \; b = (-512 + t)(-8 + t)(64 + t), \; d = 2985984t^3$$

Then ω satisfies the Picard-Fuchs equation:

$$y + (26t + 512)y' + (36t^2 + 1536t)y'' + (8t^3 + 512t^2)y''' = 0,$$

with $' = \frac{\partial}{\partial t}$. In a similar way, the curve for $n = 3$ is parametrized by

$$a = (t^2 + 246t + 729)(t + 27)^2, \; b = (t^2 - 486t - 19683)(t^2 + 18t - 27)(t + 27)^2,$$
$$d = 2^{12}3^6t^4(t + 27)^4.$$

It satisfies:

$$(t + 15)y(7t^2 + 192t + 729)y' + (6t^3 + 243t^2 + 2187t)y'' + (t^4 + 54t^3 + 729t^2)y''' = 0.$$

These computations confirm that a third order Picard-Fuchs ordinary differential equation is obtained from our methods, as must occur for a modular curve in the M-polarized locus.

6.8. N-polarized K3 surfaces. In this section we analyze the Gauss-Manin connection for N-polarized K3 surfaces using the full family given in Equation 6.2. By Proposition 6.4, we have 5 relations between the differential forms given in Equation 6.6. Let $\mathsf{R} = \mathbb{Q}[a, b, c, d]$. We consider the R-module generated by

$$\alpha dx \wedge dy \wedge dw, \quad \alpha = 1, w, w^2, w^3, xw \tag{6.10}$$

and calculate the Gauss-Manin connection on this module:

$$\nabla(\omega) = A\omega, \quad A = \frac{1}{\Delta}(Ada + Bdb + Cdc + Ddd).^3 \tag{6.11}$$

Here Δ and all the entries of the 5×5 matrices A, B, C and D are in R. For instance,

$$\begin{aligned}\Delta :=& c(34992a^7c^3d + 23328a^6bc^4 - 11664a^6cd^3 + 3888a^5c^5 - 69984a^4b^2c^3d\\ &-71928a^4c^3d^2 - 46656a^3b^3c^4 + 23328a^3b^2cd^3 - 184680a^3bc^4d + 23328a^3cd^4\\ &-97200a^2b^2c^5 + 46656a^2bc^2d^3 - 37125a^2c^5d + 34992ab^4c^3d - 68040ab^2c^3d^2\\ &-33750abc^6 + 48600ac^3d^3 + 23328b^5c^4 - 11664b^4cd^3 - 48600b^3c^4d + 23328b^2cd^4\\ &+27000bc^4d^2 - 3125c^7 - 11664cd^5)\end{aligned}$$

To check our method, we compute the Picard-Fuchs equation for ω restricted to the *decagon curve*. The decagon curve lies on the Humbert surface $\mathcal{H}_5$, but is non-arithmetic by a construction of McMullen (see [**McM06, McM05**]). Thus, we expect to obtain a fourth order Picard-Fuchs ordinary differential equation.

We calculate the Picard-Fuchs equation of $dx \wedge dy \wedge dw$ restricted to the decagon curve using the parametrization:

$$\begin{aligned} a &= 625(-3+t)^2,\\ b &= -(625/2)(-1134 + 1458t - 504t^2 + 23t^3),\\ c &= -(759375/4)(-2+t)^2(2+t)^4,\\ d &= 18984375/4(-2+t)^2(2+t)^4(9+2t).\end{aligned}$$

It is the fourth order differential equation:

$$504y + 9000ty' + 500(31t^2 - 44)y'' + 6250t(t-2)(t+2)y''' + 625(t-2)^2(t+2)^2y'''' = 0$$

In contrast, if we had chosen a curve that was a component of the intersection of two Humbert surfaces, the curve would have been either a modular or a Shimura curve, and the resulting Picard-Fuchs ordinary differential equation would have been third order. If we had chosen a curve which does not lie on a Humbert surface, the Picard-Fuchs equation would have been fifth order.

We have replicated this computation using the restriction of Picard-Fuchs differential equations obtained in § 5.

6.9. A canonical basis. The choice of the differential forms described in Equation 6.10 is not canonical. Moreover, it is not compatible with the Hodge filtration. In this section, we describe a basis which is compatible with the Hodge filtration. (The compatibility follows from Griffiths transversality.)

[3]The data of $A, B, \cdots$ as a text file is around 50KB.

For any function I in a, b, c, d let us define:

$$\partial_a I = I_a := -4a\frac{\partial I}{\partial a}, \ \partial_b I = I_b := -6b\frac{\partial I}{\partial b},$$
$$\partial_c I = I_c := -10c\frac{\partial I}{\partial c}, \ \partial_d I = I_d := -12d\frac{\partial I}{\partial d}.$$

Note that our notation for I_a and ω_a is different from the notation used in § 2.2. Let k be the fractional field of R. The k-vector V space generated by $\omega := dx \wedge dy \wedge dw$ and all its derivatives is at most 5 dimensional. We have

$$\omega_a + \omega_b + \omega_c + \omega_d - \omega = 0$$

and so we cannot take $\omega, \omega_a, \cdots$ as a basis of V. Compare the equation above to Equation 5.4. Our calculations show that we can take

$$\omega, \omega_b, \omega_c, \omega_d, \omega_{dd}$$

as a basis of V. We calculate the Gauss-Manin connection in this basis. In other words, we calculate all 5×5 matrices M^x, $x = a, b, c, d$ yielding equalities of the form:

$$\begin{pmatrix} \omega_x \\ \omega_{bx} \\ \omega_{cx} \\ \omega_{dx} \\ \omega_{ddx} \end{pmatrix} = M^x \begin{pmatrix} \omega, \\ \omega_b \\ \omega_c \\ \omega_d \\ \omega_{dd} \end{pmatrix}, \quad x = a, b, c, d. \tag{6.12}$$

Some of the information contained in the matrices M^x is not very interesting; for instance, the first row in the matrix M^b is simply $(0, 1, 0, 0, 0)$, which just encodes the fact that $\omega_b = \omega_b$.

6.10. The differential algebra annihilating the holomorphic 2-form. Let I be the left ideal of $\mathbb{Q}(a, b, c, d)[\partial_a, \partial_b, \partial_c, \partial_d]$ containing all differential operators which annihilate $\omega := dx \wedge dy \wedge dw$. The ideal I is generated by the differential operators obtained from the 20 equalities in Equation 6.12. Of these 20 equalities, 4 are trivial equalities such as $\omega_b = \omega_b$ and so on. There are 3 pairs of equations which are repeated. For instance, we have two equations of the type $\omega_{bc} = \cdots$, $\omega_{cb} = \cdots$. The 13 non-trivial differential operators have the following form. One operator is $P\omega = 0$, where P is

$$\partial_a + \partial_b + \partial_c + \partial_d - 1 \tag{6.13}$$

We have 8 second-order differential equations with a differential operator of the form:

$$p_1 + p_2\partial_b + p_3\partial_c + p_4\partial_d + p_5\partial_d\partial_d - X \tag{6.14}$$

$$X = \partial_b\partial_b, \ \partial_c\partial_c, \ \partial_b\partial_c, \ \partial_c\partial_d, \ \partial_b\partial_d, \ \partial_b\partial_a, \ \partial_c\partial_a, \ \partial_d\partial_a.$$

Here, the rational coefficients $p_i \in \mathsf{k}$ depend on the choice of X. Four of the operators are third-order differential equations with the differential operators given in Equation 6.14, where X is

$$X = \partial_d\partial_d\partial_a, \ \partial_d\partial_d\partial_b, \ \partial_d\partial_d\partial_c, \ \partial_d\partial_d\partial_d.$$

Together, these $13 = 1+8+4$ differential operators generate the differential module which annihilates the holomorphic differential form ω.

6.11. Differential equations for Humbert surfaces. Let us consider the system described in Equation 6.11. We consider a, b, c, d as functions of three other parameters u, v and w. In this way $da = \frac{\partial a}{\partial u}du + \frac{\partial a}{\partial v}dv + \frac{\partial a}{\partial w}dw$ and so on. More generally, we consider three vector fields ∂_u, ∂_v, ∂_w on the (a, b, c, d)-space. Therefore, we may have $\partial_u\partial_v \neq \partial_v\partial_u$. What we first calculate is the 3×5 matrix R in the equality:

$$\begin{pmatrix} \omega_u \\ \omega_{uv} \\ \omega_{uvw} \end{pmatrix} = R \begin{pmatrix} \omega_1 \\ \omega_2 \\ \omega_3 \\ \omega_4 \\ \omega_5 \end{pmatrix}$$

where ω_i, $i = 1, 2, \ldots, 5$ are the differential forms given in Equation 6.10. For instance, the first row of R is the first row of the matrix

$$\frac{1}{\Delta}(Aa_u + Bb_u + Cc_u + Dd_u).$$

The entries of the second row are in

$$\mathbb{Q}(a, b, c, d)[a_u, b_u, c_u, d_u, a_v, b_v, c_v, d_v, a_{uv}, b_{uv}, c_{uv}, d_{uv}]$$

and the entries of the third row are in [4]

$$\mathbb{Q}(a, b, c, d)[a_u, b_u, c_u, d_u, a_v, b_v, c_v, d_v, a_{uv}, b_{uv}, c_{uv}, d_{uv}, a_w, \\ b_w, c_w, d_w, a_{uw}, b_{uw}, c_{uw}, d_{uw}, a_{vw}, b_{vw}, c_{vw}, d_{vw}, a_{uvw}, b_{uvw}, c_{uvw}, d_{uvw}].$$

We then calculate the partial differential equations

$$\det(R^x) = 0, \quad x = uu,\ vu,\ uvu,\ vv,\ uvv \tag{6.15}$$

where

$$\begin{pmatrix} \omega \\ \omega_u \\ \omega_v \\ \omega_{uv} \\ \omega_x \end{pmatrix} = R^x \begin{pmatrix} \omega_1 \\ \omega_2 \\ \omega_3 \\ \omega_4 \\ \omega_5 \end{pmatrix}.$$

We assume that $\omega, \omega_u, \omega_v, \omega_{uv}$ are linearly independent. If not, we may replace ω_{uv} with ω_{vv}, ω_{vu} or ω_{uu}. This can be checked computationally becuase we have to write these in terms of ω_i, $i = 1, .., 5$ and check that the resulting 4×5 matrix has rank 4, see the example in the next section. Equations (6.15) imply that the subspace of the de Rham cohomology $H^2_{\mathrm{dR}}(X(a, b, c, d))$ spanned by $\omega, \omega_u, \omega_v, \omega_{uv}$ is closed under derivations with respect to u and v. Since in our list of four elements we do not have ω_{vv}, $\det(R^{vvv}) = 0$ does not appear in (6.15). If $\partial_u\partial_v = \partial_v\partial_u$ then Equations 6.15 are reduced to:

$$\det(R^x) = 0, \ x = uu, vv, uuv, uvv \tag{6.16}$$

Note that in (6.16) two equations corresponding to $x = uu, uuv$ are obtained from the other two by changing the role of u and v.

[4]The size of the matrix R stored in a computer is 5.9 MB.

6.12. Checking well-known Humbert surfaces. Suppose we are given a hypersurface

$$Z(H) := \{H(a,b,c,d) = 0\}.$$

We want to check that whether $Z(H)$ is tangent to the partial differential equations given in Equation 6.15. Let us take one of a, b, c or d as a constant. If the compactification of $Z(H)$ is birational to $\mathbb{P}^1 \times \mathbb{P}^1$, then in theory we could parametrize $Z(H)$ with algebraic coordinates (u, v) and check whether such a parametrization satisfies the differential equations given in Equation 6.15. In practice, however, such a parametrization might be difficult to find. There is another algebraic method which we explain below:

We consider two linearly independent algebraic vector fields

$$U = \sum_{x=a,b,c,d} U^x \frac{\partial}{\partial x}, \ V = \sum_{x=a,b,c,d} V^x \frac{\partial}{\partial x}, \ U^x, V^x \in \mathbb{Q}[a,b,c,d].$$

in the parameter space which are tangent to $Z(H)$. For instance, take

$$U = \frac{\partial H}{\partial b}\frac{\partial}{\partial a} - \frac{\partial H}{\partial a}\frac{\partial}{\partial b}$$

$$V = \frac{\partial H}{\partial d}\frac{\partial}{\partial c} - \frac{\partial H}{\partial c}\frac{\partial}{\partial d}$$

Let u and v be a (transcendental) coordinate system in the domain of a solution of U and V, respectively. All the parameters a, b, c, d become functions of u and v, but not simultaneously:

$$x_u = U^x, x_v = V^x, \quad x = a, b, c, d.$$

Note that $[\frac{\partial}{\partial u}, \frac{\partial}{\partial u}]$ may not be zero and so we may have $x_{uv} \neq x_{vu}$. Now, we can use the chain rule and calculate the derivatives a_{uv}, etc. The remainder of the computation is just substitution and checking the equalities given in Equation 6.15. Using this method, we have checked that the differential equations described in Equation 6.15 are tangent to the zero set of

$$H := (d - b^2 - a^3)^2 - 4a(c - ab)^2.$$

References

[BCP97] Wieb Bosma, John Cannon, and Catherine Playoust, *The Magma algebra system. I. The user language*, J. Symbolic Comput. **24** (1997), no. 3-4, 235–265, DOI 10.1006/jsco.1996.0125. Computational algebra and number theory (London, 1993). MR1484478

[Bel97] sarah-marie belcastro, *Picard lattices of families of K3 surfaces*, ProQuest LLC, Ann Arbor, MI, 1997. Thesis (Ph.D.)–University of Michigan. MR2695492

[CD07] Adrian Clingher and Charles F. Doran, *Modular invariants for lattice polarized K3 surfaces*, Michigan Math. J. **55** (2007), no. 2, 355–393, DOI 10.1307/mmj/1187646999. MR2369941 (2009a:14049)

[CD12] Adrian Clingher and Charles F. Doran, *Lattice polarized K3 surfaces and Siegel modular forms*, Adv. Math. **231** (2012), no. 1, 172–212, DOI 10.1016/j.aim.2012.05.001. MR2935386

[CDLW09] Adrian Clingher, Charles F. Doran, Jacob Lewis, and Ursula Whitcher, *Normal forms, K3 surface moduli, and modular parametrizations*, Groups and symmetries, CRM Proc. Lecture Notes, vol. 47, Amer. Math. Soc., Providence, RI, 2009, pp. 81–98. MR2500555 (2011c:14104)

[CK99] David A. Cox and Sheldon Katz, *Mirror symmetry and algebraic geometry*, Mathematical Surveys and Monographs, vol. 68, American Mathematical Society, Providence, RI, 1999. MR1677117 (2000d:14048)

[DGJ08] Charles Doran, Brian Greene, and Simon Judes, *Families of quintic Calabi-Yau 3-folds with discrete symmetries*, Comm. Math. Phys. **280** (2008), no. 3, 675–725, DOI 10.1007/s00220-008-0473-x. MR2399610 (2009g:14044)

[Dol82] Igor Dolgachev, *Weighted projective varieties*, Group actions and vector fields (Vancouver, B.C., 1981), Lecture Notes in Math., vol. 956, Springer, Berlin, 1982, pp. 34–71, DOI 10.1007/BFb0101508. MR704986 (85g:14060)

[Dol96] I. V. Dolgachev, *Mirror symmetry for lattice polarized $K3$ surfaces*, J. Math. Sci. **81** (1996), no. 3, 2599–2630, DOI 10.1007/BF02362332. Algebraic geometry, 4. MR1420220 (97i:14024)

[EK14] Noam Elkies and Abhinav Kumar, *K3 surfaces and equations for Hilbert modular surfaces*, Algebra Number Theory **8** (2014), no. 10, 2297–2411, DOI 10.2140/ant.2014.8.2297. MR3298543

[Fre83] E. Freitag, *Siegelsche Modulfunktionen* (German), Grundlehren der Mathematischen Wissenschaften [Fundamental Principles of Mathematical Sciences], vol. 254, Springer-Verlag, Berlin, 1983. MR871067 (88b:11027)

[GN97] Valeri A. Gritsenko and Viacheslav V. Nikulin, *Siegel automorphic form corrections of some Lorentzian Kac-Moody Lie algebras*, Amer. J. Math. **119** (1997), no. 1, 181–224. MR1428063 (98g:11056)

[Gro66] A. Grothendieck, *On the de Rham cohomology of algebraic varieties*, Inst. Hautes Études Sci. Publ. Math. **29** (1966), 95–103. MR0199194 (33 #7343)

[Gun90] Robert C. Gunning, *Introduction to holomorphic functions of several variables. Vol. I*, The Wadsworth & Brooks/Cole Mathematics Series, Wadsworth & Brooks/Cole Advanced Books & Software, Pacific Grove, CA, 1990. Function theory. MR1052649 (92b:32001a)

[KO68] Nicholas M. Katz and Tadao Oda, *On the differentiation of de Rham cohomology classes with respect to parameters*, J. Math. Kyoto Univ. **8** (1968), 199–213. MR0237510 (38 #5792)

[Kum08] Abhinav Kumar, *$K3$ surfaces associated with curves of genus two*, Int. Math. Res. Not. IMRN **6** (2008), Art. ID rnm165, 26, DOI 10.1093/imrn/rnm165. MR2427457 (2009d:14044)

[McM05] Curtis T. McMullen, *Teichmüller curves in genus two: the decagon and beyond*, J. Reine Angew. Math. **582** (2005), 173–199, DOI 10.1515/crll.2005.2005.582.173. MR2139715 (2006a:32017)

[McM06] Curtis T. McMullen, *Teichmüller curves in genus two: torsion divisors and ratios of sines*, Invent. Math. **165** (2006), no. 3, 651–672, DOI 10.1007/s00222-006-0511-2. MR2242630 (2007f:14023)

[Mov] Hossein Movasati, *Gauss-Manin connection of a family*, `http://w3.impa.br/~hossein/k3surfaces`.

[Mov08] Hossein Movasati, *On differential modular forms and some analytic relations between Eisenstein series*, Ramanujan J. **17** (2008), no. 1, 53–76, DOI 10.1007/s11139-006-9009-1. MR2439525 (2009k:11075)

[Mov11] Hossein Movasati, *Multiple integrals and modular differential equations*, Publicações Matemáticas do IMPA. [IMPA Mathematical Publications], Instituto Nacional de Matemática Pura e Aplicada (IMPA), Rio de Janeiro, 2011. 28° Colóquio Brasileiro de Matemática. [28th Brazilian Mathematics Colloquium]. MR2827610 (2012f:37104)

[Nag] Atsuhira Nagano, *A period differential equation for a family of k3 surfaces and the hilbert modular orbifold for the field $\mathbb{Q}(\sqrt{5})$*, arXiv:1009.5725[math.AG].

[Ram00] S. Ramanujan, *On certain arithmetical functions [Trans. Cambridge Philos. Soc.* **22** *(1916), no. 9, 159–184]*, Collected papers of Srinivasa Ramanujan, AMS Chelsea Publ., Providence, RI, 2000, pp. 136–162. MR2280861

[SY89] Takeshi Sasaki and Masaaki Yoshida, *Linear differential equations modeled after hyperquadrics*, Tohoku Math. J. (2) **41** (1989), no. 2, 321–348, DOI 10.2748/tmj/1178227829. MR996019 (90h:32050)

Department of Mathematical and Statistical Sciences, University of Alberta, Edmonton AB T6G 2G1
E-mail address: doran@math.ualberta.ca

Department of Mathematical and Statistical Sciences, University of Alberta, Edmonton AB T6G 2G1
E-mail address: aharder@ualberta.ca

Instituto de Matemática Pura e Aplicada, IMPA, Estrada Dona Castorina, 110, 22460-320, Rio de Janeiro, RJ, Brazil
E-mail address: hossein@impa.br

Department of Mathematics, Hibbard Humanities Hall 508, University of Wisconsin-Eau Claire, Eau Claire Wisconsin 54702
E-mail address: whitchua@uwec.edu

Proceedings of Symposia in Pure Mathematics
Volume **93**, 2016
http://dx.doi.org/10.1090/pspum/093/01575

Superconformal field theories and cyclic homology

Richard Eager

Abstract. One of the predictions of the AdS/CFT correspondence is the matching of protected operators between a superconformal field theory and its holographic dual. We review the spectrum of protected operators in quiver gauge theories that flow to superconformal field theories at low energies. The spectrum is determined by the cyclic homology of an algebra associated to the quiver gauge theory. Identifying the spectrum of operators with cyclic homology allows us to apply the Hochschild-Kostant-Rosenberg theorem to relate the cyclic homology groups to deRham cohomology groups. The map from cyclic homology to deRham cohomology can be viewed as a mathematical avatar of the passage from open to closed strings under the AdS/CFT correspondence.

1. Introduction

One of the most basic predictions of the AdS/CFT correspondence is the equivalence between the operators in a conformal field theory and its dual gravitational theory. Since a conformal field theory is determined by its collection of operators and their correlation functions, the matching of operators and their scaling dimensions on both sides of the duality is a natural starting point to understand the duality. However the spectrum of operators can vary with the coupling and it is often impossible to determine the scaling dimensions of operators at strong coupling without resorting to the AdS/CFT correspondence. For supersymmetric gauge theories that flow to $\mathcal{N} = 1$ superconformal field theories (SCFTs) at low energies, there is a class of BPS operators with protected scaling dimensions. We review how these protected operators can be determined using $\mathcal{Q}$-cohomolgy or equivalently cyclic homology.

The key new idea is re-writing the $\mathcal{Q}$-cohomology groups in terms of cyclic homology. This allows many techniques and theorem developed in mathematics to be applied to the analysis of quiver gauge theories. In particular for quiver gauge theories that are dual to type IIB string theory on AdS_5 times a Sasaki-Einstein manifold, the equality of the gauge theory and gravity superconformal indices was proved in [**5**] using the Hochschild-Kostant-Rosenberg (HKR) theorem relating cyclic homology of an algebra associated to the quiver gauge theory to the deRham cohomology of the Calabi-Yau cone over the Sasaki-Einstein manifold.

Quiver gauge theories dual to Sasaki-Einstein manifolds have been intensely studied, but they are at very special points in the moduli space of the SCFT that

2010 *Mathematics Subject Classification.* Primary 14J81; Secondary 16E40.

Letter	(j_1, j_2)	$\mathcal{I}$
ϕ	$(0,0)$	t^{3r}
$\overline{\psi}_{\dot{2}}$	$(0,1/2)$	$-t^{3(2-r)}$
$\partial_{\pm -}$	$(\pm 1/2, 1/2)$	$t^3 y^{\pm 1}$

Letter	(j_1, j_2)	$\mathcal{I}$
λ_1	$(1/2,0)$	$-t^3 y$
λ_2	$(-1/2,0)$	$-t^3 y^{-1}$
$\overline{f}_{\dot{2}\dot{2}}$	$(0,1)$	t^6
$\partial_{\pm -}$	$(\pm 1/2, 1/2)$	$t^3 y^{\pm 1}$

TABLE 1. Fields contributing to the $\mathcal{Q}$-cohomology groups, from a chiral multiplet (left) and from a vector multiplet (right). The equation of motion $\partial_{22}\lambda_1 = \partial_{12}\lambda_2$ must also be accounted for [**4**, **8**].

they flow to at low energy. In this paper we take the first steps to apply cyclic homology to determine the spectrum of general quiver gauge theories. While the superconformal index is constant over the SCFT moduli space, the indivual $\mathcal{Q}$-cohomolgy groups can vary over the moduli space. We illustrate this phenomena with the β-deformation of $\mathcal{N} = 4$ supersymmetric Yang-Mills in section 3.2. In this case, the HKR theorem no longer applies, but we expect the matching of operators in the AdS/CFT correspondence can still be viewed as the passage from cyclic homology to Poisson homology.

This paper is based on a talk the author gave at String-Math 2014 and reviews joint work with J. Schmude and Y. Tachikawa that appeared in [**5**]. Section 2 reviews the identification of protected BPS operators with elements of cyclic homology. Section 3 applies these techniques to two familiar examples. The first example is $\mathcal{N} = 4$ supersymmetric Yang-Mills theory and is covered in section 3.1. The second example in section 3.2 is a new application of cylic homology to analyze the β-deformation of $\mathcal{N} = 4$ supersymmetric Yang-Mills, extending the original analysis in [**2**]. Cyclic homology can also be used to analyze the other marginal and relevant deformations of $\mathcal{N} = 4$ supersymmetric Yang-Mills theory [**2**, **14**]. Some preliminary results in this direction will be briefly discussed in the final section and are part of work-in-progress.

2. Cyclic Homology of Quiver Gauge Theories

2.1. BPS Operators in SCFTs. We consider $\mathcal{N} = 1$ gauge theories on $S^1 \times S^3$ with a Lagrangian description that flow to $\mathcal{N} = 1$ SCFTs at low energies. In these theories, a special role is played by BPS operators, which are annihilated by one of the supercharges $\mathcal{Q} = \overline{Q}_{\dot{1}}$. The quantum numbers of BPS operators saturate the following inequality

$$\{\mathcal{Q}, \mathcal{Q}^\dagger\} = E - 2j_2 - \frac{3}{2}r \geq 0,$$

where E is the energy, r is the R-charge, and (j_1, j_2) are the Lorentz spins. This is the unitary bound for the $\mathcal{N} = 1$ superconformal algebra. States saturating this bound are called BPS. Our goal is to compute the $\mathcal{Q}$-cohomology for quiver gauge theories in the large N limit. The superconformal primaries of the $\mathcal{N} = 1$ superconformal algebra are by definition annihilated by $\mathcal{Q}^\dagger$. Since states in $\mathcal{Q}$-cohomology are $\mathcal{Q}$-closed, these states satisfy the unitary bound. In the quiver gauge theories that we are considering, BPS operators are formed by traces of products of letters that satisfy the bound $E - 2j_2 - \frac{3}{2}r \geq 0$. We list these letters in Table 1.

The $\mathcal{Q}$-cohomology groups are labeled by the $SU(2)_r$ spin j_2. The zeroth cohomology group consists of the elements of the chiral ring [**19**]. The higher cohomology groups correspond to short representations of the superconformal algebra known as 1/4 BPS operators[1]. Detailed information about these short representations can be found in [**4**, **8**].

The single-trace index is the Euler characteristic of these cohomology groups, and it is defined by

$$\mathcal{I}_{s.t.} = \mathrm{Tr}_{single-trace\ op.}(-1)^F t^{2(E+j_2)} y^{2j_1}.$$

From the single-trace index, the full superconformal index can be determined [**12**, **19**]. In the large-N limit, the contribution to the index from the letters ∂_α and λ_α cancels [**5**]. We therefore restrict our attention to the operators constructed from gauge invariant operators formed from words in the letters $\phi, \overline{\psi}_2$, and $\overline{f}_{22}$. Since the superconformal index is a more robust and easier to calculate quantity than the individual $\mathcal{Q}$-cohomology groups, it serves as an important guide and check on our understanding of $\mathcal{Q}$-cohomology.

While the index is invariant under exactly marginal deformations, the individual $\mathcal{Q}$-cohomology groups can jump. As the values of the coupling constants approach special loci in moduli space, the anomalous dimension of a long multiplet can decrease until it saturates a BPS bound. At this loci in moduli space the long multiplet can decompose into a direct sum of short multiplets. The short multiplets will contribute to the $\mathcal{Q}$-cohomology groups, so the $\mathcal{Q}$-cohomology will generically jump at the special loci. However, the index is constructed so that the contributions to the index from short multiplets that can form a long multiplet cancel. Therefore, the index is invariant under exactly marginal deformations. We will see an explicit example of this phenomena when we examine the β-deformation in section 3.2. Another important example is the change in the $\mathcal{Q}$-cohomology groups when a gauge coupling vanishes.

2.2. Quiver Gauge Theories. A gauge theory consists of a gauge group G along with matter fields in a representation V of G and interactions encoded by a Lagrangian. Theories with $\mathcal{N} = 1$ supersymmetry can be efficiently described using superspace. A large class of $\mathcal{N} = 1$ supersymmetric gauge theories can be specified by a quiver and a superpotential. A quiver $Q = (Q_0, Q_1, h, t)$ is a collection of vertices Q_0 and arrows Q_1 along with maps $h, t : Q_1 \to Q_0$ which specify the head and tail of an arrow. Given a quiver Q and superpotential W, we can define a gauge theory by setting the gauge group to be

$$G = \prod_{v \in Q_0} U(N_v).$$

To each arrow $a \in Q_1$ we assign a chiral superfield Φ_a which transforms in the fundamental representation of $U(N_{h(a)})$ and the anti-fundamental representation of $U(N_{t(a)})$. If a is a closed loop, then the superfield transforms in the adjoint representation of $U(N_{h(a)})$. Finally the superpotential W is a sum of gauge invariant operators. Gauge invariance requires that the superpotential W is a linear combination of cyclic words in the quiver. A fancier way of writing this is

[1]These are 1/16 BPS operators in $\mathcal{N} = 4$ SYM.

Letter	Ginzburg DG	grading
ϕ	x_e	0
$\overline{\psi}_{\dot{2}}$	x_e^*	1
$\overline{f}_{\dot{2}\dot{2}}$	t_v	2
$\overline{Q}_{\dot{1}}$	d	-1

TABLE 2. The dictionary between physical fields and generators of Ginzburg's DG algebra.

$W \in \mathbb{C}Q_{cyc} = \mathbb{C}Q/[\mathbb{C}Q, \mathbb{C}Q]$. From a quiver with potential (Q, W) we can construct its superpotential algebra $\mathcal{A}_{Q,W} = \mathbb{C}Q/(\partial W)$.

The action of the supercharge $\mathcal{Q}$ on a quiver gauge theory is

$$\begin{aligned}
\mathcal{Q}\phi_e &= 0 \\
\mathcal{Q}\overline{\psi}_{e,\dot{2}} &= \frac{\partial W(\phi)}{\partial \phi_e} \\
\mathcal{Q}\overline{f}_{v,\dot{2}\dot{2}} &= \sum_{h(e)=v} \phi_e \overline{\psi}_{e,\dot{2}} - \sum_{t(e)=v} \overline{\psi}_{e,\dot{2}} \phi_e.
\end{aligned}$$

A direct calculation shows that $\mathcal{Q}$ is nilpotent

$$\begin{aligned}
\mathcal{Q}^2 \overline{f}_{v,\dot{2}\dot{2}} &= \sum_{h(e)=v} \phi_e \frac{\partial W(\phi)}{\partial \phi_e} - \sum_{t(e)=v} \frac{\partial W(\phi)}{\partial \phi_e} \phi_e \\
&= 0.
\end{aligned}$$

While physically obvious, this relationship is a syzygy in the superpotential algebra $\mathcal{A}$ [**13**]. Thus we have uncovered a direct physical explanation for the appearance of this syzygy in the pioneering definition of Calabi-Yau algebras in [**1**].

2.3. Ginzburg's DG Algebra. The algebra generated by the letters $\phi_e, \overline{\psi}_{e,\dot{2}}$, and $\overline{f}_{v,\dot{2}\dot{2}}$ with the differential $\mathcal{Q}$ is known as Ginzburg's DG algebra [**9**]. In the DG algebra, the generators are denoted by x_e, x_e^*, and t_v and the differential is denoted by d. This dictionary is shown in Table 2. To the quiver Q we associate the quiver $\widehat{Q}$ with the same vertices as Q in the following way. For each arrow $e \in Q$ there is a corresponding arrow x_e in $\widehat{Q}$ and an arrow x_e^* with the opposite orientation. For each vertex v of Q there is a loop t_v based at vertex v in $\widehat{Q}$. The grading is simply twice the j_2 spin. In particular the letters x_e, x_e^*, and t_v are assigned charges 0,1, and 2 under the grading. The differential d has charge -1 under the grading. Let $\mathfrak{D} = \mathbb{C}\langle x_e, x_e^*, t_v \rangle$ be the free DG algebra generated by the paths x_e, x_e^*, and t_v. The degree zero homology is $HH_0(\mathfrak{D}) = \mathbb{C}\langle x_e \rangle / \left(\partial W(x)/\partial x_e\right)$. Thus, the degree zero homology $HH_0(\mathfrak{D})$ is isomorphic to the superpotential algebra $\mathcal{A}$. If the algebra $\mathcal{A}$ is a Calabi-Yau algebra of dimension three, then all positive degree homology groups $HH_{>0}(\mathfrak{D})$ vanish [**9**]. In particular, there is a quasi-isomorphism of chain complexes $\mathfrak{D} \twoheadrightarrow \mathcal{A}$. Here $\mathcal{A}$ represents the complex with the algebra $\mathcal{A}$ in degree zero and zero in all other degrees.

2.4. Cyclic Homology. Let $[\mathfrak{D}, \mathfrak{D}]$ be the $\mathbb{C}$-linear space spanned by the commutators of elements in $\mathfrak{D}$. Then $\mathfrak{D}_{cyc} := \mathfrak{D}/(\mathbb{C} + [\mathfrak{D}, \mathfrak{D}])$ is the space of cyclic words in $\mathfrak{D}$ up to scalar multiplication, or in other words, $\mathfrak{D}_{cyc}$ is the $\mathbb{C}$-vector space generated by the space of paths in $\widehat{Q}$. Equivalently, $\mathfrak{D}_{cyc}$ is the $\mathbb{C}$-vector space of single-trace operators formed by $\phi_e, \overline{\psi}_{e,\dot{2}}$, and $\overline{f}_{v,\dot{2}\dot{2}}$. For large-N quiver gauge theories with unitary groups, the gauge-invariant operators correspond to the closed paths in the Ginzburg quiver $\widehat{Q}$. The BPS operators are then elements of the homology groups $H_\bullet(\mathfrak{D}_{cyc}, d)$ where the differential d in the DG algebra is the SUSY differential $\mathcal{Q}$. If $\mathcal{A}$ is a Calabi-Yau algebra of dimension three, then since $\mathfrak{D} \twoheadrightarrow \mathcal{A}$ is a quasi-isomorphism there is an isomorphism [**6**]

$$\overline{HC}_\bullet(\mathcal{A}) \cong H_\bullet(\mathfrak{D}_{cyc}, d).$$

Thus the physical $\mathcal{Q}$-cohomology obtained from the homology of Ginzburg's DG algebra is isomorphic to the reduced (negative) cyclic homology of the superpotential algebra $\mathcal{A}$.

As an example, we consider the spin zero chiral primary operators corresponding to elements of the zeroth cyclic homology group $\overline{HC}_0(\mathcal{A}) \cong \mathcal{A}/[\mathcal{A},\mathcal{A}] \cong \mathcal{A}_{cyc}$. These operators can be easily understood as follows [**2, 3**]. To each element a of the path algebra $\mathcal{A}$, a representation associates an operator $\mathcal{O}(a)$. By the cyclic property of the trace

$$\mathcal{O}(ab) = \mathrm{Tr}(ab) = \mathrm{Tr}(ba) = \mathcal{O}(ba),$$

so the representation must factor through the map

$$\mathcal{A} \to \mathcal{A}/[\mathcal{A},\mathcal{A}] \cong \mathcal{A}_{cyc}.$$

In other words, if the F-term equations imply that $\mathrm{Tr}(ab) = \gamma\,\mathrm{Tr}(ba)$ for a constant $\gamma \neq 1$, then the operator $\mathcal{O}(ab)$ must be $\mathcal{Q}$-exact. Therefore the operator cannot be a chiral primary. We will use this argument to find the chiral primaries in the β-deformation in section 3.2.

3. Examples

3.1. $\mathcal{N}=4$ Super Yang-Mills. In this section we determine the spectrum of protected operators in $\mathcal{N}=4$ super Yang-Mills theory with respect to a $\mathcal{N}=1$ superconformal subgroup of the $\mathcal{N}=4$ superconformal group. The matter content of the $\mathcal{N}=4$ super Yang-Mills theory consists of three $\mathcal{N}=1$ adjoint chiral multiplets x, y, z and a vector multiplet. The superpotential is $W = xyz - xzy$. The F-term relations imply that the three variables x, y, and z commute and so the superpotential algebra

$$\mathcal{A} \cong \mathbb{C}\langle x,y,z\rangle/\langle xy-yx, yz-zy, zx-xz\rangle \cong \mathbb{C}[x,y,z] \cong A$$

is commutative. In this example, $X = \operatorname{Spec} A$ can be thought of a the variety $\mathbb{C}^3$.

The Hochschild homology of a commutative algebra A is isomorphic to the space of algebraic differential forms,

$$\Omega^n_A \cong HH_n(A).$$

In this example, $HH_0(A)$ is spanned by the polynomials in x, y, z and $HH_1(A)$ is spanned by the differential forms $f dx + g dy + h dz$, where f, g, h are polynomials in x, y, z. This follows from the Hochschild-Kostant-Rosenberg theorem. Using the

	1	t^2	t^4	t^6	t^8	t^{10}	t^{12}	...
HC_0	1	3	6	10	15	21	28	...
HC_1	0	0	3	8	15	24	35	...
HC_2	0	0	0	1	3	6	10	...
$\mathcal{I}_{s.t}(t)$	1	3	3	3	3	3	3	...

TABLE 3. Cyclic homology group dimensions for $\mathcal{N} = 4$ SYM

relation between Hochschild and cyclic homology groups [**5**]

$$\begin{aligned} HH_0 &= HC_0, \\ HH_1 &= HC_0 \oplus HC_1, \\ HH_2 &= HC_1 \oplus HC_2, \\ HH_3 &= \overline{HC}_2, \end{aligned}$$

we determine the cyclic homology groups of A. The cyclic homology groups and their contribution to the single-trace index are displayed in table 3. The states reproduce the supergravity states contributing to the superconformal index [**5**, **10**, **12**]. Note that all higher cyclic homology groups vanish. This corresponds to the absence of higher-spin particles in the gravity theory.

As an example we explain the determination of $\overline{HC}_0(\mathcal{A})$ in more detail. Following the original discussion in [**21**], we consider the operators

$$\mathcal{O} = \operatorname{Tr} \Phi^{z_1} \Phi^{z_2} \dots \Phi^{z_k}$$

where Φ is an x, y, or z field. If the operator is symmetric in its indices then it is in a short representation. If not, then (part of) the operator is a descendant. The descendants all arise from F-term equations, which is equivalent to being $\mathcal{Q}$-exact or d-exact in Ginzburg's DG algebra. Therefore, the degree 0 piece of $H_\bullet(\mathfrak{D}_{cyc}, d)$ consists of the chiral primary operators.

In this example it is possible to match all of the protected operators between the gauge theory and the protected Kaluza-Klein operators in type IIB supergravity on $AdS_5 \times S^5$. Of course this is much easier using $\mathcal{N} = 4$ superconformal representations, but we will illustrate the method only using a $\mathcal{N} = 1$ subalgebra. The operator $\mathcal{O}$ has conformal dimension k and transforms in the k-th symmetric representation, $\mathrm{Sym}^k \mathbf{3}$, of $SU(3)$. By the usual AdS/CFT relation, this operator is dual to a supergravity state of spin zero and mass

$$m^2 = k(k-4).$$

These operators precisely correspond to the Kaluza-Klein modes labeled by $h^\alpha_\alpha - a_{\alpha\beta\gamma\delta}$ originally determined in [**11**] and reproduced in figure 3.1. There, the operators $\mathbf{20}, \mathbf{50}, \mathbf{105}$ are in the symmetric-traceless representations of $SO(6)$. However, only the operators in the symmetric representations of $SU(3)$ are primaries with respect to a fixed $\mathcal{N} = 1$ subalgebra of the $\mathcal{N} = 4$ superconformal group.

In the examples in [**5**] where X is a cone over a Sasaki-Einstein manifold, the algebra $\mathcal{A}$ will have the same Hochschild and cyclic homology as a (commutative) variety X. Even though $\mathcal{A}$ and A are not isomorphic in this case, our general strategy can still be applied. We can again use the equality of the cyclic homology

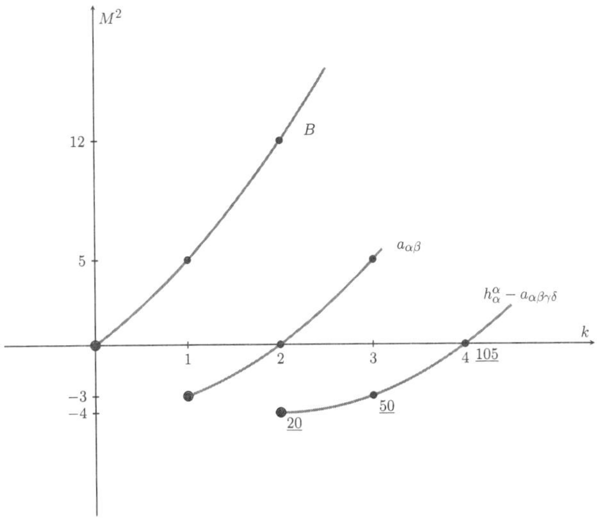

FIGURE 1. Kaluza-Klein mass spectrum of low-lying scalar excitations on $AdS_5 \times S^5$ [**11**].

of $\mathcal{A}$ and X to related the spectrum of protected operators in the gauge theory described by $\mathcal{A}$ to the protected spectrum of Kaluza-Klein modes on X.

3.2. β-deformation. We now determine the spectrum of protected operators in the β-deformation of $\mathcal{N} = 4$ super Yang-Mills theory. The β-deformation is a quiver gauge theory with potential $W = qxyz - q^{-1}xzy$ where $q = e^{i\beta}$. The F-term relations are

$$\begin{aligned} xy &= q^{-2}yx \\ yz &= q^{-2}zy \\ zx &= q^{-2}xz \end{aligned}$$

which shows that the matrices parametrizing the moduli space of vacua are non-commutative. The cyclic homology groups were computed in [**20**]. From the graded components of the cylic homology group, we extract the number of BPS operators with fixed j_2 spin and their contribution to the index. The operators and their contribution to the index are listed in table 4. We see that while the individual cohomology groups differ from those in $\mathcal{N} = 4$ super Yang-Mills, the index is the same. This is as expected, since the index is invariant under exactly marginal deformations. The operators corresponding to $\overline{HC}_0(\mathcal{A}) \cong \mathcal{A}/[\mathcal{A},\mathcal{A}] \cong \mathcal{A}_{cyc}$ were first determined in [**2**] using the techniques reviewed in section 2.4. The derivation is instructive, so we briefly recall it. Consider an operator $\mathcal{O} = \operatorname{Tr} l_1 l_2 \ldots l_n$, where l_i is one of the letters x, y, or z. Suppose that l_1 is an x. The F-term conditions imply that

$$\mathcal{O} = \operatorname{Tr} l_1 l_2 \ldots l_{n-1} l_n = q^{2(|z|-|y|)} \operatorname{Tr} l_n l_1 l_2 \ldots l_{n-1},$$

where $|x|, |y|$, and $|z|$ are the total number of x's, y's, and z's in the operator $\mathcal{O}$. Thus for q not a primitive root of unity, $|z| - |y| = 0$. Repeating the argument with the letter y or z, we find that the single-trace chiral primaries must have have

	1	t^2	t^4	t^6	t^8	t^{10}	t^{12}	...
HC_0	1	3	3	4	3	3	4	...
HC_1	0	0	0	2	0	0	2	...
HC_2	0	0	0	1	0	0	1	...
$\mathcal{I}_{s.t}(t)$	1	3	3	3	3	3	3	...

TABLE 4. Cyclic homology group dimensions for the β-deformation

charges $(k,0,0),(0,k,0),(0,k,0)$, or (k,k,k) [**2, 3, 15**]. These operators exactly correspond to the graded components of $\overline{HC}_0$.

For $G = SU(N)$ there are additional chiral primaries $\operatorname{Tr} xy, \operatorname{Tr} xz$ and $\operatorname{Tr} yz$. This agrees with the perturbative one-loop spectrum of chiral operators found in [**7,16**]. For q a root of unity, the cyclic homology groups jump [**3**]. It would be very interesting to reproduce the spectrum of protected operators from a Kaluza-Klein analysis of the dual supergravity solution [**15**].

4. Conclusion and Future Directions

We have seen how the spectrum of protected operators in a quiver gauge theory is determined by the cyclic homology groups $\overline{HC}_\bullet(\mathcal{A})$ of its superpotential algebra $\mathcal{A}$. This is a powerful new technique for analyzing this class of theories. Using off-the-shelf mathematical results, we have easily derived new predictions for the spectrum of intensely studied theories, such as the β-deformation and more general quiver gauge theories.

For theories dual to Freund-Rubin compactifications of the form AdS_5 times a Sasaki-Einstein manifold, the spectrum of protected operators has already been matched by independent calculations in both the gauge and gravity theories. However for more general compactifications it would be interesting to develop techniques to determine the spectrum of protected operators directly in supergravity. For example, it would be a strong check of our proposal to derive the spectrum of protected operators directly from the gravity dual of the β-deformation [**15**] and match it to the cyclic homology groups obtained in section 3.2. For a massive deformation of $\mathcal{N} = 4$ supersymmetric Yang-Mills theory, the cyclic homology groups were determined in [**17**]. The matching of the spectrum of protected spin-2 excitations to the short Kaluza-Klein modes in the gravity solution [**18**] will appear in a future publication. However, it is desirable to determine the full spectrum in more general compactifications.

Passing from cyclic homology to deRham or Poisson homology is a mathematical analog of the passage from open to closed strings in AdS/CFT. We have shown the utility of this approach by applying it to match the spectrum of protected operators in a gauge theory and its gravity dual. This new interpretation of AdS/CFT should hopefully allow for further insight into the mathematical underpinnings of the duality.

Acknowledgements

The author would like to thank Y. Tachikawa and J. Schmude for a previous collaboration whose results were revisited in this paper. The author also thanks

M. van den Bergh for helpful correspondence. Finally, the author would also like to thank D. Berenstein for inspiration and lending him a copy of Loday's book on cyclic homology in his formative years. The work of R. E. is supported in part by World Premier International Research Center Initiative (WPI Initiative), MEXT, Japan through the Institute for the Physics and Mathematics of the Universe, the University of Tokyo.

References

[1] David Berenstein and Michael R. Douglas, *Seiberg duality for quiver gauge theories*, (2002).

[2] David Berenstein, Vishnu Jejjala, and Robert G. Leigh, *Marginal and relevant deformations of $N = 4$ field theories and non-commutative moduli spaces of vacua*, Nuclear Phys. B **589** (2000), no. 1-2, 196–248, DOI 10.1016/S0550-3213(00)00394-1. MR1788803 (2002c:81150)

[3] David Berenstein and Robert G. Leigh, *Discrete torsion, AdS/CFT and duality*, J. High Energy Phys. **1** (2000), Paper 38, 22, DOI 10.1088/1126-6708/2000/01/038. MR1743285 (2001a:81174)

[4] F. A. Dolan and H. Osborn, *Applications of the superconformal index for protected operators and q-hypergeometric identities to $\mathcal{N} = 1$ dual theories*, Nuclear Phys. B **818** (2009), no. 3, 137–178, DOI 10.1016/j.nuclphysb.2009.01.028. MR2518083 (2010k:81264)

[5] Richard Eager, Johannes Schmude, and Yuji Tachikawa, *Superconformal Indices, Sasaki-Einstein Manifolds, and Cyclic Homologies*, (2012).

[6] B. L. Feĭgin and B. L. Tsygan, *Cyclic homology of algebras with quadratic relations, universal enveloping algebras and group algebras*, K-theory, arithmetic and geometry (Moscow, 1984), Lecture Notes in Math., vol. 1289, Springer, Berlin, 1987, pp. 210–239, DOI 10.1007/BFb0078369. MR923137 (89c:17020)

[7] Daniel Z. Freedman and Umut Gürsoy, *Comments on the β-deformed $\mathcal{N} = 4$ SYM theory*, J. High Energy Phys. **11** (2005), 042, 15 pp. (electronic), DOI 10.1088/1126-6708/2005/11/042. MR2187524 (2006g:81184)

[8] Abhijit Gadde, Leonardo Rastelli, Shlomo S. Razamat, and Wenbin Yan, *On the superconformal index of $\mathcal{N} = 1$ IR fixed points. A holographic check*, J. High Energy Phys. **3** (2011), 041, 26, DOI 10.1007/JHEP03(2011)041. MR2821147

[9] Victor Ginzburg, *Calabi-Yau algebras*, (2006).

[10] M. Günaydin and N. Marcus, *The spectrum of the S^5 compactification of the chiral $N = 2$, $D = 10$ supergravity and the unitary supermultiplets of* $\mathrm{U}(2, 2/4)$, Classical Quantum Gravity **2** (1985), no. 2, L11–L17. MR786555 (86j:83060)

[11] H. J. Kim, L. J. Romans, and P. van Nieuwenhuizen, *Mass spectrum of chiral ten-dimensional $N = 2$ supergravity on S^5*, Phys. Rev. D (3) **32** (1985), no. 2, 389–399, DOI 10.1103/PhysRevD.32.389. MR797203 (86h:83066)

[12] Justin Kinney, Juan Maldacena, Shiraz Minwalla, and Suvrat Raju, *An index for 4 dimensional super conformal theories*, Comm. Math. Phys. **275** (2007), no. 1, 209–254, DOI 10.1007/s00220-007-0258-7. MR2335774 (2009h:81297)

[13] Maxim Kontsevich, *Formal (non)commutative symplectic geometry*, The Gel′fand Mathematical Seminars, 1990–1992, Birkhäuser Boston, Boston, MA, 1993, pp. 173–187. MR1247289 (94i:58212)

[14] Robert G. Leigh and Matthew J. Strassler, *Exactly marginal operators and duality in four-dimensional $N = 1$ supersymmetric gauge theory*, Nuclear Phys. B **447** (1995), no. 1, 95–133, DOI 10.1016/0550-3213(95)00261-P. MR1343672 (96i:81274)

[15] Oleg Lunin and Juan Maldacena, *Deforming field theories with* $\mathrm{U}(1) \times \mathrm{U}(1)$ *global symmetry and their gravity duals*, J. High Energy Phys. **5** (2005), 033, 37, DOI 10.1088/1126-6708/2005/05/033. MR2155387 (2006e:81304)

[16] Kallingalthodi Madhu and Suresh Govindarajan, *Chiral primaries in the Leigh-Strassler deformed $\mathcal{N} = 4$ SYM—a perturbative study*, J. High Energy Phys. **5** (2007), 038, 31 pp. (electronic), DOI 10.1088/1126-6708/2007/05/038. MR2318089 (2008f:81250)

[17] Philippe Nuss, *L'homologie cyclique des algèbres enveloppantes des algèbres de Lie de dimension trois* (French, with English summary), J. Pure Appl. Algebra **73** (1991), no. 1, 39–71, DOI 10.1016/0022-4049(91)90105-B. MR1121630 (92i:19004)

[18] Krzysztof Pilch and Nicholas P. Warner, *A new supersymmetric compactification of chiral IIB supergravity*, Phys. Lett. B **487** (2000), no. 1-2, 22–29, DOI 10.1016/S0370-2693(00)00796-6. MR1779285 (2001j:83103)

[19] Christian Römelsberger, *Counting chiral primaries in $\mathcal{N} = 1$, $d = 4$ superconformal field theories*, Nuclear Phys. B **747** (2006), no. 3, 329–353, DOI 10.1016/j.nuclphysb.2006.03.037. MR2241553 (2007h:81197)

[20] Marc Wambst, *Complexes de Koszul quantiques* (French, with English and French summaries), Ann. Inst. Fourier (Grenoble) **43** (1993), no. 4, 1089–1156. MR1252939 (95a:17023)

[21] Edward Witten, *Anti de Sitter space and holography*, Adv. Theor. Math. Phys. **2** (1998), no. 2, 253–291. MR1633012 (99e:81204c)

DEPARTMENT OF PHYSICS, MCGILL UNIVERSITY MONTRÉAL, QC, CANADA
Current address: Mathematisches Institut, Universität Heidelberg, Heidelberg, Germany
E-mail address: `reager@physics.mcgill.ca`

Proceedings of Symposia in Pure Mathematics
Volume **93**, 2016
http://dx.doi.org/10.1090/pspum/093/01552

Differential K-characters and D-branes

Fabio Ferrari Ruffino

ABSTRACT. Starting from the definition of Cheeger-Simons K-character given in Benameur and Maghfoul (2006) and Ruffino (*Flat pairing and generalized Cheeger-Simons characters*, forthcoming Journal of Homotopy and Related Structures, DOI 10.1007/s40062-015-0124-9; arXiv:1208.1288), we show how to describe D-brane world-volumes, the Wess-Zumino action and topological D-brane charges within the K-theoretical framework in type II superstring theory. We stress in particular how each feature of the old cohomological classification can be reproduced using the K-theoretical language.

1. Introduction

In the framework of type II superstring theory, there are two fundamental pictures that describe and classify D-brane charges and the Ramond-Ramond fields. The first one relies on classical cohomology. In particular, a D-brane world-volume is a submanifold, which becomes a singular cycle via a suitable triangulation, and the Poincaré dual of the underlying homology class is the topological charge. The Ramond-Ramond fields are classified by ordinary differential cohomology, for which the Deligne cohomology provides a concrete model [**2**]. The Wess-Zumino action turns out to be the holonomy of a differential cohomology class along the world-volume. The other fundamental classification scheme relies on K-theory [**5, 14**]. In particular, the Ramond-Ramond fields are classified by a differential K-theory class [**13, 15**], while the topological charge of the D-brane is the corresponding K-theory class. What we try to clarify in this paper is how to correctly define the world-volume in this picture, in order to get a suitable generalization of the holonomy map to differential K-theory. In this way we are able to give a correct definition of the Wess-Zumino action. Considering the world-volume as a topological K-cycle is not enough, thus we have to define a suitable differential extension of K-cycles, on which we are able to compute the holonomy. We see that such a definition leads to differential K-characters, as defined in [**1**] and [**6**]. In this way we can draw a complete parallel between the two classification schemes. Since we consider ordinary K-theory, we suppose that the B-field is vanishing; otherwise, we must develop an analogous construction for twisted K-theory and its differential extension.

The paper is organized as follows. In section 2 we describe the classification scheme via ordinary homology. In section 3 we describe the classification scheme

2010 *Mathematics Subject Classification.* Primary 81T50; Secondary 19L50, 53C08.
Key words and phrases. Differential K-characters, D-branes.
The author was supported by FAPESP, processo 2014/03721-3.

via K-theory. In section 4 we recall the definition of differential K-character given in [**6**]. In section 5 we apply such a definition in order to describe the world-volume and the Wess-Zumino action in the K-theoretical framework, drawing a complete parallel between the two classification schemes.

2. Ordinary differential cohomology and Ramond-Ramond fields

If we consider the classical magnetic monopole in 3+1 space-time dimensions, it is well-known that, because of the Dirac quantization condition, the field strength $F_{\mu\nu}$ can be considered as the curvature of a connection on a gauge bundle on $\mathbb{R}^3 \setminus \{0\}$ (or $\mathbb{R}^4 \setminus (\{0\} \times \mathbb{R})$), whose first Chern class, belonging to $H^2(\mathbb{R}^3 \setminus \{0\}; \mathbb{Z}) \simeq \mathbb{Z}$, corresponds to the magnetic charge fixed in the origin. If we argue in the same way for a monopole in a generic space-time dimension $n+1$, we need a gauge invariant integral $(n-1)$-form $F_{\mu_1 \ldots \mu_{n-1}}$, whose integral on an $(n-1)$-dimensional sphere around the origin of $\mathbb{R}^n$ is the magnetic charge (up to a normalization constant). Hence, because of the Dirac quantization condition, such a field strength can be thought of as the curvature of a connection on an abelian $(n-3)$-gerbe, whose first Chern class, belonging to $H^{n-1}(\mathbb{R}^n \setminus \{0\}, \mathbb{Z}) \simeq \mathbb{Z}$, corresponds to the charge fixed in the origin. That's why p-gerbes naturally arise when dealing with monopoles in a space-time of generic dimension. Since a D-brane, at a semiclassical level, can be thought of as a generalized magnetic monopole whose charge is measured by the Ramond-Ramond field strength, it follows that the Ramond-Ramond potentials $C_{\mu_1 \ldots \mu_{p+1}}$ and field strength $G_{\mu_1 \ldots \mu_{p+2}}$ can be thought of respectively as a connection and its curvature on an abelian p-gerbe. A concrete way to describe abelian p-gerbes with connection is provided by the Deligne cohomology [**2**].

Given a compact smooth manifold X, we consider the complex of sheaves:

$$S^p_X = \underline{U}(1) \xrightarrow{\tilde{d}} \Omega^1_{\mathbb{R}} \xrightarrow{d} \cdots \xrightarrow{d} \Omega^p_{\mathbb{R}}, \tag{2.1}$$

where $\underline{U}(1)$ is the sheaf of smooth $U(1)$-valued functions, $\Omega^k_{\mathbb{R}}$ is the sheaf of real k-forms, d is the exterior differential and $\tilde{d} = \frac{1}{2\pi i} d \circ \log$. The Deligne cohomology group of degree p on X is the Čech hypercohomology group of the complex (2.1), i.e., $\check{H}^p(X, S^p_X)$. It can be concretely described via a good cover $\mathfrak{U} = \{U_\alpha\}_{\alpha \in I}$ of X: by definition, we consider the double complex whose columns are the Čech complexes of the sheaves involved in (2.1), and we consider the cohomology of the associated total complex. This means that a p-cocycle is defined by a sequence $(g_{\alpha_0 \cdots \alpha_{p+1}}, (C_1)_{\alpha_0 \cdots \alpha_p}, \ldots, (C_p)_{\alpha_0 \alpha_1}, (C_{p+1})_{\alpha_0})$, satisfying the conditions:

$$\begin{aligned}
&(C_{p+1})_\beta - (C_{p+1})_\alpha = (-1)^{p+1} d(C_p)_{\alpha\beta} \\
&(C_p)_{\alpha\beta} + (C_p)_{\beta\gamma} + (C_p)_{\gamma\alpha} = (-1)^p \, d(C_{p-1})_{\alpha\beta\gamma} \\
&\cdots \\
&\check{\delta}^p (C_1)_{\alpha_0 \ldots \alpha_p} = \tfrac{1}{2\pi i} d \log g_{\alpha_0 \ldots \alpha_{p+1}} \\
&\check{\delta}^{p+1} g_{\alpha_0 \ldots \alpha_{p+1}} = 1.
\end{aligned} \tag{2.2}$$

The local forms dC_{p+1} glue to a gauge-invariant one G_{p+2}, which is the curvature of the p-gerbe. We stress that, with respect to this model, the datum of the superstring background must include a complete equivalence class, not only the top-forms C_{p+1}. As for line bundles, the correspondence $[G_{p+2}]_{dR} \simeq c_1(\mathcal{G}) \otimes_{\mathbb{Z}} \mathbb{R}$ holds, in particular the Dirac quantization condition applies for any p. From a physical point of view, Deligne cohomology describes gauge transformations. Conditions (2.2) specify how the local potentials glue on the intersections, and this concerns a single representative of the equivalence class. There are also possible gauge transformations consisting in the addition of a coboundary. The real datum is the cohomology class, since it is determined by the two real physical observables: the field strength (corresponding to the field F in electromagnetism) and the holonomy of the connection or Wess-Zumino action (corresponding in electromagnetism to the phase difference measured in the context of the Aranhov-Bohm effect). The holonomy is the exponential of the Wilson loop; it can be defined for any p generalizing the definition of the Wilson loop for line bundles. A line bundle with connection is described by a Deligne cohomology class of degree 1, i.e., by $[(g_{\alpha\beta}, A_\alpha)] \in \check{H}^1(X, S^1_X)$. The Wilson loop is usually described as the minimal coupling between the potentials A and the loop γ, that's why it is usually written as $\int_\gamma A$. Actually the correct definition must also take into account the transition functions. In particular, we divide the loop γ in intervals $\gamma_1, \ldots, \gamma_m$, such that γ_i is contained in a chart U_{α_i}. Then we integrate the local potential A_{α_i} on γ_i and we compute the logarithm of the transition function $g_{\alpha_i\alpha_{i+1}}$ on the intersection point between γ_i and γ_{i+1}. The sum is the Wilson loop, its exponential the holonomy along γ. Such a definition can be generalized to any p, even if the explicit formula is much more complicated to write down concretely [**10**]. The basic idea is the following: given a Deligne cohomology class $[(g_{\alpha_0\cdots\alpha_{p+1}}, (C_1)_{\alpha_0\cdots\alpha_p}, \ldots, (C_p)_{\alpha_0\alpha_1}, (C_{p+1})_{\alpha_0})]$ of degree $p+1$ and a smooth $(p+1)$-submanifold Γ, we choose a suitable triangulation of Γ, such that each simplex is contained in a chart. Then we integrate the potentials C_{p+1} on the $(p+1)$-simplicies, the potentials C_p on the p-simplicies, and so on until the transition functions on the vertices. A suitable formula joining these data gives the Wilson loop, which is the Wess-Zumino action in string theory. The result depends on the cycle, not only on the homology class, except when the curvature vanishes. This is coherent with the fact that the world-volume is a cycle, not only a homology class. Only in the flat case is the holonomy a morphism from $H_{p+1}(X;\mathbb{Z})$ to $U(1)$, hence flat abelian p-gerbes are classified by the group $H^{p+1}(X;\mathbb{R}/\mathbb{Z})$. This is due to a Stokes-type formula for the holonomy on a trivial cycle: the holonomy over a boundary ∂A is the exponential of the integral of the curvature on A.

Calling $\hat{H}^p(X)$ the Deligne cohomology group of degree $p-1$, i.e. $\hat{H}^p(X) := \check{H}^{p-1}(X, S^{p-1}_X)$, we get the following commutative diagram [**11**]:

$$
\begin{array}{ccc}
\hat{H}^\bullet(X) & \xrightarrow{c_1} & H^\bullet(X;\mathbb{Z}) \\
{\scriptstyle curv}\downarrow & & \downarrow{\scriptstyle \otimes_{\mathbb{Z}}\mathbb{R}} \\
\Omega^\bullet_{int}(X) & \xrightarrow{dR} & H^\bullet_{dR}(X).
\end{array}
\tag{2.3}
$$

Here c_1 is the first Chern class, $curv$ is the curvature, dR is the de-Rham cohomology class and $\Omega^\bullet_{int}(X)$ is the group of closed real forms that represent an integral cohomology class. Diagram (2.3) shows that $\hat{H}^\bullet(X)$ is a differential refinement of $H^\bullet(X;\mathbb{Z})$, adding the piece of information due to the connection. Moreover, one can prove that, given a class $\alpha \in \hat{H}^{p+2}(X)$, if $c_1(\alpha) = 0$, then α can be represented by a cocycle of the form $(1, 0, \ldots, 0, C_{p+1})$, where C_{p+1} is a globally defined $(p+1)$-form. In this case the Wilson loop on a $(p+1)$-submanifold Γ is simply given by $\int_\Gamma C_{p+1}$. Such a global potential is unique up to large gauge transformation, i.e., up to the addition of a closed integral form.

With respect to this model, the local Ramond-Ramond potentials C_{p+1} are (a part of) a connection on an abelian p-gerbe, whose curvature is the field strength G_{p+2}. In this case a Dp-brane world-volume is thought of as a $(p+1)$-dimensional submanifold W of the space-time X. The world-volume W, via a suitable triangulation, defines a singular $(p+1)$-cycle, that we also call W. When the numerical charge is $q \in \mathbb{Z}$, we think of a stack of q D-branes (anti-branes if $q < 0$), whose underlying cycle is qW. The topological charge of the D-brane is the Poincaré dual of the underlying homology class $[qW] \in H_{p+1}(X;\mathbb{Z})$. The Wess-Zumino action, usually written as $\int_W C_{p+1}$, is the holonomy of the connection on W. Moreover, calling $n := \dim W$, the violated Bianchi identity is:

$$dG_{n-p-2} = q \cdot \delta(W) \qquad dG_{p+2} = 0.$$

This implies that G_{n-p-2} is a closed form in the complement of W and, if L is a linking manifold of W, with linking number l, we get $\frac{1}{l}\int_L G_{n-p-2} = q \in \mathbb{Z}$. That's why field strengths are quantized and can be thought of as the curvature of a connection.

3. Differential K-theory and Ramond-Ramond fields

It is well known that K-theory is a better tool than ordinary cohomology in order to classify D-brane charges [**5**, **7**]. We first introduce some technical tools about K-theory and K-homology, then we recall the advantages of the K-theoretical classification.

3.1. K-homology. We consider a variant of the usual definition of topological K-homology, that will be more suitable for our purposes later: we replace the "vector bundle modification" with the Gysin map, which is the natural push-forward in cohomology. We briefly recall the definition. Given an embedding $\iota : Y \to X$ of compact manifolds of codimension r, we consider the following data:

- a K-orientation of the normal bundle $N_Y X$, i.e., a Thom class $u \in K^r_{\text{cpt}}(N_Y X)$;
- a tubular neighbourhood U of Y and a diffeomorphism $\varphi_U : N_Y X \to U$;
- the open embedding $i : U \hookrightarrow X$, inducing a push-forward in compactly-supported cohomology. Such a push-forward is defined as the pull-back via the map $i' : X \to U^+$, which is the identity on U and sends $X \setminus U$ to the point at infinity.

There is a natural $K(X)$-module structure on $K_{\text{cpt}}(N_Y X)$, hence we define $\iota_! : K^\bullet(X) \to K^{\bullet+r}(Y)$ as follows: $\iota_!(\alpha) := i_*(\varphi_U)_*(\alpha \cdot u)$. The Gysin map turns out to be independent of the choices involved in the construction, except for the orientation of the normal bundle. If X and Y are K-oriented manifolds and ι respects the orientations, since $TX|_Y \simeq TY \oplus N_Y X$, we get an induced orientation on $N_Y X$. This implies that the Gysin map is well-defined for an embedding of K-oriented manifolds. If $f : Y \to X$ is a generic smooth map between compact manifolds, we consider an embedding $\iota : Y \hookrightarrow X \times \mathbb{R}^N$ such that $\pi_X \circ \iota = f$. Then we define $f_!(\alpha) := \int_{\mathbb{R}^N} \iota_!(\alpha)$. Again, if f is a map of K-oriented manifolds, we get an induced orientation on $N_Y(X \times \mathbb{R}^N)$, hence the Gysin map is well-defined.

We now come back to K-homology. On a smooth compact manifold X, we define the group of *n-precycles* as the free abelian group generated by the quadruples (M, u, α, f) such that:

- M is a smooth compact manifold (without boundary) with K-orientation u (i.e., with Thom class u on the tangent bundle), whose connected components $\{M_i\}$ have dimension $n + q_i$, being q_i arbitrary;
- $\alpha \in K^\bullet(M)$, such that $\alpha|_{M_i} \in K^{q_i}(M)$;
- $f : M \to X$ is a smooth map.

We define the group of *n-cycles*, denoted by $z_n(X)$, as the quotient of the group of n-precycles by the free subgroup generated by elements of the form:

- $(M, u, \alpha + \beta, f) - (M, u, \alpha, f) - (M, u, \beta, f)$;
- $(M, u, \alpha, f) - (M_1, u|_{M_1}, \alpha|_{M_1}, f|_{M_1}) - (M_2, u|_{M_2}, \alpha|_{M_2}, f|_{M_2})$, for $M = M_1 \sqcup M_2$;
- $(M, u, \varphi_!\alpha, f) - (N, v, \alpha, f \circ \varphi)$ for $\varphi : (N, v) \to (M, u)$ a smooth map.

We define the group of *n-boundaries*, denoted by $b_n(X)$, as the subgroup of $z_n(X)$ generated by the cycles which are representable by a pre-cycle (M, u, α, f) with the following property: there exists a quadruple (W, U, A, F) such that W is a manifold and $M = \partial W$, U is a K-orientation of W and $U|_M = u$, $A \in K^\bullet(W)$ and $A|_M = \alpha$, $F : W \to X$ is a smooth map satisfying $F|_M = f$. We define $K_n(X) := z_n(X)/b_n(X)$. It seems to be more natural to use the Gysin map in the definition, since it is the natural push-forward in cohomology, the vector bundle modification being just a particular case. Moreover, we do not have to quotient out explicitly up to diffeomorphism the first component of the quadruple (M, u, α, f), since the pull-back via a diffeomorphism is again a particular case of the Gysin map.

Let us consider a Dp-brane world-volume W in the space-time X. As before we call $n = \dim X$. The $U(q)$-gauge theory on W lives on a complex vector bundle $E \to W$ of rank q, being q the number of D-branes in the stack. Hence there is a well-defined K-theory class $[E] \in K^0(W)$. Moreover, because of the Freed-Witten anomaly [**9**], W is a spin^c-manifold, which is the condition in order to admit a K-theoretical orientation u (that we fix as a part of the world-volume datum). Finally, we consider the embedding in the space-time $\iota : W \hookrightarrow X$. In this way we get a K-homology class $[(W, u, E, \iota)] \in K_{p+1}(X)$. Since also X is K-orientable (because it is a spin manifold, hence, in particular, spin^c), we can apply Poincaré duality and describe the topological charge as a K-theory class of X, which is precisely $\iota_![E] \in K^{n-p-1}(X)$.

We can now recall some advantages of the K-theoretical classification. First of all, it rules out Freed-Witten anomalous world-volumes, which are precisely the non-K-orientable ones. On the contrary, the classification via singular cohomology is unable to detect this anomaly. Moreover, in the K-theoretical charge we also take into account the presence of the Chan-Patton bundle and of the embedding in the space-time; this fact will lead to the presence of the gauge and gravitational couplings in the Wess-Zumino action, therefore we get more complete information. Finally, since the D-brane charge is a K-theory class of the space-time, it can be thought of as the formal difference between two space-filling D-brane stacks of equal rank: this is compatible with the Sen conjecture, stating that any D-brane configuration in the space-time can be obtained from a pair made by a D9-brane and a D9-antibrane, via the process of annihilation due to tachyon condensation.

3.2. Ramond-Ramond fields. Since the D-brane charge is described by K-theory, the Ramond-Ramond fields, that measure such a charge, must be quantized with respect to K-theory, not with respect to ordinary cohomology. In order to make this concept more precise, we consider a diagram analogous to (2.3), but with respect to K-theory instead of ordinary cohomology [**4**]. This means that we look for a graded group $\hat{K}^\bullet(X)$ fitting into the following diagram:

$$\begin{array}{ccc} \hat{K}^\bullet(X) & \xrightarrow{c_1} & K^\bullet(X) \\ \downarrow{\scriptstyle curv} & & \downarrow{\scriptstyle \mathrm{ch}} \\ \Omega^\bullet_{Kint}(X;\mathfrak{k}^\bullet_{\mathbb{R}}) & \xrightarrow{dR} & H^\bullet_{dR}(X;\mathfrak{k}^\bullet_{\mathbb{R}}). \end{array} \tag{3.1}$$

Here ch is the Chern character, $\mathfrak{k}^\bullet_{\mathbb{R}}$ is the K-theory ring of the point and $\Omega^\bullet_{Kint}(X;\mathfrak{k}^\bullet_{\mathbb{R}})$ is the graded group of closed (poly)forms with integral K-periods, in the sense that we now specify. Given a form $\omega \in \Omega^n_{cl}(X;\mathfrak{k}^\bullet_{\mathbb{R}})$ and a K-homology class $[(M,u,\alpha,f)] \in K_n(X)$, we can consider the following pairing:

$$\langle \omega, [(M,u,\alpha,f)] \rangle := \int_M f^*\omega \wedge \mathrm{ch}(\alpha) \wedge \hat{A}_K(M). \tag{3.2}$$

We say that ω is K-integral or has integral K-periods if such a pairing gives an integral value for any K-homology class. One can prove that a form is K-integral if and only if its cohomology class belongs to the image of the Chern character.

There are various models for $\hat{K}^\bullet(X)$; here we consider the Freed-Lott model of $\hat{K}^0(X)$ [**8**], that can be extended to any degree (actually only the parity of the degree is meaningful, since Bott periodicity holds even for the differential extension). Given two connections ∇ and ∇' on the same vector bundle E, there is a natural equivalence class $\mathrm{CS}(\nabla,\nabla')$ of odd-dimensional forms up to exact ones, called *Chern-Simons class*, such that $\mathrm{ch}(\nabla) - \mathrm{ch}(\nabla') = d\mathrm{CS}(\nabla,\nabla')$. We define a *differential vector bundle* on X as a quadruple (E,h,∇,ω) where:

- E is a complex vector bundle on X;
- h is an Hermitian metric on E;
- ∇ is a connection on E compatible with h;
- $\omega \in \Omega^{\mathrm{odd}}(X)/\mathrm{Im}(d)$ is a class of real odd-dimensional differential forms up to exact ones.

The *direct sum* between differential vector bundles is defined as $(E,h,\nabla,\omega) \oplus (E',h',\nabla',\omega') := (E \oplus E', h \oplus h', \nabla \oplus \nabla', \omega + \omega')$. An *isomorphism of differential vector bundles* $\Phi : (E,h,\nabla,\omega) \to (E',h',\nabla',\omega')$ is an isomorphism of complex Hermitian vector bundles $\Phi : (E,h) \to (E',h')$ such that:

$$\omega - \omega' \in \mathrm{CS}(\nabla, \Phi^*\nabla'). \tag{3.3}$$

The isomorphism classes of differential vector bundles form an abelian semigroup, hence we can consider its Grothendieck group $\hat{K}^0(X)$. By definition an element of $\hat{K}^0(X)$ is a difference $[(E,h,\nabla,\omega)] - [(E',h',\nabla',\omega')]$, where $[(E,h,\nabla,\omega)]$ is the class up to the stable equivalence relation.

The group that we have defined fits into the diagram (3.1) considering the two maps:

$$c_1[(E,h,\nabla,\omega)] := [E] \qquad curv[(E,h,\nabla,\omega)] := \mathrm{ch}(\nabla) - d\omega,$$

where $\mathrm{ch}(\nabla) = \mathrm{Tr}\exp(\frac{i}{2\pi}\Omega)$, Ω being the curvature of ∇. The curvature is K-integral since:

$$\langle \mathrm{ch}(\nabla) - d\omega, [(M, u, \alpha, f)] \rangle = \int_M \mathrm{ch}(f^*E \cdot \alpha) \wedge \hat{A}_K(M) \in \mathbb{Z}$$

because of the index theorem. As we have already pointed out, there is an analogous model for $\hat{K}^1(X)$.

We have seen that an abelian p-gerbe with vanishing first Chern class can always be represented by a cocycle of the form $(1, 0, \ldots, 0, C_{p+1})$, where C_{p+1} is globally defined and unique up to the addition of an integral form. An analogous consideration holds for differential K-theory: a class $\alpha \in \hat{K}^p(X)$ with vanishing first Chern class can be represented in the form $[(0, 0, 0, C_{p-1})]$, where $C_{p-1} \in \Omega^{p-1}(X; \mathfrak{k}^\bullet_{\mathbb{R}})$ is globally defined and unique up to the addition of a K-integral form.

We are now able to describe the Ramond-Ramond fields in the K-theoretical framework. Because of the Bott periodicity, the two meaningful groups are $\hat{K}^0(X)$ and $\hat{K}^1(X)$, corresponding respectively to type IIB and type IIA theory. The Ramond-Ramond fields with even-degree field strength are jointly classified by a class $\alpha \in \hat{K}^0(X)$, while the ones with odd-degree field strength are classified by $\beta \in \hat{K}^1(X)$. We discuss the features of α, the discussion about β being analogous. The curvature of α is a form $G_{ev} \in \Omega^0_{cl}(X; \mathfrak{k}^\bullet_{\mathbb{R}}) \simeq \bigoplus_{p\in\mathbb{Z}} \Omega^{2p}_{cl}(X)$. The component of degree $2p$ is the field-strength G_{2p}. If we consider a local chart U of X, then $\alpha|_U$ is topologically trivial, hence it can be represented in the form $(0, 0, 0, C_{odd})$, with $C_{odd} \in \Omega^{-1}(U; \mathfrak{k}^\bullet_{\mathbb{R}}) \simeq \bigoplus_{p\in\mathbb{Z}} \Omega^{2p-1}(U)$, unique up to the addition of an exact form (on a contractible chart U, any K-integral form, being closed, is exact). The component of degree $2p-1$ is the local potential C_{2p-1}. This means that the potentials are a local expression of a global differential K-theory class, which is the complete datum encoded in the space-time.

Now the main point is the following. How do we have to think of a D-brane world-volume in the K-theoretical framework, in order to correctly define the Wess-Zumino action? Comparing with the framework of ordinary cohomology, it seems natural to think of it as a K-homology cycle, representing a class whose Poincaré dual is the topological charge. This is possible, but we will see that it is not enough in order to define the Wess-Zumino action.

3.3. Comparing the two frameworks. Let us start from the mathematics. In table 1 we compare the features of ordinary differential cohomology with the ones of differential K-theory. We can see that there is a complete analogy between the two pictures, except for the holonomy, since we have to clarify on which cycles it must be computed in the case of K-theory (in the table, $Z^{sm}_\bullet$ denotes the smooth singular cycles).

Physically, Ramond-Ramond fields in type II superstring theory are classified by an abelian p-gerbe or by a differential K-theory class (line 1 of table 1). The field strength is the curvature in each case, hence it obeys the corresponding quantization condition (line 3 of table 1). Any class is locally topologically trivial, hence we get the local Ramond-Ramond potentials up to gauge transformations (line 4 of table 1). The world-volume is a singular cycle in the first picture, and the Poincaré dual of the underlying homology class is the topological charge; the Wess-Zumino action

	Abelian p-gerbe with c.	**Diff. K-theory class**
Classified by	$\hat{H}^\bullet(X)$	$\hat{K}^\bullet(X)$
First Chern class	$c_1 \in H^\bullet(X;\mathbb{Z})$	$c_1 \in K^\bullet(X)$
Curvature	$curv \in \Omega^\bullet_{int}(X)$ $[curv]_{dR} \simeq c_1 \otimes_\mathbb{Z} \mathbb{R}$	$curv \in \Omega^\bullet_{Kint}(X; \mathfrak{k}^\bullet_\mathbb{R})$ $[curv]_{dR} \simeq \mathrm{ch}(c_1)$
Top. trivial classes	$\Omega^{\bullet-1}(X)/\Omega^{\bullet-1}_{int}(X)$	$\Omega^{\bullet-1}(X;\mathfrak{k}^\bullet_\mathbb{R})/\Omega^{\bullet-1}_{Kint}(X;\mathfrak{k}^\bullet_\mathbb{R})$
Flat classes	$H^{\bullet-1}(X;\mathbb{R}/\mathbb{Z})$	$K^{\bullet-1}(X;\mathbb{R}/\mathbb{Z})$
Holonomy	$Z^{sm}_{\bullet-1}(X) \to U(1)$	$?? \to U(1)$

TABLE 1. Comparison

is the holonomy of the Ramond-Ramond fields on the world-volume (line 6 of table 1). What is the Wess-Zumino action in the K-theoretical framework?

We have seen that the topological D-brane charge is measured by the K-theory class of the space-time Poincaré dual to $[(W, u, E, \iota)] \in K_{p+1}(X)$, where W is the world-volume as a sub-manifold, u is a fixed Thom class of the tangent bundle of W, E is the Chan-Patton bundle and ι is the embedding of W in the space-time. This class is $\iota_![E]$. Hence, we could consider as the world-volume the K-cycle (W, u, E, ι), but we do not know how to define the holonomy of the class $\alpha \in \hat{K}^{p+2}(X)$, representing the Ramond-Ramond fields. Usually the pairing is written supposing that α is topologically trivial, hence described by a global form C. It has the following form [**12**]:

$$\langle \alpha, (W, u, E, \iota) \rangle = \int_W C \wedge \mathrm{ch}(E) \wedge \hat{A}_K(W) \wedge \hat{A}_K(X)^{-\frac{1}{2}}. \tag{3.4}$$

We denote by $\hat{A}_K$ the $\hat{A}$-genus of K-theory, i.e., $\hat{A} \wedge e^{\frac{d}{2}}$, where $d \in H^2(W;\mathbb{Z})$ is a suitable class whose $\mathbb{Z}_2$-reduction is $w_2(W)$ [**12**]. Equation (3.4) has some problems. The most evident one is what we have already said: it holds only when α is topologically trivial. Actually, even in this case, we can make some more remarks. The form C in general is not-closed, hence the integral on W depends on the specific representatives of $\mathrm{ch}(E)$ and $\hat{A}_K(W)$ (we neglect for the moment $\hat{A}_K(X)$, since it does not depend on the D-brane). How do we choose them? It is not difficult to reply for $\mathrm{ch}(E)$: since $\mathrm{ch}(E) = [\mathrm{Tr}\exp(\frac{i}{2\pi}\Omega)]$, Ω being the curvature of a connection on E, we have to fix a connection on E in order to fix a representative of $\mathrm{ch}(E)$. We choose the connection defining the $U(q)$-gauge theory on the D-brane, q being the rank of E. This fact shows that, even when α is topologically trivial, we cannot consider as the world-volume the topological K-cycle (W, u, E, ι): at least we need to include the connection on E as a part of the datum. Moreover, what about the $\hat{A}$-genus? It does not seem so trivial to find a natural representative, hence we need some information more. In the next sections we try to fill this gap.

There is an element missing in the previous list: in the framework of ordinary cohomology, the numerical charge of a D-brane is measured by the integral of the dual field-strength on a linking manifold. It is not difficult to find the analogous property in the K-theoretical framework, actually we could do this even considering the world-volume just as a topological K-cycle, but we postpone the discussion to the last section.

4. Differential K-characters

We try to reply to the previous questions looking for a suitable definition of differential K-cycle and differential K-character. The idea we presented in [**6**] is the following. Let us consider a K-cycle (M, u, α, f) of degree p on X and a differential K-theory class $\hat{\beta} \in \hat{K}^{p+1}(X)$ (of course only the parity of p is meaningful). We have that $\alpha \in K^q(X)$, where q satisfies $\dim M = p + q$. If we refine α to a differential class $\hat{\alpha}$, then we can consider the product $\hat{\alpha} \cdot f^*\hat{\beta} \in \hat{K}^{p+q+1}(M)$. There is a unique map from M to the point, that we call p_M. If we are able to define the differential refinement of the Gysin map, via a suitable differential refinement of the orientation u (that we call $\hat{u}$), we can calculate $(p_M)_!(\hat{\alpha} \cdot f^*\hat{\beta}) \in \hat{K}^1(pt)$. We now prove that $\hat{K}^1(pt) \simeq \mathbb{R}/\mathbb{Z}$ canonically, hence we can define the holonomy of $\hat{\beta}$ on $(M, \hat{u}, \hat{\alpha}, f)$ as $\exp((p_M)_!(\hat{\alpha} \cdot f^*\hat{\beta}))$. This shows that, in order to define the holonomy, we must consider a suitable differential refinement of the topological K-cycles, that will lead us to define differential K-characters. We have to show that $\hat{K}^1(pt) \simeq \mathbb{R}/\mathbb{Z}$ canonically. Since $K^1(pt) = 0$, a class $\gamma \in \hat{K}^1(pt)$ is topologically trivial, hence it can be represented by a form $\omega \in \Omega^0(pt; \mathfrak{k}^\bullet_{\mathbb{R}})/\Omega^0_{Kint}(pt; \mathfrak{k}^\bullet_{\mathbb{R}})$. On a point there are non-zero forms only in degree 0, and they are real numbers. The K-integral ones are precisely the integer numbers, since, in the pairing (3.2), $f^*\omega$ is constant and $\int_M \mathrm{ch}(\alpha) \wedge \hat{A}_K(M)$ is integral because of the index theorem. This shows that $\Omega^0(pt; \mathfrak{k}^\bullet_{\mathbb{R}})/\Omega^0_{Kint}(pt; \mathfrak{k}^\bullet_{\mathbb{R}}) \simeq \mathbb{R}/\mathbb{Z}$.

Let us present the precise definition of differential K-character. We have shown above that we must consider suitable differential refinements of the components of a topological K-cycle. The main point is that, when dealing with differential classes, the curvature is meaningful as a single form, not only as a cohomology class, therefore it is not homotopy invariant. Thus, we need suitable definitions in order to recover classical topological tools as the 2x3 rule about the orientation of the bundles E, F, $E \oplus F$. In particular, we have to correctly define the concept of orientation of a smooth map with respect to differential K-theory [**11**], which encodes the data that we need to fix. First of all, following [**3**], we define a $\hat{K}$-*orientation* of a smooth vector bundle as a differential extension[1] of a Thom class of the bundle. Then we define a *representative of a* $\hat{K}$*-orientation* of a smooth map

[1]A differential extension of a class $\alpha \in K^n(X)$ is a class $\hat{\alpha} \in \hat{K}^n(X)$ such that $c_1(\hat{\alpha}) = \alpha$.

$f : Y \to X$ between compact manifolds (neat if X and Y have boundary) as the datum of:

- a (neat) embedding $\iota : Y \hookrightarrow X \times \mathbb{R}^N$ for any $N \in \mathbb{N}$, such that $\pi_X \circ \iota = f$;
- a $\hat{K}$-orientation $\hat{u}$ of the normal bundle $N_Y(X \times \mathbb{R}^N)$;
- a (neat) tubular neighbourhood U of Y in $X \times \mathbb{R}^N$ with a diffeomorphism $\varphi : N_Y(X \times \mathbb{R}^N) \to U$.

Using a definition similar to the topological one, it turns out that the Gysin map $f_! : \hat{K}^\bullet(Y) \to \hat{K}^{\bullet+r}(X)$ is well defined if f is endowed with a representative of a $\hat{K}$-orientation. We can suitably define *homotopy* and *equivalence by stabilization* in the set of representatives of $\hat{K}$-orientations, and we call *$\hat{K}$-orientation* of f an equivalence class. Moreover, a smooth manifold M is *$\hat{K}$-oriented* if the unique map from M to a point is $\hat{K}$-oriented.

With this definition, as in the topological case, if f is a *proper submersion* between $\hat{K}$-oriented manifolds, then it automatically inherits an orientation. Actually the technical details are more complicated. We just sketch the problems. First of all, one fundamental property of the Gysin map is that it is compatible with the composition, i.e., $(g \circ f)_! = g_! \circ f_!$. Moreover, it satisfies $f_!(\alpha \cdot f^*\beta) = f_!\alpha \cdot \beta$. In order to maintain these properties in the differential case, we need the hypothesis that f is a submersion, because, in this case, considering the embedding $\iota : Y \hookrightarrow X \times \mathbb{R}^N$, we can choose the tubular neighbourhood of Y in such a way that the image of the fibre of the normal bundle on $y \in Y$ is contained in $\{\iota(y)\} \times \mathbb{R}^N$. In this way, when we consider $\alpha \cdot f^*\beta$ and we apply $f_!$, the multiplication by β acts as a multiplication by a constant class on each fibre of the tubular neighbourhood, therefore it factorizes in the integral with respect to $\mathbb{R}^N$. A similar argument holds in order to prove that $(g \circ f)_! = g_! \circ f_!$. Moreover, thanks to the equivalence relation we introduced among the representatives of orientations, the embedding ι is meaningful only up to homotopy and stabilization, and the choice of the tubular neighborhood is immaterial. This is important by a physical point of view, since a fixed embedding and a fixed tubular neighbourhood would have no physical meaning.

Now we can come back to the definition of differential K-character. On a smooth compact manifold X, we define the group of *differential n-precycles* as the free abelian group generated by the quadruples $(M, \hat{u}, \hat{\alpha}, f)$ such that:

- M is a smooth compact manifold (without boundary) with $\hat{K}^\bullet$-orientation $\hat{u}$,[2] whose connected components $\{M_i\}$ have dimension $n + q_i$, with q_i arbitrary;
- $\hat{\alpha} \in \hat{K}^\bullet(M)$, such that $\hat{\alpha}|_{M_i} \in \hat{K}^{q_i}(M)$;
- $f : M \to X$ is a smooth map.

The group of *differential n-cycles*, denoted by $\hat{z}_n(X)$, is the quotient of the group of n-precycles by the free subgroup generated by elements of the form:

- $(M, \hat{u}, \hat{\alpha} + \hat{\beta}, f) - (M, \hat{u}, \hat{\alpha}, f) - (M, \hat{u}, \hat{\beta}, f)$;
- $(M, \hat{u}, \hat{\alpha}, f) - (M_1, \hat{u}|_{M_1}, \hat{\alpha}|_{M_1}, f|_{M_1}) - (M_2, \hat{u}|_{M_2}, \hat{\alpha}|_{M_2}, f|_{M_2})$, for $M = M_1 \sqcup M_2$;
- $(M, \hat{u}, \varphi_!\hat{\alpha}, f) - (N, \hat{v}, \hat{\alpha}, f \circ \varphi)$ for $\varphi : N \to M$ a submersion, oriented via the 2x3 principle.

[2]Here we denote by $\hat{u}$ the whole differential orientation, not only the differential refinement of the Thom class u.

The group of *differential n-boundaries*, denoted by $\hat{b}_n(X)$, is the subgroup of $\hat{z}_n(X)$ generated by the cycles which are representable by a pre-cycle $(M, \hat{u}, \hat{\alpha}, f)$ with the following property. There exists a quadruple $(W, \hat{U}, \hat{A}, F)$ such that W is a manifold and $M = \partial W$, $\hat{U}$ is an $\hat{K}^{\bullet}$-orientation of W and $\hat{U}|_M = \hat{u}$, $\hat{A} \in \hat{K}^{\bullet}(W)$ and $\hat{A}|_M = \hat{\alpha}$, $F : W \to X$ is a smooth map satisfying $F|_M = f$. We define $K_n(X) := \hat{z}_n(X)/\hat{b}_n(X)$.

The homology groups, defined in this way, are isomorphic to the ones defined via topological cycles, as shown above. We have defined the differential cycles in such a way that it is possible to integrate a differential cohomology class on such a cycle. When the class is flat and only the homology class is meaningful, we need no differential information, since the group of flat classes is $\mathrm{Hom}(K_{p-1}(X), \mathbb{R}/\mathbb{Z})$; that's why we do not need a non-trivial differential extension of the homology classes. We will see in the following the physical meaning of this fact.

Given a class $\hat{\beta} \in \hat{K}^{p+1}(X)$ and a differential p-cycle $(M, \hat{u}, \hat{\alpha}, f)$, with $\dim M = p+q$ and $\hat{\alpha} \in \hat{K}^q(M)$, we can compute the holonomy as we sketched at the beginning of this paragraph: we consider the class $\hat{\alpha} \cdot f^*\hat{\beta} \in \hat{K}^{p+q+1}(M)$ and, since M is $\hat{K}$-oriented, i.e., the map $p_M : M \to pt$ is $\hat{K}$-oriented, we can calculate $(p_M)_!(\hat{\alpha} \cdot f^*\hat{\beta}) \in \hat{K}^1(pt) \simeq \mathbb{R}/\mathbb{Z}$. The exponential of the result is the holonomy. One can show that the holonomy completely characterizes the differential K-theory class, as in the case of ordinary cohomology. When the cycle is a boundary, a Stokes-type formula holds even in the K-theoretical framework: if $(M, \hat{u}, \hat{\alpha}, f) = \partial(W, \hat{U}, \hat{A}, F)$, then

$$\mathrm{Hol}_{(M,\hat{u},\hat{\alpha},f)}(\hat{\beta}) = \exp \int_W F^* curv(\hat{\beta}) \wedge curv(\hat{A}) \wedge \hat{A}_{\hat{K}}(W). \tag{4.1}$$

Here $\hat{A}_{\hat{K}}(W)$ is a representative of $\hat{A}_K(W)$, which is defined as $\int_{N_W\mathbb{R}^N/W} curv(\hat{u})$, where the embedding of W in $\mathbb{R}^N$ is provided by the differential orientation of W. Formula (4.1) implies that, if α is flat, its holonomy over a trivial cycle is zero. Hence, in this case, the holonomy only depends on the K-homology class.

Thanks to differential K-characters we can complete table 1: in the K-theoretical framework, the holonomy is a group morphism $\hat{z}_{\bullet-1}(X) \to U(1)$.

5. Differential K-characters, D-branes and Ramond-Ramond fields

We can now complete the K-theoretical description of D-branes in type II superstring theory. We describe a D-brane world-volume as a differential K-cycle. In particular, we consider the topological K-cycle (W, u, E, ι), where (we recall) u is a Thom class of W, E is the Chan-Patton bundle and ι is the embedding of W in the space-time. On E there is the $U(q)$-gauge theory of the D-brane, hence E is endowed with an Hermitian metric h and a compatible connection ∇. Therefore, we can consider the differential K-theory class $[(E, h, \nabla, 0)]$, using the Freed-Lott model. We call $\hat{E}$ such a class. Moreover, we refine u to a differential orientation $\hat{u}$ of W, that must be fixed as a part of the datum. We get a differential K-theory class $(W, \hat{u}, \hat{E}, \iota)$, that is the world-volume in the K-theoretical framework. Actually, we consider one cycle made by all the even-dimensional world-volumes or one made by all the odd-dimensional ones, depending whether we are considering the type IIA or type IIB theory. In this way, we can correctly define the Wess-Zumino action: it is the holonomy of the differential K-theory class, representing the Ramond-Ramond fields, on the world-volume. How do we compute the topological charge? Here we see the physical importance of the fact that the K-homology groups, defined via differential cycles and boundaries, are isomorphic to the ones defined via topological cycles and boundaries: the Poincaré dual of the underlying K-homology class of the world-volume is the topological charge that we have already defined.

We show that, when the class is topologically trivial, the holonomy coincides with (3.4). Actually, we obtain this result normalizing the class with $\hat{A}_{\hat{K}}(X)^{-\frac{1}{2}}$. In fact, let us call $a(C_{odd})$ the topologically trivial class represented by the global form C_{odd}, i.e., in the Freed-Lott model, $a(C_{odd}) = [(0,0,0,C_{odd})]$ (again, the discussion about C_{ev} is analogous). Let us compute the holonomy of $a(C_{odd} \wedge \hat{A}_{\hat{K}}(X)^{-\frac{1}{2}})$ along the differential K-cycle $(W, \hat{u}, \hat{E}, \iota)$. We have that:

$$
\begin{aligned}
(5.1)\qquad (p_W)_!(\iota^* a(C_{odd} \wedge \hat{A}_{\hat{K}}(X)^{-\frac{1}{2}}) \cdot \hat{E}) &= (p_W)_!(a(\iota^* C_{odd} \wedge \hat{A}_{\hat{K}}(X)^{-\frac{1}{2}} \\
&\wedge curv(\hat{E})) = (p_W)_!(a(\iota^* C_{odd} \wedge \hat{A}_{\hat{K}}(X)^{-\frac{1}{2}} \wedge \mathrm{ch}\nabla_E)).
\end{aligned}
$$

For the first equality we have used the relation $a(C_{odd}) \cdot \hat{E} = a(C_{odd} \wedge curv(\hat{E}))$, which is a fundamental property of differential cohomology. Now we apply the definition of the Gysin map. We consider the data provided by any representative of the differential orientation $\hat{u}$ of W: an embedding $j : W \hookrightarrow \mathbb{R}^N$, a tubular neighbourhood U of W in $\mathbb{R}^N$, the diffeomorphism $\varphi_U : N_W\mathbb{R}^N \to U$ and the open embedding $i : U \hookrightarrow \mathbb{R}^N$. From (5.1) we get:

$$
\begin{aligned}
&\int_{\mathbb{R}^N} i_*(\varphi_U)_*(a(\iota^* C_{odd} \wedge \hat{A}_{\hat{K}}(X)^{-\frac{1}{2}} \wedge \mathrm{ch}\nabla_E) \cdot \hat{u}) \\
&\qquad = \int_{\mathbb{R}^N} i_*(\varphi_U)_*(a(\iota^* C_{odd} \wedge \hat{A}_{\hat{K}}(X)^{-\frac{1}{2}} \wedge \mathrm{ch}\nabla_E \wedge curv(\hat{u}))) \\
&\qquad = a\Big(\int_{N_W\mathbb{R}^N} \iota^* C_{odd} \wedge \hat{A}_{\hat{K}}(X)^{-\frac{1}{2}} \wedge \mathrm{ch}\nabla_E \wedge curv(\hat{u})\Big) \\
&\qquad = a\Big(\int_W \int_{N_W\mathbb{R}^N/W} \iota^* C_{odd} \wedge \hat{A}_{\hat{K}}(X)^{-\frac{1}{2}} \wedge \mathrm{ch}\nabla_E \wedge curv(\hat{u})\Big) \\
&\qquad = a\Big(\int_W \iota^* C_{odd} \wedge \hat{A}_{\hat{K}}(X)^{-\frac{1}{2}} \wedge \mathrm{ch}\nabla_E \wedge \int_{N_W\mathbb{R}^N/W} curv(\hat{u})\Big) \\
&\qquad = a\Big(\int_W \iota^* C_{odd} \wedge \mathrm{ch}\nabla_E \wedge \hat{A}_{\hat{K}}(W) \wedge \hat{A}_{\hat{K}}(X)^{-\frac{1}{2}}\Big).
\end{aligned}
$$

Thus the holonomy is the exponential of $\int_W \iota^* C_{odd} \wedge \mathrm{ch}\nabla_E \wedge \hat{A}_{\hat{K}}(W) \wedge \hat{A}_{\hat{K}}(X)^{-\frac{1}{2}}$, as stated in equation (3.4). We see that, in this case, we have canonical representatives of $\mathrm{ch}E$ and $\hat{A}_K(X)$, provided by the curvatures of $\hat{E}$ and $\hat{u}$, that are two components of the world-volume thought of as a differential K-cycle. Since it is necessary to normalize with $\hat{A}_{\hat{K}}(X)^{-\frac{1}{2}}$ the K-theory class whose holonomy we are calculating, we have to fix a representative of such a class as a part of the background. This would follow automatically refining the space-time manifold to a differential K-cycle too, but it is not necessary, we just choose a representative of the $\hat{A}$-genus as a normalization constant.

Using classical cohomology, the integral of the field-strength along a linking manifold is the numerical charge of the D-brane. A linking manifold L of W is the boundary of a manifold S that intersects W transversely in a finite number of points of the interior. The number of such points is the linking number. Within the K-theoretical framework, we can generalize this concept. First of all, when we consider a D-brane world-volume W with Chan-Patton bundle E, there is not only that charge of W itself, but there are sub-brane charges which are encoded in E. In particular, since in the Wess-Zumino action the Chern character $\mathrm{ch}\, E$ appears, we can interpret the Poincaré duals of the Chern characters as sub-branes of W with a charge. Because of this, a linking manifold of W is not enough. Since the field-strength is K-quantized, it is natural to consider a *linking K-cycle* (L, u, F, ι). Here L is a "generalized" linking manifold, i.e., L is the boundary of a manifold S such that S and W intersect transversely in a submanifold (without boundary) contained in the interior of S. If $S \cap W$ is 0-dimensional, we get a linking manifold in the usual sense. We consider the even-dimensional field-strengths G_{ev}, the discussion about G_{odd} being analogous. The violated Bianchi identity is [**12**]:

$$dG_{ev} = \delta(W) \wedge \mathrm{ch}\nabla_E \wedge \hat{A}_{\hat{K}}(W) \wedge \hat{A}_{\hat{K}}(X)^{-\frac{1}{2}}, \tag{5.2}$$

where W is the union of all the world-volumes with dimension of the suitable parity. Here, again, we see the importance of having representatives of the Chern character and the $\hat{A}$-genus, because dG_{ev} is a form (actually, a current) and not a cohomology class. Equation (5.2) implies that $G_{ev} \wedge \hat{A}_{\hat{K}}(X)^{-\frac{1}{2}}$ is K-quantized and the pairing with a linking K-cycle gives the corresponding charge. In fact:

$$\begin{aligned}
\langle G_{ev} \wedge \hat{A}_K(X)^{-\frac{1}{2}}, (L, u, F, \iota)\rangle &= \int_L G_{ev} \wedge \hat{A}_K(X)^{-\frac{1}{2}} \wedge \mathrm{ch}(F) \wedge \hat{A}_K(L) \\
&= \int_S dG_{ev} \wedge \hat{A}_{\hat{K}}(X)^{-\frac{1}{2}} \wedge \mathrm{ch}(F) \wedge \hat{A}_{\hat{K}}(S) \\
&= \int_S \delta(W) \wedge \mathrm{ch}(E \otimes F) \wedge \frac{\hat{A}_{\hat{K}}(W) \wedge \hat{A}_{\hat{K}}(S)}{\hat{A}_{\hat{K}}(X)} \\
&= \int_{S\cap W} \mathrm{ch}(E \otimes F) \wedge \hat{A}_K(S \cap W) \in \mathbb{Z}.
\end{aligned}$$

If L is a linking manifold and F is the trivial line bundle, then we get $\int_{S\cap W} \mathrm{ch}^0 E = ql$, as in the previous case (l is the linking number and $q = \mathrm{ch}^0 E$). Let us consider $\mathrm{ch}^1 E$. If we represent $PD_W(\mathrm{ch}^1 E)$ as a cycle qW' of codimension 2, we suppose that we can take a linking manifold of W', such that $S \cap W$ is a submanifold of dimension 2. Then the corresponding term of the integral is $\int_{S\cap W} \mathrm{ch}^1 E = \int_{qW'} 1 = ql$, i.e., we measure the charge of the sub-brane. An analogous consideration holds for the higher Chern characters, but we have to take into account the terms of the $\hat{A}$-genus. We just make two final remarks. Using ordinary cohomology, in order to compute the linking number l we must consider any solution of $dG_{n-p-2} = \delta(W)$ (with $q = 1$) and compute the integral along L. Similarly, in the K-theoretical picture, in order to compute the linking number of a cycle (L, u, F, ι), we consider any solution of $dG_{ev} = \delta(W) \wedge \hat{A}_{\hat{K}}(W) \wedge \hat{A}_{\hat{K}}(\hat{X})^{-\frac{1}{2}}$ (with E the trivial line bundle) and compute the integral along the cycle. Then, from the previous integral, we can compute q. Moreover, we remark that the fact that $G_{ev} \wedge \hat{A}_{\hat{K}}(X)^{-\frac{1}{2}}$, and not G_{ev} itself, is K-quantized, is just a normalization analogous to $\frac{1}{2\pi} G_p$ in the case of ordinary cohomology (the constant can appear depending on the conventions). Here $\hat{A}_{\hat{K}}(X)$ does not depend on W, hence it is a constant with respect to a fixed space-time background.

Now we have all the elements in order to draw a complete parallel between the two classification schemes of D-branes. Table 2 shows such a parallel.

	Singular cohomology	**K-theory**
World-vol.	Singular cycle qW	Diff. K-cycle $(W, \hat{u}, \hat{E}, \iota)$
Top. charge	Sing. coh. class $\mathrm{PD}_X[qW]$	K-th. class $\mathrm{PD}_X[(W, \hat{u}, \hat{E}, \iota)]$
RR fields	Ordinary diff. cohom. class Integral field strength	Diff. K-theory class K-Integral field strength
WZ action	Holonomy of the RR fields	K-Holonomy of the RR fields
Num. charge	$\int$ f.s. over a linking manifold	$\int$ f.s. over a linking K-cycle

TABLE 2. Comparison (physics).

References

[1] M.-T. Benameur and M. Maghfoul, *Differential characters in K-theory*, Differential Geom. Appl. **24** (2006), no. 4, 417–432, DOI 10.1016/j.difgeo.2005.12.008. MR2231056 (2008b:58031)

[2] J.-L. Brylinski, *Loop spaces, characteristic classes and geometric quantization*, Progress in Mathematics, vol. 107, Birkhäuser Boston, Inc., Boston, MA, 1993. MR1197353 (94b:57030)

[3] U. Bunke, *Differential cohomology*, course note, arXiv:1208.3961.

[4] U. Bunke and T. Schick, *Differential K-theory: a survey*, Global differential geometry, Springer Proc. Math., vol. 17, Springer, Heidelberg, 2012, pp. 303–357, DOI 10.1007/978-3-642-22842-1_11. MR3289847

[5] J. Evslin, *What Does(n't) K-theory Classify?*, arXiv:hep-th/0610328

[6] F. Ferrari Ruffino, *Flat pairing and generalized Cheeger-Simons characters*, Journal of Homotopy and Related Structures, DOI 10.1007/s40062-015-0124-9. arXiv:1208.1288

[7] F. Ferrari Ruffino and R. Savelli, *Comparing two approaches to the K-theory classification of D-branes*, J. Geom. Phys. **61** (2011), no. 1, 191–212, DOI 10.1016/j.geomphys.2010.10.001. MR2746991 (2012g:81180)

[8] D. S. Freed and J. Lott, *An index theorem in differential K-theory*, Geom. Topol. **14** (2010), no. 2, 903–966, DOI 10.2140/gt.2010.14.903. MR2602854 (2011h:58036)

[9] D. S. Freed and E. Witten, *Anomalies in string theory with D-branes*, Asian J. Math. **3** (1999), no. 4, 819–851. MR1797580 (2002e:58065)

[10] K. Gomi and Y. Terashima, *Higher-dimensional parallel transports*, Math. Res. Lett. **8** (2001), no. 1-2, 25–33, DOI 10.4310/MRL.2001.v8.n1.a4. MR1825257 (2002c:53082)

[11] M. J. Hopkins and I. M. Singer, *Quadratic functions in geometry, topology, and M-theory*, J. Differential Geom. **70** (2005), no. 3, 329–452. MR2192936 (2007b:53052)

[12] R. Minasian and G. Moore, *K-theory and Ramond-Ramond charge*, J. High Energy Phys. **11** (1997), Paper 2, 7 pp. (electronic), DOI 10.1088/1126-6708/1997/11/002. MR1606278 (2000a:81190)

[13] G. Moore and E. Witten, *Self-duality, Ramond-Ramond fields and K-theory*, J. High Energy Phys. **5** (2000), Paper 32, 32, DOI 10.1088/1126-6708/2000/05/032. MR1769467 (2001m:81254)

[14] K. Olsen and R. J. Szabo, *Constructing D-branes from K-theory*, Adv. Theor. Math. Phys. **3** (1999), no. 4, 889–1025. MR1797287 (2001k:81251)

[15] R. J. Szabo and A. Valentino, *Ramond-Ramond fields, fractional branes and orbifold differential K-theory*, Comm. Math. Phys. **294** (2010), no. 3, 647–702, DOI 10.1007/s00220-009-0975-1. MR2585983 (2011c:19014)

Departamento de Matemática - Universidade Federal de São Carlos - Rod. Washington Luís, Km 235 - C.P. 676 - 13565-905 São Carlos, SP, Brazil

E-mail address: ferrariruffino@gmail.com

Proceedings of Symposia in Pure Mathematics
Volume **93**, 2016
http://dx.doi.org/10.1090/pspum/093/01569

Integral pentagon relations for 3d superconformal indices

Ilmar Gahramanov and Hjalmar Rosengren

ABSTRACT. The superconformal index of a three-dimensional supersymmetric field theory can be expressed in terms of basic hypergeometric integrals. By comparing the indices of dual theories, one can find new integral identities for basic hypergeometric integrals. Some of these integral identities have the form of the pentagon identity which can be interpreted as the 2–3 Pachner move for triangulated 3-manifolds.

1. Introduction

The superconformal index is one of the efficient tools in the study of non-perturbative aspects of supersymmetric field theory providing the most rigorous mathematical check of supersymmetric dualities. Recent progress in superconformal index computations have significant implications for mathematics.

A rather striking example is the observation made by Dolan and Osborn [**DO**] that the superconformal index of four-dimensional theories is expressible in terms of elliptic hypergeometric integrals. Such integrals are a new class of special functions [**Ro**, **Ra**, **Sp**] and they are of interest both in mathematics and in physics. The identification of superconformal indices of supersymmetric dual theories is given by the Weyl group symmetry transformations for certain elliptic hypergeometric functions on different root systems. The computations of the superconformal indices of supersymmetric dual theories in four dimensions have led to new non-trivial integral identities for elliptic hypergeometric functions [**SV1**, **SV2**, **S**, **KL**].

2010 *Mathematics Subject Classification.* Primary 81T60, 33D60; Secondary 33E20, 33D90.
Key words and phrases. Basic hypergeometric function, q-hypergeometric function, pentagon identity, superconformal index, supersymmetric duality, mirror symmetry.

The first author is grateful to Tudor Dimofte and Grigory Vartanov for useful discussions. The author would like to thank the organizers of the String–Math 2014 conference held at the University of Alberta, Edmonton, Canada, June 9-13, 2014 for the chance to present these results at the conference. The author also would like to thank Nesin Mathematics Village (Izmir, Turkey), the Chalmers University of Technology (Gothenburg, Sweden), Perimeter Institute for Theoretical Physics (Waterloo, Canada) and the Institut des Hautes Etudes Scientifiques (Bures-sur-Yvette, France) for hospitality during his stay, where some parts of this work was done. The author was supported by the European Science Foundation through the research grant ITGP Short Visit 6454 during a visit to the Chalmers University of Technology and by the support of the Marie Curie International Research Staff Exchange Network UNIFY of the European Union's Seventh Framework Programme [FP7-People-2010-IRSES] under grant agreement No 269217 during a visit to the Perimeter Institute.

Superconformal indices of three-dimensional theories have attracted much attention recently [**HKPP**, **KKP**, **GR**, **KSV**, **GV2**]. Their exact computation yields new powerful verifications of various supersymmetric dualities as mirror symmetry, Seiberg-like duality etc. Since a three-dimensional superconformal index can be expressed in terms of basic hypergeometric integrals [**G**, **GaRa**], by studying supersymmetric dualities one can get new identities for this type of special functions. In this work we consider a special type of such identities, namely five term relations or the so-called pentagon identities. The pentagon relations are interesting from different aspects, see, for instance, [**FK**, **GR**, **K**, **KLV**]. Here we present some examples of integral pentagon relations related to the three-dimensional superconformal index.

The rest of the paper is organized in the following way. In Section 2 we make a brief review of the superconformal index in three dimensions. We present some examples of the integral pentagon identities in Sections 3 and 4 and briefly discuss some open problems in Section 5.

2. The superconformal index

In this section we give a short introduction to a three-dimensional superconformal index and refer the reader to [**IY**, **KSV**] and references therein for more details.

Before proceeding, it would be useful to recall the well-known Witten index. Consider a supersymmetric quantum mechanics

$$\{Q,Q^\dagger\} = 2H\,, \tag{2.1}$$

$$\{Q,(-1)^F\} = 0\,, \tag{2.2}$$

where Q, H and $(-1)^F$ are the supersymmetric charge, the Hamiltonian and the fermion number operator[1], respectively. In order to check whether the supersymmetry is broken or not, Witten introduced the topological invariant of a theory [**W**]

$$\mathrm{I}_W = \mathrm{Tr}(-1)^F e^{-\beta H}\,, \tag{2.3}$$

which tells us that supersymmetry is not spontaneously broken if $\mathrm{I}_W \neq 0$. The sum in the definition (2.3) runs over all physical states of the theory. The Witten index is independent of the parameter β and counts the difference between the number of bosonic and fermionic ground states. It is an analogue of the Atiyah–Singer index [**AS**].

In the case of a supersymmetric field theory one can generalize the Witten index[2] by including to the index global symmetries of a theory commuting with Q and $Q^\dagger$ [**KMMR**, **R1**, **R2**]. For a d-dimensional supersymmetric theory the superconformal index is the following partition function defined on $S^{d-1}\times S^1$,

$$\mathrm{I}(\{t_i\}) = \mathrm{Tr}(-1)^F e^{-\beta\{Q,Q^\dagger\}}\prod t_i^{F_i}\,, \tag{2.4}$$

where the trace is taken over the Hilbert space on S^{d-1}, F_i are generators for global symmetries that commute with Q and $Q^\dagger$, and t_i are additional regulators (fugacities) corresponding to the global symmetries. The superconformal index

[1] A fermion number F takes the value zero on bosons and one on fermions.

[2] The original Witten index for supersymmetric gauge theories gives the dual Coxeter number for the gauge group.

counts short BPS operators of the theory. We refer the reader to [**R2**, **SV1**, **GV1**] for more details and references.

We will restrict ourselves to three-dimensional theories with $\mathcal{N}=2$ supersymmetry[3]. These theories have four real supercharges which we organize into complex spinor Q_α and its conjugate $\bar{Q}_\alpha$ with $\alpha=1,2$. We focus on theories in the far IR, where the symmetry is enhanced to the superconformal symmetry and therefore there are four additional supercharges S_α and $\bar{S}_\alpha$ with $\alpha=1,2$. We choose one of the supercharges, say, $Q=Q_1$. For the theory on $S^2\times S^1$ we have $Q_1^\dagger=S_1$ and the following commutation relation (for the full superconformal algebra, see [**D**])

$$\{Q,Q^\dagger\}=\Delta-R-j_3\,, \tag{2.5}$$

where Δ, R, j_3 are the energy, the generator of the R-symmetry and the third component of the angular momentum on S^2, respectively. Note that, because of conformal symmetry, R-symmetry appears explicitly in the commutation relations[4]. The superconformal index of this theory is defined as [**BM**]

$$\operatorname{Tr}\left[(-1)^F e^{-\beta\{Q,Q^\dagger\}}x^{\Delta+j_3}\prod_i t_i^{F_i}\right], \tag{2.6}$$

where F_i are generators for flavor symmetries of the theory.

Using the localization technique [**Pes**] it was shown that the superconformal index of a three-dimensional theory has the form of matrix integral (see, for instance, [**Ki**, **IY**]) which is a basic hypergeometric integral[5] [**G**, **GaRa**]. For instance, the chiral multiplet with components being a complex scalar field ϕ, a Weyl fermion ψ and an auxiliary complex scalar F,

$$\Phi=\phi+\sqrt{2}\theta\psi+\theta^2F \tag{2.7}$$

(θ is a Grassman coordinate), and with R-charge r in the fundamental representation of the gauge group $U(N)$, contributes to the index as

$$\prod_{a=1}^{N_c}\frac{(x^{2-r+|m_a|}z_a^{-1};x^2)_\infty}{(x^{r+|m_a|}z_a;x^2)_\infty}\,, \tag{2.8}$$

where the q-Pochhammer symbol is defined as

$$(z;q)_\infty=\prod_{i=0}^{\infty}(1-zq^i) \tag{2.9}$$

and the integer parameters m_i stand for the magnetic charges corresponding to the gauge group $U(N)$ and run over integers; the fugacities z_i correspond to the gauge group.

One can also compute the superconformal index by using representations of the superconformal algebra [**KSV**], i.e. by the so-called Romelsberger prescription [**R2**].

---

[3]There are many interesting results in this direction also for theories with $\mathcal{N}=4$ [**KW**], $\mathcal{N}=6$ [**BM**, **Ki**] and $\mathcal{N}=8$ [**BK**] supersymmetry.

[4]In the case of $\mathcal{N}=2$ supersymmetric theories without conformal symmetry, the R-symmetry is only an automorphism group and does not appear in a direct way as in (2.5). For details see, for instance, [**AHISS**].

[5]In the literature it is also called q-hypergeometric integral.

From now on we will use the notation

$$x^2 = q, \tag{2.10}$$

and express the superconformal index via the so-called tetrahedron index [**DGG1**]

$$\mathcal{I}_q[m,z] \;=\; \prod_{i=0}^{\infty} \frac{1-q^{i-\frac{1}{2}m+1}z^{-1}}{1-q^{i-\frac{1}{2}m}z}, \quad \text{with } |q|<1 \text{ and } m\in Z. \tag{2.11}$$

The contribution of a chiral multiplet in terms of tetrahedron index has the following form

$$\prod_{i=0}^{\infty} \frac{1-q^{i+\frac{1}{2}|m|+1}z^{-1}}{1-q^{i+\frac{1}{2}|m|}z} \;=\; (-q^{\frac{1}{2}})^{-\frac{1}{2}(m+|m|)} z^{\frac{1}{2}(m+|m|)} \mathcal{I}_q[m,z]. \tag{2.12}$$

We defined the tetrahedron index for the free chiral with zero R-charge, but one can write the index for general R-charge by the shift $z \to z q^{r/2}$.

The three-dimensional index can be factorized into vertex and anti-vertex partition function [**KSV**, **Pas**, **HKP**] and all results presented in the next section can be written in this fashion; however, this subject is beyond the scope of the present work.

It is worth to mention here that we will consider theories with $U(1)$ gauge group only in order to obtain pentagon identities which we will study in the next section.

3. Integral pentagon identities

Our main interest is the five-term relation for the superconformal index. Let us consider the $d=3$ $\mathcal{N}=2$ supersymmetric quantum electrodynamics with $U(1)$ gauge group and one flavor. The superconformal index of this theory is

$$I_e \;=\; \sum_{m\in Z} \oint \frac{dz}{2\pi i z} z^{-m}\, \mathcal{I}_q[m;q^{1/6}z^{-1}]\, \mathcal{I}_q[-m;q^{1/6}z]\,, \tag{3.1}$$

where the integration is over the unit circle with positive orientation. For simplicity we switched off[6] the topological symmetry $U(1)_J$.

The dual theory is the free Wess–Zumino theory[7] [**IS**, **BHOY**, **AHISS**] with three chiral multiplets $q, \tilde{q}$, S interacting through the superpotential[8] $W = \tilde{q}Sq$. The index of this theory has more simple form, since we do not need to integrate over the gauge group,

$$I_m \;=\; \left(\mathcal{I}_q[0;q^{1/3}]\right)^3. \tag{3.2}$$

These two theories are dual under the mirror symmetry, i.e. under exchange of the Higgs and the Coulomb branches [9]. The mirror duality leads to the following integral pentagon identity

$$\sum_{m\in Z} \oint \frac{dz}{2\pi i z} z^{-m}\, \mathcal{I}_q[m;q^{1/6}z^{-1}]\, \mathcal{I}_q[-m;q^{1/6}z] \;=\; \left(\mathcal{I}_q[0;q^{1/3}]\right)^3. \tag{3.3}$$

---

[6] See, for instance, [**KSV**, **KW**]. We consider the influence of the topological $U(1)_J$ symmetry to the index in the next chapter, where we define the so-called generalized superconformal index.

[7] In the literature this theory sometimes is called the XYZ model.

[8] The permutation symmetry of the superpotential fixes the R-charges, but one can write the index for more general R-charge like in [**IY**].

[9] In three-dimenisonal supersymmetric theories the Coulomb and the Higgs branch are both hyper-Kähler manifolds.

This is the first example of a pentagon identity for the tetrahedron index. The identity (3.3) was obtained in [**KSV**] by using the superconformal index technique. In [**KSV**] the authors also presented the mathematical proof of the integral identity (3.3) by using Ramanujan summation formula for basic hypergeometric series.

The tetrahedron index can be written in the following form:

$$\mathcal{I}_q[m,z] = \sum_{e\in Z} \mathcal{I}(m,e) z^e \,, \tag{3.4}$$

where

$$\mathcal{I}(m,e) = \sum_{n=\frac{1}{2}(|e|-e)}^{\infty} \frac{(-1)^n q^{\frac{1}{2}n(n+1)-(n+\frac{1}{2}e)m}}{(q)_n (q)_{n+e}} \,. \tag{3.5}$$

This index was introduced in [**DGG1**]. It is also interesting from a mathematical point of view [**Gar**, **GHRS**]. The index $\mathcal{I}(m,e)$ obeys the following pentagon identity

$$\begin{aligned} \mathcal{I}(m_1-e_2,e_1)\mathcal{I}(m_2-e_1,e_2) \\ = \sum_{e_3} q^{e_3}\mathcal{I}(m_1,e_1+e_3)\mathcal{I}(m_2,e_2+e_3)\mathcal{I}(m_1+m_2,e_3). \end{aligned} \tag{3.6}$$

The proof of the identity (3.6) is given in the Appendix of [**Gar**]. This pentagon relation is a counterpart of the integral pentagon identity (3.3). In order to distinguish between this type relation and the identity of the form (3.3) we use the terminology "the *integral* pentagon identity" for the latter one.

As another example, we consider the following three-dimensional duality. The electric theory is the $d=3$ $\mathcal{N}=2$ superconformal field theory with $U(1)$ gauge symmetry and six chiral multiplets, half of them transforming in the fundamental representation and another half transforming in the anti-fundamental representation of the gauge group. Its mirror dual is a theory with nine chirals and without gauge degrees of freedom (the gauge symmetry is completely broken). The mirror symmetry leads to the following identity

$$\sum_{m\in Z}\oint \frac{dz}{2\pi i z}\,(-z)^{-3m}\prod_{i=1}^{3}\mathcal{I}_q[-m,q^{\frac{1}{6}}\xi_i z]\,\mathcal{I}_q[m,q^{\frac{1}{6}}\eta_i z^{-1}] = \prod_{i,j=1}^{3}\mathcal{I}_q[0,q^{\frac{1}{3}}\xi_i\eta_j]\,, \tag{3.7}$$

where the fugacities ξ_i and η_i stand for the flavor symmetry $SU(3)\times SU(3)$ and there is the balancing condition $\prod_{i=1}^{3}\xi_i = \prod_{i=1}^{3}\eta_i = 1$. Note that we again dropped the topological symmetry $U(1)_J$. The identity (3.7) was introduced in [**GR**], to where we refer the reader for the details and the mathematical proof of it.

Following [**GR**] we introduce a new function

$$\mathcal{B}[m;a,b] = \frac{\mathcal{I}_q[m,a]\,\mathcal{I}_q[-m,b]}{\mathcal{I}_q[0,ab]}\,, \tag{3.8}$$

and rewrite the equality (3.7) in terms of this function. The final result is a new integral pentagon identity in terms of $\mathcal{B}[m;a,b]$ functions

$$\sum_{m\in Z}\oint \frac{dz}{2\pi i z}\,(-z)^{-3m}\prod_{i=1}^{3}\mathcal{B}[m;\xi_i z^{-1},\eta_i z] = \mathcal{B}[0;\xi_1\eta_2,\xi_3\eta_1]\,\mathcal{B}[0;\xi_2\eta_1,\xi_3\eta_2] \tag{3.9}$$

where we have redefined the flavor fugacities $\xi_i \to q^{-1/6}\xi_i$ and $\eta_i \to q^{-1/6}\eta_i$ and the new balancing condition[10] is $\prod_{i=1}^3 \xi_i = \prod_{i=1}^3 \eta_i = q$.

4. Generalized superconformal index

One can also find a similar pentagon relation for the *generalized superconformal index* [**KW**]. Unlike four-dimensional gauge theories, in three dimensions there are no chiral anomalies, therefore there is no obstruction for considering a theory in a non-trivial background gauge field coupled to the global symmetries. Then one gets new discrete parameters for global symmetries in the expression of the superconformal index. The index with new integer parameters corresponding to the global symmetries is called the generalized superconformal index.

One can apply the above techniques similarly to the generalized superconformal index and obtain more general integral pentagon identities.

The expression (3.7) in terms of the generalized index has the following form

$$\sum_{m\in Z}\int_{\mathbb{T}} \frac{dz}{2\pi i z}\prod_{i=1}^{3}(-1)^m z^{-3m}\frac{((\xi_i z)^{-1}q^{1+m/2};q)_\infty(z/\eta_i q^{1-m/2};q)_\infty}{(\xi_i z q^{m/2+M_i};q)_\infty(\eta_i/zq^{-m/2+N_i};q)_\infty}$$

$$= \frac{1}{\prod_{j=1}^3 q^{\binom{M_j}{2}+\binom{N_j}{2}}\xi_j^{M_j}\eta_j^{N_j}}\prod_{i,j=1}^{3}\frac{(q/\xi_i\eta_j;q)_\infty}{(\xi_i\eta_j q^{M_i+N_j};q)_\infty} \tag{4.1}$$

where we switched on background fields coupled to the flavor symmetry and therefore the index has additional integer parameters M_i and N_i with the condition $\sum_i M_i = \sum_i N_i = 0$. There is also the balancing condition $\prod_{i=1}^3 \xi_i = \prod_{i=1}^3 \eta_i = q$ for flavor fugacities. The new discrete parameters are analogous to the magnetic charge m for the gauge symmetry.

The analogue of the first pentagon identity (3.3) in terms of the generalized superconformal index is the following pentagon identity

$$\sum_{s\in Z}\int \frac{dz}{2\pi i z} z^{2n-s}\omega^m\alpha^{-m}q^{\frac{1}{4}m}\mathcal{I}_q[m+s;q^{\frac{1}{4}}\alpha z^{-1}]\mathcal{I}_q[m-s;\alpha z q^{\frac{1}{4}}]$$

$$= \omega^{-m}\alpha^{n+2m}q^{\frac{1}{4}n}\mathcal{I}_q[m;q^{\frac{1}{4}}\alpha^{-1}\omega^{-1}]\mathcal{I}_q[-m;q^{\frac{1}{4}}\alpha^{-1}\omega]\mathcal{I}_q[2m;q^{\frac{1}{2}}\alpha^2],$$

where we switched on the background gauge field coupled to the topological $U(1)_J$ global symmetry. Here α and m denote the parameters for the axial $U(1)_A$ symmetry, ω and n denote the parameters for the topological $U(1)_J$ symmetry and the discrete parameter s stands for magnetic charge. Note that this identity was proven only for the case $m = 0$ [**KW**].

5. Concluding remarks

There is a recently proposed relation called *3d − 3d correspondence* [**DGG1, DGG2**] (see also [**TM1, TM2, TM3, GMMS, DGaGo, GKLP, CDGS**]) that connects $d = 3$ $\mathcal{N} = 2$ supersymmetric theories and triangulated 3-manifolds. Namely, the independence of the invariant of the corresponding 3-manifold on the choice of triangulation corresponds to the equality of superconformal indices of mirror dual theories [**DGG1, GHRS**]. In this context the interpretation of the integral pentagon identities discussed here is the 2–3 Pachner move [**Pa1, Pa2**] for triangulated 3-manifolds, which relates different decompositions of a polyhedron

[10] We have a misprint in our previous paper [**GR**].

with five ideal vertices into ideal tetrahedra. Much work remains to be done in this direction.

Note that one can write such pentagon identities also for partition functions on S^3 [**KLV**], i.e. for hyperbolic hypergeometric integrals [**Bu**].

As an aside comment, we would like to mention that the pentagon identity (3.9) may represent the star-triangle relation for some integrable model.

References

[AHISS] O. Aharony, A. Hanany, K. Intriligator, N. Seiberg, and M. J. Strassler, *Aspects of $N = 2$ supersymmetric gauge theories in three dimensions*, Nuclear Phys. B **499** (1997), no. 1-2, 67–99, DOI 10.1016/S0550-3213(97)00323-4. MR1468698 (99b:81227)

[AS] M. F. Atiyah and I. M. Singer, *The index of elliptic operators on compact manifolds*, Bull. Amer. Math. Soc. **69** (1963), 422–433. MR0157392 (28 #626)

[BK] D. Bashkirov and A. Kapustin, *Dualities between $\mathcal{N} = 8$ superconformal field theories in three dimensions*, J. High Energy Phys. **5** (2011), 074, 17, DOI 10.1007/JHEP05(2011)074. MR2834550 (2012k:81238)

[BM] J. Bhattacharya and S. Minwalla, *Superconformal indices for $\mathcal{N} = 6$ Chern Simons theories*, J. High Energy Phys. **1** (2009), 014, 12, DOI 10.1088/1126-6708/2009/01/014. MR2480370 (2010d:81258)

[BHOY] J. de Boer, K. Hori, Y. Oz, and Z. Yin, *Branes and mirror symmetry in $N = 2$ supersymmetric gauge theories in three dimensions*, Nuclear Phys. B **502** (1997), no. 1-2, 107–124, DOI 10.1016/S0550-3213(97)00444-6. MR1477860 (99j:81179)

[Bu] F. J. van de Bult, *Hyperbolic hypergeometric functions*, PhD thesis (2007) .

[CDGS] H.-J. Chung, T. Dimofte, S. Gukov, and P. Sulkowski, *3d-3d Correspondence Revisited*, arXiv:1405.3663.

[DGaGo] T. Dimofte, M. Gabella, and A. B. Goncharov, *K-Decompositions and 3d Gauge Theories*, arXiv:1301.0192.

[DGG1] T. Dimofte, D. Gaiotto, and S. Gukov, *3-manifolds and 3d indices*, Adv. Theor. Math. Phys. **17** (2013), no. 5, 975–1076. MR3262519

[DGG2] T. Dimofte, D. Gaiotto, and S. Gukov, *Gauge theories labelled by three-manifolds*, Comm. Math. Phys. **325** (2014), no. 2, 367–419, DOI 10.1007/s00220-013-1863-2. MR3148093

[D] F. A. Dolan, *On superconformal characters and partition functions in three dimensions*, J. Math. Phys. **51** (2010), no. 2, 022301, 43, DOI 10.1063/1.3211091. MR2605030 (2011e:81123)

[DO] F. A. Dolan and H. Osborn, *Applications of the superconformal index for protected operators and q-hypergeometric identities to $\mathcal{N} = 1$ dual theories*, Nuclear Phys. B **818** (2009), no. 3, 137–178, DOI 10.1016/j.nuclphysb.2009.01.028. MR2518083 (2010k:81264)

[FK] L. D. Faddeev and R. M. Kashaev, *Quantum dilogarithm*, Modern Phys. Lett. A **9** (1994), no. 5, 427–434, DOI 10.1142/S0217732394000447. MR1264393 (95i:11150)

[GR] I. Gahramanov and H. Rosengren, *A new pentagon identity for the tetrahedron index*, JHEP **1311** (2013) 128, arXiv:1309.2195.

[GV1] I. B. Gahramanov and G. S. Vartanov, *Superconformal indices and partition functions for supersymmetric field theories*, XVIIth International Congress on Mathematical Physics, World Sci. Publ., Hackensack, NJ, 2014, pp. 695–703. MR3204521

[GV2] I. Gahramanov and G. Vartanov, *Extended global symmetries for 4D $\mathcal{N} = 1$ SQCD theories*, J. Phys. A **46** (2013), no. 28, 285403, 10, DOI 10.1088/1751-8113/46/28/285403. MR3083462

[GMMS] D. V. Galakhov, A. D. Mironov, A. Yu. Morozov, and A. V. Smirnov, *Three-dimensional extensions of the Alday-Gaiotto-Tachikawa relation*, Theoret. and Math. Phys. **172** (2012), no. 1, 939–962, DOI 10.1007/s11232-012-0088-4. Russian version appears in Teoret. Mat. Fiz. **172** (2012), no. 1, 72–99. MR3170055

[GKLP] D. Gang, E. Koh, S. Lee, and J. Park, *Superconformal Index and 3d-3d Correspondence for Mapping Cylinder/Torus*, JHEP **1401** (2014) 063, arXiv:1305.0937.

[Gar] S. Garoufalidis, *The 3D index of an ideal triangulation and angle structures*, `arXiv:1208.1663`.

[GHRS] S. Garoufalidis, C. D. Hodgson, J. H. Rubinstein, and H. Segerman, *1-efficient triangulations and the index of a cusped hyperbolic 3-manifold*,`arXiv:1303.5278`.

[G] G. Gasper, "Lecture notes for an introductory minicourse on q-series," `arXiv:math/9509223`.

[GaRa] G. Gasper and M. Rahman, *Basic hypergeometric series*, 2nd ed., Encyclopedia of Mathematics and its Applications, vol. 96, Cambridge University Press, Cambridge, 2004. With a foreword by Richard Askey. MR2128719 (2006d:33028)

[HKP] C. Hwang, H. C. Kim and J. Park, *Factorization of the 3d superconformal index*, JHEP **1408**, 018 (2014)`arXiv:1211.6023`.

[HKPP] C. Hwang, H. Kim, K.-J. Park, and J. Park, *Index computation for 3d Chern-Simons matter theory: test of Seiberg-like duality*, J. High Energy Phys. **9** (2011), 037, 18, DOI 10.1007/JHEP09(2011)037. MR2889876

[IY] Y. Imamura and S. Yokoyama, *Index for three dimensional superconformal field theories with general R-charge assignments*, J. High Energy Phys. **4** (2011), 007, 21, DOI 10.1007/JHEP04(2011)007. MR2833291 (2012i:81265)

[IS] K. Intriligator and N. Seiberg, *Mirror symmetry in three-dimensional gauge theories*, Phys. Lett. B **387** (1996), no. 3, 513–519, DOI 10.1016/0370-2693(96)01088-X. MR1413696 (97j:81261)

[KKP] A. Kapustin, H. Kim, and J. Park, *Dualities for 3d theories with tensor matter*, J. High Energy Phys. **12** (2011), 087, 24. MR2935600

[KW] A. Kapustin and B. Willett, *Generalized Superconformal Index for Three Dimensional Field Theories*, `arXiv:1106.2484`.

[K] R. Kashaev, *On beta pentagon relations*, `arXiv:1403.1298`.

[KLV] R. Kashaev, F. Luo and G. Vartanov, *A TQFT of Turaev-Viro type on shaped triangulations*, `arXiv:1210.8393`.

[Ki] S. Kim, *The complete superconformal index for $\mathcal{N} = 6$ Chern-Simons theory*, Nuclear Phys. B **821** (2009), no. 1-2, 241–284, DOI 10.1016/j.nuclphysb.2009.06.025. MR2562335 (2010k:81275)

[KMMR] J. Kinney, J. Maldacena, S. Minwalla, and S. Raju, *An index for 4 dimensional super conformal theories*, Comm. Math. Phys. **275** (2007), no. 1, 209–254, DOI 10.1007/s00220-007-0258-7. MR2335774 (2009h:81297)

[KSV] C. Krattenthaler, V. P. Spiridonov, and G. S. Vartanov, *Superconformal indices of three-dimensional theories related by mirror symmetry*, J. High Energy Phys. **6** (2011), 008, 20, DOI 10.1007/JHEP06(2011)008. MR2870861

[KL] D. Kutasov and J. Lin, *N=1 Duality and the Superconformal Index*, `arXiv:1402.5411`.

[Pa1] U. Pachner, *Konstruktionsmethoden und das kombinatorische Homöomorphieproblem für Triangulationen kompakter semilinearer Mannigfaltigkeiten* (German), Abh. Math. Sem. Univ. Hamburg **57** (1987), 69–86, DOI 10.1007/BF02941601. MR927165 (89g:57027)

[Pa2] U. Pachner, *P.L. homeomorphic manifolds are equivalent by elementary shellings*, European J. Combin. **12** (1991), no. 2, 129–145, DOI 10.1016/S0195-6698(13)80080-7. MR1095161 (92d:52040)

[Pas] S. Pasquetti, *Factorisation of $\mathcal{N} = 2$ theories on the squashed 3-sphere*, J. High Energy Phys. **4** (2012), 120, front matter+16. MR2967591

[Pes] V. Pestun, *Localization of gauge theory on a four-sphere and supersymmetric Wilson loops*, Comm. Math. Phys. **313** (2012), no. 1, 71–129, DOI 10.1007/s00220-012-1485-0. MR2928219

[Ra] E. M. Rains, *Transformations of elliptic hypergeometric integrals*, Ann. of Math. (2) **171** (2010), no. 1, 169–243, DOI 10.4007/annals.2010.171.169. MR2630038 (2011i:33046)

[R1] C. Römelsberger, *Counting chiral primaries in $\mathcal{N} = 1$, $d = 4$ superconformal field theories*, Nuclear Phys. B **747** (2006), no. 3, 329–353, DOI 10.1016/j.nuclphysb.2006.03.037. MR2241553 (2007h:81197)

[R2] C. Romelsberger, *Calculating the Superconformal Index and Seiberg Duality*, `arXiv:0707.3702`.

[Ro] H. Rosengren, *Elliptic hypergeometric series on root systems*, Adv. Math. **181** (2004), no. 2, 417–447, DOI 10.1016/S0001-8708(03)00071-9. MR2026866 (2005b:33022)

[Sp] V. P. Spiridonov, *Essays on the theory of elliptic hypergeometric functions* (Russian, with Russian summary), Uspekhi Mat. Nauk **63** (2008), no. 3(381), 3–72, DOI 10.1070/RM2008v063n03ABEH004533; English transl., Russian Math. Surveys **63** (2008), no. 3, 405–472. MR2479997 (2010c:33045)

[SV1] V. P. Spiridonov and G. S. Vartanov, *Elliptic hypergeometry of supersymmetric dualities*, Comm. Math. Phys. **304** (2011), no. 3, 797–874, DOI 10.1007/s00220-011-1218-9. MR2794548

[SV2] V. P. Spiridonov and G. S. Vartanov, *Elliptic hypergeometry of supersymmetric dualities II. Orthogonal groups, knots, and vortices*, Comm. Math. Phys. **325** (2014), no. 2, 421–486, DOI 10.1007/s00220-013-1861-4. MR3148094

[S] M. Sudano, *The Römelsberger index, Berkooz deconfinement, and infinite families of Seiberg duals*, J. High Energy Phys. **5** (2012), 051, front matter+32. MR3042974

[TM1] Y. Terashima and M. Yamazaki, *SL(2,R) Chern-Simons, Liouville, and Gauge Theory on Duality Walls*, JHEP **1108** (2011) 135, `arXiv:1103.5748`.

[TM2] Y. Terashima and M. Yamazaki, *Semiclassical Analysis of the 3d/3d Relation*, Phys.Rev. **D88** no. 2, (2013) 026011, `arXiv:1106.3066`.

[TM3] Y. Terashima and M. Yamazaki, *Emergent 3-manifolds from 4d Superconformal Indices*, Phys.Rev.Lett. **109** (2012) 091602, `arXiv:1203.5792`.

[W] E. Witten, *Constraints on supersymmetry breaking*, Nuclear Phys. B **202** (1982), no. 2, 253–316, DOI 10.1016/0550-3213(82)90071-2. MR668987 (84j:81131)

Institut für Physik und IRIS Adlershof, Humboldt-Universität zu Berlin, Zum Grossen Windkanal 6, D12489 Berlin, Germany and Institute of Radiation Problems ANAS, B.Vahabzade 9, AZ1143 Baku, Azerbaijan

E-mail address: `ilmar@physik.hu-berlin.de`

Department of Mathematical Sciences, Chalmers University of Technology and University of Gothenburg, SE-412 96 Göteborg, Sweden

E-mail address: `hjalmar@chalmers.se`

Proceedings of Symposia in Pure Mathematics
Volume **93**, 2016
http://dx.doi.org/10.1090/pspum/093/01553

Wilson Surfaces in 6D (2,0) Theory and AdS_7/CFT_6

Hironori Mori and Satoshi Yamaguchi

ABSTRACT. We investigate the AdS_7/CFT_6 correspondence based on probe M5-branes in M-theory and Wilson surfaces defined in the 6D $\mathcal{N} = (2,0)$ superconformal field theory. With the conjecture that the (2,0) theory compactified on S^1 is equivalent to the 5D maximal super Yang-Mills (MSYM), we can compute the expectation values of the Wilson surfaces in large-rank symmetric and anti-symmetric representation as those of Wilson loops in the Chern-Simons matrix model. As new evidence for AdS_7/CFT_6, We can show that the Wilson surfaces completely agree with the on-shell values of probe M5-branes wrapping submanifolds in 11-dimensional supergravity background.

1. Introduction

This note is a short version of our original paper [**1**]. We would like to establish new evidence for AdS_7/CFT_6 between M-theory on $AdS_7 \times S^4$ and the 6D $\mathcal{N} = (2,0)$ superconformal field theory [**2**], which is known much less than AdS_d/CFT_{d-1} for $d \leq 5$. The main reason for it is that we still do not have the Lagrangian description of the (2,0) theory. However, there is the conjecture [**3**–**5**] that the (2,0) theory on $S^1 \times S^5$ is equivalent to the 5D maximal super Yang-Mills (MSYM) on S^5 under the condition

$$R_6 = \frac{g_{YM}^2}{8\pi^2}, \tag{1.1}$$

where R_6 is the radius of S^1, and g_{YM} is a coupling constant in 5D theory. This is concluded by the identification of Kaluza-Klein modes in 6D with instanton particles in 5D. We try to deal with a non-local object called Wilson surface in the (2,0) theory. The Wilson surface wrapping $S^1 \times S^1$ of which one is the compactified S^1 can be described by a Wilson loop along the other S^1 under the conjecture. Furthermore, the partition function of the 5D MSYM reduces to the Chern-Simons matrix model computed by the localization [**6**, **7**]. Then, we can evaluate the expectation values of the Wilson surfaces in symmetric and anti-symmetric representation as those of the Wilson loops in the Chern-Simons matrix model.

2010 *Mathematics Subject Classification.* Primary 81T20, 81T60, 83E50.

Key words and phrases. Wilson surface, AdS/CFT, Chern-Simons matrix model.

The work of the first author was supported in part by the JSPS Research Fellowship for Young Scientists.

The work of the second author was supported in part by JSPS KAKENHI Grant No. 22740165.

On the other hand, we consider a probe M5-brane description of the Wilson surface [**9**–**13**] instead of the M2-brane [**7**, **8**, **14**–**17**]. According to the analogy of the Wilson loop in $\mathrm{AdS}_5/\mathrm{CFT}_4$ [**18**–**22**], we expect that an M5-brane wrapping on $\mathrm{AdS}_3 \times S^3 (\subset \mathrm{AdS}_7)$ corresponds to the symmetric representation and one wrapping on $\mathrm{AdS}_3 \times \tilde{S}^3$ ($\tilde{S}^3 \subset S^4$) corresponds to the anti-symmetric representation. We evaluate the on-shell action of each probe M5-brane and compare them with the expectation values of the Wilson surfaces in large-rank representations. Finally, we find new evidence supporting $\mathrm{AdS}_7/\mathrm{CFT}_6$ as the correspondences of the probe M5-branes on $\mathrm{AdS}_3 \times S^3$ and $\mathrm{AdS}_3 \times \tilde{S}^3$ to the Wilson surfaces in symmetric and anti-symmetric representation, respectively.

2. Wilson surfaces in large-rank representations

We consider a Wilson surface defined in 6D A_{N-1} type (2,0) theory on $S^1 \times S^5$. From the conjecture of the equivalence between the (2,0) theory on $S^1 \times S^5$ and 5D MSYM on S^5, the Wilson surface wrapping $S^1 \times S^1$, the first one is the great circle of S^5 and the second one is the compactified S^1, reduces to a Wilson loop along the first S^1. We can find the expectation value of this Wilson loop in Chern-Simons matrix model derived by the localization [**6**, **8**, **23**–**25**] as the partition function of the 5D MSYM. Its explicit form in the representation $\mathbf{R}$ is given by

$$\langle W_{\mathbf{R}} \rangle = \frac{1}{\mathcal{Z}} \int \prod_{i=1}^{N} d\nu_i \prod_{i,j,i\neq j} \left| \sinh \frac{N}{2}(\nu_i - \nu_j) \right| \exp\left[-\frac{N^2}{\beta} \sum_{i=1}^{N} \nu_i^2 \right] \mathrm{Tr}_R e^{N\nu}, \tag{2.1}$$

where ν_i are the eigenvalues from the matrix model, $\beta = \frac{g_{YM}^2}{2\pi r}$, and r is the radius of S^5. $\mathcal{Z}$ is the partition function defined by

$$\mathcal{Z} := \int \prod_{i=1}^{N} d\nu_i \prod_{i,j,i\neq j} \left| \sinh \frac{N}{2}(\nu_i - \nu_j) \right| \exp\left[-\frac{N^2}{\beta} \sum_{i=1}^{N} \nu_i^2 \right]. \tag{2.2}$$

To compare (2.1) with the results on the gravity side, we take the large N limit of the partition function,

$$\mathcal{Z} \sim \int \prod_{i=1}^{N} d\nu_i \exp\left[-\frac{N^2}{\beta} \sum_{i=1}^{N} \nu_i^2 + \frac{N}{2} \sum_{i,j,i\neq j} |\nu_i - \nu_j| \right]. \tag{2.3}$$

Because both terms in the exponential of (2.3) are order $O(N^3)$, we can solve the eigenvalue equations of this reduced matrix model by

$$\nu_i = \frac{\beta}{2N}(N - 2i) \tag{2.4}$$

under the assumption $\nu_i > \nu_j$ for $i < j$. Note that instanton factors are independent of the integration variables in the limit $N \to \infty$ (see [**8**] for details), and the final results for the Wilson surfaces do not have the instanton contributions because those should be canceled by the normalization factor in (2.1).

2.1. Symmetric representation. Firstly, we consider the Wilson surface in symmetric representation S_k where we take the rank k to be $O(N)$. The trace in this representation is expressed as

$$\mathrm{Tr}_{S_k} e^{N\nu} = \sum_{1\le i_1 \le \cdots \le i_k \le N} \exp\left[N \sum_{l=1}^{k} \nu_{i_l}\right]. \tag{2.5}$$

The expectation value with this insertion is given by

$$\langle W_{S_k} \rangle \sim \int \prod_{i=1}^{N} d\nu_i \exp\left[-\frac{N^2}{\beta}\sum_{i=1}^{N}\nu_i^2 + \frac{N}{2}\sum_{i,j,i\neq j}|\nu_i - \nu_j| + Nk\nu_1\right]. \tag{2.6}$$

In fact, the third term in (2.6) affects the eigenvalue distribution of ν_1, and again we find the new solution of ν_1

$$\nu_1 = \frac{\beta}{2N}(N+k). \tag{2.7}$$

We substitute it into (2.6) and then obtain the leading value depending on k

$$\langle W_{S_k} \rangle \sim \exp\left[\frac{\beta}{2}Nk\left(1+\frac{k}{2N}\right)\right]. \tag{2.8}$$

Here, we use the fact that $\langle W_{S_k} \rangle = 1$ when $k = 0$. This expression (2.8) reproduces the result of the fundamental case when $k = 1$ [**7**, **8**].

2.2. Anti-symmetric representation. As the second example, we take anti-symmetric representation A_k with $k = O(N)$ boxes in the Young diagram. The trace in this representation is written as

$$\mathrm{Tr}_{A_k} e^{N\nu} = \sum_{1\le i_1 < \cdots < i_k \le N} \exp\left[N \sum_{l=1}^{k} \nu_{i_l}\right]. \tag{2.9}$$

When we insert this expression into the partition function, the largest contribution in the large N limit is in the case of $i_l = l$ because of our ordering $\nu_1 > \nu_2 > \cdots > \nu_N$. Accordingly, the leading one in (2.1) is given by

$$\langle W_{A_k} \rangle \sim \int \prod_{i=1}^{N} d\nu_i \exp\left[-\frac{N^2}{\beta}\sum_{i=1}^{N}\nu_i^2 + \frac{N}{2}\sum_{i,j,i\neq j}|\nu_i - \nu_j| + N\sum_{l=1}^{k}\nu_l\right]. \tag{2.10}$$

Unlike S_k case, this insertion does not affect the eigenvalue distribution. As a result, the leading contribution with the solutions (2.4) becomes

$$\langle W_{A_k} \rangle \sim \exp\left[\frac{\beta}{2}Nk\left(1-\frac{k}{N}\right)\right]. \tag{2.11}$$

The expression is invariant under the exchange of k and $(N-k)$ as expected and reproduces the result of the fundamental case when $k = 1$ [**7**, **8**].

3. Probe M5-branes in 11D supergravity

In this section, we will compute the expectation values of probe M5-branes in 11-dimensional supergravity. We use usual conventions [**2**] for the AdS radius L and the M5-brane tension T_5 given by

$$L = 2\,(\pi N)^{\frac{1}{3}}\,\ell_{\mathrm{P}}, \qquad T_5 = \frac{1}{(2\pi)^5 \ell_{\mathrm{P}}^6}, \tag{3.1}$$

where $\ell_{\rm P}$ is the 11-dimensional Planck length. The metric of Euclidean $\mathrm{AdS}_7 \times S^4$ in the global coordinates is

$$
\begin{aligned}
& \mathrm{d}s^2 = L^2\left(\cosh^2\rho \mathrm{d}\tau^2 + \mathrm{d}\rho^2 + \sinh^2\rho \mathrm{d}\Omega_5^2\right) + \frac{L^2}{4}\mathrm{d}\Omega_4^2, \\
& \mathrm{d}\Omega_5^2 = \mathrm{d}\eta^2 + \sin^2\eta \mathrm{d}\phi^2 + \cos^2\eta \mathrm{d}\Omega_3^2, \qquad \mathrm{d}\Omega_4^2 = \mathrm{d}\theta^2 + \sin^2\theta \mathrm{d}\tilde{\Omega}_3^2, \\
& \rho \geq 0, \quad 0 \leq \phi < 2\pi, \quad 0 \leq \eta \leq \frac{\pi}{2}, \quad 0 \leq \theta \leq \pi,
\end{aligned} \tag{3.2}
$$

where $\mathrm{d}\Omega_3^2$ and $\mathrm{d}\tilde{\Omega}_3^2$ are metrics of units S^3 and $\tilde{S}^3$, respectively. For $\mathrm{AdS}_7/\mathrm{CFT}_6$, we set the boundary as $S^1 \times S^5$ where S^1 is parametrized by τ with the periodicity

$$
\tau \sim \tau + \frac{2\pi R_6}{r}. \tag{3.3}
$$

To be precise, the identification (3.3) is accompanied by the rotation of the isometry in the $\tilde{S}^3$ direction in order to compare the results from 5D MSYM [**6**, **8**]. Another useful set of coordinates is the $\mathrm{AdS}_3 \times S^3$ foliation whose metric is expressed as

$$
\begin{aligned}
& \mathrm{d}s^2 = L^2\left(\cosh^2 u \mathrm{d}\check{\Omega}_3^2 + \mathrm{d}u^2 + \sinh^2 u \mathrm{d}\Omega_3^2\right) + \frac{L^2}{4}\mathrm{d}\Omega_4^2, \\
& \mathrm{d}\check{\Omega}_3^2 = \cosh^2 w \mathrm{d}\tau^2 + \mathrm{d}w^2 + \sinh^2 w \mathrm{d}\phi^2,
\end{aligned} \tag{3.4}
$$

where (u, w) are related to (ρ, η) with the transformations

$$
\sinh u = \sinh\rho\cos\eta, \qquad \tanh w = \tanh\rho\sin\eta. \tag{3.5}
$$

Let us denote the vielbein for the spacetime by E^a: (E^0, E^1, E^2) for AdS_3; $E^3 = L\mathrm{d}u$, (E^4, E^5, E^6) for S^3 in AdS_7; $E^7 = L\mathrm{d}\theta$, and $(E^8, E^9, E^\natural)$ ($\natural = 10$) for $\tilde{S}^3$ in S^4.

There is the 4-form field strength B_4 as a bosonic field in addition to the metric. When the background geometry is $\mathrm{AdS}_7 \times S^4$, B_4 is given by

$$
B_4 = \frac{6}{L}E^{789\natural}, \tag{3.6}
$$

where $E^{a_1\cdots a_p}$ is a shorthand for $E^{a_1}\wedge\cdots\wedge E^{a_p}$. In the rest of the paper, we represent all indices of field variables as the ones in the local Lorentz frame.

In order to evaluate the on-shell action of a probe M5-brane, we utilize the so-called Pasti-Sorokin-Tonin (PST) action proposed by [**26**–**28**] as the covariant action on a single M5-brane. The bosonic fields consist of an auxiliary scalar field a and a 2-form gauge field $A_2 = \frac{1}{2}A_{mn}\mathrm{d}\zeta^m\wedge\mathrm{d}\zeta^n$ as well as the spacetime coordinates. The bosonic part of the PST action $S_{\rm M5}$ with the Wess-Zumino term is given by

$$
S_{\rm M5} = T_5\int \mathrm{d}^6\zeta\sqrt{-g_{\rm ind}}\left[\mathcal{L} + \frac{1}{4}\tilde{H}^{mn}H_{mn}\right] + T_5\int\left(C_6 - \frac{1}{2}C_3\wedge H_3\right), \tag{3.7}
$$

where $\zeta^m (m = 0, 1, \ldots, 5)$ are the worldvolume coordinates, $C_{3/6}$ is background 3-form/6-form gauge field, and

$$
\mathcal{L} = \sqrt{\det\left(\delta_m^{\ n} + i\tilde{H}_m^{\ n}\right)}, \qquad F_3 = \mathrm{d}A_2, \tag{3.8}
$$

$$
H_3 = F_3 - C_3, \qquad H_{mn} = H_{mnp}v^p, \qquad \tilde{H}^{mn} = (*_6 H)^{mnp}v_p, \tag{3.9}
$$

$$
v_p = \frac{\partial_p a}{\sqrt{-g^{mn}\partial_m a\partial_n a}}. \tag{3.10}
$$

The Hodge star $*_6$ is defined with the induced metric on the M5-brane, and the indices are used in raising or lowering by the induced metric. Moreover, the action possesses the gauge symmetry δ_g

$$\delta_g A_{mn} = \partial_{[m}\phi_{n]}(\zeta), \tag{3.11}$$

and the following local symmetries δ_φ and δ_ψ:

$$\begin{cases} \delta_\varphi a = 0, \\ \delta_\varphi A_{mn} = \dfrac{1}{2}\partial_{[m}a\ \varphi_{n]}(\zeta), \end{cases} \tag{3.12}$$

$$\begin{cases} \delta_\psi a = \psi(\zeta), \\ \delta_\psi A_{mn} = -\dfrac{\psi(\zeta)}{2\sqrt{-g^{pq}\partial_p a \partial_q a}}\left(H_{mn} - \mathcal{V}_{mn}\right), \end{cases} \tag{3.13}$$

where $\varphi_m(\zeta)$ and $\psi(\zeta)$ are infinitesimal parameters for each transformation, and

$$\mathcal{V}_{mn} := -2\frac{\delta\mathcal{L}}{\delta\tilde{H}^{mn}}. \tag{3.14}$$

We will use the above symmetries to fix the gauge of the background gauge fields.

3.1. Wrapping $\mathrm{AdS}_3 \times S^3$**.** Now, we consider the probe M5-brane wrapping $\mathrm{AdS}_3 \times S^3 \subset \mathrm{AdS}_7$. First of all, we would like to determine the boundary term to regularize the on-shell PST action because there are ambiguities to choose local counter terms. For this purpose, we begin with the Poincaré coordinates given by

(3.15)

$$\mathrm{d}s^2 = \frac{L^2}{y^2}\left(\mathrm{d}y^2 + \mathrm{d}r_1^2 + r_1^2\mathrm{d}\phi^2 + \mathrm{d}r_2^2 + r_2^2\mathrm{d}\Omega_3^2\right) + \frac{L^2}{4}\mathrm{d}\Omega_4^2, \qquad y > 0, \quad r_1, r_2 \geq 0.$$

As an analogy of the string theory case, we consider the ansatz

$$r_2 = \kappa y, \qquad y = y(\lambda), \tag{3.16}$$

where κ is a constant. The induced metric on the probe M5-brane is written by

$$\begin{aligned} \mathrm{d}s_{\mathrm{ind}}^2 &= \frac{L^2}{y^2}\left[\left(1+\kappa^2\right)y'^2\mathrm{d}\lambda^2 + \mathrm{d}r_1^2 + r_1^2\mathrm{d}\phi^2 + (\kappa y)^2\mathrm{d}\Omega_3^2\right], \\ \sqrt{g_{\mathrm{ind}}} &= \frac{\kappa^3 L^6}{y^3}|y'|r_1\sqrt{1+\kappa^2}\sqrt{g_{S^3}}, \end{aligned} \tag{3.17}$$

where $y' := \mathrm{d}y/\mathrm{d}\lambda$, and g_{S^3} is the determinant of the metric of unit S^3.

There is the 7-form field strength B_7 which is the Hodge dual to B_4 and a certain combination of C_3 and C_6 to satisfy the equation of motion for B_4:

$$B_7 = *B_4 = \frac{6L^6}{y^7}r_1 r_2^3 dy\wedge\mathrm{d}r_1\wedge\mathrm{d}\phi\wedge\mathrm{d}r_2\wedge\omega_3, \tag{3.18}$$

where ω_3 is the volume form of unit S^3. Because $C_3 = 0$ on this M5-brane world-volume, the fact that the derivative of C_6 becomes to be equal to B_7 leads to

$$C_6 = \frac{\kappa^4 L^6}{y^3}r_1 y'\mathrm{d}r_1\wedge\mathrm{d}\phi\wedge\omega_3\wedge\mathrm{d}\lambda. \tag{3.19}$$

We have the flux quantization through S^3 for the coupling of a probe M5-brane involving S^3 to an open M2-brane electrically. Considering how the open M2-brane can couple to the M5-brane leads to the flux quantization condition [**29**]

$$F_3 = \frac{k}{\pi T_2}\omega_3 = \frac{k}{2N}L^3\omega_3, \tag{3.20}$$

where $T_2 = \frac{1}{(2\pi)^2\ell_{\rm P}^3}$ is the tension of the M2-brane. This quantization condition implies that a non-trivial component of H_3 is

$$H_{456} = \frac{k}{2N\kappa^3}. \tag{3.21}$$

We use the gauge symmetry (3.12) and set

$$H_{012} = 0. \tag{3.22}$$

To find the field strength $\tilde{H}_3$ dual to H_3, we must fix an auxiliary field a. Through our discussion, we choose

$$a = \phi \Rightarrow v_2 = 1. \tag{3.23}$$

With this gauge fixing of a, the non-trivial component of $\tilde{H}^{ab}$ is given by

$$\tilde{H}^{01} = H_{456}. \tag{3.24}$$

After the Wick rotation $\tilde{H}_{t1} = i\tilde{H}_{\tau 1}$ in the original PST action (3.7), the PST action with nonzero C_6 can be computed as

$$S_{\rm M5} = \mathcal{K}\int_{\lambda_{\rm min}}^{\lambda_0} {\rm d}\lambda \frac{|y'|}{y^3}\left[\sqrt{(1+\kappa^2)\left(\kappa^6 + \left(\frac{k}{2N}\right)^2\right)} - \kappa^4\right], \tag{3.25}$$

where

$$\mathcal{K} = 2\pi^2 T_5 L^6 \int_0^{2\pi} {\rm d}\phi \int_0^\infty {\rm d}r_1 r_1. \tag{3.26}$$

We assume $y' < 0$ and introduce the cutoff denoted by λ_0 and the lower bound $\lambda_{\rm min}$. The equation of motion for κ obtained by (3.25) results in the relation of κ to k given by

$$\kappa = \sqrt{\frac{k}{2N}}. \tag{3.27}$$

Putting this into (3.25), the PST action is rewritten as

$$S_{\rm M5} = \mathcal{K}\frac{k}{2N}\int_{\lambda_{\rm min}}^{\lambda_0} {\rm d}\lambda \frac{|y'|}{y^3}. \tag{3.28}$$

Here, we change the bulk coordinate y to z such that

$$z = \frac{1}{y^2}. \tag{3.29}$$

Since $\lambda_{\rm min}$ is mapped into $z_{\rm min} = 0$, we can evaluate the on-shell PST action in the simple form

$$S_{\rm M5} = \frac{k}{4N}\mathcal{K}z_0 = \frac{k}{4N}\mathcal{K}\int_{\lambda_{\rm min}}^{\lambda_0} {\rm d}\lambda z' =: \int_{\lambda_{\rm min}}^{\lambda_0} {\rm d}\lambda \mathcal{L}, \tag{3.30}$$

where a new cutoff is defined as $z_0 := z(\lambda_0)$. Then, the conjugate momentum for z is defined by

$$P_z = \frac{\partial \mathcal{L}}{\partial z'} = \frac{k}{4N}\mathcal{K} \tag{3.31}$$

with the boundary condition that the variation of P_z on the boundary vanishes. The boundary term with P_z based on the Legendre transformation is given by

$$S_{\mathrm{bdy}} = -P_z z_0. \tag{3.32}$$

Indeed, the final result does not depend on the gauge choice of a as far as we use the Legendre transformation prescription for the 2-form gauge field as in [**19**, **30**]. Thus, the regularized on-shell action $S_{\mathrm{M5}}^{\mathrm{reg}}$ of the probe M5-brane becomes

$$S_{\mathrm{M5}}^{\mathrm{reg}} = S_{\mathrm{M5}} + S_{\mathrm{bdy}} = 0. \tag{3.33}$$

This result is expected since the plane Wilson surface preserves a part of the Poincaré supersymmetry. Namely, we conclude that the boundary term is proportional to the volume of the boundary with the gauge choice (3.19) and (3.22).

To obtain the non-trivial value for the probe M5-brane dual to the Wilson surface wrapping on $S^1 \times S^1$, we transform the Poincaré coordinates into the $\mathrm{AdS}_3 \times S^3$ foliation coordinates (3.4) with identification (3.3). The former are associated with the latter by

$$y = \frac{e^\tau}{\cosh u \cosh w}, \qquad r_1 = e^\tau \tanh w, \qquad r_2 = \frac{e^\tau \tanh u}{\cosh w}. \tag{3.34}$$

Now, the M5-brane is wrapping $\mathrm{AdS}_3 \times S^3$ expressed by $u = u_k =$(constant) related to κ as

$$\kappa = \sinh u_k. \tag{3.35}$$

Also, we should rewrite the explicit form of C_6 and H_3 in the foliation coordinates. Then, substituting all factors in the foliation coordinates into the PST action results in

$$S_{\mathrm{M5}} = \frac{2\pi R_6}{r} k\,(2N + k) \sinh^2 w_0, \tag{3.36}$$

where w_0 is a cutoff. From our strategy to regulate the on-shell action (3.32), the boundary term is given by

$$S_{\mathrm{bdy}} = -\frac{2\pi R_6}{r} k\,(2N + k) \sinh w_0 \cosh w_0. \tag{3.37}$$

Summing them up and taking the limit $w_0 \to \infty$, we get the expectation value of the Wilson surface as the on-shell action of the M5-brane

$$\exp\left[-S_{\mathrm{M5}}^{\mathrm{reg}}\right] = \exp\left[-(S_{\mathrm{M5}} + S_{\mathrm{bdy}})\right] = \exp\left[\frac{\beta}{2} Nk \left(1 + \frac{k}{2N}\right)\right], \tag{3.38}$$

where we used the relation (1.1). This result completely matches the value of the Wilson surface in symmetric representation (2.8).

3.2. Wrapping $\mathrm{AdS}_3\times\tilde{S}^3$**.** Let us consider a probe M5-brane wrapping $\mathrm{AdS}_3\times\tilde{S}^3\subset\mathrm{AdS}_7\times S^4$ in the global coordinates (3.2). We take the ansatz

$$\eta=\pi/2,\quad \theta=\theta_k=(\text{constant}). \tag{3.39}$$

The induced metric on this M5-brane is written by

$$\begin{aligned} \mathrm{d}s_{\text{ind}}^2&=L^2\left(\cosh^2\rho\mathrm{d}\tau^2+\mathrm{d}\rho^2+\sinh^2\rho\mathrm{d}\phi^2\right)+\frac{L^2}{4}\sin^2\theta_k\mathrm{d}\tilde{\Omega}_3^2,\\ \sqrt{g_{\text{ind}}}&=\frac{L^6}{8}\cosh\rho\sinh\rho\sin^3\theta_k\sqrt{g_{\tilde{S}^3}}, \end{aligned} \tag{3.40}$$

where a constant θ_k is associated with integer k parametrizing the flux quantization condition. Then, we can acquire B_4 in terms of the derivative of C_3

$$B_4=\mathrm{d}C_3=\frac{3}{8}L^3\sin^3\theta\mathrm{d}\theta\wedge\tilde{\omega}_3, \tag{3.41}$$

where $\tilde{\omega}_3$ is the volume form of unit $\tilde{S}^3$. Integrating this over θ gives us the explicit form of C_3

$$C_3=-\frac{L^3}{8}\left(3\cos\theta-\cos^3\theta-2\right)\tilde{\omega}_3=:-L^3f(\theta)\tilde{\omega}_3. \tag{3.42}$$

We choose the gauge in which $C_3=0$ at $\theta=0$ because $\tilde{S}^3$ shrinks at that point. From this C_3 and the flux quantization condition (3.20), H_3 is given by

$$H_{89\natural}=\left(\frac{k}{2N}+f(\theta_k)\right)\frac{8}{\sin^3\theta_k}, \tag{3.43}$$

and the remaining component of $\tilde{H}^{ab}$ is

$$\tilde{H}^{01}=H_{89\natural} \tag{3.44}$$

under the gauge choice $H_{012}=0$. Consequently, we can calculate the PST action of this M5-brane as

$$S_{\text{M5}}=T_5\frac{\pi^2L^6}{4}\int\mathrm{d}^3\zeta\cosh\rho\sinh\rho\sqrt{\sin^6\theta_k+64\left(\frac{k}{2N}+f(\theta_k)\right)^2}. \tag{3.45}$$

We can obtain the solution to the equation of motion for θ_k as the relation to k

$$\cos\theta_k=1-\frac{2k}{N}. \tag{3.46}$$

Putting this solution back into (3.45) leads to the on-shell action

$$S_{\text{M5}}=\frac{4\pi R_6}{r}k(N-k)\sinh^2\rho_0, \tag{3.47}$$

where ρ_0 is a cutoff. As for the previous case, we can write down the boundary term proportional to the volume of the M5-brane on the boundary

$$S_{\text{bdy}}=-\frac{4\pi R_6}{r}k(N-k)\sinh\rho_0\cosh\rho_0. \tag{3.48}$$

Combining them results in the finite value of the M5-brane wrapping $\mathrm{AdS}_3\times\tilde{S}^3$

$$\exp\left[-S_{\text{M5}}^{\text{reg}}\right]=\exp\left[\frac{\beta}{2}Nk\left(1-\frac{k}{N}\right)\right]. \tag{3.49}$$

It perfectly agrees with the Wilson surface in anti-symmetric representation (2.11). As a result, we obtain nontrivial support for $\mathrm{AdS}_7/\mathrm{CFT}_6$.

4. Summary

We have established new non-trivial evidence for the AdS_7/CFT_6 correspondence where the probe M5-branes wrapping $AdS_3 \times S^3 \subset AdS_7$ and $AdS_3 \times \tilde{S}^3 \subset AdS_7 \times S^4$ perfectly match the Wilson surfaces in large-rank symmetric and anti-symmetric representation, respectively. Note that we also obtain the same results of the Wilson surfaces in rectangular representations by using the general formulation for the Wilson loops in the Chern-Simons matrix model [**31**].

We would like to comment on the relation between the bubbling geometry and Wilson surfaces in larger representations as one of interesting future directions. The authors [**9**, **13**, **32**, **33**] have derived a class of the bubbling solutions of the 11-dimensional supergravity dual to the Wilson surfaces as well as the bubbling geometry for local operators [**34**] and Wilson loops [**32**, **35**, **36**]. The eigenvalue distribution of these solutions as the matrix model is suggested that the real 1-dimensional line of the eigenvalue space consists of two types of the segment colored by black and white: the eigenvalue density is a positive constant on the black segment and zero on the white segment. The unit length of a black segment is twice that of a white segment. Actually, this picture of the bubbling solutions is compatible with the eigenvalue distribution of the Chern-Simons matrix model, which is other evidence for AdS_7/CFT_6. Now, we are working on evaluating the expectation values of the Wilson surfaces from the bubbling geometry point of view and comparing them to the computation in the Chern-Simons matrix model.

References

[1] H. Mori and S. Yamaguchi, "M5-branes and Wilson Surfaces in AdS_7/CFT_6 Correspondence," `arXiv:1404.0930 [hep-th]`.

[2] J. Maldacena, *The large N limit of superconformal field theories and supergravity*, Adv. Theor. Math. Phys. **2** (1998), no. 2, 231–252. MR1633016 (99e:81204a)

[3] E. Witten, *String theory dynamics in various dimensions*, Nuclear Phys. B **443** (1995), no. 1-2, 85–126, DOI 10.1016/0550-3213(95)00158-O. MR1334520 (96f:81104)

[4] M. R. Douglas, *On $D = 5$ super Yang-Mills theory and $(2,0)$ theory*, J. High Energy Phys. **2** (2011), 011, 18, DOI 10.1007/JHEP02(2011)011. MR2820832 (2012j:81211)

[5] N. Lambert, C. Papageorgakis, and M. Schmidt-Sommerfeld, *M5-branes, D4-branes and quantum 5D super-Yang-Mills*, J. High Energy Phys. **1** (2011), 083, 17, DOI 10.1007/JHEP01(2011)083. MR2792268 (2012c:81204)

[6] H.-C. Kim and S. Kim, *M5-branes from gauge theories on the 5-sphere*, J. High Energy Phys. **5** (2013), 144, front matter+41. MR3080517

[7] J. A. Minahan, A. Nedelin, and M. Zabzine, *5D super Yang-Mills theory and the correspondence to* AdS_7/CFT_6, J. Phys. A **46** (2013), no. 35, 355401, 19, DOI 10.1088/1751-8113/46/35/355401. MR3100880

[8] H.-C. Kim, J. Kim, and S. Kim, "Instantons on the 5-sphere and M5-branes," `arXiv:1211.0144 [hep-th]`.

[9] O. Lunin, *1/2-BPS states in M theory and defects in the dual CFTs*, J. High Energy Phys. **10** (2007), 014, 80, DOI 10.1088/1126-6708/2007/10/014. MR2357964 (2008k:81266)

[10] B. Chen, L. Zhang, W. He, and J.-B. Wu, *M5-branes and Wilson surfaces*, J. High Energy Phys. **8** (2007), 067, 31 pp. (electronic), DOI 10.1088/1126-6708/2007/08/067. MR2342046 (2008i:81137)

[11] B. Chen, C.-Y. Liu, and J.-B. Wu, *Operator product expansion of Wilson surfaces from M5-branes*, J. High Energy Phys. **1** (2008), 007, 28, DOI 10.1088/1126-6708/2008/01/007. MR2375405 (2009a:81138)

[12] B. Chen and J.-B. Wu, *Wilson-Polyakov surfaces and M-theory branes*, J. High Energy Phys. **5** (2008), 046, 26, DOI 10.1088/1126-6708/2008/05/046. MR2411313 (2009d:81282)

[13] E. D'Hoker, J. Estes, M. Gutperle, and D. Krym, *Exact half-BPS flux solutions in M-theory. II. Global solutions asymptotic to* $\mathrm{AdS}_7 \times S^4$, J. High Energy Phys. **12** (2008), 044, 21, DOI 10.1088/1126-6708/2008/12/044. MR2469896 (2009m:83147)

[14] O. J. Ganor, *Six-dimensional tensionless strings in the large N limit*, Nuclear Phys. B **489** (1997), no. 1-2, 95–121, DOI 10.1016/S0550-3213(96)00702-X. MR1443797 (98i:81200)

[15] D. Berenstein, R. Corrado, W. Fischler, and J. Maldacena, *Operator product expansion for Wilson loops and surfaces in the large N limit*, Phys. Rev. D (3) **59** (1999), no. 10, 105023, 10, DOI 10.1103/PhysRevD.59.105023. MR1709200 (2000g:81182)

[16] R. Corrado, B. Florea, and R. McNees, *Correlation functions of operators and Wilson surfaces in the* $d = 6$, $(0, 2)$ *theory in the large N limit*, Phys. Rev. D (3) **60** (1999), no. 8, 085011, 18, DOI 10.1103/PhysRevD.60.085011. MR1718504 (2000g:81167)

[17] D. Young, *Wilson loops in five-dimensional super-Yang-Mills*, J. High Energy Phys. **2** (2012), 052, front matter + 15. MR3040661

[18] S.-J. Rey and J.-T. Yee, *Macroscopic strings as heavy quarks: large-N gauge theory and anti-de Sitter supergravity*, Eur. Phys. J. C Part. Fields **22** (2001), no. 2, 379–394, DOI 10.1007/s100520100799. MR1875413 (2002j:81243)

[19] N. Drukker and B. Fiol, *All-genus calculation of Wilson loops using D-branes*, J. High Energy Phys. **2** (2005), 010, 26 pp. (electronic), DOI 10.1088/1126-6708/2005/02/010. MR2140501 (2005m:81246)

[20] S. A. Hartnoll and S. Prem Kumar, *Multiply wound Polyakov loops at strong coupling*, Phys. Rev. D (3) **74** (2006), no. 2, 026001, 11, DOI 10.1103/PhysRevD.74.026001. MR2249981 (2007c:81176)

[21] S. Yamaguchi, *Wilson loops of anti-symmetric representation and D5-branes*, J. High Energy Phys. **5** (2006), 037, 17 pp. (electronic), DOI 10.1088/1126-6708/2006/05/037. MR2231501 (2007d:81233)

[22] J. Gomis and F. Passerini, *Holographic Wilson loops*, J. High Energy Phys. **8** (2006), 074, 30 pp. (electronic), DOI 10.1088/1126-6708/2006/08/074. MR2249907 (2007d:81222)

[23] J. Källén and M. Zabzine, *Twisted supersymmetric 5D Yang-Mills theory and contact geometry*, J. High Energy Phys. **5** (2012), 125, front matter+25. MR3042942

[24] K. Hosomichi, R.-K. Seong, and S. Terashima, *Supersymmetric gauge theories on the five-sphere*, Nuclear Phys. B **865** (2012), no. 2, 376–396, DOI 10.1016/j.nuclphysb.2012.08.007. MR2967134

[25] J. Källén, J. Qiu, and M. Zabzine, *The perturbative partition function of supersymmetric 5D Yang-Mills theory with matter on the five-sphere*, J. High Energy Phys. **8** (2012), 157, front matter + 32. MR3006905

[26] P. Pasti, D. Sorokin, and M. Tonin, *Covariant action for a* $D = 11$ *five-brane with the chiral field*, Phys. Lett. B **398** (1997), no. 1-2, 41–46, DOI 10.1016/S0370-2693(97)00188-3. MR1442904 (98b:81210)

[27] I. Bandos, K. Lechner, A. Nurmagambetov, P. Pasti, D. Sorokin, and M. Tonin, *Covariant action for the super-five-brane of M theory*, Phys. Rev. Lett. **78** (1997), no. 23, 4332–4334, DOI 10.1103/PhysRevLett.78.4332. MR1456124 (98e:81127)

[28] I. Bandos, K. Lechner, A. Nurmagambetov, P. Pasti, D. Sorokin, and M. Tonin, *On the equivalence of different formulations of the M theory five-brane*, Phys. Lett. B **408** (1997), no. 1-4, 135–141, DOI 10.1016/S0370-2693(97)00784-3. MR1483585 (98j:81245)

[29] J. M. Camino, A. Paredes, and A. V. Ramallo, *Stable wrapped branes*, J. High Energy Phys. **5** (2001), Paper 11, 38, DOI 10.1088/1126-6708/2001/05/011. MR1845420 (2002c:81152)

[30] N. Drukker, D. J. Gross, and H. Ooguri, *Wilson loops and minimal surfaces*, Phys. Rev. D (3) **60** (1999), no. 12, 125006, 20, DOI 10.1103/PhysRevD.60.125006. MR1732810 (2001k:81291)

[31] N. Halmagyi and T. Okuda, *Bubbling Calabi-Yau geometry from matrix models*, J. High Energy Phys. **3** (2008), 028, 27, DOI 10.1088/1126-6708/2008/03/028. MR2391092 (2009d:81329)

[32] S. Yamaguchi, *Bubbling geometries for half-BPS Wilson lines*, Internat. J. Modern Phys. A **22** (2007), no. 7, 1353–1374, DOI 10.1142/S0217751X07035070. MR2309814 (2008d:83198)

[33] E. D'Hoker, J. Estes, M. Gutperle, and D. Krym, *Exact half-BPS flux solutions in M-theory. I. Local solutions*, J. High Energy Phys. **8** (2008), 028, 55, DOI 10.1088/1126-6708/2008/08/028. MR2434558 (2009i:83125)

[34] H. Lin, O. Lunin, and J. Maldacena, *Bubbling AdS space and 1/2 BPS geometries*, J. High Energy Phys. **10** (2004), 025, 68, DOI 10.1088/1126-6708/2004/10/025. MR2116025 (2005j:83122)

[35] O. Lunin, *On gravitational description of Wilson lines*, J. High Energy Phys. **6** (2006), 026, 44 pp. (electronic), DOI 10.1088/1126-6708/2006/06/026. MR2233826 (2007f:83094)

[36] E. D'Hoker, J. Estes, and M. Gutperle, *Gravity duals of half-BPS Wilson loops*, J. High Energy Phys. **6** (2007), 063, 52 pp. (electronic), DOI 10.1088/1126-6708/2007/06/063. MR2326588 (2009e:81176)

Department of Physics, Graduate School of Science, Osaka University, Toyonaka, Osaka 560-0043, Japan
E-mail address: hiromori@het.phys.sci.osaka-u.ac.jp

Department of Physics, Graduate School of Science, Osaka University, Toyonaka, Osaka 560-0043, Japan
E-mail address: yamaguch@het.phys.sci.osaka-u.ac.jp

Proceedings of Symposia in Pure Mathematics
Volume **93**, 2016
http://dx.doi.org/10.1090/pspum/093/01558

Motivic zeta functions of the quartic and its mirror dual

Johannes Nicaise, D. Peter Overholser, and Helge Ruddat

ABSTRACT. We use a formula of Bultot to compute the motivic zeta function for the toric degeneration of the quartic K3 and its Gross-Siebert mirror dual degeneration. We check for this explicit example that the identification of the logarithm of the monodromy and the mirror dual Lefschetz operator works at an integral level.

1. Introduction

Motivic integration was introduced by Kontsevich in a famous lecture at Orsay in 1995. Kontsevich invented this theory in order to prove that birationally equivalent complex Calabi-Yau varieties have the same Hodge numbers.

Motivic integrals take values in a suitable Grothendieck ring of varieties and can be specialized to various additive invariants of algebraic varieties, such as Hodge-Deligne polynomials.

An early application of motivic integration was Denef and Loeser's definition of the *motivic zeta function* of a hypersurface singularity, designed as a motivic upgrade of Igusa's p-adic local zeta function. This motivic zeta function is a very rich invariant of the singularity and contains several interesting classical invariants, such as the Hodge spectrum.

We refer to [**12**] for an elementary introduction. In [**10**], an analogous motivic zeta function was defined for a smooth and proper variety X over $\mathbb{C}((t))$ with trivial canonical sheaf. The case of abelian varieties was studied in detail in [**9**], but only very few examples are known when X is Calabi-Yau.

A large class of Calabi-Yau varieties over $\mathbb{C}((t))$ can be constructed and described by means of the theory of *toric degenerations* of Gross and Siebert. Using tropical and logarithmic geometry, Gross and Siebert showed how to construct a Calabi-Yau variety over $\mathbb{C}((t))$ from combinatorial data: a topological manifold with an integral affine structure with singularities and a piecewise linear function on this manifold [**8**]. This construction explains mirror symmetry between Calabi-Yau varieties over $\mathbb{C}((t))$ as a duality between the combinatorial data.

2010 *Mathematics Subject Classification.* Primary 11G42; Secondary 14M25, 14D05, 14E18, 14G10.

Key words and phrases. Motivic zeta function, mirror symmetry, toric degenerations, Gross-Siebert program.

The research leading to these results has received funding from the European Research Council under the European Union's Seventh Framework Programme (FP/2007-2013) / ERC Grant Agreement n. 306610.

We aim to use this construction to compute motivic zeta functions. As a first step, and to illustrate the techniques that come into play, we work out the details in the case of a mirror pair of $K3$ surfaces, and we compare the result with the formula of Stewart and Vologodsky [**17**].

Notation. We will denote by K the field of complex Laurent series $\mathbb{C}((t))$ and by R its valuation ring $\mathbb{C}[\![t]\!]$. If $\mathcal{X}$ is a scheme over R then its special fiber will be denoted by $\mathcal{X}_0$. If X is a separated K-scheme of finite type, then an R-model of X is a flat separated R-scheme $\mathcal{X}$ of finite type, endowed with an isomorphism of K-schemes $\mathcal{X}_K \to X$ where $\mathcal{X}_K$ denotes the base change to K.

We denote by $K_0(\mathrm{Var}_\mathbb{C})$ the Grothendieck ring of complex varieties, by $\mathbb{L} = [\mathbb{A}^1_\mathbb{C}] \in K_0(\mathrm{Var}_\mathbb{C})$ the class of the affine line, and by $\mathcal{M}_\mathbb{C} = K_0(\mathrm{Var}_\mathbb{C})[\mathbb{L}^{-1}]$ the localized Grothendieck ring. We recall that $K_0(\mathrm{Var}_\mathbb{C})$ is the quotient of the free abelian group on isomorphism classes $[X]$ of complex varieties X modulo the scissor relations $[X] = [Y] + [X \setminus Y]$ whenever Y is a closed subvariety of X. The product on $K_0(\mathrm{Var}_\mathbb{C})$ is induced by the fibered product of algebraic varieties over $\mathbb{C}$.

All the logarithmic structures in this paper are defined with respect to the Zariski topology.

2. Motivic zeta functions of Calabi-Yau varieties and the formula of Stewart-Vologodsky

2.1. Motivic zeta functions of Calabi-Yau varieties. Let X be a smooth, proper, geometrically connected variety over K with trivial canonical line bundle, and assume that $X(K)$ is non-empty. The motivic zeta function $Z_X(T)$ is a formal power series in T with coefficients in the localized Grothendieck ring $\mathcal{M}_\mathbb{C}$. It was introduced in [**10**], and measures the sets of rational points on X over the finite extensions of K. In his PhD thesis, Bultot gave a formula for $Z_X(T)$ in terms of a log-smooth R-model of X. We will now recall a particular case of his formula.

Let $\mathcal{X}$ be a proper R-model of X. We endow $\operatorname{Spec} R$ with its standard log structure and we endow $\mathcal{X}$ with the divisorial log structure induced by its special fiber $\mathcal{X}_0$. The sheaf of monoids defining the log structure on $\mathcal{X}$ will be denoted by $M_\mathcal{X}$. We assume that $\mathcal{X}$ is log-smooth over R, that $\mathcal{X}_0$ is reduced, and that the relative canonical line bundle $\omega^{\log}_{\mathcal{X}/R}$ (that is, the sheaf of relative log-differentials of maximal degree) is trivial. These assumptions substantially simplify Bultot's formula for the zeta function and they will be sufficient for the applications in this paper.

We denote by $F(\mathcal{X})$ the *Kato fan* of $\mathcal{X}$. This is a topological space endowed with a sheaf of monoids. The underlying space of $F(\mathcal{X})$ is the subspace of $\mathcal{X}_0$ that consists of the generic points of intersections of sets of irreducible components of $\mathcal{X}_0$, and the monoid sheaf on $F(\mathcal{X})$ is the restriction of the sheaf of characteristic monoids $\overline{M}_\mathcal{X} = M_\mathcal{X}/\mathcal{O}^\times_\mathcal{X}$ on $\mathcal{X}$.

The special fiber $\mathcal{X}_0$ has a natural stratification, indexed by the points of the fan $F(\mathcal{X})$. For every ξ in $F(\mathcal{X})$, we denote by $U(\xi)$ the subset of the closure of $\{\xi\}$ obtained by removing the closures of the sets $\{\xi'\}$ with ξ' a point of $F(\mathcal{X})$ that does not specialize to ξ. The set $U(\xi)$ is a locally closed subset of $\mathcal{X}_0$, and we endow it with its reduced induced structure.

For every point ξ of $F(\mathcal{X})$, the stalk $\overline{M}_{\mathcal{X},\xi}$ is a toric monoid. The dimension of the monoid $\overline{M}_{\mathcal{X},\xi}$, which coincides with the rank of its groupification, will be denoted by $r(\xi)$. The monoid $\overline{M}_{\mathcal{X},\xi}$ contains a distinguished element e_ξ, which is the image of the residue class of the local parameter t under the morphism

$$\overline{M}_{\mathrm{Spec}\,R,0} = R/R^\times \cong \mathbb{N} \to \overline{M}_{\mathcal{X},\xi}$$

where we wrote 0 for the closed point of $\mathrm{Spec}\,R$. We denote by $\overline{M}^{\vee,\mathrm{loc}}_{\mathcal{X},\xi}$ the set of local morphisms of monoids

$$\phi : \overline{M}_{\mathcal{X},\xi} \to \mathbb{N}.$$

Here, "local" simply means that $\phi^{-1}(0) = \{1\}$.

THEOREM 2.1 (Bultot). *With the above notations and assumptions, we have*

$$Z_X(T) = \sum_{\xi \in F(\mathcal{X})} (\mathbb{L}-1)^{r(\xi)-1}[U(\xi)] \sum_{\phi \in \overline{M}^{\vee,\mathrm{loc}}_{X\mathcal{X}\xi}} T^{\phi(e_\xi)} \quad \in \mathcal{M}_{\mathbb{C}}[[T]].$$

PROOF. This is a particular case of [**1**, 3.1]. □

2.2. The formula of Stewart-Vologodsky. Here we will focus on the case where X is a $K3$-surface with a strictly semistable R-model, i.e., a regular proper R-model $\mathcal{X}$ such that $\mathcal{X}_0$ is a reduced divisor with strict normal crossings. If we enlarge our category of models to include algebraic spaces over R instead of only schemes, then we can find such a strictly semistable model $\mathcal{X}$ with the additional property that the relative canonical line bundle $\omega_{\mathcal{X}/R}$ is trivial. Such a model is called a *Kulikov model* for X. The possible special fibers of Kulikov models have been classified by Kulikov [**11**] and Persson-Pinkham [**14**]. The shape of the special fiber is closely related to the limit mixed Hodge structure of X, and this allowed Stewart and Vologodsky to obtain an explicit formula for the motivic zeta function $Z_X(T)$ of X in terms of the limit mixed Hodge structure [**17**].

If X is of type III and $\mathcal{X}$ is a Kulikov model, then the special fiber $\mathcal{X}_0$ is a union of smooth rational surfaces that intersect along cycles of smooth rational curves and the dual intersection complex of $\mathcal{X}_0$ is a triangulation of the 2-sphere. Stewart and Vologodsky have shown that $Z_X(T)$ only depends on the number $t(X)$ of triple points in $\mathcal{X}_0$, i.e., the number of 2-simplices in the dual intersection complex (this number is denoted by $r_2(X,K)$ in [**17**]). The invariant $t(X)$ be read from the limit mixed Hodge structure on the degree 2 cohomology of X using [**2**, 7.1]; see Section 4.

THEOREM 2.2 (Stewart-Vologodsky). *If X is a $K3$-surface over K of type III, then*

$$Z_X(T) = \frac{t(X)}{2}(\mathbb{L}-1)^2\frac{T(1+T)}{(1-T)^3} + (1+10\mathbb{L}+\mathbb{L}^2)\frac{2T}{1-T} \quad \in \mathcal{M}_{\mathbb{C}}[[T]].$$

In particular, it only depends on $t(X)$, and it fully determines the invariant $t(X)$.

PROOF. This is an immediate consequence of Stewart and Vologodsky's formula (Theorem 1 in [**17**]) for the coefficients of the generating series $Z_X(T)$. □

We can also use Theorem 2.2 to compute the *motivic volume* or *motivic nearby fiber* of X, which encodes the geometry of a general fiber of X viewed as a degenerating family of complex varieties over a formal punctured disc. It is formally defined by taking the limit of $-Z_X(T)$ for $T \to +\infty$; see [**13**, §8]. This leads to the following result.

COROLLARY 2.3. *If X is a $K3$-surface over K of type III, then the motivic volume of X is equal to*

$$2 + 20\mathbb{L} + 2\mathbb{L}^2 \quad \in \mathcal{M}_{\mathbb{C}}.$$

This expression nicely reflects the cohomological structure of X: subtracting the contributions 1 and $\mathbb{L}^2$ for the cohomology spaces of degree 0 and 4, respectively, we obtain the motivic decomposition $1 + 20\mathbb{L} + \mathbb{L}^2$ for the degree 2 cohomology of X. The fact that the motivic volume lies in $\mathbb{Z}[\mathbb{L}]$ reflects the property that the limit mixed Hodge structure is of Hodge-Tate type.

3. The quartic and its mirror

We begin by describing the affine manifolds which will serve as the dual intersection complexes of the degenerations of the quartic and its mirror.

3.1. Affine manifolds and subdivisions. The affine manifold (see [**7**]) B that is the intersection complex of a degeneration of the quartic can be constructed on the boundary of the Newton polytope of the quartic. This Newton polytope P is four times a standard 3-simplex. Translating it so that its unique interior lattice point becomes the origin yields the tetrahedron in $\mathbb{R}^3$ whose vertices are given by

$$(-1,-1,-1), (3,-1,-1), (-1,3,-1), \text{ and } (-1,-1,3). \tag{3.1}$$

We define a subdivision $\mathcal{P}$ of $B := \partial P$ by cutting each facet of the tetrahedron P into 16 elementary two-simplices as shown in Figure 3.1. We introduce a distinguished set Δ given by the barycenters of the 24 one-polytopes of $\mathcal{P}$ that are contained in an edge of the tetrahedron. These are marked in Figure 3.1 by dotted circles. The standard affine structure on the interior of each maximal simplex and a fan structure at each vertex, which is given as shown in Figure 3.2, yield an integral affine structure on $B \setminus \Delta$ (a variation on example 2.10.4 of [**7**]). We equip $(B, \mathcal{P})$ with a piecewise linear function φ, defined by the Newton polytopes given in Figure 3.3. Applying the discrete Legendre transform (see [**7**, §4]), we obtain another integral tropical manifold $(\check{B}, \check{\mathcal{P}})$ with multi-valued piecewise linear function $\check{\varphi}$. A chart of $(\check{B}, \check{\mathcal{P}})$ is pictured in Figure 3.4.

3.2. Degenerations and log smooth models. A general description of how to obtain toric degenerations from Batyrev-Borisov Calabi-Yau manifolds was given in [**4**]. We follow the slight variant of this given in [**15**].

3.2.1. *A degeneration by compactifying a pencil.* Let P be the Newton polytope of some hypersurface in $(\mathbb{C}^*)^n$ that we wish to degenerate. Assume the origin is the unique interor point of P and that all facets have integral distance one to it, i.e. P is reflexive. Let f be a Laurent polynomial defining the hypersurface. Let $\bar{C}(P)$ be the sideways truncated cone over P, i.e.

$$\bar{C}(P) = \{(rm, r) \in \mathbb{R}^{n+1} \mid r \geq 0, m \in P, rm \in P\}.$$

Let $\mathcal{Y}'$ denote the toric variety associated to $\bar{C}(P)$. We have a proper map $\mathcal{Y}' \to \mathbb{A}^1$ given by the monomial $z^{e_{n+1}}$ that is associated to the last unit vector e_{n+1} in $\mathbb{R}^{n+1}$. We assume that the coefficients of f are generic. Next consider the Newton polytope $\tilde{P}$ of the pencil in $(\mathbb{C}^*)^n$ generated by the original hypersurface and the empty set, i.e.

$$\tilde{P} = \text{Newton}(f z^{e_{n+1}} + 1) = \text{conv}(P \times \{e_{n+1}\}, 0).$$

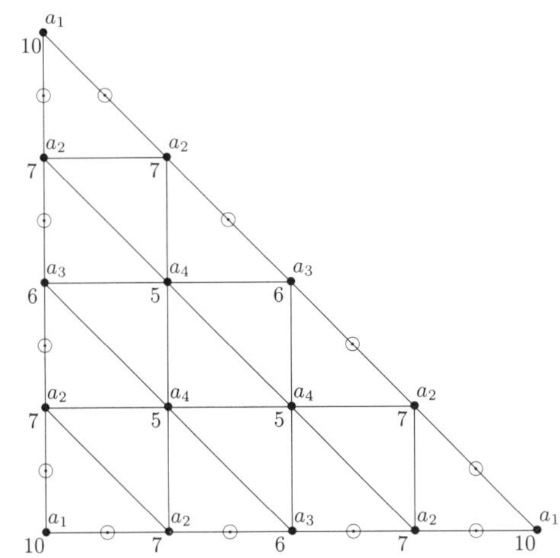

FIGURE 3.1. The subdivision of a facet of the Newton polytope of the quartic. The union of facets gives the *integral tropical manifold* B with polyhedral decomposition $\mathcal{P}$. The pair $(B, \mathcal{P})$ is the intersection complex ("cone picture") of a degeneration $\mathcal{X}$ of a quartic K3 surface and the dual intersection complex ("fan picture") of a *mirror* degeneration $\check{\mathcal{X}}$. The symbol $\odot$ marks affine singularities, while the vertex labelings a_i correspond to different fan structures (see Figure 3.2). We label edges between vertices of type a_2 and a_3 by b_1, and the remaining edges by b_2. All two-dimensional polytopes are of the same type, which we label by c.

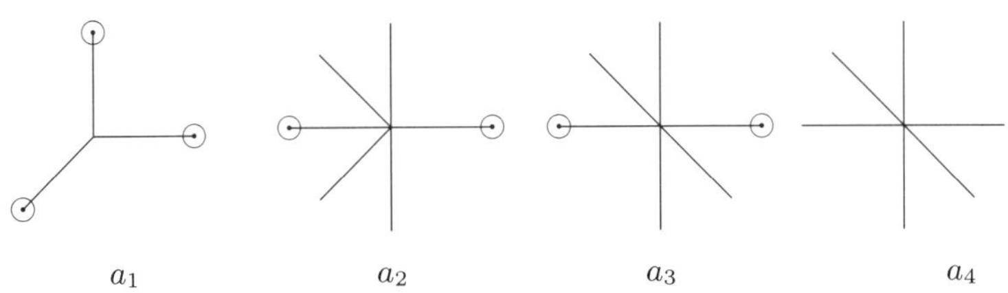

FIGURE 3.2. Fan structures.

This is the convex hull of the origin and a copy of P placed at height one (with respect to e_{n+1}), $z^{e_{n+1}}$ is the pencil parameter. Since $\tilde{P}$ coincides with the cut-off of $\bar{C}(P)$ at $e^*_{n+1} \leq 1$, the hypersurface $\mathcal{X}'$ defined by $fz^{e_{n+1}}+1$ doesn't contain any zero-dimensional orbits of $\mathcal{Y}'$. The restriction of $\mathcal{Y}' \to \mathbb{A}^1$ to $\mathcal{X}'$ gives a proper map $\pi' : \mathcal{X}' \to \mathbb{A}^1$. This is a degeneration of the compactified hypersurface associated to f (set $z^{e_1} = \infty$ to obtain the original hypersurface).

3.2.2. *Resolving singularities in codimension two.* Note that $\mathcal{Y}'$ and $\mathcal{X}'$ are typically very singular. Let D' denote the horizontal divisor in $\mathcal{Y}'$ given by the vertical facets of $\bar{C}(P)$ (those that arise from the sideways truncation). A resolution of singularities in codimension two can be done by choosing another lattice polytope Q whose normal fan is a refinement of P's such that the resulting toric variety $\mathcal{Y}$, i.e. the family $\mathcal{Y} \to \mathbb{A}^1$, has the properties

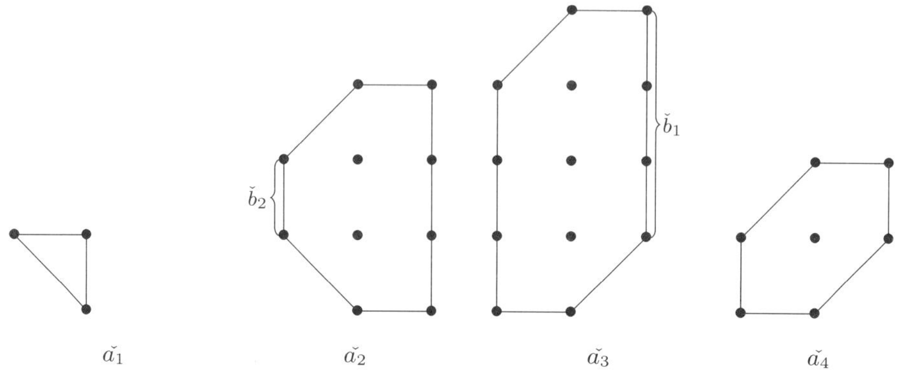

FIGURE 3.3. Newton polytopes

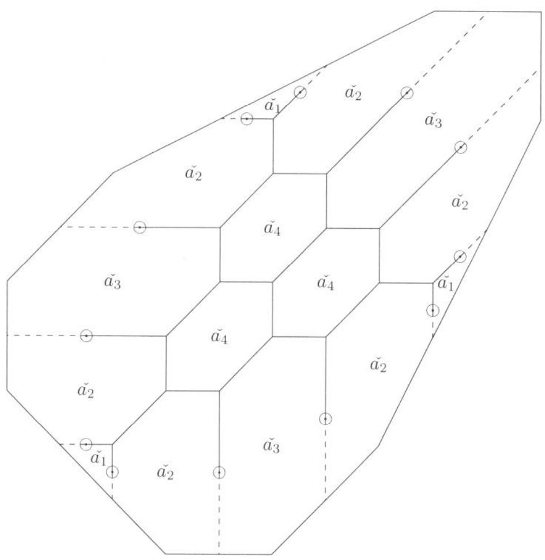

FIGURE 3.4. A chart of the pair $(\check{B}, \check{\mathcal{P}})$ dual to the patch of $(B, \mathcal{P})$ depicted in Figure 3.1. The labelings $\check{a}_i$ refer to two-dimensional polyhedra dual to the vertices in Figure 3.1. Edges of type $\check{b}_1$ separate cells of type $\check{a}_2$ and $\check{a}_3$, while all other edges are of type $\check{b}_2$. Each vertex is of type $\check{c}$ with identical fan structure given by the fan of $\mathbb{P}^2$. The markings $\odot$ again indicate affine singularities, while the dashed segments are "cut lines" across which the affine structure changes in this chart.

(1) the general fibre of $\mathcal{Y}$ is regular,
(2) away from D, the strict transform of D', the family $\mathcal{Y} \to \mathbb{A}^1$ is semistable, i.e. the central fibre $\mathcal{Y}_0$ is a normal crossing divisor in Y.

We will give such degenerations explicitly in the examples that interest us. We then restrict the family to the strict transform of $\mathcal{X}'$ to obtain a degenerating family $\pi : \mathcal{X}'' \to \mathbb{A}^1$ for the quartic and its mirror dual.

3.2.3. *Resolving the remaining log singularities.* We now assume $\dim \mathcal{X}'' = 3$, i.e. $\mathcal{X}''$ is a degeneration of surfaces. Denote by $\mathcal{X}_0 = \pi^{-1}(0)$ the special fibre. It is not convenient to deal with the remaining log singularities in $\mathcal{X}''$ torically. If there are some, these lie in the intersection of D and $\operatorname{Sing}\mathcal{X}_0$, hence they are of codimension three in $\mathcal{X}''$. Note that $\mathcal{Y}$ is smooth at the generic points of $D\cap\mathcal{X}_0$ by the reflexivity assumption. In the quartic and its mirror, the set of points where $\mathcal{X}''\to\mathbb{A}^1$ is not log smooth is $D\cap\operatorname{Sing}\mathcal{X}_0$ and each singularity is an ordinary double point. At such a singularity, $\mathcal{X}''\to\mathbb{A}^1$ takes the shape

$$xy-tw=0$$

where t is the coordinate of $\mathbb{A}^1$. We resolve all of these simultaneously by a projective small resolution $\mathcal{X}\to\mathcal{X}''$ simply by blowing up components of $\mathcal{X}_0$ in $\mathcal{X}''$ until all singularities are gone. Each singularity in the resolution is replaced by a $\mathbb{P}^1$ that is contained in the strict(=proper) transform of the component of $\mathcal{X}_0$ that was blown up to resolve this singularity. Let $\tilde{\mathcal{X}_0}$ be the strict transform of $\mathcal{X}_0$ in $\mathcal{X}$. What is most important for us is that motivically

$$[\tilde{\mathcal{X}_0}]=[\mathcal{X}_0]+m\mathbb{L} \tag{3.2}$$

where m is the number of singularities. We will also need that the copies of $\mathbb{L}$ don't lie in any proper intersection of two strata of $\tilde{\mathcal{X}_0}$. In other words, they are added to the part of a component of $\mathcal{X}_0$ that is the complement of the proper intersections with other components. We will have $m=24$ for the quartic and its mirror-dual. We denote the resulting semistable family by $\pi:\mathcal{X}\to\mathbb{A}^1$.

3.2.4. *Degeneration of the quartic.* The Newton polytope P of the quartic was given in (3.1). We construct the family $\mathcal{X}'\to\mathbb{A}^1$ as in the previous subsections. The general fibre is already smooth. It remains to make the central fibre semistable. For this, we choose a piecewise linear function $\hat{\varphi}:P\to\mathbb{R}$ that is uniquely determined by its values at the lattice points in the boundary of P which we choose on each facet to be as given in Figure 3.1. We then define the polytope Q to be the convex hull of the graph of $\hat{\varphi}$, i.e.

$$Q=\{(m,r)\in\mathbb{R}^{n+1}\mid r\geq\hat{\varphi}(m)\}.$$

We leave it to the interested reader to check that

(1) The restriction of $\hat{\varphi}$ to each facet induces the subdivision given in Figure 3.1,
(2) Let τ be a simplex in the subdivision of a facet then the piecewise linear function induced by $\hat{\varphi}$ on the quotient fan of P by τ is up to addition of a linear function given by the Newton polytope $\check{\tau}$ in Fig. 3.3.

Using Q, we obtain $\mathcal{X}\to\mathbb{A}^1$ as in the previous subsections. Since the quartic meets the singular locus of the union of hyperplanes in $\mathbb{P}^3$ in 24 points, we find $m=24$.

3.2.5. *Degeneration of the quartic mirror dual.* To obtain the mirror dual degeneration $\check{\mathcal{X}}\to\mathbb{A}^1$ we proceed analogously. Let $\check{P}$ denote the reflexive dual of P. The Batyrev quartic mirror dual is the crepant resolution of a general hypersurface in the projective toric variety associated to $\check{P}$. Using $\check{P}$ we obtain $\check{C}(\check{P})$, $\check{\mathcal{Y}}\to\mathbb{A}^1$ and $\check{\mathcal{X}}'\to\mathbb{A}^1$ similar to before. Let $\check{Q}_1$ denote the Newton polytope of the piecewise linear function $\hat{\varphi}$ that we constructed for the resolution of the quartic degeneration. Let $\hat{\varphi}_0:P\to\mathbb{R}$ be the piecewise linear function that takes value $\hat{\varphi}(v)-1$ at a lattice point v in ∂P and let $\check{Q}_0$ be its Newton polytope. This results in a Minkowski sum.

$$\check{Q}_1=\check{Q}_0+\check{P}.$$

Finally, we set

$$\check{Q} = \{(q_0, 0) + r(p, 1) \in \mathbb{R}^{n+1} \mid q_0 \in \check{Q}_0, p \in \check{P}, r \geq 0\}$$

We leave it to the interested reader to check that this satisfies the two items in §3.2.2. Again there are $m = 24$ double points to resolve to obtain a semistable family $\check{\mathcal{X}} \to \mathbb{A}^1$ and the intersection complex of the central fibre is the one depicted in Figure 3.4.

3.3. Applying Bultot's theorem. With the aid of Bultot's formula for log smooth models (Theorem 2.1), we use the combinatorial data of $(B, \mathcal{P})$ and $(\check{B}, \check{\mathcal{P}})$ to compute the motivic zeta function of the quartic, the general fiber of $\mathcal{X}$. The same argument can be applied (after applying the appropriate dualization) to the mirror quartic.

The polyhedral complex $\mathcal{P}$ is the intersection complex of $\mathcal{X}_0$, encoding its toric strata, while $\check{\mathcal{P}}$ encodes the discrete part of its log structures. When translated into the language of our construction, incorporating (3.2), Theorem 2.1 yields

$$Z_{\text{quartic}}(T) = 24\mathbb{L}\frac{T}{1-T} + \sum_{\tau \in \mathcal{P}} (\mathbb{L}-1)^{r(\tau)}[U(\tau)] \sum_{\phi \in \overline{M}^{\vee,\text{loc}}_{\mathcal{X},\tau}} T^{\phi(e_\xi)} \tag{3.3}$$

$$= 24\mathbb{L}\frac{T}{1-T} + \sum_{\tau \in \mathcal{P}} (\mathbb{L}-1)^{r(\tau)}[U(\tau)] \sum_{(\mathbf{m},t) \in C(\check{\tau})^\circ \cap \mathbb{Z}^{r(\tau)} \times \mathbb{N}} T^n. \tag{3.4}$$

Here we've identified $\check{\tau}$ with a polyhedron in $\mathbb{Z}^{r(\tau)}$, $C(\check{\tau})^\circ \subseteq \mathbb{R}^{r(\tau)} \times \mathbb{R}_{\geq 0}$ is the interior of the cone over $\check{\tau}$, $r(\tau)$ is the codimension of τ in B, and

$$U(\tau) := \mathbb{P}_\tau \setminus \bigcup_{\substack{\sigma \in \mathcal{P} \\ \sigma \subsetneq \tau}} \mathbb{P}_\sigma.$$

The equality above follows from the identification of $\overline{M}_{\mathcal{X},\tau}$ with the set of points $(\mathbf{n}, t) \in \mathbb{Z}^{r(\tau)} \times \mathbb{N}$ with $t \geq \varphi_\tau(\mathbf{n})$, where φ_τ is a representative of φ on the *fan structure* along τ (see [**8**]). Then, by definition of $\check{\tau}$, $C(\check{\tau}) \cap \mathbb{Z}^{r(\tau)} \times \mathbb{N}$ can be identified with $\overline{M}^\vee_{\mathcal{X},\tau}$, with $C(\check{\tau})^\circ \cap \mathbb{Z}^{r(\tau)} \times \mathbb{N} = \overline{M}^{\vee,\text{loc}}_{\mathcal{X},\tau}$. If $(\mathbf{m}, t)$ corresponds to $\phi \in \overline{M}^{\vee,\text{loc}}_{\mathcal{X},\tau}$, then $t = \phi(e_\tau)$. Then

$$Z_{\text{quartic}}(T) = 24\mathbb{L}\frac{T}{1-T} + \sum_{\tau \in \mathcal{P}} (\mathbb{L}-1)^{r(\tau)}[U(\tau)] \sum_{n \geq 1} l_{\check{\tau}}(n) T^n$$

where

$$l_{\check{\tau}}(n) := \#\left\{(\mathbf{m}, t) \in C(\check{\tau})^\circ \cap \mathbb{Z}^{r(\tau)} \times \mathbb{N} | t = n\right\}.$$

As noted in [**5**, Proof of Proposition 2.2], one can compute the function $l_{\check{\tau}}$ by counting simplices in a unimodular triangulation of $\check{\tau}$. Let $\mathcal{P}_{\check{\tau}}$ be the set of simplices of such a subdivision (which always exists for $\dim \check{\tau} \leq 2$). The number of integral lattice points in the interior of the jth scaling of the standard k-simplex is given by the coefficient of T^j in the expansion about 0 of $\left(\frac{T}{1-T}\right)^{k+1}$. We obtain

$$\sum_{n \geq 1} l_{\check{\tau}}(n) T^n = \sum_{\substack{\sigma \in \mathcal{P}_{\check{\tau}} \\ \sigma \not\subseteq \partial\check{\tau}}} \left(\frac{T}{1-T}\right)^{\dim(\sigma)+1}.$$

Combining this result with the observation that $U(\tau) = (\mathbb{L}-1)^{2-r(\tau)}$, we can rewrite 3.4. Consider $\tilde{\check{\mathcal{P}}}$, a unimodular triangulation of $\check{\mathcal{P}}$ As before, let $t(\text{quartic})$ be the number of triangles appearing in $\tilde{\check{\mathcal{P}}}$. Then

$$\begin{aligned}
Z_{\text{quartic}}(T) =& 24\mathbb{L}\frac{T}{1-T} + \sum_{\tau\in\mathcal{P}} (\mathbb{L}-1)^2 \sum_{\substack{\sigma\in\tilde{\check{\mathcal{P}}} \\ \sigma\subset\check{\tau} \\ \sigma\not\subseteq\partial\check{\tau}}} \left(\frac{T}{1-T}\right)^{3-r(\sigma)} \\
=& 24\mathbb{L}\frac{T}{1-T} + (\mathbb{L}-1)^2 \sum_{\sigma\in\tilde{\check{\mathcal{P}}}} \left(\frac{T}{1-T}\right)^{3-r(\sigma)} \\
=& 24\mathbb{L}\frac{T}{1-T} + (\mathbb{L}-1)^2 \left[t(\text{quartic}) \left(\frac{T}{1-T}\right)^3 \right. \\
&\left. + \frac{3t(\text{quartic})}{2}\left(\frac{T}{1-T}\right)^2 + \left(\frac{t(\text{quartic})}{2}+2\right)\left(\frac{T}{1-T}\right) \right] \\
=& \frac{t(\text{quartic})}{2}(\mathbb{L}-1)^2\frac{T(1+T)}{(1-T)^3} + (1+10\mathbb{L}+\mathbb{L}^2)\frac{2T}{1-T}.
\end{aligned}$$

Thus we see agreement with Theorem 2.2.

Each facet of the tetrahedron P can be triangulated as in Fig. 3.1. This yields $4\cdot 16$ triangles giving $t(\text{mirror quartic}) = 64$. Letting $t(\tau)$ denote the count of lattice triangles in an elementary triangulation of a polygon τ, using the labelling as in Fig. 3.3, we obtain $t(\text{quartic})$ as

$$4t(\check{a}_1) + 2\cdot 6t(\check{a}_2) + 6t(\check{a}_3) + 3\cdot 4t(\check{a}_4) = 4 + 2\cdot 6\cdot 6 + 6\cdot 14 + 3\cdot 4\cdot 10 = 280.$$

Thus,

$$\begin{aligned}
Z_{\text{quartic}}(T) =& 140(\mathbb{L}-1)^2\frac{T(1+T)}{(1-T)^3} + (1+10\mathbb{L}+\mathbb{L}^2)\frac{2T}{1-T} \\
Z_{\text{mirror quartic}}(T) =& 32(\mathbb{L}-1)^2\frac{T(1+T)}{(1-T)^3} + (1+10\mathbb{L}+\mathbb{L}^2)\frac{2T}{1-T}.
\end{aligned}$$

4. Arithmetic of the monodromy

Let us assume now that $\mathcal{X}$ is a type III degeneration of K3 surfaces, $\mathcal{X}_s$ the nearby fibre and $\mathcal{X}_0$ the central fibre. We follow [**2**]. Let $\mathcal{T} \in \text{Aut}(H^2(\mathcal{X}_s;\mathbb{Z}))$ be the Picard Lefschetz transformation (i.e. monodromy). One sets

$$N = \log\mathcal{T} = (\mathcal{T}-1) - \frac{1}{2}(\mathcal{T}-1)^2.$$

Denote by

$$W_0 \subset W_2 \subset W_4 = H^2(\mathcal{X}_s;\mathbb{Q})$$

the weight filtration induced by N, i.e. $W_0 = \operatorname{im} N^2$, $W_2 = \ker N^2 = W_0^{\perp}$. We have that $W_0 \cap H^2(\mathcal{X}_s;\mathbb{Z}) \cong \mathbb{Z}$, let γ be a generator. Let $(x\cdot y)$ denote the pairing of $x, y \in H^2(X_s;\mathbb{Q})$. By unimodularity, there is a $\gamma' \in H^2(\mathcal{X}_s;\mathbb{Z})$ such that $(\gamma\cdot\gamma') - 1$. Set

$$\delta = N\gamma' \in W_2.$$

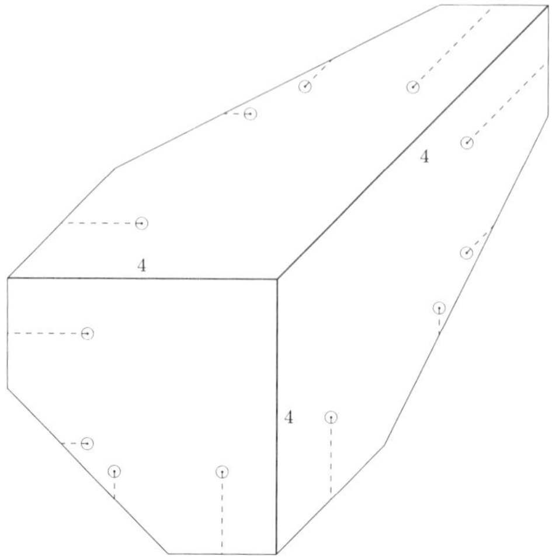

FIGURE 4.1. The tropical curve given by the 1-skeleton in Fig. 3.4 is homologous to the tropical curve depicted here

PROPOSITION 4.1 (Friedman and Scattone). *We have for* $x \in H^2(\mathcal{X}_s;\mathbb{Q})$,

$$Nx = (x \cdot \gamma)\delta - (x \cdot \delta)\gamma$$

and $\delta \in H^2(\mathcal{X}_s;\mathbb{Z})$, *so* N *preserves* $H^2(\mathcal{X}_s;\mathbb{Z})$.

PROOF. [**2**, Lemma 1.1]. □

There is a pair of integers (t,k) that determines N up to conjugation [**2**, Prop. 1.7]. One defines

$$t \overset{\text{def}}{=} (\delta \cdot \delta) = \#(\text{triangles in a unimodular triangulation of } (\check{B}, \check{\mathcal{P}}))$$

where the second equality is [**2**, Prop 7.1] where it appears as the count of triple points of a semistable birational model for $\mathcal{X}_0$. The index k is the maximal integer such that N/k is integral, i.e. by the proposition this is the maximal integer such that δ/k is integral. Note that k^2 divides t.

Let $f : \mathcal{X}_s \to B$ denote the continuous map that is the SYZ fibration [**3**]. It follows from [**3**] that γ is a fibre and γ' is a section.

LEMMA 4.2. *We have that* $\delta = N\gamma'$ *is given by the tropical 1-cycle that is the 1-skeleton of* $\mathcal{P}$ *supporting sections of* $\check{\Lambda}$ *that are given by the kinks of* φ.

PROOF. We leave this to the upcoming paper [**6**] and refer to [**15**] that explains how one arrives at this. □

This Lemma allows us to compute the index k, i.e. the maximal integer such that δ/k is integral. For the computation of t, it is a nice exercise to compute $(\delta \cdot \delta)$ using the tropical intersection product but for us it is easier to go by the triangle count.

THEOREM 4.3. *We have*

$$t(\textit{quartic}) = 280 = 2^3 \cdot 5 \cdot 7, \qquad k(\textit{quartic}) = 1,$$
$$t(\textit{quartic-mirror}) = 64 = 2^6, \quad k(\textit{quartic-mirror}) = 4.$$

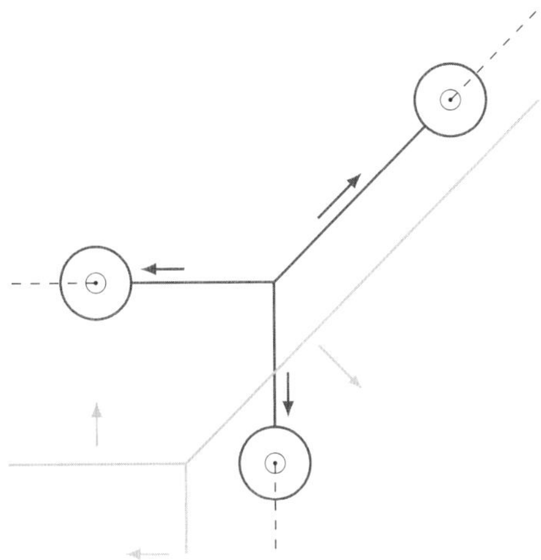

FIGURE 4.2. Simple intersection of a tropical cycle and cocycle

PROOF. The quantities t were computed in the previous section. Consider Fig. 3.4. The tropical curve α obtained from the 1-skeleton in the intersection complex of the quartic-mirror is the tropical curve part of which is shown in Fig. 3.4 and the sections of $\check{\Lambda}$ are primitive along each edge. Furthermore, the sections are invariant under local monodromy wherever an edge passes through a singularity. This allows us to deform the edges over the singularities by adding a suitable tropical 2-cycle. This way, it can be seen that α is homologous to the curve depicted in Fig. 4.1. Let α' be the reduced tropical curve, i.e. $\alpha = 4\alpha'$. We need to show that $k = 4$ is the maximum, i.e. that α' is not a homologous to a proper multiple of another curve. This follows from looking at $t = t(\text{quartic}) = 64$. Since k^2 divides t, the only other options would be $k = 8$ but then, by [**2**, Fig. B], $\mathcal{P}$ would need to be the refinement of a triangulation of the sphere with a single triangle which is impossible, hence $k = 4$. It remains to show that $k(\text{quartic}) = 1$. It suffices to exhibit a cocycle that pairs to 1 with the tropical cycle in given by the 1-skeleton in the intersection complex of the quartic, see Fig. 3.1. One such cycle is shown in black in Fig 4.2. It is Y-shaped where each end encircles a singularity that neighbours a fixed corner of the tetrahedron. Let's denote this one by α. We perturb the 1-skeleton cycle slightly so that it intersects α transversely in a single point, let's call this β. The pairing of the sections of Λ and $\check{\Lambda}$ at the stalk of the intersection point is ± 1, see Fig 4.2. This pairing coincides with the homology-cohomology pairing of the usual (co-)cycles in $\hat{\alpha} \in H^1(\mathcal{X}_s, \mathbb{Z})$ and $\hat{\beta} \in H_1(\mathcal{X}_s, \mathbb{Z})$ reconstructed from α and β respectively as in [**16**]. This implies that $\hat{\beta}$ is primitive and thus $k(\text{quartic}) = 1$. □

Let us consider the Lefschetz operator $L \in \operatorname{End}(H^\bullet(\mathcal{X}_s, \mathbb{Z}))$ on the cohomology of the quartic, i.e. L is the cup product with $4H$ for H the hyperplane class. We may define $\check{k}(\text{quartic})$, $\check{t}(\text{quartic})$ using L analogous to how we defined k, t using N. This means $\check{k}(\text{quartic})$ is the maximum such that $H/\check{k}$ is integral, i.e. $\check{k}(\text{quartic}) = 4$. We find that $\check{t}(\text{quartic}) := |\operatorname{coker} L^2| = 64$ as this can be computed from the ambient $\mathbb{P}^3$ where it is $(4H)^2.(\text{quartic}) = (4H)^3 = 64H^3$.

We deduce the following mirror symmetry result.

COROLLARY 4.4. *$\check{t}$(quartic) = t(quartic-mirror), $\check{k}$(quartic) = k(quartic-mirror).*

References

[1] Emmanuel Bultot, *Computing zeta functions on log smooth models* (English, with English and French summaries), C. R. Math. Acad. Sci. Paris **353** (2015), no. 3, 261–264, DOI 10.1016/j.crma.2014.11.014. MR3306495

[2] Robert Friedman and Francesco Scattone, *Type* III *degenerations of* $K3$ *surfaces*, Invent. Math. **83** (1986), no. 1, 1–39, DOI 10.1007/BF01388751. MR813580 (87k:14044)

[3] Mark Gross, *Topological mirror symmetry*, Invent. Math. **144** (2001), no. 1, 75–137, DOI 10.1007/s002220000119. MR1821145 (2002c:14062)

[4] Mark Gross, *Toric degenerations and Batyrev-Borisov duality*, Math. Ann. **333** (2005), no. 3, 645–688, DOI 10.1007/s00208-005-0686-7. MR2198802 (2007b:14086)

[5] Mark Gross, Ludmil Katzarkov, and Helge Ruddat, *Towards mirror symmetry for varieties of general type* (201202), available at 1202.4042.

[6] Mark Gross and Bernd Siebert, *Torus fibrations and toric degenerations.* In preparation.

[7] Mark Gross and Bernd Siebert, *Affine manifolds, log structures, and mirror symmetry*, Turkish J. Math. **27** (2003), no. 1, 33–60. MR1975331 (2004g:14041)

[8] Mark Gross and Bernd Siebert, *From real affine geometry to complex geometry*, Ann. of Math. (2) **174** (2011), no. 3, 1301–1428, DOI 10.4007/annals.2011.174.3.1. MR2846484

[9] Lars Halvard Halle and Johannes Nicaise, *Motivic zeta functions of abelian varieties, and the monodromy conjecture*, Adv. Math. **227** (2011), no. 1, 610–653, DOI 10.1016/j.aim.2011.02.011. MR2782205 (2012c:14050)

[10] Lars Halvard Halle and Johannes Nicaise, *Motivic zeta functions for degenerations of abelian varieties and Calabi-Yau varieties*, Zeta functions in algebra and geometry, Contemp. Math., vol. 566, Amer. Math. Soc., Providence, RI, 2012, pp. 233–259, DOI 10.1090/conm/566/11223. MR2858926

[11] Vik. S. Kulikov, *Degenerations of* $K3$ *surfaces and Enriques surfaces* (Russian), Izv. Akad. Nauk SSSR Ser. Mat. **41** (1977), no. 5, 1008–1042, 1199. MR0506296 (58 #22087b)

[12] Johannes Nicaise, *An introduction to p-adic and motivic zeta functions and the monodromy conjecture*, Algebraic and analytic aspects of zeta functions and L-functions, MSJ Mem., vol. 21, Math. Soc. Japan, Tokyo, 2010, pp. 141–166. MR2647606 (2011g:11223)

[13] Johannes Nicaise and Julien Sebag, *Motivic Serre invariants, ramification, and the analytic Milnor fiber*, Invent. Math. **168** (2007), no. 1, 133–173, DOI 10.1007/s00222-006-0029-7. MR2285749 (2009c:14040)

[14] Ulf Persson and Henry Pinkham, *Degeneration of surfaces with trivial canonical bundle*, Ann. of Math. (2) **113** (1981), no. 1, 45–66, DOI 10.2307/1971133. MR604042 (82f:14030)

[15] Helge Ruddat, Nicolò Sibilla, David Treumann, and Eric Zaslow, *Skeleta of affine hypersurfaces*, Geom. Topol. **18** (2014), no. 3, 1343–1395, DOI 10.2140/gt.2014.18.1343. MR3228454

[16] Helge Ruddat, *Mirror duality of Landau-Ginzburg models via discrete Legendre transforms*, Homological mirror symmetry and tropical geometry, Lect. Notes Unione Mat. Ital., vol. 15, Springer, Cham, 2014, pp. 377–406, DOI 10.1007/978-3-319-06514-4_9. MR3330791

[17] Allen J. Stewart and Vadim Vologodsky, *Motivic integral of K3 surfaces over a non-archimedean field*, Adv. Math. **228** (2011), no. 5, 2688–2730, DOI 10.1016/j.aim.2011.07.015. MR2838055

IMPERIAL COLLEGE, DEPARTMENT OF MATHEMATICS, SOUTH KENSINGTON CAMPUS, LONDON SW72AZ, UK

E-mail address: j.nicaise@imperial.ac.uk

IMPERIAL COLLEGE, DEPARTMENT OF MATHEMATICS, SOUTH KENSINGTON CAMPUS, LONDON SW72AZ, UK

E-mail address: d.overholser@imperial.ac.uk

MATHEMATISCHES INSTITUT, JGU MAINZ, STAUDINGERWEG 9, D-55128 MAINZ, GERMANY

E-mail address: ruddat@uni-mainz.de

Proceedings of Symposia in Pure Mathematics
Volume **93**, 2016
http://dx.doi.org/10.1090/pspum/093/01574

Semistability and Instability in Products and Applications

Alexander H. W. Schmitt

Abstract. We study asymptotic properties of the notions of semistability and instability from geometric invariant theory for the action of a reductive affine algebraic group on the product of two projective spaces. These findings have interesting applications to relative geometric invariant theory and the theory of quiver sheaves.

Introduction

Let G be a reductive group and $\rho\colon G \longrightarrow \mathrm{GL}(V)$, $\sigma\colon G \longrightarrow \mathrm{GL}(W)$ two rational representations of G. These data induce an action of G on the projective variety $P(V) \times P(W)$ and, for every pair m, n of positive integers, a linearization of this action in the ample line bundle $\mathscr{O}(m,n)$. In this note, we will study semistable and unstable points in $P(V) \times P(W)$ in the case that the ratio n/m is very large.

First, we will recall the characterization of (semi)stable points in $P(V) \times P(W)$ from [**15**]. As an application, we will give an elementary proof of the relative Hilbert–Mumford criterion of Gulbrandsen, Halle, and Hulek [**6**] in the quasi-projective setting.

For an unstable point $x = ([v],[w]) \in P(V) \times P(W)$, Kempf's theory [**9**] yields a so-called instability one parameter subgroup which destabilizes x as much as possible in a certain sense. The characterization of semistable points from the first section suggests an interesting property of the instability flag of x with respect to the linearization in $\mathscr{O}(m,n)$ under the assumptions that

a) $[w] \in P(W)$ is semistable and
b) $n/m \gg 0$.

This criterion was originally proved in [**15**] with a view toward boundedness results for decorated sheaves. We will present it in the second section and discuss some refinements from the papers [**19**] and [**20**].

In the last section, we will review recent results of the author on regions of stability parameters for quiver sheaves. These are obtained by combining the result on unstable points from the second section with ideas of Ramanan and Ramanathan [**14**]. We will include a new example which illustrates these results.

2010 *Mathematics Subject Classification.* Primary 14L24, 14D20; Secondary 16G20.
Key words and phrases. Hilbert–Mumford criterion, instability flag, quiver sheaf, boundedness.

Conventions. The ground field k is assumed to be algebraically closed. For the applications to quiver sheaves, we will need to assume that it has characteristic zero. All objects occurring will be defined over k. The symbols $P(V)$ and $P(E)$ will stand for the classical projectivization $(V \setminus \{0\})/\mathbb{G}_m(k)$ of a k-vector space V and $(E \setminus \{\text{ zero section }\})/\mathbb{G}_m(k)$ of a vector bundle E on an algebraic variety X, respectively. We will use $\mathbb{P}(V)$ and $\mathbb{P}(E)$ to denote projectivization according to Grothendieck. If not otherwise specified, the tensor product of two $\mathscr{O}_X$-modules on a scheme X will be taken over $\mathscr{O}_X$, i.e., we will usually write "$\otimes$" instead of "$\underset{\mathscr{O}_X}{\otimes}$".

Acknowledgments. This research as well as the author's participation in String-Math 2014 were funded by the German Research Council (DFG) via project C3 "Algebraic Geometry: Deformations, Moduli and Vector Bundles" within CRC 647 "Space-Time-Matter". Final revisions were made during the author's visit to the IHÉS. I would like to express my gratitude to the directors and the staff for the possibility to work there and the great hospitality. Special thanks go to the Klaus-Tschira-Stiftung for funding the visit.

1. Semistable and Stable Points

Let G be a reductive affine algebraic group, X a quasi-projective variety, $\overline{\alpha}\colon G\times X \longrightarrow X$ an action of G on X, $\mathscr{L}$ a line bundle on X, and $\alpha\colon G \times \mathscr{L} \longrightarrow \mathscr{L}$ a linearization of $\overline{\alpha}$ in $\mathscr{L}$.

Given these data, a one parameter subgroup $\lambda\colon \mathbb{G}_m(k) \longrightarrow G$, and a point $x \in X$, we look at the morphism

$$\begin{array}{rcl} \mathbb{G}_m(k) & \longrightarrow & X \\ z & \longmapsto & \overline{\alpha}(\lambda(z), x). \end{array}$$

We view $\mathbb{G}_m(k)$ as a subset of $\mathbb{P}_1$ and ask whether the above morphism can be extended to the point $\infty \in \mathbb{P}_1$. If not, we set

$$\mu_\alpha(\lambda, x) := \infty.$$

If it can be extended, we let $x_\infty \in X$ be the value of the extension at ∞ and write

$$\lim_{z\to\infty} \lambda(z)\cdot x := x_\infty.$$

Now, x_∞ is a fixed point for the $\mathbb{G}_m(k)$-action on X induced by $\overline{\alpha}$ and λ. Via α, $\mathbb{G}_m(k)$ acts on the fiber of $\mathscr{L}$ over x_∞. Let $\gamma \in \mathbb{Z}$ be the weight of that action and set

$$\mu_\alpha(\lambda, x) := -\gamma.$$

Remark 1.1. i) If X is projective and $\mathscr{L}$ is ample, then x_∞ always exists and the Hilbert–Mumford criterion ([**12**], Theorem 2.1) asserts that $x \in X$ is (semi)stable with respect to α if and only if

$$\mu_\alpha(\lambda, x)(\geq)0$$

holds for every non-constant one parameter subgroup $\lambda\colon \mathbb{G}_m(k) \longrightarrow G$.

ii) In the situation outlined in the introduction, ρ induces an action $\overline{\alpha}$ of G on $P(V)$ and a linearization α of that action in $\mathscr{O}_{P(V)}(1)$. Likewise, σ yields an action $\overline{\beta}$ of G on $P(W)$ and a linearization β of $\overline{\beta}$ in $\mathscr{O}_{P(W)}(1)$. Given positive integers m, n, we get the linearization $\alpha^{\otimes m} \boxtimes \beta^{\otimes n}$ of the G-action on $P(V) \times P(W)$ in

$\mathscr{O}(m,n)$. For a point $x = ([v],[w]) \in P(V) \times P(W)$ and a one parameter subgroup $\lambda\colon \mathbb{G}_m(k) \longrightarrow G$, the formula

$$\mu_{\alpha^{\otimes m} \boxtimes \beta^{\otimes n}}(\lambda, x) = m \cdot \mu_\alpha\big(\lambda, [v]\big) + n \cdot \mu_\beta\big(\lambda, [w]\big) \tag{1}$$

holds.

PROPOSITION 1.2. *There is a positive rational number l_∞, such that, for positive integers m, n with*

$$\frac{n}{m} \geq l_\infty$$

and a point $x = ([v],[w]) \in P(V) \times P(W)$, the following conditions are equivalent:

i) *The point x is (semi)stable with respect to the linearization $\alpha^{\otimes m} \boxtimes \beta^{\otimes n}$.*

ii) *The point $[w] \in P(W)$ is semistable with respect to the linearization β and, for every non-constant one parameter subgroup $\lambda\colon \mathbb{G}_m(k) \longrightarrow G$ with $\mu_\beta(\lambda,[w]) = 0$, one has*

$$\mu_\alpha\big(\lambda, [v]\big)(\geq)0.$$

PROOF. This is Proposition 2.9 in [**15**] or Proposition 1.7.3.1 in [**17**]. The statement is intuitively clear from (1), and the proof rests on basic finiteness properties. □

We now turn to [**6**], Corollary 1.1.

THEOREM 1.3 (Gulbrandsen, Halle, Hulek). *Let R be a finitely generated k-algebra, $S := \operatorname{Spec}(R)$, $f\colon X \longrightarrow S$ a projective morphism, and $\mathscr{L}$ an ample line bundle on X. Suppose G is a reductive affine algebraic group, $\overline{\delta}\colon G \times X \longrightarrow X$, $\overline{\beta}\colon G \times S \longrightarrow S$ are actions, such that f is equivariant for these actions, and $\delta\colon G \times \mathscr{L} \longrightarrow \mathscr{L}$ is a linearization of $\overline{\delta}$. Then, a point $x \in X$ is (semi)stable with respect to the linearization δ if and only if*

$$\mu_\delta(\lambda, x)(\geq)0$$

holds true for every non-trivial one parameter subgroup $\lambda\colon \mathbb{G}_m(k) \longrightarrow G$.

PROOF. **Step 1.** By definition, there are an $r > 0$ and a G-equivariant closed embedding

$$X \hookrightarrow P\big(f_\star(\mathscr{L}^{\otimes r})\big).$$

Since S is affine and noetherian, $f_\star(\mathscr{L}^{\otimes r})$ is generated by a finite dimensional subspace of $H^0(S, f_\star(\mathscr{L}^{\otimes r})) = H^0(X, \mathscr{L}^{\otimes r})$. Furthermore, by [**12**], p. 25, Lemma, the action of G on $H^0(X, \mathscr{L}^{\otimes r})$ is locally finite. Altogether, we may find a natural number $r \gg 0$, a G-invariant finite dimensional subspace $V \subset H^0(X, \mathscr{L}^{\otimes r})$, and a G-equivariant closed embedding

$$\iota\colon X \hookrightarrow P(V) \times S$$

relative to S. The formation of the set of (semi)stable points commutes with closed embeddings ([**12**], Theorem 1.19, Lemma A.1.2). Hence, we may replace X by $P(V) \times S$ and assume $\mathscr{L} = \pi^\star_{P(V)}\mathscr{O}_{P(V)}(1)$.

Step 2. According to [**5**], Chapter I, 1.12 Proposition, there exist a finite dimensional k-vector space U on which G acts linearly and a G-equivariant closed embedding $\kappa\colon S \hookrightarrow U$. So, for the same reason as before, we may replace S by U.

Step 3. Set $W := U \oplus k$. The representation of G on U and the trivial representation of G on k provide us with a representation $\sigma\colon G \longrightarrow \mathrm{GL}(W)$, an action $\overline{\beta}\colon G \times P(W) \longrightarrow P(W)$, and a linearization β of this action in $\mathscr{O}_{P(W)}(1)$. Now, $U \subset P(W)$ is a G-equivariant compactification.

Note. All points in U are semistable for the linearization β, and one has the compactification
$$U/\!\!/G \subset P(W)/\!\!/G.$$
Let
$$x = \big([v], u\big) = \big([v], [u, 1]\big) \in X \hookrightarrow P(V) \times U \subset P(V) \times P(W).$$
The one parameter subgroups $\lambda\colon \mathbb{G}_m(k) \longrightarrow G$ with $\mu_\beta(\lambda, [u, 1]) > 0$ are precisely those with $\mu_\delta(\lambda, x) = \infty$.

Step 4. Let $(P(V) \times P(W))^{\infty\text{-(s)s}}$ be the G-invariant open set of points that satisfy the second condition in Proposition 1.2. For any pair m, n of positive integers with $n/m \geq l_\infty$, it is the set of points that are (semi)stable with respect to the linearization in $\alpha^{\otimes m} \boxtimes \beta^{\otimes n}$.

Next, we look at the G-invariant open subset $(P(V) \times U)^{\delta\text{-(s)s}}$ of points which are (semi)stable with respect to the linearization δ. There are a positive integer r and G-invariant sections
$$s_i \in S^r(V^\vee) \otimes S^*(U^\vee) = H^0(P(V) \times U, \mathscr{L}^{\otimes r}), \quad i = 1, ..., t,$$
such that the G-invariant open subsets $U_1, ..., U_t$ with
$$U_i := \Big\{ x \in P(V) \times U \,\big|\, s_i(x) \neq 0 \Big\}, \quad i = 1, ..., t,$$
are affine and cover $(P(V) \times U)^{\delta\text{-ss}}$. Since the G-action on the symmetric algebra $S^*(U^\vee)$ of $U^\vee$ preserves the grading, we may assume that there are positive integers $l_1, ..., l_t$ with
$$s_i \in S^r(V^\vee) \otimes S^{l_i}(U^\vee), \quad i = 1, ..., t.$$
Let $s \in H^0(P(W), \mathscr{O}_{P(W)}(1))$ be the G-invariant section corresponding to the last coordinate. Pick $l \geq l_\infty$, such that $r \cdot l > l_i$, $i = 1, ..., t$. The section $s_i \cdot s^{r \cdot l - l_i}$ extends to a G-invariant section of $\mathscr{O}(r, r \cdot l)$, $i = 1, ..., t$. This discussion shows
$$(P(V) \times U)^{\delta\text{-(s)s}} \subset \big(P(V) \times P(W)\big)^{\infty\text{-(s)s}}.$$
To see the converse inclusion, note that, for $x = ([v], u) = ([v], [u, 1]) \in P(V) \times U$,
$$\overline{G \cdot x} \cap \big(P(V) \times P(W)\big)^{\infty\text{-ss}} \subset P(V) \times U. \tag{2}$$
This follows from the Hilbert–Mumford criterion in the form [**12**], p. 58, or [**17**], Theorem 1.5.1.4. There exists a one parameter subgroup $\lambda\colon \mathbb{G}_m(k) \longrightarrow G$ with $\mu_{\alpha^{\otimes m} \boxtimes \beta^{\otimes n}}(\lambda, x) = 0$, such that $x_\infty = \lim_{z \to \infty} \lambda(z) \cdot x$ lies in the unique orbit in $\overline{G \cdot x}$ that is closed in $(P(V) \times P(W))^{\infty\text{-ss}}$. For this one parameter subgroup, we must have $\mu_\beta(\lambda, [u, 1]) = 0$.

Using (2) and arguments of Mumford ([**12**], Amplification 1,8), the assertion follows. We refer to [**17**], Theorem 1.4.3.7, for more details. □

REMARK 1.4. i) We have the cartesian diagram
$$\begin{array}{ccc} \big(P(V) \times U\big)^{\delta\text{-ss}} /\!\!/ G & \longrightarrow & \big(P(V) \times P(W)\big)^{\infty\text{-ss}} /\!\!/ G \\ \big\downarrow & & \big\downarrow \\ \mathrm{Spec}(A^G) = U/\!\!/G & \longrightarrow & P(W)^{\mathrm{ss}} /\!\!/ G. \end{array}$$
Since the right hand map is obviously proper, so is the left hand map.

ii) In the case $X = S$, S is a finite dimensional k-vector space on which G acts linearly, and $\mathscr{L} = \mathscr{O}_X$, this is Proposition 2.5 in [**10**]. It corresponds to the case where V is a one dimensional representation, i.e., a character of G, in Proposition 1.2.

2. The Instability Flag

In this section, we will use the theory of the instability flag. We refer to [**9**] and [**14**] for accounts of the theory. Section 1.7 in [**17**] contains a brief summary adapted to the situation we are dealing with here.

THEOREM 2.1. *There is a positive rational number $l'_\infty \geq l_\infty$, such that, for positive integers m, n with*

$$\frac{n}{m} \geq l'_\infty$$

and a point $x = ([v],[w]) \in P(V) \times P(W)$ which is unstable with respect to the linearization $\alpha^{\otimes m} \boxtimes \beta^{\otimes n}$, but for which $[w] \in P(W)$ is semistable with respect to the linearization β, an instability one parameter subgroup λ of x satisfies

$$\mu_\beta(\lambda, [v]) = 0.$$

PROOF. This is Theorem 2.10 in [**15**]. A slightly simplified proof is contained in [**4**], proof of Proposition 2.9. The property is suggested by Proposition 1.2. □

REMARK 2.2. i) Suppose G acts linearly on the vector space U. As in the proof of Theorem 1.3, we form the G-equivariant compactification $U \subset P(W)$, $W := U \oplus k$. The G-action on $P(W)$ has a natural linearization β in $\mathscr{O}_{P(W)}(1)$. Given a character $\chi \in X(G)$ of G, we may twist the linearization β by χ and obtain a new linearization β^χ in $\mathscr{O}_{P(W)}(1)$. The resulting notion of (semi)stability makes sense for any real character $\chi \in X_{\mathbb{R}}(G) := X(G) \otimes_{\mathbb{Z}} \mathbb{R}$. Suppose $\Upsilon \subset X_{\mathbb{R}}(G)$ is a compact subset and $[u, 1] \in P(W)$ is a point which is unstable for every linearization of the form β^χ, for $\chi \in \Upsilon$. For applications to decorated principal bundles, we need to investigate how the instability one parameter subgroup varies with $\lambda \in \Upsilon$. There are two techniques for doing so: The first one is based on the observation that the instability one parameter subgroup depends more or less continuously on χ (see [**19**], Theorem 3.4). The second one is to slightly modify the definition of the instability flag in the specific situation in order to obtain improved estimates [**20**]. In both cases, one can derive a uniform version of Theorem 2.1 ([**19**], Theorem 3.9). The fact that Υ is compact is, of course, essential for this.

ii) The theory of the instability flag depends on the choice of a maximal torus $T \subset G$ and a scalar product on $X_{\mathbb{R}}(T) := X(T) \otimes_{\mathbb{Z}} \mathbb{R}$ which is invariant under the action of the Weyl group. The choice of T is not an issue, but the choice of the scalar product matters, at least for applications to decorated principal bundles. For example, in the setting of quiver representations, the structure group takes the form $G = \mathrm{GL}_{n_1}(k) \times \cdots \times \mathrm{GL}_{n_t}(k)$. Denote by $D_i \subset \mathrm{GL}_{n_i}(k)$ the subgroup of diagonal matrices, $i = 1, ..., t$. There is a standard choice of a basis for $X_{\mathbb{R}}(D_i)$, and we let $(\cdot,\cdot)_i$ be the standard scalar product with respect to this basis, $i = 1, ..., t$. For $\underline{\kappa} \in (\mathbb{R}_{>0})^{\times t}$,

$$(\cdot,\cdot)_{\underline{\kappa}} := \kappa_1 \cdot (\cdot,\cdot)_1 + \cdots + \kappa_t \cdot (\cdot,\cdot)_t$$

is a scalar product on $X_{\mathbb{R}}(D)$, for the maximal torus $D = D_1 \times \cdots \times D_t$ of G. Let $\|\cdot\|_\infty$ be the maximum norm on $\mathbb{R}^t$ and

$$\Xi := \Big\{ \underline{\nu} \in (\mathbb{R}_{>0})^{\times t} \,\big|\, \|\underline{\nu}\|_\infty = 1 \Big\}.$$

The set Ξ is not compact and it is difficult to analyze what happens when we approach the boundary of Ξ. In [**20**], we derive some subtle estimates which help to obtain results which are independent of $\underline{\kappa} \in \Xi$.

3. Quiver Sheaves

Let $(X, \mathscr{O}_X(1))$ be a polarized smooth projective variety. With the help of the ample line bundle $\mathscr{O}_X(1)$, we define the degree of a coherent $\mathscr{O}_X$-module. Suppose we are also given a quiver $Q = (U, A, t, h)$ and a tuple $\underline{M} = (M_a, a \in A)$ of locally free $\mathscr{O}_X$-modules of finite rank. An *$\underline{M}$-twisted Q-sheaf* is a tuple $(\mathscr{E}_u, u \in U, \varphi_a, a \in A)$ with $\mathscr{E}_u$ a coherent $\mathscr{O}_X$-module, $u \in U$, and $\varphi_a\colon M_a \otimes \mathscr{E}_{t(a)} \longrightarrow \mathscr{E}_{h(a)}$ a twisted homomorphism, $a \in A$. If $M_a = \mathscr{O}_X$, $a \in A$, we will speak of an *untwisted Q-sheaf.*

EXAMPLE 3.1. Suppose X is a curve, $Q = \bullet\!\circlearrowleft$ is the quiver with one vertex and one arrow connecting this vertex to itself, and the tangent bundle T_X is the twisting bundle. So, a T_X-twisted Q-sheaf is a pair $(\mathscr{E}, \vartheta)$ in which $\mathscr{E}$ is a coherent $\mathscr{O}_X$-module and $\vartheta\colon T_X \otimes \mathscr{E} \longrightarrow \mathscr{E}$ is a twisted endomorphism. The datum of ϑ is equivalent to the datum of a twisted endomorphism $\varphi\colon \mathscr{E} \longrightarrow \mathscr{E} \otimes \omega_X$. A pair $(E, \varphi\colon E \longrightarrow E \otimes \omega_X)$ in which E is a locally free $\mathscr{O}_X$-module of finite rank is known as a *Higgs bundle.*

In this note, we will be concerned with slope (semi)stability of quiver sheaves. This notion depends on several parameters. These are a tuple $\underline{\kappa} = (\kappa_u, u \in U)$ of positive real numbers and a tuple $\underline{\chi} = (\chi_u, u \in U)$ of real numbers.

For a collection $(\mathscr{F}_u, u \in U)$ of coherent $\mathscr{O}_X$-modules, the *$(\underline{\kappa}, \underline{\chi})$-degree* is

$$\deg_{\underline{\kappa},\underline{\chi}}(\mathscr{F}_u, u \in U) := \sum_{u\in U} \big(\kappa_u \cdot \deg(\mathscr{F}_u) + \chi_u \cdot \mathrm{rk}(\mathscr{F}_u)\big),$$

the *$\underline{\kappa}$-rank*

$$\mathrm{rk}_{\underline{\kappa}}(\mathscr{F}_u, u \in U) := \sum_{u\in U} \kappa_u \cdot \mathrm{rk}(\mathscr{F}_u),$$

and, if the $\underline{\kappa}$-rank is positive, i.e., not all the $\mathscr{F}_u$ are torsion sheaves, the *$(\underline{\kappa}, \underline{\chi})$-slope* is

$$\mu_{\underline{\kappa},\underline{\chi}}(\mathscr{F}_u, u \in U) := \frac{\deg_{\underline{\kappa},\underline{\chi}}(\mathscr{F}_u, u \in U)}{\mathrm{rk}_{\underline{\kappa}}(\mathscr{F}_u, u \in U)}.$$

An $\underline{M}$-twisted Q-sheaf $(\mathscr{E}_u, u \in U, \varphi_a, a \in A)$ is *$(\underline{\kappa}, \underline{\chi})$-slope (semi)stable,* if $\mathscr{E}_u$ is torsion free, $u \in U$, and the inequality

$$\mu_{\underline{\kappa},\underline{\chi}}(\mathscr{F}_u, u \in U)(\le)\mu_{\underline{\kappa},\underline{\chi}}(\mathscr{E}_u, u \in U)$$

holds for every non-trivial, proper $\underline{M}$-twisted Q-subsheaf $(\mathscr{F}_u, u \in U)$.[1]

REMARK 3.2. These notions are due to Álvarez-Cónsul and García-Prada ([**2**], Definition 2.2).

[1]This means that $\mathscr{F}_u$ is a subsheaf of $\mathscr{E}_u$, $u \in U$, such that $\varphi_a(M_a \otimes \mathscr{F}_{t(a)}) \subset \mathscr{F}_{h(a)}$, $a \in A$.

Global Boundedness. We will discuss a very basic problem on semistable quiver bundles and their moduli spaces.

PROBLEM 3.3 (Boundedness). Fix a tuple $\underline{n} = (n_u, u \in U)$ of positive integers and a tuple $\underline{d} = (d_u, u \in U)$ of integers. Is there a constant C, such that, for every stability parameter $\underline{\kappa} \in (\mathbb{R}_{>0})^{\times \# U}$, every stability parameter $\underline{\chi} \in \mathbb{R}^{\# U}$, every $(\underline{\kappa}, \underline{\chi})$-slope semistable $\underline{M}$-twisted Q-sheaf $(\mathscr{E}_u, u \in U, \varphi_a, a \in A)$ with $\operatorname{rk}(\mathscr{E}_u) = n_u$, $\deg(\mathscr{E}_u) = d_u$, $u \in U$, and every index $u_0 \in U$

$$\mu_{\max}(\mathscr{E}_{u_0}) \leq C?$$

REMARK 3.4. i) There are various options for normalizing the stability parameters [**20**]. We will use the following: Suppose we are given a tuple $\underline{n} = (n_u \in \mathbb{Z}_{>0}, u \in U)$ of positive integers. Then, we may assume

$$\sum_{u \in U} n_u \cdot \chi_u = 0.$$

Let $\|\cdot\|_\infty$ be the maximum norm on $\mathbb{R}^{\# U}$. We will suppose

$$\|\underline{\kappa}\|_\infty = 1.$$

ii) Fix $\underline{n} = (n_u \in \mathbb{Z}_{>0}, u \in U)$ and set

$$N := \Big\{ \underline{\eta} \in \mathbb{R}^{\# U} \mid \sum_{u \in U} n_u \cdot \eta_u = 0 \Big\}, \quad \Xi := \Big\{ \underline{\nu} \in (\mathbb{R}_{>0})^{\times \# U} \mid \|\underline{\nu}\|_\infty = 1 \Big\}.$$

In addition, we need to fix a tuple $\underline{P} = (P_u, u \in U)$ of Hilbert polynomials. Assume that Problem 3.3 has a positive answer. Then, there is a subdivision of $\Xi \times N$ into finitely many locally closed subsets, called chambers, such the concepts of slope stability and semistability for $\underline{M}$-twisted Q-sheaves $(\mathscr{E}_u, u \in U, \varphi_a, a \in A)$ with $P(\mathscr{E}_u) = P_u$, $u \in U$, are constant within each chamber. In particular, there are only finitely many distinct notions of slope stability and semistability. We refer to [**20**] for the details.

iii) Suppose X is a curve. Then, for $\underline{n}$ and $\underline{d}$ as in Problem 3.3 and $(\underline{\kappa}, \underline{\chi}) \in \Xi \times N$, there is a quasi-projective moduli space for $(\underline{\kappa}, \underline{\chi})$-semistable $\underline{M}$-twisted quiver bundles $(E_u, u \in U, \varphi_a, a \in A)$ with $\operatorname{rk}(E_u) = n_u$, $\deg(E_u) = d_u$, $u \in U$ [**16**]. If Problem 3.3 has a positive answer, then there are only finitely many distinct moduli spaces, and the chamber decomposition mentioned in Part ii) is an important tool for understanding the relation between different moduli spaces. We refer to [**3**], [**11**] for some examples. If $\dim(X) > 1$, then one has to use the more delicate notion of Gieseker (semi)stability for constructing moduli spaces ([**16**], [**1**], [**18**]). A positive answer to Problem 3.3 is still an important boundedness statement and has consequences similar to those mentioned for slope (semi)stability and moduli spaces on curves.

Semistable Representations of *Q*. Let $\mathbb{K}$ be a field. In this paper, it will be either the ground field k or the function field $k(X)$ of the variety X. A *representation* of Q is a tuple $R = (V_u, u \in U, f_a, a \in A)$ in which V_u is a finite dimensional $\mathbb{K}$-vector space, $u \in U$, $f_a \colon V_{t(a)} \longrightarrow V_{h(a)}$ is a linear map, $a \in A$. The tuple $(\dim(V_u), u \in U)$ is the *dimension vector* of R. Let $\underline{n} = (n_u, u \in \mathbb{Z}_{>0})$ be a dimension vector and $\underline{\chi} \in N$ a tuple of real numbers. We say that R is $\underline{\chi}$*-(semi)stable*, if

$$\sum_{u \in U} \chi_u \cdot \dim_{\mathbb{K}}(W_u)(\leq)0$$

holds for every subrepresentation $(W_u, u \in U)$ of R. We say that R is *geometrically (semi)stable*, if the extension of R to the algebraic closure $\overline{\mathbb{K}}$ is $\underline{\chi}$-(semi)stable.

REMARK 3.5. As in the case of sheaves ([**8**], Theorem 1.3.7), the Harder–Narasimhan filtration ([**7**], Theorem 2.5) and its properties imply that semistability and geometric semistability are equivalent.

As an example. we look at the quiver

(3)

PROPOSITION 3.6. *Assume that $\mathbb{K}$ is algebraically closed. Let n_1 and n_2 be two positive integers and $R = (V_1, V_2, f_1, f_2)$ a representation of Q with* $\dim_{\mathbb{K}}(V_i) = n_i$, $i = 1, 2$.

i) *If both f_1 and f_2 are isomorphisms, then R is semistable with respect to $\chi \cdot (1, -1)$, $\chi \in \mathbb{R}$. It is stable if and only if $n_1 = n_2 = 1$.*

ii) *If f_1 is an isomorphism and f_2 isn't, then R is semistable with respect to $\chi \cdot (1, -1)$, $\chi \in \mathbb{R}_{>0}$. It is stable if and only if $n_1 = n_2 = 1$ and $\chi > 0$.*

iii) *If f_1 is an isomorphism and f_2 isn't, then R is semistable with respect to $\chi \cdot (-1, 1)$, $\chi \in \mathbb{R}_{>0}$. It is stable if and only if $n_1 = n_2 = 1$ and $\chi > 0$.*

iv) *In all the other cases, R is only semistable with respect to the stability parameter $(0, 0)$ and not stable.*

PROOF. i) A subrepresentation (W_1, W_2) of R satisfies $\dim_{\mathbb{K}}(W_1) = \dim_{\mathbb{K}}(W_2)$, so that $\chi \cdot \dim_{\mathbb{K}}(W_1) - \chi \cdot \dim_{\mathbb{K}}(W_2) = 0$. This shows that R is semistable with respect to $\chi \cdot (1, -1)$, $\chi \in \mathbb{R}$. If $n_1 = n_2 = 1$, then R has no non-zero, proper subrepresentation and is, therefore, stable. If $n_1 = n_2 \geq 2$, let $w \in W_1 \setminus \{0\}$ be an eigenvector of $f_2 \circ f_1$. Set $W_1 := \langle w \rangle$ and $W_2 := \langle f(w) \rangle$. Then, (W_1, W_2) is a destabilizing subrepresentation.

iv) Suppose f_2 is not injective, and let W_2 be the kernel of f_2. Then, $(0, W_2)$ is a non-zero, proper subrepresentation. So, if R is (χ_1, χ_2)-semistable, we must have $\chi_2 \leq 0$.

If f_1 isn't injective either, then also $\chi_1 \leq 0$. Using $\chi_1 \cdot n_1 + \chi_2 \cdot n_2 = 0$, we infer $\chi_1 = \chi_2 = 0$.

For the remaining argument, we assume that f_1 is injective. Note that our assumption implies $1 \leq n_1 < n_2$. Assume that R is semistable with respect to $\underline{\chi} = (\chi_1, \chi_2)$. The same argument as above implies $\chi_2 \leq 0$. As before, we let w be a non-zero eigenvector of $f_2 \circ f_1$ and construct the subrepresentation (W_1, W_2). For this subrepresentation, we have

$$\chi_1 \cdot \dim_{\mathbb{K}}(W_1) + \chi_2 \cdot \dim_{\mathbb{K}}(W_2) = \chi_1 + \chi_2 \geq \frac{1}{n_1} \cdot (\chi_1 \cdot n_1 + \chi_2 \cdot n_2) = 0.$$

If $\underline{\chi} \neq (0, 0)$, the inequality would be strict. This would contradict $\underline{\chi}$-semistability.

To conclude, recall that $(0, \ker(f_2))$ is a non-zero, proper subrepresentation. Thus, R is not $(0, 0)$-stable.

ii) and iii) We prove ii). Since f_2 is not injective, the argument presented in iv) shows that (χ_1, χ_2)-semistability implies $\chi_2 \leq 0$. Moreover, $\chi_1 = -\chi_2$, because $n_1 = n_2$ and $\chi_1 \cdot n_1 + \chi_2 \cdot n_2 = 0$. For a subrepresentation (W_1, W_2), we have $\dim_{\mathbb{K}}(W_1) \leq \dim_{\mathbb{K}}(W_2)$, so that

$$\chi \cdot \dim_{\mathbb{K}}(W_1) - \chi \cdot \dim_{\mathbb{K}}(W_2) \leq \chi \cdot \dim_{\mathbb{K}}(W_2) - \chi \cdot \dim_{\mathbb{K}}(W_2) = 0.$$

As in iv), it follows that there is a subrepresentation (W_1, W_2) with $\dim_{\mathbb{K}}(W_1) = \dim_{\mathbb{K}}(W_2) = 1$. This is destabilizing, if $n_1 = n_2 > 1$.

Finally, for $n_1 = n_2 = 1$, $(0, V_2)$ is the only non-trivial, proper subrepresentation. From this, the rest follows. □

Recent Results. As a warm up, we will discuss Theorem 2.7 from [**3**] which is based on the paper [**15**].

PROPOSITION 3.7. *Assume* $\operatorname{char}(k) = 0$. *Fix a tuple* $\underline{n} = (n_u, u \in U)$ *of positive integers, a tuple* $\underline{d} = (d_u, u \in U)$ *of integers, stability parameters* $\underline{\kappa} \in \Xi$, $\underline{\zeta}, \underline{\vartheta} \in N$, *and set, for* $\lambda \in \mathbb{R}_{>0}$,

$$\underline{\chi}^\lambda := \underline{\zeta} + \lambda \cdot \underline{\vartheta}.$$

Then, there is a value $\lambda_\infty \in \mathbb{R}_{>0}$, *such that, for* $\lambda \geq \lambda_\infty$, *an* $\underline{M}$*-twisted* Q*-sheaf* $(\mathscr{E}_u, u \in U, \varphi_a, a \in A)$ *with* $\operatorname{rk}(\mathscr{E}_u) = n_u$, $\deg(\mathscr{E}_u) = d_u$, $u \in U$, *is* $(\underline{\kappa}, \underline{\chi}^\lambda)$*-slope (semi)stable if and only if, for every* Q*-subsheaf* $(\mathscr{F}_u, u \in U)$, *the following two conditions are satisfied:*

a) $\displaystyle\sum_{u \in U} \vartheta_u \cdot \operatorname{rk}(\mathscr{F}_u) \leq 0.$

b) *If "=" holds in* a), *then* $\mu_{\underline{\kappa},\underline{\zeta}}(\mathscr{F}_u, u \in U)(\leq)\mu_{\underline{\kappa},\underline{\zeta}}(\mathscr{E}_u, u \in U)$.

REMARK 3.8. Let $\mathbb{K}$ be the function field of X and $R = (V_u, u \in U, f_a, a \in A)$ the representation obtained by restricting $(\mathscr{E}_u, u \in U, \varphi_a, a \in A)$ to the generic point of X. Condition a) says that R is $\underline{\vartheta}$-semistable. By Remark 3.5, it is also geometrically $\underline{\vartheta}$-semistable.

SKETCH OF PROOF. Let m_a be the rank of M_a, $a \in A$. To compare quiver sheaves with quiver representations over the function field $\mathbb{K}$ of X, one has to work with the quiver $Q' = (U, A', t', h')$ in which the arrow a is replaced by m_a copies of it. To simplify the situation, we assume $m_a = 1$, $a \in A$. In this situation, it is easy to reduce to the case in which there is a line bundle M on X with $M_a = M$, $a \in A$.

"$\Longrightarrow$": It suffices to prove that Condition a) holds. If not, the point

$$R \in \operatorname{Rep}_{\underline{n}}(Q) := \bigoplus_{u \in U} \operatorname{Hom}_{\mathbb{K}}(\mathbb{K}^{n_{t(a)}}, \mathbb{K}^{n_{h(a)}})$$

is geometrically $\underline{\vartheta}$-unstable. Since we assume characteristic zero, there is an instability one parameter subgroup of R which is defined over $\mathbb{K}$. We apply Theorem 2.1 in the set-up described in Remark 1.4, ii). It implies that the instability one parameter subgroup is built from subrepresentations of R (see [**17**], Theorem 1.5.1.25). Let us assume that it just corresponds to a subrepresentation $(W_u, u \in U)$. This can be extended to a Q-subsheaf $(\mathscr{F}_u, u \in U)$ of $(\mathscr{E}_u, u \in U, \varphi_a, a \in A)$. It satisfies

$$\sum_{u \in U} \vartheta_u \cdot \operatorname{rk}(\mathscr{F}_u) > 0.$$

To conclude, we have to find a lower bound for $\mu_{\underline{\kappa},\underline{\zeta}}(\mathscr{F}_u, u \in U)$. Fix an index $u_0 \in U$ and let $r(u_0)$ be the rank of $\mathscr{F}_{u_0}$. We may find an open subset $\mathscr{U} \subset X$ whose complement has codimension at least two, such that $\mathscr{E}_{u_0|\mathscr{U}}$ is locally free and $\mathscr{F}_{u_0|\mathscr{U}}$ is a subbundle of $\mathscr{E}_{u_0|\mathscr{U}}$. We get a morphism

$$\sigma\colon \mathscr{U} \longrightarrow \mathbb{P}\left(\bigwedge^{r(u_0)} \mathscr{E}_{u_0|\mathscr{U}}\right) \times P\left(\bigoplus_{a \in A} \underline{\operatorname{Hom}}_{\mathscr{O}_X}(\mathscr{E}_{t(a)|\mathscr{U}}, \mathscr{E}_{h(a)|\mathscr{U}})\right).$$

Restricting to the generic point yields a point

$$s \in \mathbb{P}\left(\bigwedge^{r(u_0)} \mathbb{K}^{\mathrm{rk}(\mathscr{E}_{u_0})}\right) \times P\left(\bigoplus_{a\in A} \mathrm{Hom}_{\mathbb{K}}(\mathbb{K}^{\mathrm{rk}(\mathscr{E}_{t(a)})}, \mathbb{K}^{\mathrm{rk}(\mathscr{E}_{t(a)})})\right).$$

Look at the line bundle $\mathscr{O}(1,l)$ with $l \geq l_\infty$, l_∞ as in Theorem 2.1. If s is semistable, then, by [**14**], Proposition 3.10,

$$\deg\bigl(\det(\mathscr{E}_{u_0|\mathscr{U}}/\mathscr{F}_{u_0|\mathscr{U}})\bigr) - l\cdot\deg(M) \geq 0,$$

so that

$$\mu(\mathscr{F}_{u_0}) \leq \max\left\{ \frac{\deg(d_{u_0}) - l\cdot\deg(M)}{s} \,\middle|\, s = 1, ..., \mathrm{rk}(\mathscr{E}_{u_0}) - 1 \right\}.$$

If not, one invokes the instability one parameter subgroup for s. By the same token as before, it is built from subrepresentations. Now, one uses the method of associating with s a point which is semistable with respect to some other representation ([**14**], Proposition 1.12). Then, as before, one gets an upper bound for $\mu(\mathscr{F}_{u_0})$.

"$\Longleftarrow$": Here, the crucial point is to prove that the family of $\underline{M}$-twisted Q-sheaves $(\mathscr{E}_u, u\in U, \varphi_a, a\in A)$ with $\mathrm{rk}(\mathscr{E}_u) = n_u$, $\deg(\mathscr{E}_u) = d_u$, $u\in U$, that satisfy Condition a) and b) is bounded. This proceeds along the lines of the second part of proof of the implication "$\Longrightarrow$". □

We use the notation of the proof, in particular, $Q' = (U, A', t', h')$. Let $\underline{\vartheta} \in N$. Call an $\underline{M}$-twisted Q-sheaf $(\mathscr{E}_u, u\in U, f_a, a\in A)$ *generically $\underline{\vartheta}$-unstable*, if the representation of Q' obtained by restricting $(\mathscr{E}_u, u\in U, f_a, a\in A)$ to the generic point of X is $\underline{\vartheta}$-unstable, and *generically totally unstable*, if it is generically $\underline{\vartheta}$-unstable for all $\underline{\vartheta} \in N\setminus\{0\}$.

REMARK 3.9. i) Fix $\underline{n}$ and $\underline{d}$ as usual. Proposition 3.7 suggests that there is a bounded subset $S \subset \Xi\times N$, such that, for a stability parameter $(\underline{\kappa}, \underline{\chi}) \in \Xi \times N$, the existence of a generically totally unstable $(\underline{\kappa}, \chi)$-slope semistable quiver sheaf $(\mathscr{E}_u, u\in U, f_a, a\in A)$ with invariants $\underline{n}$, $\underline{d}$ implies

$$(\underline{\kappa}, \underline{\chi}) \in S.$$

This is the basic idea for subsequent investigations.

ii) Proposition 3.6 and the papers [**3**], [**11**], and [**20**] give examples of quivers and dimension vectors, such that any $\underline{M}$-twisted Q-sheaf (with $\mathrm{rk}(M_a) = 1$, $a\in A$) is generically totally unstable.

The first general statement we obtained concerns the variation of the stability parameter $\underline{\chi}$.

THEOREM 3.10. *Fix a tuple $\underline{n} = (n_u, u\in U)$ of positive integers, a tuple $\underline{d} = (d_u, u\in U)$ of integers, and a stability parameter $\underline{\kappa}_0 \in (\mathbb{R}_{>0})^{\times \# U}$. Then, there is a constant C, such that, for every stability parameter $\underline{\chi} \in \mathbb{R}^{\#U}$, every $(\underline{\kappa}_0, \underline{\chi})$-slope semistable $\underline{M}$-twisted Q-sheaf $(\mathscr{E}_u, u\in U, \varphi_a, a\in A)$ with* $\mathrm{rk}(\mathscr{E}_u) = n_u$, $\deg(\mathscr{E}_u) = d_u$, *$u\in U$, and every index $u_0\in U$*

$$\mu_{\max}(\mathscr{E}_{u_0}) \leq C.$$

IDEA OF PROOF. This is [**19**], Main Theorem 1.5. The result is also a consequence of Theorem 3.11.

It is possible to reduce to (a variant of) generically totally unstable quiver sheaves ([**19**], Page 481ff). Generic total instability is constant on positive half lines. Set

$$\Omega := \Big\{ \underline{\eta} \in N \,\big|\, \|\underline{\eta}\|_\infty = 1 \Big\}.$$

We need a uniform version of Proposition 3.7 as $\underline{\vartheta}$ varies over Ω. This was achieved by the methods sketched in Remark 2.2. □

In general, the following result from [**20**] holds.

THEOREM 3.11. *Assume* $\mathrm{char}(k) = 0$. *Fix the* $\underline{n}$ *and* $\underline{d}$ *as before. Then, there is a constant* K, *depending only on* $\underline{n}$ *and* $\underline{d}$, *such that, for stability parameters* $\underline{\kappa} \in \Xi \cap (\mathbb{R}_{>0} \cdot \mathbb{Q}^{\#U})$, $\underline{\chi} \in N \cap (\mathbb{R}_{>0} \cdot \mathbb{Q}^{\#U})$, *the existence of a* $(\underline{\kappa}, \underline{\chi})$*-slope semistable* $\underline{M}$*-twisted* Q*-sheaf* $(\mathscr{E}_u, u \in U, \varphi_a, a \in A)$ *with* $\mathrm{rk}(\mathscr{E}_u) = n_u$, $\deg(\mathscr{E}_u) = d_u$, $u \in U$, *which is generically totally unstable implies*

$$\|\underline{\chi}\|_\infty \leq K.$$

REMARK 3.12. By [**20**], for stability parameters $\underline{\kappa} \in (\mathbb{R}_{>0})^{\times \#U}$, $\underline{\chi} \in \mathbb{R}^{\#U}$, there are stability parameters $\underline{\kappa}' \in (\mathbb{Q}_{>0})^{\times \#U}$, $\underline{\chi}' \in \mathbb{Q}^{\#U}$, such that the notions of $(\underline{\kappa}, \underline{\chi})$-slope (semi)stability and $(\underline{\kappa}', \underline{\chi}')$-slope (semi)stability are equivalent for $\underline{M}$-twisted Q-sheaves $(\mathscr{E}_u, u \in U, \varphi_a, a \in A)$ with $\mathrm{rk}(\mathscr{E}_u) = n_u$, $\deg(\mathscr{E}_u) = d_u$, $u \in U$.

Unfortunately, Theorem 3.11 is not sufficient for completely solving Problem 3.3. The issue is that expressions of the form χ_u/κ_u may appear in the estimates for some vertex $u \in U$ and, in principle, κ_u can be arbitrarily small. These difficulties arise already for quite simple quivers such as

$$\bullet \longrightarrow \bullet \longrightarrow \cdots \longrightarrow \bullet \longrightarrow \bullet.$$

In the next section, we will give the first example based on the above theorem.

An Example. We return to the quiver (3). To simplify the notation and the arguments a little bit, we will assume that X is a smooth projective curve. We fix line bundles L_1 and L_2 on X, set $l_i := \deg(L_i)$, $i = 1, 2$, and look at quadruples $(E_1, E_2, \varphi_1, \varphi_2)$ in which E_1 and E_2 are vector bundles on X and $\varphi_1 \colon E_1 \longrightarrow E_2 \otimes L_2$ and $\varphi_2 \colon E_2 \longrightarrow E_1 \otimes L_1$ are twisted homomorphisms.

REMARK 3.13. It seems hard to analyze the regions of stability parameters for which there do exist semistable $\underline{L}$-twisted Q-bundles explicitly, because there are no obvious subrepresentations one can check. However, we have two Higgs bundles, namely,

$$\big(E_1, \psi \colon E_1 \longrightarrow E_1 \otimes L\big), \quad \psi := (\varphi_2 \otimes \mathrm{id}_{L_2}) \circ \varphi_1,$$

and

$$\big(E_2, \psi' \colon E_2 \longrightarrow E_2 \otimes L\big), \quad \psi' := (\varphi_1 \otimes \mathrm{id}_{L_1}) \circ \varphi_2, \quad L := L_1 \otimes L_2.$$

This is essential for our reasoning.

THEOREM 3.14. *Assume* $\mathrm{char}(k) = 0$. *Fix positive integers* n_1, n_2 *and integers* d_1, d_2. *There exists a constant* C, *such that, for every stability parameter* $\underline{\kappa} = (\kappa_1, \kappa_2) \in \mathbb{R}_{>0} \times \mathbb{R}_{>0}$ *with* $\max\{\kappa_1, \kappa_2\} = 1$, *every stability parameter* $\underline{\chi} = (\chi_1, \chi_2) \in \mathbb{R}^2$ *with* $\chi_1 \cdot n_1 + \chi_2 \cdot n_2 = 0$, *and every* $(\underline{\kappa}, \underline{\chi})$*-slope semistable*

$\underline{L}$-twisted Q-bundle $(E_1, E_2, \varphi_1, \varphi_2)$ with $\mathrm{rk}(E_i) = n_i$, $\deg(E_i) = d_i$, $i = 1, 2$, one has

$$\mu_{\max}(E_i) \le C, \quad i = 1, 2.$$

PROOF. We will use the following notation and terminology: For a subbundle $F \subset E_1$, $F' := \varphi_1(F) \otimes L_2^\vee$ is a subsheaf of E_2. We call F *invariant*, if (F, F') is a Q-subsheaf.

Case 1. We first assume that φ_1 is generically an isomorphism. Then, $n_1 = n_2$ and the stability parameter is of the form $\chi \cdot (-1, 1)$ with $\chi \in \mathbb{R}$.

For an invariant subbundle F of E_1, semistability yields

$$\mu(F) - \frac{\kappa_2 \cdot l_2}{\kappa_1 + \kappa_2} \le \frac{\kappa_1 \cdot d_1 + \kappa_2 \cdot d_2}{\kappa_1 \cdot n_1 + \kappa_2 \cdot n_2}.$$

We deduce that there is a constant C_1 with

$$\mu(F) \le C_1, \tag{4}$$

for every invariant subbundle of E_1. Now, we can use Nitsure's proof for Higgs bundles ([**13**], Proposition 3.2, [**17**], Example 2.5.6.7): Let

$$\{0\} = F_0 \subsetneq F_1 \subsetneq \cdots \subsetneq F_{s-1} \subsetneq F_s := E_1$$

be the Harder–Narasimhan filtration of E_1. If F_1 is invariant, we are done, by (4). Otherwise, let $t \ge 2$ be the smallest index for which F_t is invariant. So, $(F_t, \psi\colon F_t \longrightarrow F_t \otimes L)$, $\psi := ((\varphi_2 \otimes \mathrm{id}_{L_2}) \circ \varphi_1)_{|F_t}$, $L := L_1 \otimes L_2$, is a Higgs bundle, $F_0 \subsetneq F_1 \subsetneq \cdots \subsetneq F_t$ is the Harder–Narasimhan filtration of F_t, and $F_1, ..., F_{t-1}$ are not invariant under the Higgs field ψ. Proposition 3.2 in [**13**] implies

$$\mu(F_1) \le \mu(F_t) + \max\{\, 0, l_1 + l_2 \,\} \le C_1 + \max\{\, 0, l_1 + l_2 \,\}.$$

We infer that E_1 belongs to a bounded family. We have the short exact sequence

$$0 \longrightarrow E_1 \xrightarrow{\varphi_1} E_2 \otimes L_2 \longrightarrow \mathscr{T} \longrightarrow 0$$

in which $\mathscr{T}$ is a torsion sheaf of length $d_2 + r_2 \cdot l_2 - d_1$. This shows that $E_2 \otimes L_2$ and, thus, E_2 also belongs to a bounded family.

Case 2. Here, we assume that neither φ_1 nor φ_2 is a generic isomorphism. In this case, $(E_1, E_2, \varphi_1, \varphi_2)$ is generically totally unstable, by Proposition 3.6. We have $\max\{\, \kappa_1, \kappa_2 \,\} = 1$. Without loss of generality, we assume $\kappa_1 = 1$. For every subbundle F of E_1 which is contained in the kernel of φ_1, semistability implies

$$\mu(F) + \chi_1 \le \frac{d_1 + \kappa_2 \cdot d_2}{n_1 + \kappa_2 \cdot n_2}.$$

By Theorem 3.11 and Remark 3.12, there is a constant C_2 with

$$\mu(F) \le C_2. \tag{5}$$

An arbitrary subbundle F of E_1 can be written as an extension

$$0 \longrightarrow F \cap \ker(\varphi_1) \longrightarrow F \longrightarrow \varphi_1(F) \longrightarrow 0.$$

By (5), there is a constant C_3 with

$$\deg\bigl(\varphi_1(F)\bigr) \ge \deg(F) + C_3.$$

For an invariant subbundle F of E_1, we find $\mu(F) \le 0$ or, with $\mathrm{rk}(\varphi_1(F)) \le \mathrm{rk}(F) \le r_1 - 1$, $\varepsilon := 0$, if $l_2 \le 0$, and $\varepsilon := r_1 - 1$, if $l_2 > 0$,

$$\mu(F) + \frac{\kappa_2 \cdot (C_3 - \varepsilon \cdot l_2)}{\mathrm{rk}(F) + \kappa_2 \cdot \mathrm{rk}(\varphi_1(F))} + \frac{\chi_1 \cdot \mathrm{rk}(F) + \chi_2 \cdot \mathrm{rk}(\varphi_1(F))}{\mathrm{rk}(F) + \kappa_2 \cdot \mathrm{rk}(\varphi_1(F))} \le \frac{d_1 + \kappa_2 \cdot d_2}{n_1 + \kappa_2 \cdot n_2}.$$

So, there is a constant C_4 with

$$\mu(F) \leq C_4,$$

for every invariant subbundle F of E_1. Now, we continue as in Case 1 to prove that E_1 belongs to a bounded family.

Since E_1 lives in a bounded family and $\varphi_1(E_1)$ is a quotient bundle of E_1, the slope of $\varphi_1(F)$ is bounded from below. If φ_1 is generically surjective, then it follows readily that E_2 also belongs to a bounded family. If not, $\mathrm{Coker}(\varphi_1)$ has positive rank. Furthermore, there is a constant C_5 with

$$\mu\big(\mathrm{Coker}(\varphi_1)\big) \leq C_5.$$

Since $(E_1, \varphi_1(E_1) \otimes L_2^\vee)$ is a non-zero, proper Q-subsheaf, $(0, \mathrm{Coker}(\varphi_1) \otimes L_2^\vee)$ is a non-trivial, proper quotient Q-sheaf. Semistability implies

$$\frac{d_1 + \kappa_2 \cdot d_2}{n_1 + \kappa_2 \cdot n_2} \leq \mu\big(\mathrm{Coker}(\varphi_1) \otimes L_2^\vee\big) + \frac{\chi_2}{\kappa_2} \leq C_5 - l_2 + \frac{\chi_2}{\kappa_2}.$$

We conclude that there is a constant C_6 with

$$\frac{\chi_2}{\kappa_2} \geq C_6.$$

For a subbundle G of E_2 which lies in the kernel of φ_2, $(0, G)$ is a Q-subbundle. The semistability assumption gives

$$\mu(G) \leq \frac{d_1 + \kappa_2 \cdot d_2}{n_1 + \kappa_2 \cdot n_2} - \frac{\chi_2}{\kappa_2} \leq \frac{d_1 + \kappa_2 \cdot d_2}{n_1 + \kappa_2 \cdot n_2} - C_6.$$

From here on, we may proceed as before to prove that E_2 moves in a bounded family, too. □

References

[1] L. Álvarez-Cónsul, *Some results on the moduli spaces of quiver bundles*, Geom. Dedicata **139** (2009), 99–120, DOI 10.1007/s10711-008-9327-0. MR2481840 (2010b:14015)

[2] L. Álvarez-Cónsul and O. García-Prada, *Hitchin-Kobayashi correspondence, quivers, and vortices*, Comm. Math. Phys. **238** (2003), no. 1-2, 1–33, DOI 10.1007/s00220-003-0853-1. MR1989667 (2005b:32027)

[3] L. Álvarez-Cónsul, O. García-Prada, and A. H. W. Schmitt, *On the geometry of moduli spaces of holomorphic chains over compact Riemann surfaces*, IMRP Int. Math. Res. Pap. (2006), Art. ID 73597, 82. MR2253535 (2007e:14014)

[4] N. Beck, *Stable parabolic Higgs bundles as asymptotically stable decorated swamps*, Journal of Geometry and Physics **104** (2016), June, 229–241. DOI 10.1016/j.geomphys.2016.02.014. `http://www.sciencedirect.com/science/article/pii/S0393044016300407`

[5] A. Borel, *Linear algebraic groups*, 2nd ed., Graduate Texts in Mathematics, vol. 126, Springer-Verlag, New York, 1991. MR1102012 (92d:20001)

[6] M. G. Gulbrandsen, L. H. Halle, and K. Hulek, *A relative Hilbert-Mumford criterion*, Manuscripta Math. **148** (2015), no. 3-4, 283–301, DOI 10.1007/s00229-015-0744-8. MR3414477

[7] L. Hille and J. A. de la Peña, *Stable representations of quivers*, J. Pure Appl. Algebra **172** (2002), no. 2-3, 205–224, DOI 10.1016/S0022-4049(01)00167-0. MR1906875 (2003e:16014)

[8] D. Huybrechts and M. Lehn, *The geometry of moduli spaces of sheaves*, 2nd ed., Cambridge Mathematical Library, Cambridge University Press, Cambridge, 2010. MR2665168 (2011e:14017)

[9] G. R. Kempf, *Instability in invariant theory*, Ann. of Math. (2) **108** (1978), no. 2, 299–316, DOI 10.2307/1971168. MR506989 (80c:20057)

[10] A. D. King, *Moduli of representations of finite-dimensional algebras*, Quart. J. Math. Oxford Ser. (2) **45** (1994), no. 180, 515–530, DOI 10.1093/qmath/45.4.515. MR1315461 (96a:16009)

[11] A. Laudin and A. Schmitt, *Recent results on quiver sheaves*, Cent. Eur. J. Math. **10** (2012), no. 4, 1246–1279, DOI 10.2478/s11533-012-0007-9. MR2925600

[12] D. Mumford, J. Fogarty, and F. Kirwan, *Geometric invariant theory*, 3rd ed., Ergebnisse der Mathematik und ihrer Grenzgebiete (2) [Results in Mathematics and Related Areas (2)], vol. 34, Springer-Verlag, Berlin, 1994. MR1304906 (95m:14012)

[13] N. Nitsure, *Moduli space of semistable pairs on a curve*, Proc. London Math. Soc. (3) **62** (1991), no. 2, 275–300, DOI 10.1112/plms/s3-62.2.275. MR1085642 (92a:14010)

[14] S. Ramanan and A. Ramanathan, *Some remarks on the instability flag*, Tohoku Math. J. (2) **36** (1984), no. 2, 269–291, DOI 10.2748/tmj/1178228852. MR742599 (85j:14017)

[15] A. H. W. Schmitt, *Global boundedness for decorated sheaves*, Int. Math. Res. Not. **68** (2004), 3637–3671, DOI 10.1155/S1073792804141652. MR2130049 (2006b:14017)

[16] A. Schmitt, *Moduli for decorated tuples of sheaves and representation spaces for quivers*, Proc. Indian Acad. Sci. Math. Sci. **115** (2005), no. 1, 15–49, DOI 10.1007/BF02829837. MR2120597 (2006e:14015)

[17] A. H. W. Schmitt, *Geometric invariant theory and decorated principal bundles*, Zurich Lectures in Advanced Mathematics, European Mathematical Society (EMS), Zürich, 2008. MR2437660 (2010c:14038)

[18] A. Schmitt, *A remark on semistability of quiver bundles*, Eurasian Math. J. **3** (2012), no. 1, 110–138. MR3024113

[19] A. Schmitt, *Global boundedness for semistable decorated principal bundles with special regard to quiver sheaves*, J. Ramanujan Math. Soc. **28A** (2013), 443–490. MR3115203

[20] A.H.W. Schmitt, in preparation.

Freie Universität Berlin, Institut für Mathematik, Arnimallee 3, D-14195 Berlin, Deutschland

E-mail address: alexander.schmitt@fu-berlin.de

Proceedings of Symposia in Pure Mathematics
Volume **93**, 2016
http://dx.doi.org/10.1090/pspum/093/01557

Local and relative BPS state counts for del Pezzo surfaces

Michel van Garrel

Abstract. Relative BPS state counts for log Calabi-Yau surface pairs were introduced by Gross-Pandharipande-Siebert. We describe how in the case of del Pezzo surfaces they are linearly related to local BPS state counts by means of generalized Donaldson-Thomas invariants of loop quivers.

1. Introduction

In this survey paper, we describe a fascinating relationship between local and relative BPS state counts of del Pezzo surfaces, as established by the author in [**3**]. These A-model local and relative BPS numbers are related via a linear transformation. This correspondence is intriguing because the coefficients of the corresponding matrix are generalized Donaldson-Thomas invariants of loop quivers, which are B-model invariants. We as of now do not know how to explain this phenomenon. In [**3**], these entries were computed and found to coincide with the calculation of said Donaldson-Thomas invariants as carried out by Reineke in [**13, 14**]. Our description is purely mathematical, we however expect that it fits into a natural physics setting. Let us note that to our knowledge, there is as of now no mention of relative BPS state counts in the physics literature.

2. Local BPS state counts

Let S be a smooth del Pezzo surface. Local BPS state counts, as defined in definition 2.2 below, are the A-model invariants of local mirror symmetry, which was developed in [**2**]. See also [**8**] of a description in terms of Yukawa couplings. Denote by D a smooth effective anticanonical divisor on S, by K_S the total space of the canonical bundle $\mathcal{O}_S(-D)$, and let $\beta \in \mathrm{H}_2(S,\mathbb{Z})$. Denote moreover by $\overline{M}_{0,0}(S,\beta)$, resp. by $\overline{M}_{0,1}(S,\beta)$, the moduli stack of genus 0 stable maps

$$f : C \to S$$

with no, resp. one, marked point and such that $f_*([C]) = \beta$. We have the forgetful morphism

$$\pi : \overline{M}_{0,1}(S,\beta) \to \overline{M}_{0,0}(S,\beta)$$

and the evaluation map

$$ev : \overline{M}_{0,1}(S,\beta) \to S.$$

2010 *Mathematics Subject Classification.* 05A15, 14J33, 14J45, 14N35.

Key words and phrases. BPS state counts, Gromov-Witten invariants, Generalized Donaldson-Thomas invariants, Mirror Symmetry, log Calabi-Yau surfaces, del Pezzo surfaces.

Consider the obstruction bundle $R^1\pi_*ev^*K_S$. Its fibre over a stable map $f: C \to S$ is $\mathrm{H}^1(C, f^*K_S)$. Denote by e the Euler class and by $[\,]^{vir}$ the virtual fundamental class.

DEFINITION 2.1. The *genus 0 degree β local Gromov-Witten invariant of S* is

$$I_{K_S}(\beta) := \int_{[\overline{M}_{0,0}(S,\beta)]^{vir}} \mathrm{e}\left(R^1\pi_*ev^*K_S\right) \in \mathbb{Q}.$$

We proceed with the definition of the local BPS numbers.

DEFINITION 2.2 (See Gopakumar-Vafa in [**4, 5**] and Bryan-Pandharipande in [**1**]). Assume that β is primitive and let $d \geq 1$. The *local BPS state counts $n_{d\beta} \in \mathbb{Q}$* are defined through the following equality of generating functions.

$$\sum_{l=1}^{\infty} I_{K_S}(l\beta)\, q^l = \sum_{d=1}^{\infty} n_{d\beta} \sum_{k=1}^{\infty} \frac{1}{k^3}\, q^{dk}.$$

That the BPS state counts (of any genus and for any Calabi-Yau threefold) are integers is a conjecture stated in [**1**] and attributed to Gopakumar-Vafa. In genus 0, the conjecture is proven in the following case relevant to us.

THEOREM 2.3 (Peng in [**12**]). *If S is a toric del Pezzo surface, then the BPS numbers $n_{d\beta}$ of K_S are integers $\forall d \geq 1$.*

More generally for toric Calabi-Yau threefolds, the analogous result was proven by Konishi in [**7**].

3. Relative BPS state counts

Relative BPS numbers are defined for the geometry of log Calabi-Yau surface pairs:

DEFINITION 3.1 (See section 6 of [**6**]). Let X be a smooth surface, let $D \subseteq X$ be a smooth divisor and let $\gamma \in \mathrm{H}_2(S, \mathbb{Z})$ be non-zero. If

$$D \cdot \gamma = c_1(S) \cdot \gamma,$$

then (X, D) is said to be *log Calabi-Yau with respect to γ*. Moreover, (S, D) is said to be simply *log Calabi-Yau* provided that the above equation holds for all γ.

The pairs (S, D) consisting of a del Pezzo surface and a smooth anticanonical divisor form a class of examples of log Calabi-Yau surface pairs. We will proceed to defining the relative BPS state counts for this class of examples, though we note that the definition in [**6**] is formulated for all log Calabi-Yau surface pairs. Their definition is entirely parallel to the definition of the local BPS state counts of section 2.

We are concerned with relative stable maps, that is, in addition to considering stable maps we also prescribe the tangencies of how the maps meet D. Let $\beta \in \mathrm{H}_2(S, \mathbb{Z})$ be the class of a curve and set $w = D \cdot \beta$ to be the total intersection multiplicity of β with D. Denote by $\overline{M}(S/D, w)$ the moduli space of genus 0 degree β relative stable maps meeting D in one point of maximal tangency w. Then $\overline{M}(S/D, w)$ is of virtual dimension 0. Indeed, a generic stable map of degree β meets D in w points. The moduli space of stable maps of degree β is of virtual dimension $\int_\beta c_1(\mathrm{T}_S) + (\dim S - 3)(1 - g) = w - 1$. Identifying two points of intersection cuts it by one. Repeating this process for all but one of the w intersection points drops the virtual dimension to 0.

DEFINITION 3.2. The *genus 0 degree β relative Gromov-Witten invariant of maximal tangency* is

$$N_S[w] := \int_{[\overline{M}(S/E,w)]^{vir}} 1 \in \mathbb{Q}.$$

In the notation, β is hidden for simplicity. Relative BPS numbers are defined by extracting multiple cover contributions under the idealized assumption that all embedded curves are rigid (and that there are only finitely many in each degree). Let $\iota : P \to S$ be such a rigid element[1] of $\overline{M}(S/E, w)$ and denote by $M_P[k]$, for $k \geq 1$, the contribution of k-fold multiple covers of P to $N_S[kw]$. Then:

PROPOSITION 3.3 (Proposition 6.1 in [**6**]).

$$M_P[k] = \frac{1}{k^2}\binom{k(w-1)-1}{k-1}.$$

This leads to the following definition.

DEFINITION 3.4 (Paragraph 6.3 in [**6**]). For $d \geq 1$, the *relative BPS state counts* $n_S[dw] \in \mathbb{Q}$ of class $d\beta$ are defined as the unique numbers making the following equation true:

$$\sum_{l=1}^{\infty} N_S[lw]\, q^l = \sum_{d=1}^{\infty} n_S[dw] \sum_{k=1}^{\infty} \frac{1}{k^2}\binom{k(dw-1)-1}{k-1} q^{dk}. \tag{3.1}$$

CONJECTURE 3.5 (Conjecture 6.2 in [**6**]). For $\beta \in \mathrm{H}_2(S,\mathbb{Z})$ an effective curve class, $w = \beta \cdot E$ and $d \geq 1$ as above,

$$n_S[dw] \in \mathbb{Z}.$$

THEOREM 3.6 (Corollary 10 in [**3**]). *Assume the same notation as in conjecture* 3.5 *and suppose furthermore that S is toric. Then*

$$n_S[dw] \in \mathbb{Z}$$

for all $d \geq 1$.

Note that conjecture 6.2 of [**6**] is stated for any Calabi-Yau surface pair.

4. Generalized Donaldson-Thomas invariants of loop quivers

We discuss the definition and state the explicit computation of the generalized Donaldson-Thomas invariants of loop quivers. Their definition is motivated by the framework of Kontsevich-Soibelman of [**9**] and they were studied by Reineke in [**14**]. Their calculation as stated in [**14**] and reproduced in theorem 4.2 below is a special case of a result by Reineke from [**13**].

Let $m \geq 1$, which will be fixed. The m-loop quiver consists of one vertex and m loops. The framed m-loop quiver L_m has an additional vertex with an arrow connecting the two vertices:

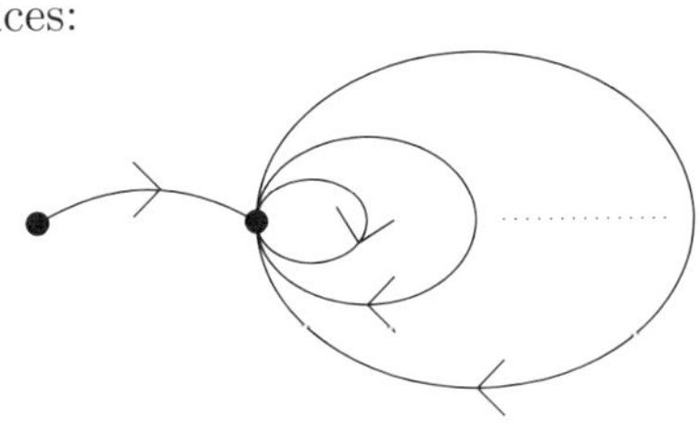

[1]See [**6**] for precise definitions.

For $n \geq 0$, $\mathbb{C}$-representations of L_m of dimension $(1, n)$ consist of the data

$$(V_0, V, \alpha_0, \alpha_1, \ldots, \alpha_n),$$

where V_0 and V are vector spaces of dimension 1, resp. n, where $\alpha_0 \in \mathrm{Hom}(V_0, V)$ and where $\alpha_i \in \mathrm{End}(V)$ for $i = 1, \cdots, n$. A morphism (γ_0, γ) between two representations is a commutative diagram

$$\begin{array}{ccc} V_0 & \xrightarrow{\alpha_0} & V \circlearrowleft \alpha_i \\ \gamma_0 \downarrow & & \downarrow \gamma \\ V_0' & \xrightarrow{\alpha_0'} & V' \circlearrowleft \alpha_i' \end{array}$$

such that $\gamma \circ \alpha_i = \alpha_i' \circ \gamma$. A morphism (γ_0, γ) is an isomorphism if both γ_0 and γ are. The space of all representations of L_m up to isomorphism is parametrized by

$$\mathbb{C}^n \oplus \mathrm{M}_n(\mathbb{C})^{\oplus m},$$

where α_0 corresponds to $v \in \mathbb{C}^n$ and the α_i to m-tuples of $n \times n$ matrices.

Denote by $\mathbb{C}\langle x_1, \ldots, x_m \rangle$ the free $\mathbb{C}$-algebra on m elements and let $(\phi_i) \in \mathrm{M}_n(\mathbb{C})^{\oplus m}$. Then (ϕ_i) determines a representation of $\mathbb{C}\langle x_1, \ldots, x_m \rangle$ on $\mathbb{C}^m$ via $x_i \mapsto \phi_i$. Moreover, $v \in \mathbb{C}^m$ is said to be *cyclic* for such a representation if its image generates all of $\mathbb{C}^m$, i.e. if

$$\mathbb{C}\langle \phi_1, \ldots, \phi_m \rangle \, v = \mathbb{C}^n.$$

The open subset of *stable* representations

$$U \subseteq \mathbb{C}^n \oplus \mathrm{M}_n(\mathbb{C})^{\oplus m}$$

consists of those (v, ϕ_i) such that v is cyclic for (ϕ_i). Consider the action of $\mathrm{GL}_n(\mathbb{C})$ on U given by:

$$g \cdot (v, \phi_i) = (gv, g\phi_i g^{-1}).$$

There is a geometric quotient of this action, called the *noncommutative Hilbert scheme for* $\mathbb{C}\langle x_1, \ldots, x_m \rangle$, and denoted by $\mathrm{Hilb}_n^{(m)}$. We package the Euler characteristics of these spaces into a generating function:

$$F(t) := \sum_{n \geq 0} \chi\left(\mathrm{Hilb}_n^{(m)}\right) t^n \in \mathbb{Z}[[t]].$$

Note that $F(0) = 1$, so that $F(t)$ admits a product expansion.

DEFINITION 4.1 (Definition 3.1 in [**14**], after [**9**]). For $n \geq 1$, the *generalized Donaldson-Thomas invariants* $\mathrm{DT}_n^{(m)} \in \mathbb{Q}$ *of the m-loop quiver* L_m are defined by means of the following product expansion:

$$F((-1)^{m-1}t) = \prod_{n \geq 1} (1 - t^n)^{-(-1)^{(m-1)n} n \, \mathrm{DT}_n^{(m)}}.$$

Recall that the Möbius function μ, for $n \geq 1$, is defined as:

$$\mu(n) = \begin{cases} 1 & \text{if } n \text{ is square-free with an even number of prime factors,} \\ -1 & \text{if } n \text{ is square-free with an odd number of prime factors,} \\ 0 & \text{if } n \text{ is not square-free.} \end{cases}$$

THEOREM 4.2 (Reineke, theorem 3.2 in [**14**]). $\mathrm{DT}_n^{(m)} \in \mathbb{N}$ *and*

$$\mathrm{DT}_n^{(m)} = \frac{1}{n^2} \sum_{d|n} \mu\left(\frac{n}{d}\right) (-1)^{(m-1)(n-d)} \binom{mn-1}{d-1}.$$

5. The correspondence

Recall that S stands for a del Pezzo surface, D for an anticanonical divisor on it, β for a primitive curve class and that we set $w = D \cdot \beta$. We define an infinite-dimensional matrix C with entries generalized Donaldson-Thomas invariants of loop quivers. If $t|s$, set

$$C_{st} := \mathrm{DT}_{s/t}^{(tw-1)}.$$

If $t \nmid s$, set $C_{st} = 0$. Since lower triangular, each row of C has only a finite number of non-zero entries. Hence applying to C an infinite-dimensional vector does not yield convergence issues. The diagonal entries of C are 1, so that $\det(C) = 1$ and its inverse is integer-valued.

We come to the interplay between the local and relative BPS state counts of S that were introduced in sections 2 and 3. The following theorem states that the matrix C provides a linear invertible relationship of the relative and local BPS numbers of S. Note that theorem 3.6, the integrality of relative BPS state counts of toric del Pezzo surfaces, follows from the integrality of C^{-1} and the theorems 2.3 and 5.1.

THEOREM 5.1 (Lemma 12 in [**3**]).

$$C \cdot [n_S[dw]]_{d\geq 1} = \left[(-1)^{dw+1}\, dw\, n_{d\beta}\right]_{d\geq 1}.$$

Why relative and local BPS state counts of del Pezzo surfaces should be related via Donaldson-Thomas invariants of loop quivers remains unclear.

References

[1] J. Bryan and R. Pandharipande, *BPS states of curves in Calabi-Yau 3-folds*, Geom. Topol. **5** (2001), 287–318 (electronic), DOI 10.2140/gt.2001.5.287. MR1825668 (2002h:14092)

[2] T.-M. Chiang, A. Klemm, S.-T. Yau, and E. Zaslow, *Local mirror symmetry: calculations and interpretations*, Adv. Theor. Math. Phys. **3** (1999), no. 3, 495–565. MR1797015 (2002e:14064)

[3] M. van Garrel, T. W. H. Wong, and G. Zaimi, *Integrality of relative BPS state counts of toric del Pezzo surfaces*, Commun. Number Theory Phys. **7** (2013), no. 4, 671–687, DOI 10.4310/CNTP.2013.v7.n4.a3. MR3228298

[4] R. Gopakumar and C. Vafa, *Topological gravity as large N topological gauge theory*, Adv. Theor. Math. Phys. **2** (1998), no. 2, 413–442. MR1633024 (2000c:81254)

[5] R. Gopakumar and C. Vafa, *Topological gravity as large N topological gauge theory*, Adv. Theor. Math. Phys. **2** (1998), no. 2, 413–442. MR1633024 (2000c:81254)

[6] M. Gross, R. Pandharipande, and B. Siebert, *The tropical vertex*, Duke Math. J. **153** (2010), no. 2, 297–362, DOI 10.1215/00127094-2010-025. MR2667135 (2011f:14093)

[7] Y. Konishi, *Integrality of Gopakumar-Vafa invariants of toric Calabi-Yau threefolds*, Publ. Res. Inst. Math. Sci. **42** (2006), no. 2, 605–648. MR2250076 (2007c:14059)

[8] Y. Konishi and S. Minabe, *Local B-model and mixed Hodge structure*, Adv. Theor. Math. Phys. **14** (2010), no. 4, 1089–1145. MR2821394 (2012h:14106)

[9] M. Kontsevich and Y. Soibelman, Stability structures, motivic Donaldson-Thomas invariants and cluster transformations, arxiv.org/abs/0811.2435, 2008.

[10] R. Pandharipande, *Hodge integrals and degenerate contributions*, Comm. Math. Phys. **208** (1999), no. 2, 489–506, DOI 10.1007/s002200050766. MR1729095 (2002k:14087)

[11] R. Pandharipande, *Three questions in Gromov-Witten theory*, Proceedings of the International Congress of Mathematicians, Vol. II (Beijing, 2002), Higher Ed. Press, Beijing, 2002, pp. 503–512. MR1957060 (2003k:14069)

[12] P. Peng, *A simple proof of Gopakumar-Vafa conjecture for local toric Calabi-Yau manifolds*, Comm. Math. Phys. **276** (2007), no. 2, 551–569, DOI 10.1007/s00220-007-0348-6. MR2346400 (2008g:14108)

[13] M. Reineke, *Cohomology of quiver moduli, functional equations, and integrality of Donaldson-Thomas type invariants*, Compos. Math. **147** (2011), no. 3, 943–964, DOI 10.1112/S0010437X1000521X. MR2801406 (2012i:16031)

[14] M. Reineke, *Degenerate cohomological Hall algebra and quantized Donaldson-Thomas invariants for m-loop quivers*, Doc. Math. **17** (2012), 1–22. MR2889742

[15] N. Takahashi, *Log mirror symmetry and local mirror symmetry*, Comm. Math. Phys. **220** (2001), no. 2, 293–299, DOI 10.1007/PL00005567. MR1844627 (2002e:14066)

KIAS, 85 Hoegiro Dongdaemun-gu, Seoul 02455, Republic of Korea
E-mail address: `vangarrel@kias.re.kr`

Proceedings of Symposia in Pure Mathematics
Volume **93**, 2016
http://dx.doi.org/10.1090/pspum/093/01590

Resurgence and topological strings

M. Vonk

Abstract. The mathematical idea of resurgence allows one to obtain nonperturbative information from the large–order behavior of perturbative expansions. This idea can be very fruitful in physics applications, in particular if one does not have access to such nonperturbative information from first principles. An important example is topological string theory, which is a priori only defined as an asymptotic perturbative expansion in the coupling constant g_s. We show how the idea of resurgence can be combined with the holomorphic anomaly equation to extend the perturbative definition of the topological string and obtain, in a model–independent way, a large amount of information about its nonperturbative structure.

1. Introduction

The vast majority of calculational problems in physics are impossible to solve exactly. For this reason, it is important to have good approximation techniques at one's disposal. One such technique is the perturbative approach: one identifies a (preferably small) parameter x in the problem, such that the problem can be solved exactly in the special case where $x = 0$. Then, one tries to construct the full solution $f(x)$ to the problem order by order in a perturbative expansion:

$$f(x) = \sum_{n=0}^{\infty} a_n x^n. \tag{1.1}$$

Of course, it will in general not be possible to find a closed form for all the coefficients a_n (that would essentially amount to finding an exact solution to the problem), but often one can calculate the individual coefficients one by one, up to arbitrarily high n. One may then calculate a partial sum of the form

$$f_N(x) = \sum_{n=0}^{N} a_n x^n \tag{1.2}$$

and, for large N, view such a sum as an approximation to the true answer $f(x)$.

There are two well–known problems that may arise in this approach. The first one is that the partial sums $f_N(x)$ may not converge when one takes the limit

2010 *Mathematics Subject Classification.* Primary 81T30; Secondary 81T45, 40G10.

Key words and phrases. Resurgence, topological strings, transseries, instantons, holomorphic anomaly equation.

The research of the author was supported by the European Research Council Advanced Grant EMERGRAV.

$N \to \infty$. The canonical example where this happens is the Taylor expansion

$$\frac{1}{1-x} \approx 1 + x + x^2 + x^3 + \dots \tag{1.3}$$

Here, the partial sums of the right hand side diverge when $|x| > 1$, even though for any $x \neq 1$, the left hand side is well–defined. This example still has a finite radius of convergence, but there are other examples where the perturbation series is *asymptotic*: even though the partial sums become better and better approximations as $x \to 0$, for any $x \neq 0$ they diverge when $N \to \infty$. A famous example is Stirling's approximation to the logarithm of the Gamma function:

$$\log \Gamma(z) - \frac{1}{2}\log(2\pi) - \left(z - \frac{1}{2}\right)\log z + z \approx \frac{1}{12z} - \frac{1}{360z^3} + \frac{1}{1260z^5} - \dots \tag{1.4}$$

where now z^{-1} is the small parameter, and we have moved all terms that do not involve positive powers of this parameter to the left hand side. One can show that the coefficients of z^{1-2n} on the right hand side grow like $\frac{(2n-2)!}{(2\pi)^{2n}}$. From this factorial growth of the coefficients, one then easily shows that the partial sums diverge for any nonzero value of z^{-1}. This factorial growth of perturbative coefficients, and the resulting asymptotic behavior, is ubiquitous in quantum mechanics and quantum field theory.

A second problem that often arises when one uses perturbative methods is that some nontrivial functions have a vanishing perturbative expansion. Here, the canonical example is

$$f(x) = e^{-\frac{1}{x^2}} \tag{1.5}$$

which is a well–defined and smooth function on the real axis, but which has a vanishing Taylor series around $x = 0$. More generically, instanton and soliton effects in physics often cannot be "seen" in perturbation theory.

An important observation is that the above two problems are not at all independent, as can be seen e.g. from Borel resummation. In section 2, we will review how this comes about, and how this relation between asymptotic perturbative series and nonperturbative effects takes its full form in the theory of resurgence. The idea of resurgence has great potential in physical applications, since it allows us to obtain nonperturbative information from a purely perturbative expansion. In this contribution, based on the work [**1**, **2**] with R. Couso-Santamaría, J. D. Edelstein and R. Schiappa, we want to work out this idea for the example of topological strings. This example is particularly interesting, as topological strings are a priori only *defined* as an asymptotic perturbative expansion. Finding a generic way to extend their partition sums into full, nonperturbative functions, is therefore a very interesting open question. In section 3, we explain how the holomorphic anomaly equation provides a window into this problem. In section 4, we then present some explicit results for the example of B–model toplogical strings on local $\mathbb{P}^2$. We end with a conclusion and outlook in section 5.

2. Resurgence

In this section, we very briefly review some of the basic ideas of resurgence. The reader is referred to [**3**] and references therein for a more thorough introduction.

2.1. Borel transforms and asymptotic behavior of coefficients. If a quantity $f(x)$ has a divergent perturbation series whose coefficients grow as

$$a_n \sim \frac{n!}{A^n} \tag{2.1}$$

for some (generically complex) A, then there is a simple trick due to Borel that often allows one to find a well–defined function for which the given series is an asymptotic perturbative approximation. One defines the *Borel transform* $\mathcal{B}[f]$ as the formal power series

$$\mathcal{B}[f](s) = \sum_{n=0}^{\infty} \frac{a_n}{n!} s^n. \tag{2.2}$$

This turns the asymptotic series for f into a new series which has a singularity at $s = A$ but is convergent when $|s| < A$. If we now assume that A is not a positive, real number and that $\mathcal{B}[f](s)$ can be analytically continued to the positive real axis, then one easily checks that

$$\mathcal{S}_0 f(x) \equiv \int_{s=0}^{\infty} \mathcal{B}[f](sx)\, e^{-s} ds \tag{2.3}$$

gives back a function $\mathcal{S}_0 f(x)$, called the *Borel resummation* of $f(x)$, which has the same asymptotic expansion as the original function. Thus, this Laplace–type transform can be thought of as a formal inverse of the Borel transform.

Of course, when A is real and positive, the above procedure does not work, since for positive x the integral in (2.3) runs into the singularity of $\mathcal{B}[f]$. One may define two alternative Borel resummations, $\mathcal{S}_{\pm} f(x)$, by using integration contours in the complex plane which circumvent the singularity either above or below. Using Cauchy's theorem, one then sees that the difference between those two resummations is of the order

$$(\mathcal{S}_- - \mathcal{S}_+) f(x) \approx e^{-A/x}. \tag{2.4}$$

This result is perhaps not too surprising: the difference between the two Borel resummations — which each have the same asymptotic expansion as the original function — is a function of the "instanton type", which itself has a vanishing perturbative expansion. This is the relation between asymptotic series and nonperturbative functions that we alluded to in the introduction.

Often, for example for reasons of reality (see e.g. [**4**]), one can show that neither of the two Borel resummations gives back the original function $f(x)$, but that the true function $f(x)$ lies "in the middle", in the sense that

$$\begin{aligned} f(x) &= \mathcal{S}_+ f(x) + \frac{1}{2} e^{-A/x} (\ldots) \\ &= \mathcal{S}_- f(x) - \frac{1}{2} e^{-A/x} (\ldots) \end{aligned} \tag{2.5}$$

where the dots indicate the resummation of a further expansion that we will make more precise in the next subsection.

A crucial observation to make about this discussion is that the coefficient A (usually called the "instanton action") which can be read off from the asymptotic growth of the *perturbative* coefficients in (2.1), also appears in the *nonperturbative* contributions in (2.5). Somehow, the perturbative solution to our problem already knows something about its nonperturbative completion. As we will see, resurgence makes this statement precise, and extends it immensely.

2.2. Transseries. As we have seen in the previous subsection, a perturbative power series contains some information (like the instanton action A) about the nonperturbative contributions to the function one wants to describe. At the same time, this information is not encoded in a very straightforward way. To make the nonperturbative content of a function more transparent, it turns out to be very useful to describe it using a *transseries*. A simple example of a transseries is an expression of the form

$$f(x) \approx \sum_{k=0}^{\infty} \sum_{n=0}^{\infty} a_n^{(k)} e^{-kA/x} x^n. \tag{2.6}$$

In this case, the transseries is a formal expansion in the variable x and in a "nonperturbative building block" $e^{-A/x}$. More general transseries may have many more of such building blocks: they can be expansions in several different instanton factors, $e^{-A_i/x}$, expansions in $\log(x)$ or other nonanalytic functions of x, etc. We will not discuss the general theory of transseries here — the reader can find several good references in [**3**]. For most of this paper, formal transseries of the above form will be sufficient.

Transseries solutions to physics problems often arise in a very natural way. For example, if the problem is described by a differential or finite difference equation, one can often simply insert an ansatz of the form (2.6) into this equation and solve it order by order to find the coeffients $a_n^{(k)}$ and the instanton action A — in the same way that one would construct a formal power series solution. In fact, this is exactly what we will do in this paper.

Of course, given a transseries, one may again ask how to obtain an actual *function* from it. To this end, let us write the transseries expansion (2.6) as

$$f(x) \approx \sum_{k=0}^{\infty} e^{-kA/x} \Phi^{(k)}(x) \qquad \text{with} \qquad \Phi^{(k)}(x) = \sum_{n=0}^{\infty} a_n^{(k)} x^n. \tag{2.7}$$

In general, all of the pertrubative series $\Phi^{(k)}(x)$ may be asymptotic, divergent series. However, as was the case for ordinary power series, one can turn a formal transseries into an actual function by choosing an integration contour and Borel resumming each sector. That is, one extends the definition of Borel resummation in the natural way to be

$$\mathcal{S}_+ f(x) = \sum_{k=0}^{\infty} e^{-kA/x} \int_{s=0^+}^{\infty} \mathcal{B}\left[\Phi^{(k)}\right](sx)\, e^{-s} ds \tag{2.8}$$

where we have chosen the $+$-contour to be specific.

Note that expressions of the form (2.7) appear very naturally in physics. In many problems, e.g. in quantum field theory, one expands a solution around a trivial "vacuum" background. However, there are in general other, nonperturbative backgrounds, such as instantons and solitons, that one wants to take into account as well. Each of these backgrounds is suppressed by a nonperturbative factor, often of the form $e^{-A/x}$, and one can construct a new, perturbative solution around such a background. One then wants to "add up" all of the different nonperturbative sectors into a single solution to the problem. The prescription (2.8) makes this procedure mathematically precise.

2.3. Resurgence. Even though we can formally write down any transseries we like, the transseries solutions that arise from physical or mathematical problems usually have a lot of extra structure. We saw an example of this in equation (2.1) and (2.5) where, essentially just from the requirement of analyticity, we derived that the instanton action A could be read off from the large–order behavior of the perturbative coefficients $a_n^{(k)}$. The theory of resurgence, first developed by J. Écalle in [**5**], extends this example immensely, and makes the resulting structure very precise.

In [**5**], a class of functions called *resurgent functions* is defined. The defining property is that the Borel transform of a resurgent function only has a discrete set of singularities and any analytic continuation along a path avoiding these singularities can be defined. For our purposes, the precise definition is not very relevant; all that matters is that most of the functions one encounters in (toy model) physical examples belong to the class of resurgent functions.

One can show that for resurgent functions, not only the instanton action A can be derived from the perturbative coefficients $a_n^{(0)}$, but in fact *all* coefficients $a_n^{(k)}$ in *all* other instanton sectors $\Phi^{(k)}$ can be obtained from the perturbative sector. This is the origin of the name "resurgence": the instanton sectors can be "resurrected" from the perturbative sector alone. Actually, the choice of the perturbative "vacuum" sector is somewhat arbitrary: one could also reconstruct the vacuum sector (and all other sectors) from an arbitrary given instanton sector.

The way to obtain this nonperturbative information from perturbative information is essentially through a huge generalization of (2.1). The details of this depend on the problem at hand (and the derivation requires several new mathematical techniques that are explained in [**5**]), but in the end it turns out that one can derive large–order relations which are schematically of the form

$$a_n^{(0)} \sim \sum_{k=1}^{\infty} \frac{S_1^k}{2\pi i} \sum_{m=0}^{\infty} \frac{\Gamma(n-m)}{(kA)^{n-m}} a_m^{(k)}. \tag{2.9}$$

Here, S_1 is an unknown problem–dependent constant known as the *Stokes constant*. We stress that the above expression is just schematic: in actual computations (see e.g. [**3**]), there may be many different instanton actions A_i, many different Stokes constants, and arguments of the form $n-m$ may be shifted by problem–dependent constants. However, the general structure of these large–order relations and the way they can be used is always the same. Note for example that taking the leading $(k=1, m=0)$ contribution in the above equation, we recover the fact that the leading growth of the perturbative coefficients $a_n^{(0)}$ is determined by A as in (2.1). Taking higher m terms into account, we see that $\frac{1}{m}$–corrections to this leading growth determine the one–instanton coefficients $a_n^{(1)}$. Then from the $k=2$ terms, we see that nonperturbative corrections of order $(2)^{-m}$ to this growth determine the two–instanton coefficients $a_n^{(2)}$, and so on.

The use of large–order relations in physical applications is twofold. First of all, there are problems where the perturbative sector can be calculated, but where the nonperturbative contributions are unknown — either due to computational difficulties, or for more fundamental reasons. In fact, the example of topological strings that we will discuss is a case where the theory is only defined perturbatively, and there is no generally defined nonperturbative completion. In such cases, one may use large–order relations to calculate nonperturbative contributions from consistency conditions alone.

Of course, in doing so, one *assumes* that the function one is investigating belongs to the class of resurgent functions. This can in general not be proven, but here, the large–order relation is again useful. If one can obtain conjectured nonperturbative contributions by other means — for example by plugging a transseries ansatz into a differential or finite difference equation — one can then use the large–order relation to *test* whether the function indeed has resurgent behavior. One may perform these tests for several nonperturbative contributions to gain confidence in the resurgent properties, and then use the large–order relation to calculate further nonperturbative terms. In what follows, we will illustrate this approach using the example of the topological string.

3. The holomorphic anomaly

There are several equivalent ways to define topological string theories and their partition functions. Physically, a topological string theory be obtained by twisting a two–dimensional $\mathcal{N} = 2$ supersymmetric field theory and coupling the resulting theory to two–dimensional gravity. Mathematically, topological string theory partition functions can be defined as generating functions of Gromov–Witten invariants, or they can be obtained from studying the complex structure deformations of Calabi–Yau manifolds. It would go too far to review any of these definitions here; we refer the interested reader to the many available reviews of the topic, such as the extensive book [**6**].

For our purposes, all that matters is that all of these definitions depend on a parameter g_s, called the topological string coupling constant, and that they all lead to a partition function which is a perturbative expansion in g_s. More precisely, the free energy (which is the logarithm of the partition function) is an expansion of the form

$$F(t; g_s) \approx \sum_{g=0}^{\infty} F_g(t)\, g_s^{2g-2}. \tag{3.1}$$

Here, we denoted any additional parameters that the problem may have by t; these can for example take the form of couplings in the two–dimensional field theory, or of moduli of the Calabi–Yau manifold.

Calculating the coefficients $F_g(t)$ from one of the definitions of the topological string theory is often very complicated, especially if one wants to go beyond the first few values of g. Fortunately, in the incarnation where the topological string is defined from the complex structure deformations of Calabi–Yau manifolds (the so–called B–model), a shortcut was found in [**7**, **8**]. It turns out that the $F_g(t)$ are almost holomorphic in the parameters t^i, where the "almost" means that there is a recursion relation, called the *holomorphic anomaly equation*, of the form

$$\partial_{\bar{\imath}} F_g = \frac{1}{2} \overline{C}_{\bar{\imath}}^{jk} \left(D_j D_k F_{g-1} + \sum_{h=1}^{g-1} D_j F_{g-h} D_k F_h \right). \tag{3.2}$$

The definition of the coefficients $\overline{C}_{\bar{\imath}}^{jk}$ and of the covariant derivatives D_i can be found e.g. in [**6**]. The crucial point is that the derivative with respect to the anti–holomorphic variable $\bar{t}^{\bar{\imath}}$ depends *only* on the F_h with $h < g$. This allows one to determine the F_g recursively, up to an integration constant which is purely holomorphic in the t^i. It turns out that this integration constant can often be determined exactly from the behavior of $F(t)$ at special points in the moduli space; see e.g. [**9**]. This so–called "direct integration" method is very efficient, and can be used to compute the F_g up to high values of g.

The direct integration technique allows one to construct a formal, perturbative expression for the topological string theory free energy, but the power series one constructs in this way are asymptotic and diverge. This is a serious problem, even more so because, as we have mentioned, the usual definitions of the topological string theory are *only* perturbative in nature. To obtain an actual function $F(t)$, we need to somehow identify the nonperturbative contributions and construct a nonperturbative completion of the asymptotic series. While this has been done in specific instances (usually through the use of dualities), a generic procedure does not exist. Clearly, the theory of resurgence can be a major asset to fill this gap.

Thus, one would first of all like to extend the power series solution (3.1) to a transseries solution. An immediate problem presents itself, as the holomorphic anomaly equation (3.2) is an equation for the *individual coefficients* $F_g(t)$, not for the full $F(t)$. Thus, we cannot simply insert a transseries ansatz into this equation. Fortunately, as was already pointed out in [**8**], it is not too hard to find an equation that the full partition function $Z = e^{F(t)}$ satisfies. Roughly, Z satisfies a heat kernel equation of the form

$$\left(\partial_{\bar{\imath}} - \frac{1}{2} g_s^2 \overline{C}_{\bar{\imath}}^{\;jk} D_j D_k\right) Z = 0. \tag{3.3}$$

By "roughly", we mean that some additional terms must be included in the equation to correct the resulting anomaly equations at low–lying g. The details of this, and the precise equation that replaces (3.3), can be found in our paper [**1**]. In what follows, we will simply refer to the above equation, but the reader should keep in mind that it is only a schematic representation of the true holomorphic anomaly equation for $Z(t)$.

Now, the plan of attack is clear: one can make a transseries ansatz for $F(t)$, plug $Z = e^F$ into the above equation, and recursively solve for all nonperturbative coefficients. As was the case for the ordinary power series solution, this requires the fixing of holomorphic integration constants at every order, which can be done using boundary behavior of $F(t)$ at special points in moduli space. Thus, we can describe this procedure as "nonperturbative direct integration".

Following this procedure, we find a conjectural form of the nonperturbative topological string free energy expressed as a transseries. One can then compare these results to existing conjectures for nonperturbative topological strings, and test, as described in the previous section, whether the $F(t)$ one finds is indeed a resurgent function. We have carried this out for general models in [**1**], and for the specific example of B–model topological strings on local $\mathbb{P}^2$ in [**2**]. In the next section, we report some of the most interesting results from these papers.

4. Resurgence and the holomorphic anomaly

We will present the results of the procedure outlined above in a mostly graphical manner. For the analytical derivations and numerics behind the images, and a much

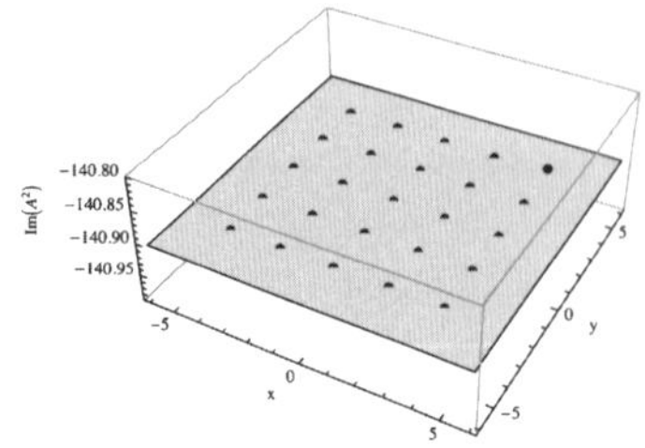

FIGURE 1. The instanton action is independent of the anti-holomorphic modulus.

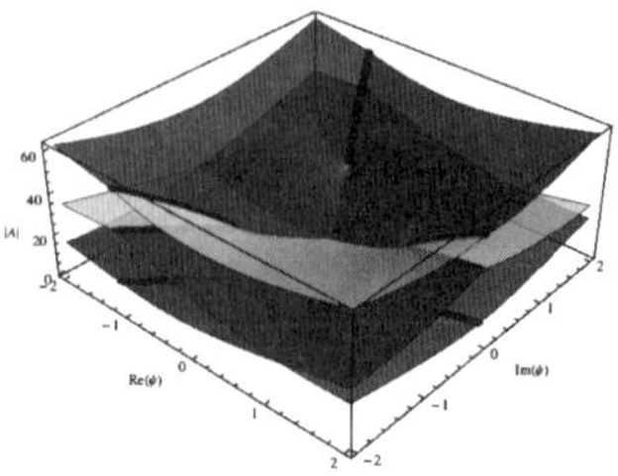

FIGURE 2. Local $\mathbb{P}^2$ has three different conifold instanton actions.

more detailed description, we refer the reader to [**1**, **2**]. The figures in this section are originally in full–color; the reader can find the colored version of the images in the online version of this paper.

The most basic prediction of resurgence is that the instanton actions A, calculated using a transseries ansatz for the equation (3.3), can also be obtained from the large–order behavior of the perturbative $F_g(t)$ using the analogue of (2.1). We have found that for topological strings, this is indeed the case. Perhaps more interestingly, using resurgence one can show on general ground that the $\bar{t}$–dependence disappears in the large g limit, and that the instanton actions are all *purely holomorphic* in t. This can also be shown numerically in explicit examples. In figure 1, for example, we see in the local $\mathbb{P}^2$ example that an instanton action (of which for technical reasons we plot the imaginary part of the square) is independent of the value of an anti–holomorphic modulus $\overline{\psi} = x + iy$. This result is consistent with a conjecture made in [**10**], which states that the instanton actions of topological string theories quite generally are given by periods of holomorphic three–forms — which naturally depend in a holomorphic manner on the moduli.

Of course, for a generic topological string theory, we expect to find more than one instanton action. In fact, in the case of local $\mathbb{P}^2$, one easily finds three different conifold periods which can all appear as instanton actions. (There is in fact a fourth period associated to the large–volume limit of local $\mathbb{P}^2$ which we will ignore in this paper.) In figure 2, we plot the absolute value of those three periods as functions of the modulus ψ of the model. At each point in moduli space, one can check that the large–order behavior of the F_g is indeed determined by the period A_i with smallest absolute value.

Next, one can use a transseries ansatz and (3.3) to calculate nonperturbative coefficients $F_g^{(k)}$ in several instanton sectors. One can then check whether these coefficients match the large–order behavior of the perturbative F_g. In figure 3, we show these tests for three different values of the modulus (from left to right) for the first four leading coefficients $F_g^{(1)}$ (from top to bottom). The dependence on the anti–holomorphic modulus $\overline{\psi}$ is plotted in the figures. The lines show the predictions (real and imaginary part) from the transseries ansatz; the dots show the results from the large–order behavior of F_g. We see that the dots match the lines perfectly, meaning that all of this nonperturbative information is indeed captured by the perturbative F_g. As an interesting side remark: we see that for holomorphic moduli (0 on the horizontal axis in the plots) all $F_{g\geq 1}^{(1)}$ vanish. This is consistent

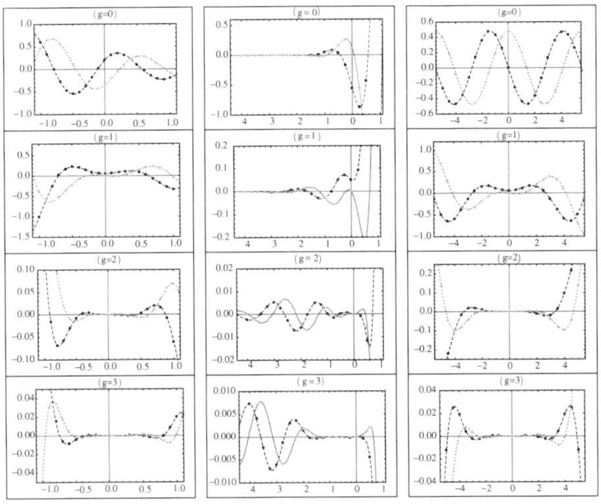

FIGURE 3. Comparison between transseries results and large–order results for several nonperturbative coefficients.

with the fact that in this limit, from the direct integration procedure one expects to find back the exact conifold results. In [**11**], these coefficients were indeed shown to vanish.

As can already be seen from figure 2, at different points in moduli space, different instanton actions have the smallest absolute value, meaning that different instanton actions determine the dominant large–order behavior of the perturbative coefficients. One can now try to Borel resum the contributions coming from a given instanton sector, and measure the contributions in the large–order relations subleading to all of those. In fact, two things can happen: those subdominant instanton contributions can come from a *different* instanton, or they can come from *two–instanton* effects of the same type as the resummed one–instanton sector. Which of these is the case depends on the absolute values of the relevant instanton actions. This is plotted for a specific slice in moduli space in figure 4. For example, for values of ψ in the left of the figure, we see that the instanton action A_3 dominates, whereas the instanton action A_1 determines the subdominant large–order behavior of F_g. For values of ψ near the middle of the plot, the A_1–sector dominates, and the subdominance comes from two–instanton effects corresponding to the same instanton.

We have checked all of these predictions against large–order behavior, and found perfect agreement. For example, in figure 5 we zoom in to a region in moduli space where the subdominant behavior is caused by two different one–instanton sectors. After resumming the leading large–order behavior, we see that what remains (the dots in the figure) indeed perfectly matches the predictions from the analytic transseries solution. In particular, a jump in the large–order data occurs exactly where one would expect to see it. A similar plot in figure 6 shows a region where first a one–instanton sector is subdominant, then a two–instanton sector takes over, and finally a different one–instanton sector determines the subdominant behavior. Once again, the dots coming from the large–order perturbative data perfectly match the predictions from the resurgent transseries ansatz.

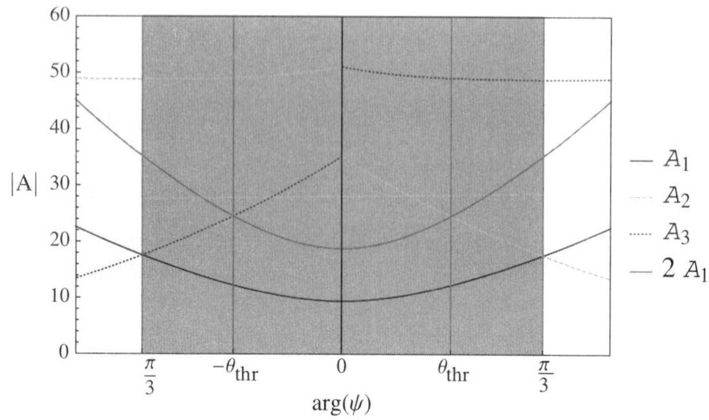

FIGURE 4. Dominant and subdominant instanton actions.

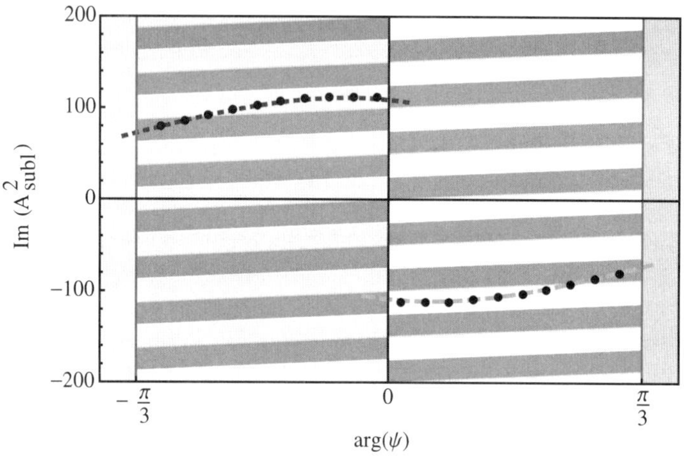

FIGURE 5. A crossover between two different subdominant one–instanton sectors.

Finally, one may wonder what happens at points in moduli space where two instanton actions have exactly the same absolute value, so that each of the corresponding sectors gives an equal size contribution to the large–order behavior. It is not too hard to show that at those points, resurgence predicts that the large–order behavior of the F_g obtains an oscillatory component. In figure 7, we isolate the expected oscillatory behavior from the transseries prediction (the continuous line) and see that for large g (plotted horizontally), the data again nicely matches the prediction.

5. Conclusion and outlook

All of the results in the previous section (and many more reported in [**1**, **2**]) seem to indicate that the nonperturbative topological string free energy is indeed a resurgent function, and that therefore, resurgent transseries techniques based on the holomorphic anomaly equation can be used to give a proper nonperturbative

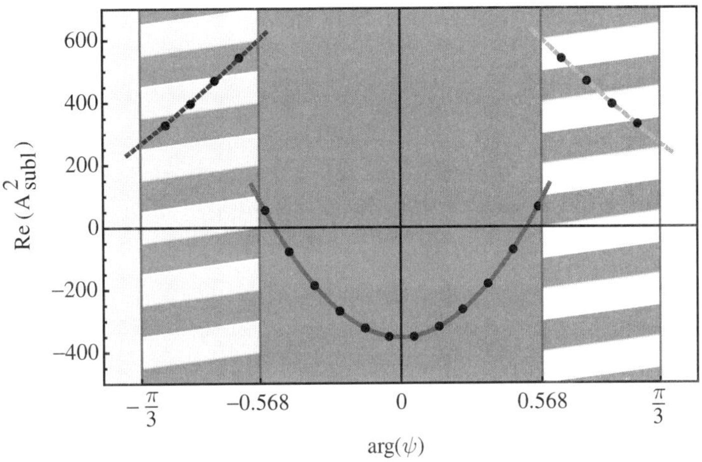

FIGURE 6. A crossover between subdominant one– and two–instanton sectors.

completion of the topological string. While these results are very encouraging, a number of open questions remain, of which we mention the two most pressing ones:

- Even though one can now *calculate* nonperturbative contributions to the topological string free energy, it would be nice to have a more physical *interpretation* of the effects that these contributions describe. In certain examples, through dualities, these effects turn out to match D–brane or tunneling effects, but a generic description from a purely topological string point of view is still missing.
- A crucial technical observation is that the holomorphic anomaly equation (3.3) is a differential equation in t, and *not* in the transseries variable g_s. As a result (due to the lack of a so–called "bridge equation", as explained in [**1**]), one cannot fully derive the large–order relations and the transseries structure from first principles, but one has to resort to information obtained from either dual desciptions or large–order analysis. At the level discussed in this paper, this is not an issue, but on a more fundamental level it is. For example, one cannot a priori determine the number of different instanton actions, and once one goes to higher than leading subdominant contributions, it is no longer fully clear which transseries coefficients determine the large–order behavior. In fact, we cannot say with certainty whether the topological free energy is a so–called "simple resurgent function" (meaning that its Borel transform only has simple poles and logarithmic branch cuts) or whether it is of some more complicated type. It would be good to have a better mathematical handle on these issues.

Apart from these fundamental open questions, it would be good to study more examples — for example, models with more moduli, or models with different (non–conifold) types of special points in moduli space — in order to develop further the technical tools presented here. Undoubtedly, this will shed a lot of new light on the so far mysterious issue of the nonperturbative definition of the topological string.

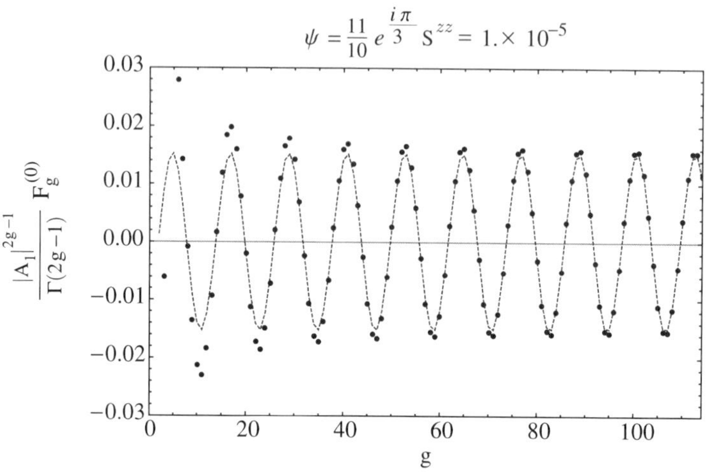

FIGURE 7. Oscillatory large–order behavior when two instanton actions dominate.

References

[1] R. C. Santamaría, J. D. Edelstein, R. Schiappa and M. Vonk, "Resurgent Transseries and the Holomorphic Anomaly," arXiv:1308.1695 [hep-th].

[2] ——— "Resurgent Transseries and the Holomorphic Anomaly: Nonperturbative Closed Strings in Local $\mathbb{CP}^2$," arXiv:1407.4821 [hep-th].

[3] I. Aniceto, R. Schiappa, and M. Vonk, *The resurgence of instantons in string theory*, Commun. Number Theory Phys. **6** (2012), no. 2, 339–496, DOI 10.4310/CNTP.2012.v6.n2.a3. MR2993121

[4] I. Aniceto and R. Schiappa, *Nonperturbative ambiguities and the reality of resurgent transseries*, Comm. Math. Phys. **335** (2015), no. 1, 183–245, DOI 10.1007/s00220-014-2165-z. MR3314503

[5] J. Écalle, "Les Fonctions Résurgentes," Prépub. Math. Univ. Paris-Sud **81-05** (1981), **81-06** (1981), **85-05** (1985)

[6] K. Hori, S. Katz, A. Klemm, R. Pandharipande, R. Thomas, C. Vafa, R. Vakil, and E. Zaslow, *Mirror symmetry*, Clay Mathematics Monographs, vol. 1, American Mathematical Society, Providence, RI; Clay Mathematics Institute, Cambridge, MA, 2003. With a preface by Vafa. MR2003030 (2004g:14042)

[7] M. Bershadsky, S. Cecotti, H. Ooguri, and C. Vafa, *Holomorphic anomalies in topological field theories*, Nuclear Phys. B **405** (1993), no. 2-3, 279–304, DOI 10.1016/0550-3213(93)90548-4. MR1240687 (94j:81254)

[8] M. Bershadsky, S. Cecotti, H. Ooguri, and C. Vafa, *Kodaira-Spencer theory of gravity and exact results for quantum string amplitudes*, Comm. Math. Phys. **165** (1994), no. 2, 311–427. MR1301851 (95f:32029)

[9] B. Haghighat, A. Klemm, and M. Rauch, *Integrability of the holomorphic anomaly equations*, J. High Energy Phys. **10** (2008), 097, 37, DOI 10.1088/1126-6708/2008/10/097. MR2452948 (2010e:81229)

[10] N. Drukker, M. Mariño, and P. Putrov, *Nonperturbative aspects of ABJM theory*, J. High Energy Phys. **11** (2011), 141, 29, DOI 10.1007/JHEP11(2011)141. MR2913213

[11] S. Pasquetti and R. Schiappa, *Borel and Stokes nonperturbative phenomena in topological string theory and $c = 1$ matrix models*, Ann. Henri Poincaré **11** (2010), no. 3, 351–431, DOI 10.1007/s00023-010-0044-5. MR2671567 (2011j:81305)

INSTITUTE FOR THEORETICAL PHYSICS, UNIVERSITY OF AMSTERDAM, SCIENCE PARK 904, POSTBUS 94485, 1090 GL AMSTERDAM, THE NETHERLANDS

E-mail address: `m.l.vonk@uva.nl`

Proceedings of Symposia in Pure Mathematics
Volume **93**, 2016
http://dx.doi.org/10.1090/pspum/093/01556

Chern-Simons splitting of 2+1D gauge theories

Tuna Yildirim

Abstract. Geometric quantization of topologically massive and pure Yang-Mills theories is studied in 2+1 dimensions. Analogous to the topologically massive AdS gravity model, both topologically massive Yang-Mills and pure Yang-Mills theories are shown to exhibit a Chern-Simons splitting behavior at large distances. It is also shown that, this large distance topological behavior can be further used to incorporate link invariants.

1. Introduction

In 2+1 dimensions, Yang-Mills(YM) theory is known to have a mass gap, which makes the theory trivial at very large distances. However, this is not the case for topologically massive Yang-Mills theory(TMYM). In TMYM theory, as Yang-Mills contribution decays exponentially at large distances, the Chern-Simons(CS) term becomes dominant, which leads to a topological theory. For a mass gap of m, the scale is given in comparison to $1/m$. This work focuses on the large but finite distance behavior of TMYM and pure YM theories by taking the first order contributions in the $1/m$ expansion into account, neglecting all higher order terms. At this limited scale, both theories have interesting topological behaviors, very similar to the topologically massive AdS-gravity model at corresponding limits.

1.1. Topologically Massive AdS Gravity. For a dynamical metric $\gamma_{\mu\nu}$, the action for the topologically massive AdS gravity model is given by[**1**, **2**]

$$S = \int d^3x \left[-\sqrt{-\gamma}(R - 2\Lambda) + \frac{1}{2\mu}\epsilon^{\mu\nu\rho}\left(\Gamma^{\alpha}_{\mu\beta}\partial_\nu \Gamma^{\beta}_{\rho\alpha} + \frac{2}{3}\Gamma^{\alpha}_{\mu\gamma}\Gamma^{\gamma}_{\nu\beta}\Gamma^{\beta}_{\rho\alpha}\right)\right]. \tag{1.1}$$

This action naturally splits into two CS terms[**3**–**6**], by defining

$$A^{\pm}{}_{\mu}{}^{a}{}_{b}[e] = \omega_{\mu}{}^{a}{}_{b}[e] \pm \epsilon^{a}{}_{bc}e_{\mu}{}^{c}, \tag{1.2}$$

where $e_\mu{}^a$ is the dreibein and $\omega_\mu{}^a{}_b[e]$ is the torsion-free spin connection. Then, the action (1.1) can be written as

$$S[e] = -\frac{1}{2}\left(1 - \frac{1}{\mu}\right) S_{CS}\left[A^+[e]\right] + \frac{1}{2}\left(1 + \frac{1}{\mu}\right) S_{CS}\left[A^-[e]\right] \tag{1.3}$$

2010 *Mathematics Subject Classification.* Primary 81T13, 81T45, 53D50; Secondary 81T40, 57M27.

Key words and phrases. Topological field theory, Wilson loop, link invariants, geometric quantization.

The author was supported in part by NSF Grant #1067889.

where

$$S_{CS}[A] = \frac{1}{2}\int \epsilon^{\mu\nu\rho}\left(A_{\mu}{}^{a}{}_{b}\partial_{\nu}A_{\rho}{}^{b}{}_{a} + \frac{2}{3}A_{\mu}{}^{a}{}_{c}A_{\nu}{}^{c}{}_{b}A_{\rho}{}^{b}{}_{a}\right). \tag{1.4}$$

For our interests, the main difference between this gravity model and TMYM theory is that the latter has a mass gap, therefore it has a topological behavior only at large distances. But the gravity model is given by CS theory irrespective of the value of μ. In TMYM theory, large distances are obtained by taking large values of m. Large m, in the sense of near CS limit of TMYM theory, corresponds to small values of μ in the gravitational analogue. On the other hand, the $\mu \to \infty$ limit(pure Einstein-Hilbert limit) of the gravity model is analogous to pure YM theory.

Now, let us focus on the two important limits of (1.3). For small values of μ, (1.3) can be written as a *sum* of two half CS theories as

$$S[e] \approx \frac{1}{2\mu}S_{CS}\big[A^{+}[e]\big] + \frac{1}{2\mu}S_{CS}\big[A^{-}[e]\big]. \tag{1.5}$$

In the $\mu \to \infty$ limit, it is equal to the *difference* between two half CS theories, as

$$S[e] = \frac{1}{2}S_{CS}\big[A^{-}[e]\big] - \frac{1}{2}S_{CS}\big[A^{+}[e]\big]. \tag{1.6}$$

The main goal of this work is to investigate whether or not a similar CS+CS type splitting appears in TMYM theory and a CS–CS type splitting in pure YM theory at large distances.

1.2. Holomorphic Quantization of Chern-Simons Theory. Before we tackle pure and topologically massive Yang-Mills theories, it would be beneficial to review the geometric quantization of pure Chern-Simons theory. This section will be a quick review, following refs. **7**, **8**.

In this section and all following sections, we will choose the temporal gauge $A_0 = 0$ for all of the theories we study. We will also use the complex coordinates, where $A_z = \frac{1}{2}(A_1 + iA_2)$ and $A_{\bar{z}} = \frac{1}{2}(A_1 - iA_2)$. In these coordinates, the conjugate momenta of CS theory are given by

$$\Pi^{az} = \frac{ik}{4\pi}A^{a}_{\bar{z}} \quad \text{and} \quad \Pi^{a\bar{z}} = -\frac{ik}{4\pi}A^{a}_{z}. \tag{1.7}$$

Now, we can write the symplectic two-form for pure CS theory as

$$\Omega = \frac{ik}{2\pi}\int\limits_{\Sigma} \delta A^{a}_{\bar{z}}\delta A^{a}_{z} \tag{1.8}$$

where Σ is an orientable two-dimensional surface.

1.2.1. *The Wave-Functional.* We start with choosing the holomorphic polarization that gives $\Phi[A_z, A_{\bar{z}}] = e^{-\frac{1}{2}K}\psi[A_{\bar{z}}]$, where K is the Kähler potential, Φ and ψ are pre-quantum and quantum wave-functionals. Kähler potential is given by $K = \frac{k}{2\pi}\int\limits_{\Sigma} A^{a}_{\bar{z}}A^{a}_{z}$.

Upon quantization we can write

$$A_z\psi[A_{\bar{z}}] = \frac{2\pi}{k}\frac{\delta}{\delta A_{\bar{z}}}\psi[A_{\bar{z}}]. \tag{1.9}$$

For the gauge theories that we are interested in, the wave-functional can be factorized as $\psi = \phi\chi$, where ϕ is the part that satisfies Gauss' law of the theory and χ is the gauge invariant part. To find ϕ, one makes an infinitesimal gauge transformation, then forces the Gauss law constraint. This leads to a condition that is usually solved by some WZW action. Then, the Schrödinger's equation can be solved to find χ. The Hamiltonian of pure CS theory in the temporal gauge is zero, hence Schrödinger's equation is trivially satisfied. This makes $\chi = 1$ a sufficient solution. For the theories that are not scale invariant, χ is where the scale dependence would be hidden.

For pure CS theory, the generator of infinitesimal gauge transformations is $G^a = \frac{ik}{2\pi} F^a_{z\bar{z}}$ and the Gauss law is given by $G^a \psi = 0$.

Before we continue, we shall parametrize the gauge fields, using Karabali-Nair[**9**] parametrization, as

$$A_z = U^{\dagger -1} \partial_z U^\dagger \text{ and } A_{\bar{z}} = -\partial_{\bar{z}} U U^{-1}. \tag{1.10}$$

Here $U \in SL(N, \mathbb{C})$ and it gauge transforms as $U \to gU$, where $g \in \mathcal{G}$ and $\mathcal{G}$ is the gauge group. U can be written as

$$U(x, 0, C) = \mathcal{P}exp\left(- \int_{\substack{0\\C}}^{x} (A_{\bar{z}} d\bar{z} + \mathcal{A}_z dz) \right), \tag{1.11}$$

where $\mathcal{A}_z$ satisfies $\partial_z A_{\bar{z}} - \partial_{\bar{z}} \mathcal{A}_z + [\mathcal{A}_z, A_{\bar{z}}] = 0$. This flatness condition is what makes (1.10) a good parametrization, since it makes U invariant under small deformations of the path C on Σ. From (1.11), it follows that

$$\mathcal{A}_z = -\partial_z U U^{-1} \text{ and } \mathcal{A}_{\bar{z}} = U^{\dagger -1} \partial_{\bar{z}} U^\dagger. \tag{1.12}$$

Now, we make an infinitesimal gauge transformation on ψ, as

$$\begin{aligned} \delta_\epsilon \psi[A_{\bar{z}}] &= \int d^2 z \; \delta_\epsilon A^a_{\bar{z}} \, \frac{\delta \psi}{\delta A^a_{\bar{z}}} \\ &= \int d^2 z \; \epsilon^a \left(\partial_{\bar{z}} \frac{\delta}{\delta A^a_{\bar{z}}} + i f^{abc} A^b_{\bar{z}} \frac{\delta}{\delta A^c_{\bar{z}}} \right) \psi \\ &= -\frac{k}{2\pi} \int d^2 z \; \epsilon^a (F^a_{z\bar{z}} - \partial_z A^a_{\bar{z}}) \psi. \end{aligned} \tag{1.13}$$

After applying the Gauss law, one gets

$$\delta_\epsilon \psi[A_{\bar{z}}] = \frac{k}{2\pi} \int d^2 z \; \epsilon^a \partial_z A^a_{\bar{z}} \psi[A_{\bar{z}}], \tag{1.14}$$

which is solved by[**10, 11**]

$$\psi[A_{\bar{z}}] = exp\big(k S_{WZW}(U)\big). \tag{1.15}$$

1.2.2. *The Gauge Invariant Measure.* The metric of the space of gauge potentials ($\mathscr{A}$) is given by [**12**]

$$\begin{aligned} ds^2_{\mathscr{A}} &= \int d^2 x \; \delta A^a_i \delta A^a_i = -8 \int Tr(\delta A_{\bar{z}} \delta A_z) \\ &= 8 \int Tr[D_{\bar{z}}(\delta U U^{-1}) D_z(U^{\dagger -1} \delta U^\dagger)]. \end{aligned} \tag{1.16}$$

This metric is similar to the Cartan-Killing metric for $SL(N, \mathbb{C})$, which is given by

$$ds^2_{SL(N,\mathbb{C})} = 8 \int Tr[(\delta U U^{-1})(U^{\dagger -1} \delta U^\dagger)]. \tag{1.17}$$

The volumes of these two spaces are related by

$$d\mu(\mathscr{A}) = det(D_{\bar{z}}D_z)d\mu(U, U^\dagger). \tag{1.18}$$

To make this measure gauge invariant, we need to define a new gauge invariant matrix $H = U^\dagger U$, which is an element of the coset $SL(N,\mathbb{C})/SU(N)$. To integrate over only the gauge invariant combinations of U and $U^\dagger$, we write

$$d\mu(\mathscr{A}) = det(D_{\bar{z}}D_z)d\mu(H). \tag{1.19}$$

The determinant is given by[**7**, **8**]

$$det(D_{\bar{z}}D_z) = constant \times e^{2c_A S_{WZW}(H)} \tag{1.20}$$

where c_A is the quadratic Casimir in the adjoint representation.

Now, using the Polyakov-Wiegmann(PW)[**10**, **11**] identity, we can finally write the inner product

$$\langle\psi|\psi\rangle = \int d\mu(\mathscr{A})\ e^{-K}\ \psi^*\psi = \int d\mu(H)\ e^{(2c_A+k)S_{WZW}(H)}. \tag{1.21}$$

1.2.3. *Wilson Loops.* In the temporal gauge, finding the expectation value of a Wilson loop using geometric quantization is problematic. In this gauge, the operator is given by

$$W_R(C) = Tr_R\ \mathcal{P}\ e^{-\oint\limits_c (A_z dz + A_{\bar{z}} d\bar{z})}. \tag{1.22}$$

The derivative operator A_z in the Wilson loop makes it difficult to calculate the path ordered exponential acting on the wave-functional. To go around this problem paying almost no price, we will use a Wilson loop-like observable $\mathcal{W}(C) = Tr\ U(x,x,C)$, where U is given by (1.11). We can write $\mathcal{W}$ as

$$\mathcal{W}_R(C) = Tr_R\ \mathcal{P}\ e^{-\oint\limits_c (\mathcal{A}_z dz + A_{\bar{z}} d\bar{z})}. \tag{1.23}$$

$\mathcal{A}_z$ is defined as $\partial_z A_{\bar{z}} - \partial_{\bar{z}}\mathcal{A}_z + [\mathcal{A}_z, A_{\bar{z}}] = 0$. Since Gauss' law forces $F_{z\bar{z}} = 0$, we can say that $\mathcal{W}$ is the Wilson loop on the constraint hypersurface.

Since the theory is given by $S_{WZW}(H)$, WZW currents $J_{\bar{z}} = -\partial_{\bar{z}}HH^{-1}$ and $J_z = H^{-1}\partial_z H$ can be used to write gauge invariant observables. $\mathcal{A}_z$ and $A_{\bar{z}}$ can be written as

$$\begin{aligned} A_{\bar{z}} &= -\,\partial_{\bar{z}}UU^{-1} = U^{\dagger -1}J_{\bar{z}}U^\dagger + U^{\dagger -1}\partial_{\bar{z}}U^\dagger, \\ \mathcal{A}_z &= -\,\partial_z UU^{-1} = U^{\dagger -1}J_z U^\dagger + U^{\dagger -1}\partial_z U^\dagger, \end{aligned} \tag{1.24}$$

which are $SL(N,\mathbb{C})$ transformed WZW currents. With this information, we can write (1.23) in terms of H as

$$\mathcal{W}_R(C,H) = Tr_R\ \mathcal{P}\ e^{\oint\limits_c (\partial_z HH^{-1}dz + \partial_{\bar{z}}HH^{-1}d\bar{z})}. \tag{1.25}$$

Now that $\mathcal{W}$ is written in terms of H and it commutes with the wave-functional, we can write its expectation value as

$$\langle\mathcal{W}(C)\rangle = \int d\mu(H)\ e^{(2c_A+k)S_{WZW}(H)}\mathcal{W}(C,H). \tag{1.26}$$

2. Topologically Massive Yang-Mills Theory

The TMYM action is given by

$$
\begin{aligned}
S_{TMYM} =& S_{CS} + S_{YM} \\
=& -\frac{k}{4\pi} \int\limits_{\Sigma\times[t_i,t_f]} d^3x\ \epsilon^{\mu\nu\alpha}\ Tr\left(A_\mu\partial_\nu A_\alpha + \frac{2}{3}A_\mu A_\nu A_\alpha\right) \\
& - \frac{k}{4\pi}\frac{1}{4m} \int\limits_{\Sigma\times[t_i,t_f]} d^3x\ Tr\ F_{\mu\nu}F^{\mu\nu}.
\end{aligned} \tag{2.1}
$$

By defining

$$\tilde{A}_\mu = A_\mu + \frac{1}{2m}\epsilon_{\mu\alpha\beta}F^{\alpha\beta}, \tag{2.2}$$

the conjugate momenta are given by

$$\Pi^{az} = \frac{ik}{4\pi}\tilde{A}^a_{\bar{z}} \quad\text{and}\quad \Pi^{a\bar{z}} = -\frac{ik}{4\pi}\tilde{A}^a_z. \tag{2.3}$$

Using $E_z = \frac{i}{2m}F^{0\bar{z}}$ and $E_{\bar{z}} = -\frac{i}{2m}F^{0z}$, components of $\tilde{A}$ can be written as $\tilde{A}_z = A_z + E_z$ and $\tilde{A}_{\bar{z}} = A_{\bar{z}} + E_{\bar{z}}$.

The symplectic two-form of the theory is given by

$$\Omega = \frac{ik}{4\pi}\int\limits_\Sigma (\delta\tilde{A}^a_{\bar{z}}\delta A^a_z + \delta A^a_{\bar{z}}\delta\tilde{A}^a_z). \tag{2.4}$$

From (2.4), it can be seen that the phase space of TMYM theory consists of two CS-like halves. This becomes more clear in coordinates $B_z = \frac{1}{2}(A_1 + i\tilde{A}_2)$, $C_z = \frac{1}{2}(\tilde{A}_1 + iA_2)$. In these coordinates the sympectic two-form is in the form of $\delta B_z\delta B_{\bar{z}} + \delta C_z\delta C_{\bar{z}}$.

2.1. The Wave-Functional.

Choosing the holomorphic polarization leads to $\Phi[A_z, A_{\bar{z}}, \tilde{A}_z, \tilde{A}_{\bar{z}}] = e^{-\frac{1}{2}K}\psi[A_{\bar{z}}, \tilde{A}_{\bar{z}}]$, where $K = \frac{k}{4\pi}\int_\Sigma(\tilde{A}^a_{\bar{z}}A^a_z + A^a_{\bar{z}}\tilde{A}^a_z)$ is the Kähler potential. Upon quantization, we write

$$A^a_z\psi = \frac{4\pi}{k}\frac{\delta}{\delta\tilde{A}^a_{\bar{z}}}\psi \quad\text{and}\quad \tilde{A}^a_z\psi = \frac{4\pi}{k}\frac{\delta}{\delta A^a_{\bar{z}}}\psi. \tag{2.5}$$

2.1.1. *The Gauss Law.* An infinitesimal gauge transformation on $\psi[A_{\bar{z}}, \tilde{A}_{\bar{z}}]$ is given by

$$\delta_\epsilon\psi[A_{\bar{z}}, \tilde{A}_{\bar{z}}] = \int d^2z\ \left(\delta_\epsilon A^a_{\bar{z}}\frac{\delta\psi}{\delta A^a_{\bar{z}}} + \ \delta_\epsilon\tilde{A}^a_{\bar{z}}\frac{\delta\psi}{\delta\tilde{A}^a_{\bar{z}}}\right). \tag{2.6}$$

Using (2.5) and $\delta A^a_{\bar{z}} = D_{\bar{z}}\epsilon^a$, $\delta\tilde{A}^a_{\bar{z}} = \tilde{D}_{\bar{z}}\epsilon^a$, we get

$$
\begin{aligned}
\delta_\epsilon\psi =& \int d^2z\ \epsilon^a\left(\tilde{D}_{\bar{z}}\frac{\delta}{\delta\tilde{A}^a_{\bar{z}}} + D_z\frac{\delta}{\delta A^a_{\bar{z}}}\right)\psi \\
=& \frac{k}{4\pi}\int d^2z\ \epsilon^a\left(\partial_z\tilde{A}^a_{\bar{z}} + \partial_z A^a_{\bar{z}} - 2F_{z\bar{z}} - D_z E_{\bar{z}} + D_{\bar{z}}E_z\right)\psi
\end{aligned} \tag{2.7}
$$

For TMYM theory, Gauss' law is given by $G^a\psi = 0$, where $G^a = \frac{ik}{4\pi}(2F_{z\bar{z}} + D_z E_{\bar{z}} - D_{\bar{z}} E_z)$. After applying Gauss' law, the gauge transformation becomes

$$\delta_\epsilon \psi = \frac{k}{4\pi}\int d^2z\ \epsilon^a \left(\partial_z \tilde{A}^a_{\bar{z}} + \partial_z A^a_{\bar{z}}\right)\psi. \tag{2.8}$$

This condition is very similar to (1.14) and can be solved by $\psi = \phi\chi$, with

$$\phi[A_{\bar{z}}, \tilde{A}_{\bar{z}}] = exp\left[\frac{k}{2}(S_{WZW}(\tilde{U}) + S_{WZW}(U))\right]. \tag{2.9}$$

Here we used the parametrization

$$\tilde{U}(x, 0, C) = \mathcal{P}exp\left(-\int_{\substack{0\\C}}^{x} (\tilde{A}_{\bar{z}} d\bar{z} + \tilde{\mathcal{A}}_z dz)\right), \tag{2.10}$$

which can be followed by the tilde versions of the equations (1.10) and (1.12).

2.1.2. *Schrödinger's Equation.* With $\alpha = \frac{4\pi}{k}$, $B = \frac{ik}{2\pi}F^{z\bar{z}}$ and using Euclidean metric, the Hamiltonian for TMYM is given by

$$\mathcal{H} = \frac{m}{2\alpha}(E^a_{\bar{z}} E^a_z + E^a_z E^a_{\bar{z}}) + \frac{\alpha}{m} B^a B^a. \tag{2.11}$$

E-fields satisfy the commutator

$$[E^a_z(x), E^b_{\bar{z}}(x')] = -2\alpha\ \delta^{ab}\delta^{(2)}(x - x'). \tag{2.12}$$

This allows us to interpret E^a_z as an annihilation operator and $E^b_{\bar{z}}$ as a creation operator [**13**]. To get rid of the infinity, the Hamiltonian can be normal ordered as

$$\mathcal{H} = \frac{m}{\alpha} E^a_{\bar{z}} E^a_z + \frac{\alpha}{m} B^a B^a. \tag{2.13}$$

At large distances compared to $1/m$, it is the standard practice to neglect the B^2 term[**13**]. Since we are interested in the *near* CS limit of the theory, we take m to be large and ignore the potential energy term. In that case, the vacuum wave-functional is given by $E_z\psi = 0$. In terms of the derivative operator, $E_z = \tilde{A}_z - A_z$ can be written as

$$E^a_z \psi = \frac{4\pi}{k}\left(\frac{\delta\phi}{\delta A^a_{\bar{z}}} - \frac{\delta\phi}{\delta \tilde{A}^a_{\bar{z}}}\right)\chi + \frac{4\pi}{k}\left(\frac{\delta\chi}{\delta A^a_{\bar{z}}} - \frac{\delta\chi}{\delta \tilde{A}^a_{\bar{z}}}\right)\phi. \tag{2.14}$$

The derivatives of ϕ are given by $\mathcal{A}$ and $\tilde{\mathcal{A}}$[**8**] as

$$\tilde{A}^a_z \phi = \frac{4\pi}{k}\frac{\delta\phi}{\delta A^a_{\bar{z}}} = \mathcal{A}^a_z \phi \text{ and } A^a_z \phi = \frac{4\pi}{k}\frac{\delta\phi}{\delta \tilde{A}^a_{\bar{z}}} = \tilde{\mathcal{A}}^a_z \phi. \tag{2.15}$$

Now by defining $\mathcal{E}_z = \tilde{\mathcal{A}}_z - \mathcal{A}_z$ we can write

$$E^a_z \psi = -\mathcal{E}^a_z \psi + \frac{4\pi}{k}\left(\frac{\delta\chi}{\delta A^a_{\bar{z}}} - \frac{\delta\chi}{\delta \tilde{A}^a_{\bar{z}}}\right)\phi. \tag{2.16}$$

Solving $E_z\psi = 0$ for χ gives

$$\chi = exp\left(-\frac{k}{8\pi}\int_\Sigma (\tilde{A}^a_{\bar{z}} - A^a_{\bar{z}})\mathcal{E}^a_z\right) = exp\left(-\frac{k}{8\pi}\int_\Sigma E^a_{\bar{z}}\mathcal{E}^a_z\right). \tag{2.17}$$

Since both E and $\mathcal{E}$ are first order in $1/m$, $\chi = 1 + \mathcal{O}(1/m^2)$. Thus χ can be taken as unity at large enough distances compared to $1/m$.

2.1.3. *The Gauge Invariant Measure.* For TMYM theory, the metric of the space of gauge potentials is given by

$$\begin{aligned} ds^2_{\mathscr{A}} &= -4\int Tr(\delta\tilde{A}_{\bar{z}}\delta A_z + \delta A_{\bar{z}}\delta\tilde{A}_z) \\ &= 4\int Tr[\tilde{D}_{\bar{z}}(\delta\tilde{U}\tilde{U}^{-1})D_z(U^{\dagger-1}\delta U^{\dagger}) + D_{\bar{z}}(\delta U U^{-1})\tilde{D}_z(\tilde{U}^{\dagger-1}\delta\tilde{U}^{\dagger})]. \end{aligned} \tag{2.18}$$

The gauge invariant measure for this case is

$$d\mu(\mathscr{A}) = det(\tilde{D}_{\bar{z}}D_z)det(D_{\bar{z}}\tilde{D}_z)d\mu(\tilde{U}^{\dagger}U)d\mu(U^{\dagger}\tilde{U}). \tag{2.19}$$

For a certain choice of local counter terms, the determinants can be written as

$$det(\tilde{D}_{\bar{z}}D_z)det(D_{\bar{z}}\tilde{D}_z) = constant \times e^{2c_A\left(S_{WZW}(\tilde{U}^{\dagger}U)+S_{WZW}(U^{\dagger}\tilde{U})\right)}. \tag{2.20}$$

To simplify the notation we define a new gauge invariant matrix $N = \tilde{U}^{\dagger}U$. Now, the measure becomes

$$d\mu(\mathscr{A}) = constant \times e^{2c_A\left(S_{WZW}(N)+S_{WZW}(N^{\dagger})\right)}d\mu(N)d\mu(N^{\dagger}). \tag{2.21}$$

2.1.4. *Chern-Simons Splitting.* To find the inner product, using PW identity we write

$$e^{-K_{TMYM}}\psi^*_{TMYM}\psi_{TMYM} = e^{\frac{k}{2}\left(S_{WZW}(N)+S_{WZW}(N^{\dagger})\right)}\chi^*\chi. \tag{2.22}$$

We have shown that $\chi^*\chi = 1 + \mathcal{O}(1/m^2)$. Then the inner product for the vacuum state in the near CS limit is

$$\langle\psi|\psi\rangle_{TMYM} = \int d\mu(N)d\mu(N^{\dagger})\ e^{(2c_A+\frac{k}{2})\left(S_{WZW}(N)+S_{WZW}(N^{\dagger})\right)} + \mathcal{O}(1/m^2). \tag{2.23}$$

Using (1.21), this inner product can be written as two CS theories with half the level, as

$$\langle\psi|\psi\rangle_{TMYM_k} = \langle\psi|\psi\rangle_{CS_{k/2}}\langle\psi|\psi\rangle_{CS_{k/2}} + \mathcal{O}(1/m^2). \tag{2.24}$$

Here the labels k and $k/2$ indicate the levels of the CS terms in the Lagrangian. It is well known that k has to be an integer to ensure gauge invariance of CS. Here, we have two half level CS parts and each piece transforms as $\frac{1}{2}S_{CS}(A^g) \to \frac{1}{2}S_{CS}(A) + \pi k\omega(g)$ where $\omega(g)$ is the winding number. Then, the sum of the two will bring an extra $2\pi k\omega(g)$ that will not change the value of the path integral, even for odd values of k. But if one wants to write operator expectation values of TMYM theory in terms of CS expectation values using CS splitting, this can only be done for even values of k.

3. Wilson Loops in TMYM Theory

Since $\tilde{A}$ transforms like a gauge field, we can define a Wilson loop-like observable with it, as

$$T_R(C) = Tr_R\ \mathcal{P}\ e^{-\oint_c \tilde{A}_\mu dx^\mu}. \tag{3.1}$$

To make a physical interpretation of this new observable, we will check and see if it satisfies a 't Hooft-like algebra with the Wilson loop. To simplify the calculation we will consider the abelian versions of the loop operators. In complex coordinates with temporal gauge, the operators are

$$W(C) = e^{i \oint_c (A_z dz + A_{\bar{z}} d\bar{z})} \quad \text{and} \quad T(C) = e^{i \oint_c (\tilde{A}_z dz + \tilde{A}_{\bar{z}} d\bar{z})}. \tag{3.2}$$

These operators satisfy the following 't Hooft-like algebra

$$T(C_1)W(C_2) = e^{\frac{2\pi i}{k} l(C_1, C_2)} W(C_2)T(C_1), \tag{3.3}$$

where $l(C_1, C_2)$ is the intersection number of C_1 and C_2 , which can only take values $0, \pm 1$. This allows us to interpret T as a 't Hooft-like operator for TMYM theory. In this work, we will only consider loops that have zero intersection number.

Instead of directly using T and W, we will use $Tr\ U(x, x, C)$ and $Tr\ \tilde{U}(x, x, C)$ or

$$\mathcal{W}_R(C) = Tr_R\ \mathcal{P}\ e^{-\oint_c (\mathcal{A}_z dz + \mathcal{A}_{\bar{z}} d\bar{z})} \quad \text{and} \quad \mathcal{T}_R(C) = Tr_R\ \mathcal{P}\ e^{-\oint_c (\tilde{\mathcal{A}}_z dz + \tilde{\mathcal{A}}_{\bar{z}} d\bar{z})} \tag{3.4}$$

to avoid the same problem we had in CS Wilson loops. Once again these can be written using gauge $SL(N, \mathbb{C})$ transformed WZW currents $-\partial_{\bar{z}} N N^{-1}$, $-\partial_z N N^{-1}$, $-\partial_{\bar{z}} N^\dagger N^{\dagger -1}$ and $-\partial_z N^\dagger N^{\dagger -1}$ as

$$\begin{aligned} \mathcal{W}_R(C, N) &= Tr_R\ \mathcal{P}\ e^{\oint_c (\partial_z N N^{-1} dz + \partial_{\bar{z}} N N^{-1} d\bar{z})}, \\ \mathcal{T}_R(C, N^\dagger) &= Tr_R\ \mathcal{P}\ e^{\oint_c (\partial_z N^\dagger N^{\dagger -1} dz + \partial_{\bar{z}} N^\dagger N^{\dagger -1} d\bar{z})}. \end{aligned} \tag{3.5}$$

Using these operators, we can calculate the following expectation value

(3.6)

$$\begin{aligned} \langle \mathcal{W}_{R_1}(C_1) \mathcal{T}_{R_2}(C_2) \rangle &= \int d\mu(\mathscr{A}) \psi_0^* \mathcal{W}_{R_1}(C_1) \mathcal{T}_{R_2}(C_2) \psi_0 \\ &= \int d\mu(N) d\mu(N^\dagger)\ e^{(2c_A + \frac{k}{2})\left(S_{WZW}(N) + S_{WZW}(N^\dagger)\right)}\ \mathcal{W}_{R_1}(C_1, N) \mathcal{T}_{R_2}(C_2, N^\dagger) \\ &\quad + \mathcal{O}(1/m^2). \end{aligned}$$

This leads to interesting equivalences between the observables of TMYM and CS theories:

$$\langle \mathcal{W}_R(C) \rangle_{TMYM_{2k}} = \langle \mathcal{W}_R(C) \rangle_{CS_k} + \mathcal{O}(1/m^2), \tag{3.7a}$$

$$\langle \mathcal{T}_R(C) \rangle_{TMYM_{2k}} = \langle \mathcal{W}_R(C) \rangle_{CS_k} + \mathcal{O}(1/m^2) \tag{3.7b}$$

and

(3.7c)

$$\langle \mathcal{W}_{R_1}(C_1) \mathcal{T}_{R_2}(C_2) \rangle_{TMYM_{2k}} = \Big(\langle \mathcal{W}_{R_1}(C_1) \rangle_{CS_k} \Big) \Big(\langle \mathcal{W}_{R_2}(C_2) \rangle_{CS_k} \Big) + \mathcal{O}(1/m^2).$$

Notice that these equivalences are written for even level number on the TMYM side. Although $\langle \mathcal{W}_R(C_1) \mathcal{T}_R(C_2) \rangle_{TMYM_k}$ is gauge invariant for all integer values of k, (3.7) can only be written for even level number on the TMYM side.

4. Pure Yang-Mills Theory

In previous sections, we have shown that at large distances, TMYM theory exhibits a CS+CS type splitting behavior, analogous to (1.5). This section is a quick summary of how CS-CS splitting can be obtained for pure YM theory at large distances, analogous to (1.6).

The action is given by

$$S_{YM} = -\frac{k}{4\pi}\frac{1}{4m}\int\limits_{\Sigma\times[t_i,t_f]} d^3x\ Tr\ (F_{\mu\nu}F^{\mu\nu}). \tag{4.1}$$

Here, the constant $\frac{k}{4\pi}$ is inserted so that the split CS terms will have level numbers at the end of our calculation. At this point there are no restrictions on k. The symplectic two-form of the theory is

$$\Omega = \frac{ik}{4\pi}\int\limits_{\Sigma}(\delta E^a_{\bar{z}}\delta A^a_z + \delta A^a_{\bar{z}}\delta E^a_z). \tag{4.2}$$

With defining

$$\tilde{A}_i = A_i + E_i \text{ and } \hat{A}_i = A_i - E_i, \tag{4.3}$$

the symplectic two-form can be written as a difference of two half CS-like parts as

$$\Omega = \frac{ik}{4\pi}\int\limits_{\Sigma}(\delta \tilde{A}^a_{\bar{z}}\delta A^a_z - \delta A^a_{\bar{z}}\delta \hat{A}^a_z). \tag{4.4}$$

Using the methods we described in the previous sections, it can be shown that[**14**] the wave-functional is given by

$$\psi[A_{\bar{z}}, \tilde{A}_{\bar{z}}] = exp\left[\frac{k}{2}\left(S_{WZW}(\tilde{U}) - S_{WZW}(U)\right)\right]\chi, \tag{4.5}$$

or equally

$$\psi[A_{\bar{z}}, \hat{A}_{\bar{z}}] = exp\left[\frac{k}{2}\left(S_{WZW}(U) - S_{WZW}(\hat{U})\right)\right]\chi. \tag{4.6}$$

Similar to TMYM theory, it can be shown that $\chi = 1+\mathcal{O}(1/m^2)$, since both TMYM theory and pure YM theory have the same Hamiltonian in the temporal gauge.

By defining new gauge invariant matrices $H_1 = U^\dagger\tilde{U}$ and $H_2 = \hat{U}^\dagger U$, the measure is given by

$$d\mu(\mathscr{A}) = e^{2c_A\left(S_{WZW}(H_1)+S_{WZW}(H_2)\right)}d\mu(H_1)d\mu(H_2). \tag{4.7}$$

Now, using PW identity, the inner product can be written as

$$\langle\psi|\psi\rangle = \int d\mu(H_1)d\mu(H_2)e^{\left(2c_A+\frac{k}{2}\right)S_{WZW}(H_1)+\left(2c_A-\frac{k}{2}\right)S_{WZW}(H_2)} + \mathcal{O}(1/m^2). \tag{4.8}$$

This inner product can also be written in terms of two CS inner products with levels $k/2$ and $-k/2$ as

$$\langle\psi_0|\psi_0\rangle_{YM_k} = \langle\psi|\psi\rangle_{CS_{k/2}}\langle\psi|\psi\rangle_{CS_{-k/2}} + \mathcal{O}(1/m^2). \tag{4.9}$$

This shows that the insight provided by the gravitational analogue theory was correct in this limit as well. Analogous to the pure Hilbert-Einstein limit (1.6), pure YM theory can also be written as a difference of two split half-CS parts at large finite distances.

Although initially there were no restrictions on k, now it may seem like it has to be an even integer to ensure gauge invariance of the split CS terms. This is not necessary, since each CS part transforms like $\frac{1}{2}S_{CS}(A^g) \to \frac{1}{2}S_{CS}(A) + \pi k\omega(g)$ and two $\pi k\omega(g)$ terms cancel each other. But if one wants to write YM observables in terms of CS Wilson loops, similar to our discussion on TMYM observables, this can only be done for even values of k.

5. Conclusions

We have shown that at large enough distances, both near CS limit and pure YM limit of TMYM theory exhibits a CS splitting behavior as predicted by the analogous gravitational theory, topologically massive AdS gravity. In the near CS limit, each split CS part has the level $k/2$. The pure YM theory however, has split parts with levels $k/2$ and $-k/2$. At very large distances, split CS parts of TMYM theory add up to give the original level number k. On the other hand, for pure YM theory, the split parts cancel at very large distances to give a trivial result, as required by the existence of a mass gap.

In section 3, we have shown that the CS splitting can be exploited to write TMYM observables in terms of CS observables. This allows us to use skein relations in order to calculate TMYM Wilson loop expectation values. A similar calculation can be done for YM observables at large enough distances.

Detailed calculations on this work can be found in refs. **14–16**.

Acknowledgements

The author thanks Vincent Rodgers and Parameswaran Nair for their support and supervision. The author would also like to thank Tudor Dimofte and Chris Baesley for helpful discussions, and conference organizers for the support they provided.

References

[1] S. Deser, R. Jackiw, and S. Templeton, *Topologically massive gauge theories*, Ann. Physics **140** (1982), no. 2, 372–411, DOI 10.1016/0003-4916(82)90164-6. MR665601 (84j:81128)

[2] S. Deser, R. Jackiw, and S. Templeton. Three-Dimensional Massive Gauge Theories. *Physical Review Letters*, 48:975–978, 1982.

[3] A. Achúcarro and P. K. Townsend, *A Chern-Simons action for three-dimensional anti-de Sitter supergravity theories*, Phys. Lett. B **180** (1986), no. 1-2, 89–92, DOI 10.1016/0370-2693(86)90140-1. MR865090 (88e:83079)

[4] E. Witten, $2+1$*-dimensional gravity as an exactly soluble system*, Nuclear Phys. B **311** (1988/89), no. 1, 46–78, DOI 10.1016/0550-3213(88)90143-5. MR974271 (90a:83041)

[5] S. Carlip, S. Deser, A. Waldron, and D. K. Wise, *Topologically massive AdS gravity*, Phys. Lett. B **666** (2008), no. 3, 272–276, DOI 10.1016/j.physletb.2008.07.057. MR2445839 (2009h:83109)

[6] S. Carlip, S. Deser, A. Waldron, and D. K. Wise, *Cosmological topologically massive gravitons and photons*, Classical Quantum Gravity **26** (2009), no. 7, 075008, 24, DOI 10.1088/0264-9381/26/7/075008. MR2512699 (2010e:83045)

[7] M. Bos and V. P. Nair, *Coherent state quantization of Chern-Simons theory*, Internat. J. Modern Phys. A **5** (1990), no. 5, 959–988, DOI 10.1142/S0217751X90000453. MR1035400 (91a:81203)

[8] V. P. Nair, *Quantum field theory*, Graduate Texts in Contemporary Physics, Springer, New York, 2005. A modern perspective. MR2124673 (2006d:81001)

[9] D. Karabali and V. P. Nair, *Gauge invariance and mass gap in* (2+1)*-dimensional Yang-Mills theory*, Proceedings of the Workshop on Low-dimensional Field Theory (Telluride, CO, 1996), Internat. J. Modern Phys. A **12** (1997), no. 6, 1161–1171, DOI 10.1142/S0217751X9700089X. MR1474287 (98k:81249)

[10] A. Polyakov and P. B. Wiegmann, *Theory of nonabelian Goldstone bosons in two dimensions*, Phys. Lett. B **131** (1983), no. 1-3, 121–126, DOI 10.1016/0370-2693(83)91104-8. MR722389 (85a:81067)

[11] A. M. Polyakov and P. B. Wiegmann, *Goldstone fields in two dimensions with multivalued actions*, Phys. Lett. B **141** (1984), no. 3-4, 223–228, DOI 10.1016/0370-2693(84)90206-5. MR750642 (85m:81104b)

[12] D. Karabali and V. P. Nair, *A gauge-invariant Hamiltonian analysis for non-abelian gauge theories in* $(2+1)$ *dimensions*, Nuclear Phys. B **464** (1996), no. 1-2, 135–152, DOI 10.1016/0550-3213(96)00034-X. MR1390947 (98b:81143)

[13] G. Grignani, G. Semenoff, P. Sodano, and O. Tirkkonen, *G/G models as the strong coupling limit of topologically massive gauge theory*, Nuclear Phys. B **489** (1997), no. 1-2, 360–384, DOI 10.1016/S0550-3213(97)00050-3. MR1443808 (98f:81300)

[14] T. Yildirim. Chern-Simons Splitting of 2+1D Pure Yang-Mills Theory at Large Distances. *arXiv preprint arXiv:1410.8593*, 2014.

[15] T. Yildirim, *Topologically massive Yang-Mills theory and link invariants*, Internat. J. Modern Phys. A **30** (2015), no. 7, 1550034, 18, DOI 10.1142/S0217751X15500347. MR3318974

[16] T. Yildirim, *Topologically massive yang-mills theory and link invariants*, ProQuest LLC, Ann Arbor, MI, 2014. Thesis (Ph.D.)–The University of Iowa. MR3322059

Physics and Astronomy Department, The University of Iowa, Iowa City, Iowa 52242
Current address: Department of Physics, Arizona State University, Tempe, Arizona 85287
E-mail address: `tuna.yildirim@asu.edu`

Proceedings of Symposia in Pure Mathematics
Volume **93**, 2016
http://dx.doi.org/10.1090/pspum/093/01568

A strange family of Calabi-Yau 3-folds

Howard J. Nuer and Patrick Devlin

ABSTRACT. We study the predictions of mirror symmetry for the 1-parameter family of Calabi-Yau 3-folds $\tilde{X}$ with hodge numbers $h^{11} = 31, h^{21} = 1$ constructed by Borisov and Nuer. We calculate the Picard-Fuchs differential equation associated to this family, and use it to predict the instanton numbers on the hypothetical mirror. These exhibit a strange vanishing in odd degrees. We also calculate the monodromy action on $H^3(\tilde{X}, \mathbb{Q})$ and find that it strangely predicts a positive Euler characteristic for its mirror. From a degenerate fiber of our family we construct a new rigid Calabi-Yau 3-fold. In an appendix we prove the expansion of the conifold period conjectured to hold for all 1-parameter families.

1. Introduction

Ever since [**COGP**] the mathematical ramifications of mirror symmetry have revolutionized algebraic geometry and have been heartily pursued by many mathematicians. The mirror symmetry of Calabi-Yau threefolds with one-dimensional complex moduli space, that is $h^{21} = 1$, are particularly interesting since predictions about the mirror Calabi-Yau threefold are easily obtained from the so-called Picard-Fuchs equation. Defined as a fourth-order differential equation

$$(A_4(z)D_z^4 + A_3(z)D_z^3 + A_2(z)D_z^2 + A_1(z)D_z + A_0(z))f(z) = 0,$$

satisfied by the periods

$$f(z) = \int_{\gamma(z)} \Omega(z)$$

associated to the one-parameter family, the Picard-Fuchs equation can be used to calculate the Yukawa coupling for the variation of Hodge structure of this one-parameter family, the monodromy action on the third cohomology group, as well as the mirror map. Using the latter, one obtains enumerative predictions for the number of rational curves on the mirror Calabi-Yau threefold.

Studying individual examples of families of Calabi-Yau threefolds and their mirror symmetry has been a crucial step in formulating mirror symmetry as a rigorous mathematical discipline. New examples shape and hone our understanding of the validity of techniques and definitions which have become standard in the industry by now, but are still only conjectures at the end of the day. It is with the hope of broadening our understanding of mirror symmetry that we present here a

2010 *Mathematics Subject Classification.* Primary 14J33; Secondary 14J32, 14J28.
Key words and phrases. Picard-Fuchs equation, Yukawa coupling.

geometric example which seems to clash at every turn with the usual prescriptions of mirror symmetry. The authors have only seen examples with such anomalies coming from Picard-Fuchs equations constructed formally but without a geometric one-parameter family attached to them.

We now give a brief overview of the contents of this paper. In Section 2 we provide the details of the construction of the new family of Calabi-Yau threefolds with hodge numbers (31,1) briefly mentioned in [**BN**]. Consider the matrices

$$\begin{pmatrix} a & b+c & 0 & 0 \\ b-c & a & 0 & 0 \\ 0 & 0 & 0 & 0 \\ 0 & 0 & 0 & 0 \end{pmatrix}, \begin{pmatrix} 0 & 0 & 0 & 0 \\ 0 & a & b+c & 0 \\ 0 & b-c & a & 0 \\ 0 & 0 & 0 & 0 \end{pmatrix},$$

$$\begin{pmatrix} 0 & 0 & 0 & 0 \\ 0 & 0 & 0 & 0 \\ 0 & 0 & a & b+c \\ 0 & 0 & b-c & a \end{pmatrix}, \begin{pmatrix} a & 0 & 0 & b-c \\ 0 & 0 & 0 & 0 \\ 0 & 0 & 0 & 0 \\ b+c & 0 & 0 & a \end{pmatrix}, a, b, c \in \mathbb{C}$$

as sections $s_1, ..., s_4$ of the bundle $Q \otimes Q$ on $G(2,4)$, the Grassmannian of planes in $\mathbb{C}^4$, where Q is the tautological quotient bundle. The family of determinantal varieties in $G(2,4)$ given by the vanishing of $\det(s_1, ..., s_4)$ turns out to be a one-parameter family of nodal Calabi-Yau threefolds, and in Theorem 2.1 we prove that they exhibit small resolutions with hodge numbers (31,1). We also investigate the singular fibers of this family and in doing so identify a good candidate for a point of maximally unipotent monodromy (MUM point for short), as well as a new rigid Calabi-Yau threefold.

In Section 3 we recall the necessary background on variation of Hodge structure to understand the Picard-Fuchs equation, and we determine it for the one-parameter family constructed in Section 2. We show that the candidate from Section 2 is indeed an MUM point. We also identify the existence of a second singular point which is not an MUM point in the strict sense but may nevertheless correspond to some interesting phenomena in mirror symmetry.

We use the results of Section 3 to make predictions in Section 4.1 about the instanton numbers of the conjectured mirror family to the one constructed in this paper by computing the A-model Yukawa coupling of the mirror. It is here that we first encounter some of the anomalies of our family as the Yukawa coupling is an even function, indicating that all odd degree Gopakumar-Vafa invariants vanish. This is the first example known to the authors of such a strange prediction, and we know of no actual varieties exhibiting this behavior. We offer in this section a possible explanation for this strange behavior coming from torsion in the second homology of the mirror. We also discuss the possibility of a second mirror variety and make similar instanton number predictions for this point. This Yukawa coupling corresponds to the one associated to equation 110 in the list of [**AESZ**] although the Picard-Fuchs equation and monodromy are different. We observe a very surprising relationship between these two Yukawa couplings, namely

$$2\kappa_{ttt}^{\infty}(q^2) = \kappa_{ttt}^{0}(q),$$

where $\kappa_{ttt}^{p}(q)$ is the Yukawa coupling at the point p considered as a power series in q.

Inspired by [**ES**] we calculate the monodromy action on the middle cohomology associated to our one-parameter family, as well as the conifold period, in Section 5.

In an attempt to glean information about the hypothetical mirror we used the predictions of homological mirror symmetry to translate these monodromy matrices and conifold period into information about the basic invariants of this mirror. Specifically, we predict the size of its fundamental group, $H^3, c_2 \cdot H$, and c_3, where H is the ample generator of the rank one Picard group of the mirror. Here we find another strange facet of our family: homological mirror symmetry predicts a positive Euler characteristic of 48, which is impossible for a Calabi-Yau threefold with $h^{11} = 1, h^{21} = 31$. In the end we have been unable to determine a mirror.

In an appendix, we present a proof of the form of the conifold period,

$$z_2(t) = \frac{H^3}{6}t^3 + \frac{c_2(Y).H}{24}t + \frac{c_3(Y)}{(2\pi i)^3}\zeta(3) + O(q),$$

which was conjectured to hold in [**ES**] for all one-parameter families of Calabi-Yau threefolds.

Acknowledgements. Our debt to the papers [**ES**] and [**R**] cannot be overstated. The techniques and arguments here were greatly influenced by those presented there. The first author would like to thank his advisor Lev Borisov for his continued guidance and support. We are also extremely happy to thank Donu Arapura, Jim Bryan, Hiroshi Iritani, Atsushi Kanazawa, Anatoly Libgober, and Wadim Zudilin for invaluable discussions and insight into different aspects of mirror symmetry. Special thanks go to Duco van Straten and his two students Michael Bogner and Joerg Hofmann for helpful discussions and their confirmation of the computer calculations obtained here. We are also grateful to Mike Stillman and Dan Grayson for the program *Macaulay2* [**GS**] which was crucial for investigating the geometry of our family. The first author was partially supported by NSF Grant DMS 1201466.

2. Construction and Basic Invariants of the (31,1) family

We recall here the construction of a family of $(31, 1)$ Calabi-Yau threefolds via the degeneration of the $(2, 32)$ family of Calabi-Yau threefolds constructed in [**BN**]. We also fill in the details that were glossed over there and describe the basic invariants of this family.

Consider the 4×4 matrices

$$s_1 = \begin{pmatrix} a & b+c & 0 & 0 \\ b-c & a & 0 & 0 \\ 0 & 0 & 0 & 0 \\ 0 & 0 & 0 & 0 \end{pmatrix}, \quad s_2 = \begin{pmatrix} 0 & 0 & 0 & 0 \\ 0 & a & b+c & 0 \\ 0 & b-c & a & 0 \\ 0 & 0 & 0 & 0 \end{pmatrix},$$

$$s_3 = \begin{pmatrix} 0 & 0 & 0 & 0 \\ 0 & 0 & 0 & 0 \\ 0 & 0 & a & b+c \\ 0 & 0 & b-c & a \end{pmatrix}, \quad s_4 = \begin{pmatrix} a & 0 & 0 & b-c \\ 0 & 0 & 0 & 0 \\ 0 & 0 & 0 & 0 \\ b+c & 0 & 0 & a \end{pmatrix},$$

with $a, b, c \in \mathbb{C}$. We may view these as global sections of $Q \otimes Q$, where Q is the universal quotient bundle on the Grassmannian $G(2, 4)$ of planes in $\mathbb{C}^4$. Consider the determinantal variety $X_{a,b,c} \subset G(2, 4)$ given by the vanishing of the determinant $D := \det(s_1, s_2, s_3, s_4)$. For the sake of clarity we write down explicitly the equations for this variety as a complete intersection in $\mathbb{P}^5$ of a quartic hypersurface with $G(2, 4)$. Pulling back D and writing it in terms of the Plücker coordinates, which we denote by $y_0, ..., y_5$, one finds this family can be written as the intersection of $G(2,4)$, defined by $h_1 = y_2y_3 - y_1y_4 + y_0y_5 = 0$, and the quartic hypersurface given by

$$h_2 = c(a^3y_0^2y_1y_2 - a^3y_0y_1y_2^2 + 2a^2by_0^2y_1y_4 + a^3y_0y_1^2y_4 + 2a^2by_1^3y_4-$$
$$2a^2by_0y_1y_2y_4 + 4ab^2y_1^2y_2y_4 + 2a^2by_1y_2^2y_4 + a^3y_0^2y_3y_4 + 2a^2by_0y_1y_3y_4+$$
$$4ab^2y_1^2y_3y_4 + a^3y_0y_3^2y_4 + 2a^2by_1y_3^2y_4 - a^3y_0y_1y_4^2 + 8b^3y_1^2y_4^2 + 4ab^2y_1y_2y_4^2$$
$$+ 4ab^2y_1y_3y_4^2 + 2a^2by_1y_4^3 + 4ab^2y_0^2y_1y_5 - a^3y_0^2y_1y_5 - 4ab^2y_0y_1y_2y_5+$$
$$a^3y_0y_1y_2y_5 - 4ab^2y_0y_1y_3y_5 + a^3y_0y_1y_3y_5 + a^3y_1y_3^2y_5 - 4ab^2y_0^2y_4y_5+$$
$$a^3y_0^2y_4y_5 - 16b^3y_0y_1y_4y_5 - a^3y_1^2y_4y_5 - 4ab^2y_0y_2y_4y_5 + a^3y_0y_2y_4y_5-$$
$$2a^2by_1y_2y_4y_5 - a^3y_2^2y_4y_5 - 4ab^2y_0y_3y_4y_5 + a^3y_0y_3y_4y_5 + 2a^2by_1y_3y_4y_5+$$
$$a^3y_1y_4^2y_5 - 4ab^2y_0y_1y_5^2 + a^3y_0y_1y_5^2 + a^3y_1y_3y_5^2 + 4ab^2y_0y_4y_5^2 - a^3y_4y_5^2+$$
$$2a^2by_1y_4y_5^2 + a^3y_2y_4y_5^2) = 0.$$

Notice from the equations that we may assume $c \neq 0$ and scale it to be 1. Moreover, scaling the two parameters a, b simultaneously doesn't change the variety, so we may view this family as being over $\mathbb{P}^1$.

We have the following result enumerating the important properties of $X := X_{a,b}$:

THEOREM 2.1. *For generic choices of a, b we have*

(1) X is an irreducible threefold whose singular locus consists of 118 ordinary double points.

(2) There is a small resolution $\pi : \tilde{X} \to X$ of the ordinary double points, with $\tilde{X}$ a non-singular Calabi-Yau threefold.

(3) $\chi(\tilde{X}) = 60$, $h^{1,1}(\tilde{X}) = 31$, and $h^{2,1}(\tilde{X}) = 1$.

PROOF. (1) We note that X is the complete intersection defined by h_1, h_2 and thus is certainly connected. Irreducibility will follow from the fact that the singular locus is zero dimensional. To see that its singular locus is as claimed for generic choice of parameters, we may check on each standard affine open subset U_i of $G(2,4)$, given by $\{y_i \neq 0\}$, which is isomorphic to $\mathbb{C}^4 = \text{Spec } \mathbb{C}[x_1, ..., x_4]$. $X \cap U_i$ then becomes a hypersurface, say $V(h)$, and the locus of worse-than-nodal points can be described by adding the Hessian of h to the Jacobian ideal of $X \cap U_i$. By eliminating the coordinate variables from this ideal for each i, for example by using *Macaulay2*, one finds that the worse-than-nodal locus on $\mathbb{P}^1$ is given by $a^2(a^2 - b^2)b^2 = 0$. By checking over all parameter values in $\mathbb{P}^1$ over various finite fields one easily sees that 118 is the generic number of nodes. (An exceptional parameter value for which X has more than 118 nodes will be discussed later in the paper).

(2) We construct one such small resolution using *Macaulay2*. First notice that we can scale b to be 1 since $\infty \in \mathbb{P}^1$ has unnodal fiber anyway. Then we consider the intersection of $Sing(X)$ with the union of the coordinate hyperplanes given by $V(y_0 \cdots y_5)$. By calculating the Hilbert polynomial of $Sing(X) \cap V(y_0 \cdots y_5)$, one finds that 114 of the singular points are contained in this locus. We consider the effective Cartier divisor defined by $y_0 \cdots y_5$ on X. Using *Macaulay2*, we can explicitly calculate the irreducible components $\{S_i\}_{i=0,...,18}$ of this Cartier divisor and find that 13 of them, say $S_0, ..., S_{12}$, are smooth surfaces. Moreover, all 114 of the above-mentioned singular points lie on the union of these 13 surfaces, and thus

these are non-Cartier Weil divisors. If we blow up X along S_0, then we obtain a small partial resolution X^0 of the nodes of X that lie on S_0. By considering the proper transform $\tilde{S}_1$ of S_1 on X^0, we proceed to blow up X^0 along $\tilde{S}_1$. Above the nodes we resolved previously X^0 is smooth and thus $\tilde{S}_1$ is Cartier there, so this blow-up has no effect there, and we obtain another small partial resolution X^1 of the nodes of X contained in $S_0 \bigcup S_1$. Proceeding in this way, we find a small partial resolution $\hat{X}$ of 114 of the 118 nodes of X. For the remaining 4 nodes, one notices that they lie on the hyperplane given by $V(y_1 - y_4 - ay_5 + ay_0)$. Again consider the Cartier divisor on X given by $V(y_1 - y_4 - ay_5 + ay_0) \cap X$. It has 4 irreducible components which are all smooth surfaces. By blowing each of these up in succession as before, we resolve the remaining 4 nodes to obtain a small resolution $\tilde{X}$. That it is a Calabi-Yau is standard, but details are given in Section 7 of [**BN**].

(3) The claim about the topological Euler characteristic follows immediately from the fact that a generic (2,4) complete intersection has Euler characteristic -176, and from the number of nodes. The calculation of the Hodge numbers follows from the results of Section 7 in [**BN**] and *Macaulay2* calculations. □

REMARK 2.2. It is worth noting that the small resolution above is not canonical. The order in which the S_i are blown up likely determines different smooth birational models of X related by flops. To obtain a simultaneous small resolution of the family over the locus of $\mathbb{P}^1$ parametrizing fibers with 118 nodes, one must take an étale cover of this locus to eliminate the possible monodromy amongst the S_i. Since the B-model is unaffected by resolution of singularities, we may work on the explicit singular family above and use it as a universal family of objects over the moduli space. This will allow us to use mirror symmetry below to make predictions about the hypothetical mirror.

2.1. Degenerate Fibers. Using *Macaulay2* one can verify that this family becomes reducible for $(a, b) = (0, 1), (\pm 1, 1), (1, 0)$. Over $(0, 1)$ it becomes the union of a degree 4 complete intersection with 6 nodes given by $2y_2y_3 - y_1y_4 = 2y_0y_5 - y_1y_4 = 0$, and two quadric cones given by $y_4 = y_2y_3 + y_0y_5 = 0$ and $y_1 = y_2y_3 + y_0y_5 = 0$, respectively. The decomposition over $(1, 1)$ is harder to see explicitly, but the Hilbert polynomial of its singular locus has degree 2, so it must certainly be reducible there. At infinity, the family becomes the union of a degree 2 component given by $y_0 - y_2 + y_3 + y_5 = -y_2y_3 + y_1y_4 - y_2y_5 + y_3y_5 + y_5^2 = 0$, which is a quadratic cone, and a degree 6 component given by $y_2y_3 - y_1y_4 + y_0y_5 = y_0y_1y_2 + y_0y_3y_4 + y_1y_3y_5 + y_2y_4y_5 = 0$ with 34 nodes.

From the proof of the above theorem, we saw that the nodal singularities of the generic fibre live along the union of the coordinate axes with another moving hyperplane given by $y_1 - y_4 - ay_5 + ay_0$. One can use elimination theory again to determine if there are any fibers which acquire additional singularities (this time by localizing at $y_0 \cdots y_5 \cdot (y_1 - y_4 - ay_5 + ay_0)$ to avoid the known singularities). We indeed find that $(\pm\sqrt{-8}, 1)$ gives such a fiber. By checking over various finite fields fiber by fiber, we became quite certain that there are no others. We'll see below from the Picard-Fuchs equation that this is indeed the case.

2.2. A New Rigid Calabi-Yau. The fiber of our family above $a = \sqrt{-8}$ is a nodal Calabi-Yau threefold with 122 nodes. One can easily verify that it has a crepant resolution, obtained similarly as before, and using the techniques of

Section 7 of [**BN**] we found that the hodge numbers of the resolution are $(34,0)$. To the best of our knowledge this is a new example of a rigid Calabi-Yau threefold. As traditional mirror symmetry does not apply to this variety, it would be interesting to understand this example in the context of generalized mirror symmetry as in [**CDP**].

2.3. Effectivity of this family. Although it is not obvious from looking at the equations, the above family is slightly ineffective in a moduli-theoretic sense since $X_{a,1} \cong X_{-a,1}$:

PROPOSITION 2.3. *There is a natural isomorphism* $X_{a,1} \cong X_{-a,1}$

PROOF. Consider the linear map on $\mathbb{C}^4$ with basis $e_1, ..., e_4$ given by $e_1 \mapsto -e_1, e_2 \mapsto e_2, e_3 \mapsto -e_3, e_4 \mapsto e_4$. Then this has the effect on the matrices s_i of negating the entries which involve b, c while leaving the rest of the matrix unchanged. Thus $X_{a,b,c} \cong X_{a,-b,-c}$ in the original notation. But negating the sign of all entries of each matrix s_i doesn't change the determinantal locus at all, so $X_{a,-b,-c} = X_{-a,b,c}$. Thus indeed $X_{a,1} \cong X_{-a,1}$. □

This tells us that a^2, not a is the natural moduli parameter for this family. We will switch to this parameter shortly.

2.4. Integral homology of $\tilde{X}$. For the sake of completeness and for later use, we record here the important facts about the integral homology of our $\tilde{X}$:

PROPOSITION 2.4. *The nonsingular small resolution $\tilde{X}$ of the complete intersection X is simply connected. Moreover, for its homology we have*
(i) $H_i(\tilde{X}) = H_i(\mathbb{P}^3)$ *for* $i \neq 2,3,4$;
(ii) $H_4(\tilde{X})$ *is torsion free;*
(iii) $\mathrm{Tor}(H_3(\tilde{X})) = \mathrm{Tor}(H_3(X))$.

PROOF. That the complete intersection X is simply connected follows from Corollary 5.2.4 in [**Dim**]. The resolution $\pi : \tilde{X} \to X$ replaces the nodes by simply connected $\mathbb{P}^1$'s, so it too is simply connected. For the statement about homology, we first note that by Theorem 5.4.3 and Corollary 5.4.4 in [**Dim**] we have $H_j(X) = H_j(\mathbb{P}^3)$ for $j \neq 3,4$, and $H_4(X)$ is torsion-free. Now consider the union U of small open balls around the nodes of X and its preimage $V = \pi^{-1}(U)$. Then by excision for the union of the nodes W (the exceptional locus E of π, respectively) we get that $H_i(X-W, U-W) \cong H_i(X,U)$ (respectively, $H_n(\tilde{X}-E, V-E) \cong H_i(\tilde{X}, V)$). Thus clearly $H_n(X,U) \cong H_n(\tilde{X}, V)$ since the excised relative homologies are obviously isomorphic. Using the long exact sequence of homology for the pair $(\tilde{X}, V)$ and the fact that V deformation retracts onto E, topologically a union of S^2's, we get that $H_i(\tilde{X}) \cong H_i(\tilde{X}, V)$ for $i > 3$. We also obtain two exact sequences:

$$0 \to H_1(\tilde{X}) \to H_1(\tilde{X}, V) \to \mathbb{Z}^{117} \to 0, \text{and}$$

$$0 \to H_3(\tilde{X}) \to H_3(\tilde{X}, V) \to \mathbb{Z}^{118} \to H_2(\tilde{X}) \to H_2(\tilde{X}, V) \to 0.$$

Doing the same for the pair (X,U) and using the fact that U deformation retracts onto 118 points, we see that $H_i(X) \cong H_i(X,U)$ for $i > 1$ and we get an exact sequence

$$0 \to H_1(X) \to H_1(X,U) \to \mathbb{Z}^{117} \to 0.$$

It follows from this that $H_i(X) \cong H_i(\tilde{X})$ for $i \neq 2,3$ which proves *(i)* and *(ii)*. Since $H_2(\tilde{X}, V) \cong H_2(X,U) \cong H_2(X) \cong \mathbb{Z}$, we get that $H_2(\tilde{X}) \cong \mathbb{Z} \oplus M$ and $H_3(X) \cong$

$H_3(\tilde{X}) \oplus K$, where $M = \text{im}\,(\mathbb{Z}^{118} \to H_2(\tilde{X}))$ and $K = \ker(\mathbb{Z}^{118} \to H_2(\tilde{X}))$. Since K is free, *(iii)* follows as well. □

3. Picard-Fuchs Equation and Maximally Unipotent Monodromy

3.1. Theory behind the approach. For details and proofs of the techniques used in this section, see [**CK**]. For notational simplicity, we assume $b = 1$ and write our family as $\tilde{X}_a$ with moduli parameter a.

Let $S = \mathbb{P}^1 - \{0, \pm 1, \pm\sqrt{-8}, \infty\}$ be the uncompactified complex moduli space of our family, and let $\mathcal{F}^0 = R^3\pi_*\mathbb{C} \otimes \mathcal{O}_S$ be the induced local system, with subbundles $\mathcal{F}^p$ which correspond to the Hodge filtration on fibers. By the nilpotent orbit theorem of Schmid in [**Sch**], we may canonically extend these bundles to a filtration of bundles

$$0 \subset \overline{\mathcal{F}}^3 \subset \ldots \subset \overline{\mathcal{F}}^0,$$

on the compactification, $\mathbb{P}^1$. Since $\overline{\mathcal{F}}^3$ is a line bundle (as π is a family of Calabi-Yau 3-folds and thus $h^{3,0} = 1$), we may choose a fixed local generator $\Omega \in \overline{\mathcal{F}}^3$ around $a = 0$. The Picard-Fuchs equation is a differential equation satisfied by the periods $f(a) = \int_{\gamma(a)} \Omega(a)$ for $\gamma(a)$ a 3-cycle on $\tilde{X}_a$ moving continuously with a.

To see where the Picard-Fuchs equation comes from, we note that if ∇ is the Gauss-Manin connection on $\mathcal{F}^0$ (we also use the same notation for the logarmithmic extension of it to the compactification, due to [**Del1**]), then according to [**BG**] $\Omega, \nabla_\delta\Omega, \nabla_\delta^2\Omega, \nabla_\delta^3\Omega$ are generically linearly independent, where $\delta = a\frac{d}{da}$ is seen as a tangent vector along S, and thus form a basis in a punctured neighborhood of the origin. Applying ∇_δ once more then gives a relation of the form

$$\nabla_\delta^4\Omega + B_3(a)\nabla_\delta^3\Omega + B_2(a)\nabla_\delta^2\Omega + B_1(a)\nabla_\delta\Omega + B_0(a)\Omega = 0,$$

where the B_i are holomorphic near 0 because Deligne's extension has only regular singular points. We may of course clear denominators to obtain a holomorphic coefficient in front of $\nabla_\delta^4\Omega$, and by restricting to algebraic differentials the coefficients must then be polynomial. Applying this operator to any period $f(a) = \int_{\gamma(a)} \Omega$ shows that it must satisfy a differential equation,

$$\delta^4 f(a) + B_3(a)\delta^3 f(a) + B_2(a)\delta^2 f(a) + B_1(a)\delta f(a) + B_0(a) f(a) = 0,$$

called the Picard-Fuchs equation.

We must also consider the monodromy action $\mathcal{T} : H^3(\tilde{X}, \mathbb{C}) \to H^3(\tilde{X}, \mathbb{C})$ obtained by going around the point $a = 0$, where $\tilde{X}$ is a smooth fiber over a point very close to the origin. According to the monodromy theorem [**Lan**], this linear action is quasi-unipotent, i.e there exist positive integers n, m such that

$$(\mathcal{T}^n - 1)^m = 0, (\mathcal{T}^n - 1)^k \neq 0 \text{ for } k < m,$$

with unipotent index m at most 4, since $\dim H^3(X, \mathbb{C}) = 2h^{3,0} + 2h^{2,1} = 4$. An MUM point is defined by the condition that $n = 1, m = 4$. Thus we can find a basis g_0, g_1, g_2, g_3 of $H^3(\tilde{X}, \mathbb{C})$ such that

$$\mathcal{T} = \begin{pmatrix} 1 & 1 & 0 & 0 \\ 0 & 1 & 1 & 0 \\ 0 & 0 & 1 & 1 \\ 0 & 0 & 0 & 1 \end{pmatrix}.$$

Then we can choose a basis in homology $\gamma_0, \gamma_1, \gamma_2, \gamma_3$ Poincare dual to this basis.

According to the nilpotent orbit theorem, g_0 extends to a single-valued flat section of $\overline{\mathcal{F}}^0$, and accordingly $f_0 = \int_{\gamma_0} \Omega$ extends to a single-valued holomorphic function at $a = 0$. If follows from the form of $\mathcal{T}$ that analytically continuing $f_i = \int_{\gamma_i} \Omega$ around $a = 0$ then gives $f_i + f_{i-1}$. It follows that at an MUM point there should be two cycles γ_0, γ_1 whose periods, f_0, f_1 respectively, satisfy $t := f_1/f_0 = g + \log a$, where g is holomorphic at 0.

We have the following proposition from [**CK**] that allows us to use the Picard-Fuchs equation to determine the type of monodromy we have around a given boundary point:

PROPOSITION 3.1. *Let the Picard-Fuchs equation for a given local section* Ω *of* $\overline{\mathcal{F}}^0$ *be written as*

$$\delta^4 f + B_3(a)\delta^3 f + B_2(a)\delta^2 f + B_1(a)\delta f + B_0(a) f = 0.$$

Then the monodromy action $\mathcal{T}$ *is unipotent if and only if the roots of the indicial equation are integers. Furthermore,* $\mathcal{T}$ *is maximally unipotent if and only if the indical equation is of the form* $(y - l)^4 = 0$ *for some integer* l.

One can check that this integer l is insignificant in that altering the given holomorphic 3-form Ω by multiplying by a meromorphic function $p(a) = c_l a^l + ...$ with $c_l \neq 0$ alters the indicial equation of the new differential equation by replacing y with $y - l$. So the significant part is that all roots of the indicial equation be equal.

3.2. Determining the Picard-Fuchs equation near $a = 0$**.** We saw above that the fiber above $a = 0$ was highly degenerate, breaking into three irreducible components. This suggests that $a = 0$ is a good candidate for an MUM point. Since the B-model (that is the Hodge theoretic side of mirror symmetry) is unchanged via desingularization, we may do all of our calculations on the complete intersection X_a. We may also restrict ourselves to the open affine $U_0 \subset \mathbb{P}^5$ given by $y_0 \neq 0$ and choose a 3-cycle inside the open subset $X_a \cap U_0$ which various continuously with a and against which we will integrate a local section of $\overline{\mathcal{F}}^3$ to obtain a period.

We may further simplify the calculation by considering $X_a \cap U_0$ as the hypersurface in $\mathbb{C}^4$ with coordinates $y_1, ..., y_4$ defined by the equation $h(y_1, ..., y_4) = h_2(1, y_1, ..., y_4, y_1y_4 - y_2y_3)$. The natural choice for a local section of the canonical bundle of X_a on U_0 is then given by the residue of

$$\Psi = (\frac{1}{2\pi i})^4 \frac{dy_1 \wedge ... \wedge dy_4}{h}.$$

We may choose a constant 4-cycle Γ on $\mathbb{C}^4 \backslash X_a \cap U_0$, and then a period can be obtained as

$$f_0(a) = \int_\Gamma \Psi(a).$$

Since all periods must satisfy the Picard-Fuchs equation, the choice of such a 4-cycle is irrelevant for our purposes. One obviously must worry about holomorphicity, but we can choose an appropriate 4-cycle in a moment.

The essence of the approach we take follows that of Rødland in [**R**]. We find that

$$\Psi = \frac{1}{16} \cdot \frac{1}{1 - \sum_i v_i} \cdot \bigwedge \frac{dy_i}{(2\pi i) y_i},$$

where this comes from writing $\frac{y_1 \cdots y_4}{h}$ as

$$\frac{1}{\left(\frac{h}{y_1 \cdots y_4}\right)},$$

and writing the denominator in terms of the Laurent monomials v_i. Then we take the 4-cycle Γ to be a topological torus given $|y_i| = \epsilon_i$ and consider the geometric series expansion of

$$\frac{1}{1 - \sum_i v_i}.$$

For this series to converge and to allow us to manipulate it freely, it suffices to ensure that $\sum |v_i| < 1$. We can do this by choosing $\epsilon_1 = \epsilon_3 = .05, \epsilon_2 = \epsilon_4 = .5$, and $|a| < 1/40$ for example. When expanding the resulting absolutely convergent series as a Laurent series in terms of the v_i, the only terms that contribute to the integral are those without any y_i's. Indeed, if a Laurent monomial has either negative powers or positive powers for some y_i, then when multiplied by $\frac{dy_i}{(2\pi i) y_i}$ the resulting function of y_i being integrated either has a removable singularity (the total exponent would then be at most -2) or is holomorphic, and thus the integral of this function over the closed circle $|y_i| = \epsilon_i$ is 0.

Now generators of the subring of $\mathbb{C}[\{v_i\}]$ generated by those monomials that are independent of the y_i's can be calculated as in the appendix to [**R**], and a closed form for the period $f_0(a)$ can be obtained as Rødland does. Unfortunately, because of the number of Laurent monomials v_i and the fact that the generators of this subring have different powers of a, the resulting form of the period is almost useless and involves an infinite sum in approximately 40 indices. Instead we may calculate iterated residues in *Maple* to compute the power series expansion of f_0 to a large number of terms. We find that

$$f_0(a) = -\frac{1}{16} - \frac{3}{64}a^2 - \frac{81}{2048}a^4 - \frac{143}{4096}a^6 - \frac{66357}{2097152}a^8 - \ldots,$$

where we notice that this function is entirely even. This is what we expect from Proposition 2.3.

Converting to the true moduli parameter $z = a^2$, we use recurrence relations on the coefficients of the power series to determine what polynomials $A_i(z)$ satisfy the Picard-Fuchs equation

$$A_4(z) D_z^4 f_0(z) + \ldots + A_1(z) D_z f_0(z) + A_0(z) f_0(z) = 0,$$

where $D_z = z\frac{d}{dz}$. We find a solution in degree 6 given by the operator

$$\begin{aligned}\mathcal{D} :=& 16(z-1)^3(z-4)^2(z+8)D_z^4 + 96z(z-1)^2(z-4)(z^2-28)D_z^3 \\ &+ 12z(z-1)(18z^4 - 129z^3 - 136z^2 + 2000z - 1024)D_z^2 \\ &+ 36z(192 - 752z + 540z^2 + 99z^3 - 58z^4 + 6z^5)D_z \\ &+ 3z(512 - 2688z + 1824z^2 + 856z^3 - 288z^4 + 27z^5).\end{aligned}$$

We notice from the leading coefficient that as expected the only singular fibers are at $z = 0, 1, \ 8, \infty \in \mathbb{P}^1$. One can check that $z = 4$ does not represent an actual singular fiber, and we'll see that it is the unique vanishing point for the B-model Yukawa coupling of this family. From calculating the indicial equation of this ODE, we immediately get the following proposition:

PROPOSITION 3.2. *The point $z = 0$ is an MUM point of the family of Calabi-Yau 3-folds constructed here.*

3.3. The Picard-Fuchs equation at other singular fibers. In the new coordinate z, we find that there are 3 other singular values to check, $z = 1, -8, \infty$. Translating our ODE accordingly around those points, one finds that the Riemann scheme for our differential operator is

$$P\left\{\begin{array}{cccc} -8 & 0 & 1 & \infty \\ \hline 0 & 0 & 0 & 3/2 \\ 1 & 0 & 0 & 3/2 \\ 1 & 0 & -1/2 & 3/2 \\ 2 & 0 & 1/2 & 3/2 \end{array}\right\}.$$

From Proposition 3.1 we see that the monodromy is only unipotent outside the origin around the point $z = -8$. This value corresponds to where the family acquires extra nodes and is known as a conifold point. That $z = -8$ has spectrum (this is the set of roots of its indicial equation) $\{0, 1, 1, 2\}$ agrees with predictions made in [**ES**] about the kind of monodromy around such a singular fiber and its spectrum.

The spectrum for ∞ is somewhat strange. It is not an MUM point in the strict sense, but if one takes a double cover of $\mathbb{P}^1$, for example branched at 1 and ∞, then the spectrum at ∞ becomes that of a genuine MUM point. At first glance this suggests the existence of a second mirror family to $\tilde{X}_z$. According to homological mirror symmetry this family should conjecturally be derived equivalent to the mirror family corresponding to the MUM point at $z = 0$. The validity of $z = \infty$ as an MUM point will be elaborated on later.

4. Mirror Symmetry and A-model Yukawa couplings

According to mirror symmetry, the B-model (Hodge-theoretic) Yukawa coupling on a Calabi-Yau threefold X gives the A-model (Gromov-Witten) Yukawa coupling on its mirror Y after applying the mirror map.

The A-model Yukawa coupling is defined in terms of the genus zero Gromov-Witten (GW) invariants, and can be seen (in the Picard rank one case) as the generating function of the genus zero Gromov-Witten invariants over all degrees. More specifically, the A-model Yukawa coupling is written as

$$\kappa_{ttt} = H^3 + \sum_{d=1}^{\infty} N_d q^d,$$

where H is the ample generator of $\mathrm{Pic}(Y)$, $N_d := \int_{[\overline{M}_{0,0}(Y,d)]^{virt}} 1$ is the degree d unpointed genus 0 Gromov-Witten invariant, defined by integrating over the virtual fundamental class $[\overline{M}_{0,0}(Y,d)]^{virt}$ of the moduli space of stable maps $\overline{M}_{0,0}(Y,d)$. Conjecturally, these are integers, and this power series can further be written as

$$\kappa_{ttt} = H^3 + \sum_{d=1}^{\infty} n_d \frac{d^3 q^d}{1 - q^d},$$

where the n_d are the Gopakumar-Vafa (GV) invariants, or instanton numbers, which naively should count the number of rational curves on Y of degree d(degree being measured against the ample generator of Pic(Y)). These numbers are also conjectured to be integral. The relation between these two invariants is given by

$$N_d = \sum_{k|d} n_{\frac{d}{k}} k^{-3}.$$

It should be noted that this relation is often the definition of the GV invariants since a rigorous mathematical definition of them does not exist at the moment. The enumerative significance of both the GW and GV invariants is a subtle issue.

If we take f_0 to be the unique holomorphic solution to the Picard-Fuchs equation around an MUM point, and choose a period f_1 such that $t = f_1/f_0 = g + \log z$ with g holomorphic at $z = 0$ and normalized so that $g(0) = 0$, the mirror map is defined by $q = e^t$. Obviously q is determined only up to a constant. Mirror symmetry then predicts that the A-model Yukawa coupling defined above is in fact equal to the B-model Yukawa coupling, that is

$$\kappa_{ttt} = (\frac{d \log z}{dt})^3 \frac{1}{f_0(z)^2} \int_{X_z} \Omega \wedge \nabla^3_{z\frac{d}{dz}} \Omega,$$

where the right hand side is the B-model Yukawa coupling calculated by Hodge theory on the mirror family.

4.1. A-model Yukawa coupling prediction. Applying these ideas to our family, we write g as a power series in z and substite the resulting formula for f_1 back into the Picard-Fuchs equation to get

$$g(z) = \frac{3}{8}z + \frac{81}{512}z^2 + \frac{187}{2048}z^3 + \frac{64797}{1048576}z^4 + ..., \text{ and}$$

$$q = c_2 z(e^g) = c_2(z + \frac{3}{8}z^2 + \frac{117}{512}z^3 + \frac{653}{4096}z^4 + ...).$$

Using the Picard-Fuchs equation and Griffith's transversality, it can be shown that $L = \int_{X_z} \Omega \wedge \nabla^3_{z\frac{d}{dz}} \Omega$ satisfies the differential equation

$$\frac{dL}{L} = -\frac{A_3(z)}{2zA_4(z)},$$

so that in our case

$$L = c_1 \frac{z-4}{(z-1)^3(z+8)}.$$

Here we notice that $z = 4$ is the unique vanishing point of L.

Finally, one can calculate the inverse series giving $z = z(q)$ as a power series in q, and using that $\frac{d \log z}{dt} = \frac{q}{z}\frac{dz}{dq}$, we put everything together to find that according to mirror symmetry

$$\begin{aligned} \kappa^0_{ttt} &= m(2 + 72q^2 - 2232q^4 + 43617q^6 - 7425720q^8 + ...) \\ &= m(2 + 9\frac{2^3q^2}{1-q^2} - 36\frac{4^3q^4}{1-q^4} + 2019\frac{6^3q^6}{1-q^6} + ...), \end{aligned}$$

where we've found the unique choice $c_2 = 1/32$ which makes all of the n_d integral. We've also let $c_1 = 4m$ to scale away the rest of the denominators. This leads to the

CONJECTURE 4.1. The genus 0 GV invariants n_d of the mirror to $\tilde{X}$ are given by the above values.

Obviously this Yukawa coupling is strange since it suggests that $n_d = 0$ for all odd d. We discuss a possible explanation in the next section. It is worth noting that the conjectured integrality of GW/GV invariants holds for our family, and moreover our family provides the first example known to the authors (and others) of such an even Yukawa coupling coming from geometry.

4.2. Torsion in homology and the vanishing of odd GV invariants. It is explained in [**AM**] that if torsion is present in $H_2(Y,\mathbb{Z})$, then the A-model Yukawa coupling takes on a more complicated form. For example, if $H_2(Y,\mathbb{Z}) \cong \mathbb{Z}\times\mathbb{Z}_2$, then we find that

$$\kappa_{ttt} = H^3 + (n_1^0 + (-1)^a n_1^1)q + O(q^2),$$

where $H^3 = \deg Y$, $a = 0$ or 1, and n_1^i are the number of lines whose torsion component in homology is in the class $i \in \mathbb{Z}_2$. Then $n_1^0 + n_1^1$ is the total number of lines. In this case if we had $n_1^0 = n_1^1$ and $a = 1$, then the corresponding coefficient of q would vanish. Assuming this paradigm continues in higher degrees, this could be a possible explanation of the vanishing of the odd coefficients in our Yukawa coupling.

It is worth noting that the predictions in [**BK**] cannot apply in total generality to all families of Calabi-Yau threefolds, as the examples of [**HT**],[**S**] of double mirrors with different fundamental groups show. Therefore torsion in homology may indeed explain the vanishing of the odd GW/GV invariants above. Moreover, as noted in [**BK**] $\mathrm{Tor}(H_2(Y,\mathbb{Z}))$ is dual to $Br(Y)$, the Brauer group of Y. Thus nontrivial torsion in the second homology would correspond to a nontrivial Brauer group as well (see [**A**], [**HT**] for examples of Calabi-Yau threefolds with nontrivial Brauer group). We note the possible relevance of this later.

4.3. Virtual GW/GV invariants at infinity. As we observed above, ∞ is not a classicaly defined MUM point, but it almost is. This type of point appears also in the families constructed by Kanazawa in [**Kan**]. As in those examples, we may transform our operator to the point at infinity and calculate the virtual GW/GV invariants that are conjectured to hold on the virtual mirror. We use the term *virtual* since it's not clear that mirror symmetry predicts the existence of a mirror or the equality of Yukawa couplings in this case. Nevertheless, changing the coordinate from z to $1/z$ and transforming the gauge by $\sqrt{z^3}$ transforms the Euler operator by $D_z \to -D_z - 3/2$. Thus the Picard-Fuchs operator at infinity becomes

$$\begin{aligned}\tilde{\mathfrak{D}} =& 16(z-1)^3(4z-1)^2(8z+1)D_z^4\\ &+ 96z(z-1)^2(4z-1)(32z^2-8z+3)D_z^3\\ &+ 12z(z-1)(2304z^4-2464z^3+992z^2-118z+15)D_z^2\\ &+ 36z(4z-1)(192z^4-304z^3+150z^2-30z+1)D_z\\ &+ 3z^2(3456z^4-5840z^3+3408z^2-933z+152).\end{aligned}$$

Now this operator has an MUM point at the new origin, so we may proceed as before, and assuming the equality of the B-model Yukawa coupling with the virtual A-model Yukawa coupling, we get that after choosing the constants of integration to be $c_1 = m$ and $c_2 = -2^{-4}$ the Yukawa coupling is

$$\begin{aligned}\kappa_{ttt}^{\infty} &= m(1 + 36q - 1116q^2 + 218088q^3 - 3712860q^4 + ...)\\ &= m(1 + 36\frac{q}{1-q} - 144\frac{2^3q^2}{1-q^2} + 8076\frac{3^3q^3}{1-q^3} - 57996\frac{4^3q^4}{1-q^4} + ...).\end{aligned}$$

It is interesting to see that this Yukawa coupling again satisfies the integrality conjecture expected of genuine GW/GV invariants. The mathematical meaning of these virtual numbers is not at the moment understood. It would be important to understand them better in the future. M. Bogner has pointed out to the first author that this Yukawa coupling is the same as that of equation 110 in the list of [**AESZ**]. The differential operator itself and the associated monodromy are different however.

Upon reviewing the sequence of instanton numbers in κ_{ttt}^{∞} and comparing with those of κ_{ttt}^{0}, it becomes clear that $n_d^{\infty} = 4n_{2d}^{0}$. In terms of power series expansions one in fact has $2\kappa_{ttt}^{\infty}(q^2) = \kappa_{ttt}^{0}(q)$, at least up to 100 terms. The meaning of these relations, geometric or otherwise, is unknown at the moment, but it suggests an intimate connection between the behavior at 0 and ∞.

5. Calabi-Yau differential equations, Monodromy, and the search for a mirror pair

As mentioned in the introduction, we have been unsuccesful in determining a mirror candidate to $\tilde{X}$. In an attempt to determine a possible mirror, we used the observation that the Picard-Fuchs equation of a one-parameter family can be used to calculate the monodromy action of the family, and these matrices can be conjecturally used to determine the fundamental numerical invariants of its mirror. In addition to the strange Yukawa coupling above, we find that our family also presents strange results in the context of these by-now standard techniques in mirror symmetry. To see this, we follow the approach in [**ES**] for calculating the monodromy matrices associated to a *Calabi-Yau differential equation.*

5.1. Calabi-Yau Differential Equations. These equations were defined in [**AESZ**] as fourth-order ODE's

$$\frac{d^4f}{dz^4} + a_3(z)\frac{d^3f}{dz^3} + a_2(z)\frac{d^2f}{dz^2} + a_1(z)\frac{df}{dz} + a_0(z)f(z) = 0,$$

satisfying the following five conditions which are expected to correspond to equations obtained as Picard-Fuchs equations of genuine one-parameter families:

(1) The singular point at $z = 0$ is an MUM point;

(2) The coefficients $a_i(z)$ satisfy the equation

$$a_1 = \frac{1}{2}a_2a_3 - \frac{1}{8}a_3^3 + a_2' - \frac{3}{4}a_3a_3' - \frac{1}{2}a_3'';$$

(3) The solutions $\lambda_1 \leq \lambda_2 \leq \lambda_3 \leq \lambda_4$ of the indicial equation at $z = \infty$ are positive rational numbers satisfying $\lambda_1 + \lambda_4 = \lambda_2 + \lambda_3$, and we suppose that the eigenvalues of the monodromy around $z = \infty$ are the zeros of a product of cyclotomoic polynomials;

(4) The power series solution near $z = 0$ has integral coefficients;

(5) The Yukawa coupling satisfies the integrality conjecture up to multiplication by a positive integer.

One can easily check that our Picard-Fuchs equations at 0 and ∞ satisfy all of the above condition except (4), but this pathology is not serious. By scaling the moduli parameters by 32 and -16, respectively, we obtain new ODE's which now satisfy all of the above conditions. Of course, this does not actually change the family or the Yukawa couplings calculated above.

5.2. Monodromy for the operator $\mathcal{D}$. The monodromy matrices for the action of $\pi_1(S,p)$ on $H^3(\tilde{X},\mathbb{Q})$ associated to our family $\tilde{X}_z$ can be calculated numerically using *Maple* as in [**ES**] ($S = \mathbb{P}^1 - \{0,1,-8,\infty\}$ as above). We obtain the monodromy matrices to be

$$T_0 = \begin{pmatrix} 1 & 0 & 0 & 0 \\ 1 & 1 & 0 & 0 \\ \frac{1}{2} & 1 & 1 & 0 \\ \frac{1}{6} & \frac{1}{2} & 1 & 1 \end{pmatrix}, \; T_{-8} = \begin{pmatrix} 1 & -8 & 0 & -192 \\ 0 & 1 & 0 & 0 \\ 0 & 0 & 1 & 0 \\ 0 & 0 & 0 & 1 \end{pmatrix}, \; T_4 = Id,$$

$$T_1 = \begin{pmatrix} 8 & -28 & 72 & -96 \\ \frac{7}{2} & -11 & 24 & -24 \\ \frac{5}{8} & -\frac{3}{2} & 2 & 0 \\ \frac{1}{48} & \frac{1}{12} & -\frac{1}{2} & 1 \end{pmatrix}, \; T_\infty = \begin{pmatrix} -8 & 28 & -120 & 96 \\ -\frac{1}{2} & 1 & 0 & -24 \\ \frac{1}{8} & -\frac{1}{2} & 2 & 0 \\ -\frac{1}{48} & \frac{1}{12} & -\frac{1}{2} & 1 \end{pmatrix},$$

where the matrices are presented in the first standard form mentioned in [**ES**]. It is certainly comforting that the monodromy around $z = 4$ calculated in this way yields the identity, as one would expect from a smooth fiber.

According to homological mirror symmetry (HMS), the relationship between the Calabi-Yau $\tilde{X}$ and its conjectural mirror Y goes beyond just the mirror duality of the hodge diamond or the equality of the A and B-model Yukawa couplings. Kontsevich expressed this symmetry by the stronger equivalence of two different derived categories associated to Calabi-Yau threefolds. The main statement of HMS is that the bounded derived category of coherent sheaves on $\tilde{X}$, $D^b(\tilde{X})$, is equivalent to the derived Fukaya category of Y, $D\mathcal{F}(Y)$, and vice-versa. Via the chern character this descends to cohomology as

$$H^{ev}(\tilde{X},\mathbb{Q}) \cong H^3(Y,\mathbb{Q}), H^{ev}(Y,\mathbb{Q}) \cong H^3(\tilde{X},\mathbb{Q}).$$

Accordingly , the monodromy action on $H^3(\tilde{X},\mathbb{Q})$ above should correspond to autoequivalences of $D^b(Y)$ which descend to automorphisms of the even cohomology. Moreover, the monodromy around the MUM point is conjectured to correspond to the action on $D^b(Y)$ given by tensoring by $\mathcal{O}(H)$, and the monodromy around the conifold point is conjectured to correspond to the spherical twist by $\mathcal{O}_Y$. These two actions on $D^b(Y)$ descend to cohomology as the matrices

$$\begin{pmatrix} 1 & 0 & 0 & 0 \\ 1 & 1 & 0 & 0 \\ \frac{1}{2} & 1 & 1 & 0 \\ \frac{1}{6} & \frac{1}{2} & 1 & 1 \end{pmatrix} \text{ and } \begin{pmatrix} 1 & -c & 0 & -d \\ 0 & 1 & 0 & 0 \\ 0 & 0 & 1 & 0 \\ 0 & 0 & 0 & 1 \end{pmatrix},$$

respectively, where we've used the basis $1, H, H^2, H^3$ for $H^{ev}(Y,\mathbb{Q})$, and $d := H^3, c := c_2 \cdot H/12$.

In our family this yields the prediction that the mirror Y has $H^3 = 192, c_2 \cdot H = 96$. At first glance this may seem very promising since using Hirzebruch-Riemann-Roch and Kodaira vanishing we may then determine the natural embedding of Y by the linear system $|H|$ and the numerics of the equations that define it. But as in

[**ES**] we performed a second check of consistency for these predictions by calculating the so-called *conifold period.*

5.3. The Conifold Period. The conifold period $f(z) = \int_{C(z)} \Omega$ is defined as the integral of our local generator $\Omega \in \overline{\mathcal{F}}^3$ against the vanishing cycle $C(z)$ near the conifold point $z = -8$, where $C(z)$ consists of the four S^3's which get collapsed in the formation of the four nodes in the fiber above $z = -8$. The monodromy around this point is the sum of the four Picard-Lefschetz transformations

$$\alpha \mapsto \alpha - \langle \alpha, C_i(z) \rangle C_i(z)$$

associated to each vanishing sphere $C_i(z)$ (See Section 3.2.1 in [**V**] for a discussion of vanishing spheres and the Picard-Lefschetz transformation). Since the $C_i(z)$ are all homologous in $H^3(\tilde{X}_z, \mathbb{Q})$, this monodromy is a symplectic reflection in $C(z)$. The Frobenius basis of the Picard-Fuchs equation at the conifold point has three holomorphic solutions and one solution of the form

$$f(z) \log z + k(z).$$

Since the monodromy around the conifold point is a symplectic involution in $C(z)$, the conifold period must be $f(z)$ up to scaling. Analytically continuing this function along a straight line to the MUM point at the origin, we can write this function as

$$z_2(s) = \frac{H^3}{6} s^3 + \frac{c_2 \cdot H}{24} s + \frac{c_3}{(2\pi i)^3} \zeta(3) + O(q),$$

where $q = e^{2\pi i s}$ and $s = \frac{1}{2\pi i} t = \frac{1}{2\pi i} \log z + \frac{1}{2\pi i} g(z)$, in the notation from above (we include a derivation of this form of the conifold period in an appendix for lack of a reference). Again following the algorithm of [**ES**], we numerically calculated the conifold period and scaled so that $H^3 = 192$ in the above expansion. This indeed gave $c_2 \cdot H = 96$, providing an internal consistency check of the above monodromy matrix calculations. Suprisingly, we also found a positive value for the topological Euler characteristic, $c_3 = 48$, which cannot correspond to a genuine mirror to our family. It is again interesting that we nevertheless get integral values for these invariants as one would expect from a genuine geometric situation. As an alternative to scaling the conifold period to give the expected H^3 in the above method, we may also scale to give the necessary Euler characteristic -60, obtaining $H^3 = -240, c_2 \cdot H = -120$. Of course, replacing H by $-H$ gives positive values for these. It is quite possible that there is no way to distinguish between $\pm H$ in this expression of the conifold period.

5.4. Monodromy for $\tilde{\mathcal{D}}$. As with the calculation of the Yukawa couplings, it seemed worth investigating the behavior near our almost MUM point at infinity. Repeating the above calculations with the virtual Picard-Fuchs operator $\tilde{\mathcal{D}}$, one obtains monodromy matrices

$$T_0 = \begin{pmatrix} 1 & 0 & 0 & 0 \\ 1 & 1 & 0 & 0 \\ \frac{1}{2} & 1 & 1 & 0 \\ \frac{1}{6} & \frac{1}{2} & 1 & 1 \end{pmatrix}, \quad T_{-1/8} = \begin{pmatrix} 1 & -4 & 0 & -24 \\ 0 & 1 & 0 & 0 \\ 0 & 0 & 1 & 0 \\ 0 & 0 & 0 & 1 \end{pmatrix}, \quad T_{1/4} = Id,$$

$$T_1 = \begin{pmatrix} 7 & 12 & 24 & 0 \\ 1 & 3 & 12 & 24 \\ -\frac{5}{2} & -5 & -13 & -12 \\ \frac{7}{6} & \frac{13}{6} & 5 & 3 \end{pmatrix}, \quad T_\infty = \begin{pmatrix} 7 & 16 & 24 & 168 \\ -8 & -17 & -36 & -168 \\ 2 & 4 & 11 & 36 \\ -\frac{1}{3} & -\frac{2}{3} & -2 & -5 \end{pmatrix},$$

giving predictions of $H^3 = 24, c_2 \cdot H = 48$. Calculating the conifold period confirms these numbers and again gives a prediction of $c_3 = 48$. We note that like the virtual instanton numbers, these invariants are off by a factor of 4 from those calculated for the operator $\mathcal{D}$.

6. Conclusion and Open Questions

Without an explicit mirror Y, we cannot even begin to verify the predictions of mirror symmetry for the family of Calabi-Yau threefolds constructed here. Nevertheless, we feel that the failed attempts to glean information about the mirror reveal the necessity to understand more deeply the prescriptions for calculating the mirror map and Yukawa couplings used above. The work here presents some important questions:

1) Does the mirror map need to be altered in some cases? Would this account for the strange vanishing of odd GV invariants on the conjectural mirror? Or does torsion in the second homology group of the mirror indeed explain this vanishing?

2) What is the meaning of the equal, but fractional, roots of the indicial equation for $\mathcal{D}$ at ∞? Does this have a Hori-Tong GLSM description (see[**HoTo**]) as suggested by [**Kan**]?

3) What is the meaning of the close relationship between invariants computed at 0 and ∞?

4) Do we need to reinterpret the correspondence between the vanishing cycle $C \in D\mathcal{F}(\tilde{X})$ and $\mathcal{O}_Y \in D^b(Y)$? Would this explain the strange positive Euler characteristic prediction?

5) Does the existence of a nontrivial Brauer group for the mirror suggest that derived categories of *twisted* sheaves must be incorporated into the picture of homological mirror symmetry? Would this explain the strange numerical invariants predicted by homological mirror symmetry?

Hopefully upon answering these questions, we may place the example constructed in this paper in its proper context in mirror symmetry.

7. Appendix: Expansion of the conifold period

It was asked in [**ES**] if the appearance of the term $\frac{c_3}{(2\pi i)^3}\zeta(3)$ in the expansion of the conifold period obtained in [**COGP**] is a mathematical theorem holding for families of Calabi-Yau threefolds other than the famous example of the quintic threefold treated there. The purpose of this appendix is to prove that indeed this expansion always holds assuming the validity of the standard conjectures of HMS. In particular, we assume that the vanishing cycle S in $D\mathcal{F}(X)$ corresponds to $\mathcal{O}_Y$ in $D^b(Y)$ for its mirror partner Y. A similar result for toric varieties in all dimensions was obtained in [**Lib**]. See also the related discussion in [**HJLM**].

THEOREM 7.1. *Under the above assumptions, the conifold period $z_2(t)$ has the following expansion up to scaling:*

$$z_2(t) = \frac{H^3}{6}t^3 + \frac{c_2(Y).H}{24}t + \frac{c_3(Y)}{(2\pi i)^3}\zeta(3) + O(q),$$

where of course H is the ample generator of $\mathrm{Pic}(Y)$ *and* $q = e^{2\pi i t}$ *is the usual parameter in the mirror map.*

PROOF. In order to relate the conifold period to anything on the mirror, we must analytically continue it to the MUM point. Then according to HMS $S \in D\mathcal{F}(X)$ should correspond to some object of $D^b(Y)$, $\mathcal{O}_Y$ by our assumption. Moreover, according to HMS the pairing $\langle S(z), \Omega\rangle = \int_{S(z)} \Omega$ at the complex moduli point z should correspond to taking the central charge of $\mathcal{O}_Y$ at the Kähler moduli point t corresponding to z (see comments on pages 2 and 10 of [**Iri**]). Of course the coordinate t corresponds to the complexified divisor tH, where H is the ample generator of $\mathrm{Pic}(Y)$.

From the fourth formula on page 12 of [**Iri**], we get that this central charge is

$$Z_{tH}(\mathcal{O}_Y) = -\frac{c_3(Y)}{(2\pi i)^3}\zeta(3) - \frac{c_2(Y).H}{24}\frac{t}{2\pi i} + \frac{1}{(2\pi i)^3}G(tH).$$

Here the function G is defined by

$$G(tH) := 2F_0(tH) - t\frac{d}{dt}F_0(tH), \text{ where } F_0(tH) := \frac{H^3}{6}t^3 + \sum_{d>0} N_d e^{dH.Ct}$$

is the genus zero potential of Y and C is the generator of $H_2(Y, Z)$ modulo torsion. Expanding out $G(tH)$, we get

$$G(tH) = -\frac{H^3}{6}t^3 - t(\sum_{d>0} dH.CN_d e^{dH.Ct}) + 2(\sum_{d>0} N_d e^{dH.Ct}).$$

Putting this back into the formula above for the central charge and noticing that the terms different from $-\frac{H^3}{6}t^3$ can be grouped into the $O(q)$ term, we get the desired form of the conifold period up to scaling by -1. $\square$

References

[A] N. Addington, "The Brauer group is not a derived invariant," arXiv:1306.6538.

[AESZ] G. Almkvist, C. van Enckevort, D. van Straten, and W. Zudilin, "Tables of Calabi-Yau equations," arXiv:0507430.

[AM] P. S. Aspinwall and D. R. Morrison, *Stable singularities in string theory*, Comm. Math. Phys. **178** (1996), no. 1, 115–134. With an appendix by Mark Gross. MR1387944 (97d:32049)

[BG] R. L. Bryant and P. A. Griffiths, *Some observations on the infinitesimal period relations for regular threefolds with trivial canonical bundle*, Arithmetic and geometry, Vol. II, Progr. Math. vol. 36, Birkhäuser Boston, Boston, MA, 1983, pp. 77–102. MR717607 (86a:32044)

[BK] V. Batyrev and M. Kreuzer, *Integral cohomology and mirror symmetry for Calabi-Yau 3-folds*, Mirror symmetry. V, AMS/IP Stud. Adv. Math. vol. 38, Amer. Math. Soc., Providence, RI, 2006, pp. 255–270. MR2282962 (2008d:14060)

[BN] L. Borisov, H. Nuer, "On (2,4) Complete Intersection threefolds that contain an Enriques surface," arXiv:1210.1903.

[CDP] P. Candelas, E. Derrick, and L. Parkes, *Generalized Calabi-Yau manifolds and the mirror of a rigid manifold*, Nuclear Phys. B **407** (1993), no. 1, 115–154, DOI 10.1016/0550-3213(93)90276-U. MR1242064 (94i:32041)

[CK] D. A. Cox and S. Katz, *Mirror symmetry and algebraic geometry*, Mathematical Surveys and Monographs, vol. 68, American Mathematical Society, Providence, RI, 1999. MR1677117 (2000d:14048)

[COGP] P. Candelas, X. C. de la Ossa, P. S. Green, and L. Parkes, *A pair of Calabi-Yau manifolds as an exactly soluble superconformal theory*, Nuclear Phys. B **359** (1991), no. 1, 21–74, DOI 10.1016/0550-3213(91)90292-6. MR1115626 (93b:32029)

[Del1] P. Deligne, *Équations différentielles à points singuliers réguliers* (French), Lecture Notes in Mathematics, Vol. 163, Springer-Verlag, Berlin-New York, 1970. MR0417174 (54 #5232)

[Dim] A. Dimca, *Singularities and topology of hypersurfaces*, Universitext, Springer-Verlag, New York, 1992. MR1194180 (94b:32058)

[ES] C. van Enckevort and D. van Straten, *Monodromy calculations of fourth order equations of Calabi-Yau type*, Mirror symmetry. V, AMS/IP Stud. Adv. Math. vol. 38, Amer. Math. Soc., Providence, RI, 2006, pp. 539–559. MR2282974 (2007m:14057)

[GS] D. Grayson, M. Stillman, "Macaulay 2: A computer program designed to support computations in algebraic geometry and computer algebra." Source and object code available from http://www.math.uiuc.edu/Macaulay2/.

[HJLM] J. Halverson, H. Jockers, J. Lapan, D. R. Morrison, "Perturbative Corrections to Kähler Moduli Spaces," arXiv:1308.2157.

[HT] S. Hosono and H. Takagi, *Determinantal quintics and mirror symmetry of Reye congruences*, Comm. Math. Phys. **329** (2014), no. 3, 1171–1218, DOI 10.1007/s00220-014-1971-7. MR3212882

[HoTo] K. Hori and D. Tong, *Aspects of non-abelian gauge dynamics in two-dimensional $\mathcal{N} = (2,2)$ theories*, J. High Energy Phys. **5** (2007), 079, 41 pp. (electronic), DOI 10.1088/1126-6708/2007/05/079. MR2318130 (2009d:81351)

[Iri] H. Iritani, *Ruan's conjecture and integral structures in quantum cohomology*, New developments in algebraic geometry, integrable systems and mirror symmetry (RIMS, Kyoto, 2008), Adv. Stud. Pure Math. vol. 59, Math. Soc. Japan, Tokyo, 2010, pp. 111–166. MR2683208 (2011h:14081)

[Kan] A. Kanazawa, *Pfaffian Calabi-Yau threefolds and mirror symmetry*, Commun. Number Theory Phys. **6** (2012), no. 3, 661–696, DOI 10.4310/CNTP.2012.v6.n3.a3. MR3021322

[Lan] A. Landman, *On the Picard-Lefschetz transformation for algebraic manifolds acquiring general singularities*, Trans. Amer. Math. Soc. **181** (1973), 89–126. MR0344248 (49 #8987)

[Lib] A. Libgober, *Chern classes and the periods of mirrors*, Math. Res. Lett. **6** (1999), no. 2, 141–149, DOI 10.4310/MRL.1999.v6.n2.a2. MR1689204 (2000h:32017)

[R] E. A. Rødland, *The Pfaffian Calabi-Yau, its mirror, and their link to the Grassmannian $G(2,7)$*, Compositio Math. **122** (2000), no. 2, 135–149, DOI 10.1023/A:1001847914402. MR1775415 (2001h:14051)

[Sch] W. Schmid, *Variation of Hodge structure: the singularities of the period mapping*, Invent. Math. **22** (1973), 211–319. MR0382272 (52 #3157)

[S] C. Schnell, *The fundamental group is not a derived invariant*, Derived categories in algebraic geometry, EMS Ser. Congr. Rep. Eur. Math. Soc., Zürich, 2012, pp. 279–285. MR3050707

[V] C. Voisin, *Hodge theory and complex algebraic geometry. II*, Cambridge Studies in Advanced Mathematics, vol. 77, Cambridge University Press, Cambridge, 2003. Translated from the French by Leila Schneps. MR1997577 (2005c:32024b)

Rutgers University, Department of Mathematics, 110 Frelinghuysen Rd., Piscataway, New Jersey 08854

E-mail address: `hjn11@math.rutgers.edu`

Rutgers University, Department of Mathematics, 110 Frelinghuysen Rd., Piscataway, New Jersey 08854

E-mail address: `prd41@math.rutgers.edu`

Proceedings of Symposia in Pure Mathematics
Volume **93**, 2016
http://dx.doi.org/10.1090/pspum/093/01572

Calabi-Yau threefolds fibred by Kummer surfaces associated to products of elliptic curves

Charles F. Doran, Andrew Harder, Andrey Y. Novoseltsev, and Alan Thompson

ABSTRACT. We study threefolds fibred by Kummer surfaces associated to products of elliptic curves, that arise as resolved quotients of threefolds fibred by certain lattice polarized K3 surfaces under a fibrewise Nikulin involution. We present a general construction for such surfaces, before specializing our results to study Calabi-Yau threefolds arising as resolved quotients of threefolds fibred by mirror quartic K3 surfaces. Finally, we give some geometric properties of the Calabi-Yau threefolds that we have constructed, including expressions for Hodge numbers.

1. Introduction

Building on earlier work by Shioda, Inose [**24**][**12**], Nikulin [**21**] and Morrison [**18**], Clingher and Doran [**2**][**3**] exhibited a duality between K3 surfaces admitting a lattice polarization by the lattice

$$M := H \oplus E_8 \oplus E_8$$

and Kummer surfaces associated to products of elliptic curves, that closely relates the geometry of the surfaces on each side. This duality is easy to describe: any M-polarized K3 surface admits a canonically defined Nikulin involution, the resolved quotient by which is a Kummer surface associated to a product of elliptic curves and, conversely, a Kummer surface associated to a product of elliptic curves also admits a Nikulin involution, the resolved quotient by which is isomorphic to an M-polarized K3 surface. Moreover, applying this process twice returns us to the surface we started with.

This duality was exploited in [**4**], to obtain certain geometric properties of a Calabi-Yau threefold admitting a fibration by M-polarized K3 surfaces. In that case, it was proven that Clingher's and Doran's construction could be performed

2010 *Mathematics Subject Classification.* Primary 14D06, Secondary 14J28, 14J30, 14J32.

Key words and phrases. K3 surface, Calabi-Yau threefold, fibration.

C. F. Doran and A. Y. Novoseltsev were supported by the Natural Sciences and Engineering Resource Council of Canada (NSERC), the Pacific Institute for the Mathematical Sciences (PIMS), and the McCalla Professorship at the University of Alberta.

A. Harder was supported by an NSERC Post-Graduate Scholarship.

A. Thompson was supported by a Fields-Ontario-PIMS Postdoctoral Fellowship with funding provided by NSERC, the Ontario Ministry of Training, Colleges and Universities, and an Alberta Advanced Education and Technology Grant.

fibrewise, giving rise to a new Calabi-Yau threefold that was fibred by Kummer surfaces associated to products of elliptic curves. As it turns out, the geometry of the Kummer fibred threefold thus obtained was easier to study, and could be used to derive geometric properties of the original threefold fibred by M-polarized K3 surfaces.

The main aim of this paper is to investigate to what extent this construction can be generalized to arbitrary threefolds fibred by M-polarized K3 surfaces.

More precisely, suppose that $\mathcal{X}$ is a threefold fibred by K3 surfaces and that the restriction $\mathcal{X}_U$ of this fibration to the locus of smooth fibres is an M-polarized family of K3 surfaces, in the sense of [**10**, Definition 2.1]. Then results of [**10**] show that the canonical Nikulin involution on the fibres of $\mathcal{X}_U$ extends to the entire threefold, so we may quotient and resolve singularities to obtain a threefold fibred by Kummer surfaces $\mathcal{Y}_U$. We may then ask whether $\mathcal{Y}_U$ can be compactified to a threefold $\mathcal{Y}$ and, if so, what properties this new threefold has.

One case that is of particular interest is when the threefold $\mathcal{X}$ is Calabi-Yau, as occurred in [**4**]. In this case, one would like to know whether $\mathcal{Y}$ (if it exists) is also Calabi-Yau and, if so, how its properties relate to those of $\mathcal{X}$.

In the latter part of this paper we address this second question in a special case, where the Calabi-Yau threefolds $\mathcal{X}$ are very well-understood. Specifically, we consider the setting where $\mathcal{X}$ is one of the Calabi-Yau threefolds $\mathcal{X}_g$ fibred by mirror quartic K3 surfaces constructed in [**9**]. Note that this is a special case of the construction above as, by definition, a mirror quartic K3 surface is polarized by the lattice $M_2 := H \oplus E_8 \oplus E_8 \oplus \langle -4 \rangle$, which clearly contains M as a primitive sublattice. These threefolds $\mathcal{X}_g$ encompass many well-known examples, including the quintic mirror threefold, and provide a useful illustration of our methods in a concrete setting.

In this special case, we show that we can explicitly construct Kummer surface fibred threefolds $\mathcal{Y}_g$ that are related to the $\mathcal{X}_g$ by a fibrewise quotient-resolution procedure as above. Moreover, the $\mathcal{Y}_g$ are Calabi-Yau in most cases and have geometric properties that are closely related to those of the $\mathcal{X}_g$. This gives a new perspective from which to study the geometry of the threefolds $\mathcal{X}_g$, amongst them the quintic mirror.

Finally, we note that the construction of the threefolds $\mathcal{Y}_g$ is somewhat interesting in its own right, as they are all constructed from a single, rigid Calabi-Yau threefold. This rigid Calabi-Yau threefold is in turn built from a well-known extremal rational elliptic surface, using a method originally due to Schoen [**22**] that was later extended by Kapustka and Kapustka [**13**].

The structure of this paper is as follows. In Section 2 we review some background material, mostly taken from [**2**] and [**3**], about M-polarized K3 surfaces, and describe the threefolds $\mathcal{X}$ that are fibred by them. Then, in Section 3, we develop the theory required to construct the associated threefolds fibred by Kummer surfaces $\mathcal{Y}$, and describe their construction in general terms. This construction proceeds by first *undoing the Kummer construction*, as originally described in [**10**, Section 4.3], then running a generalized version of the *forward construction* from [**4**, Section 7].

Finally, in Section 4, we specialize the entire discussion to the case where $\mathcal{X}$ is one of the Calabi-Yau threefolds fibred by quartic mirror K3 surfaces $\mathcal{X}_g$ constructed in [**9**]. In this case we can construct the associated threefolds fibred by Kummer surfaces $\mathcal{Y}_g$ completely explicitly as pull-backs of a special threefold $\mathcal{Y}_2$. This special threefold is constructed in turn as a resolved quotient of a rigid Calabi-Yau threefold $\mathcal{A}_2$, which is described in Section 4.2. The properties of $\mathcal{A}_2$ are then studied in Section 4.3, after which we carefully describe the quotient-resolution procedure used to obtain $\mathcal{Y}_2$ from it in Sections 4.4 and 4.5. The method for constructing the $\mathcal{Y}_g$ from $\mathcal{Y}_2$ is detailed in Section 4.6, and some of the properties of the $\mathcal{Y}_g$ are computed in Section 4.7.

2. Background material

We begin by setting up some notation. Let $\mathcal{X}$ be a smooth projective threefold that admits a fibration $\mathcal{X} \to B$ by K3 surfaces over a smooth curve. Let $N := \mathrm{NS}(X_p)$ denote the Néron-Severi group of the fibre of $\mathcal{X}$ over a general point $p \in B$. Suppose that there exists a primitive lattice embedding $M \hookrightarrow N$ of the lattice $M := H \oplus E_8 \oplus E_8$ into N (we will assume that such an embedding has been fixed in what follows).

Denote the open set over which the fibres of $\mathcal{X}$ are smooth K3 surfaces by $U \subset B$ and let $\mathcal{X}_U \to U$ denote the restriction of $\mathcal{X}$ to U. Suppose further that $\mathcal{X}_U \to U$ is an N-polarized family of K3 surfaces, in the sense of [**10**, Definition 2.1].

REMARK 2.1. Nineteen such fibrations are known on Calabi-Yau threefolds $\mathcal{X}$ with $h^{2,1}(\mathcal{X}) = 1$; these are summarized by [**10**, Table 5.1]. Moreover, a large class of additional examples of Calabi-Yau threefolds fibred by K3 surfaces polarized by the lattice $M_2 := H \oplus E_8 \oplus E_8 \oplus \langle -4 \rangle$ are constructed in [**9**]; we will return to these examples in Section 4.

2.1. M-polarized K3 surfaces.

By assumption, a general fibre X_p of $\mathcal{X}$ is an M-polarized K3 surface. We recall here some basic properties of M-polarized K3 surfaces, that will be used repeatedly in what follows. In this section we will denote an M-polarized K3 surface by (X, i), where X is a K3 surface and i is a primitive lattice embedding $i\colon M \hookrightarrow \mathrm{NS}(X)$.

Building upon work of Inose [**12**], Clingher, Doran, Lewis and Whitcher [**5**] have shown that M-polarized K3 surfaces have a coarse moduli space given by the locus $d \neq 0$ in the weighted projective space $\mathbb{WP}(2,3,6)$ with weighted coordinates (a, b, d). Thus, by normalizing $d = 1$, we may associate a pair of complex numbers (a, b) to an M-polarized K3 surface (X, i).

The first piece of structure that we need on (X, i) comes from the work of Morrison [**18**], who showed that the composition of i with the canonical embedding $E_8 \oplus E_8 \hookrightarrow M$ defines a canonical *Shioda-Inose structure* on (X, i) (named for Shioda and Inose [**24**], who were the first to study such structures). By definition, such a structure consists of a Nikulin involution β on X, such that the resolved quotient $Y = \widetilde{X/\beta}$ is a Kummer surface and there is a Hodge isometry $T_Y \cong T_X(2)$, where T_X and T_Y denote the transcendental lattices of X and Y respectively, and $T_X(2)$ indicates that the bilinear pairing on T_X has been multiplied by 2.

By [**2**, Theorem 3.13][1], we see that in our setting Y is isomorphic to the Kummer surface $\mathrm{Kum}(A)$, where $A \cong E_1 \times E_2$ is an Abelian surface that splits as a product of elliptic curves. By [**2**, Corollary 4.2][1] the j-invariants of these elliptic curves are given by the roots of the equation

$$j^2 - \sigma j + \pi = 0,$$

where σ and π are given in terms of the (a, b) values associated to (X, i) by $\sigma = a^3 - b^2 + 1$ and $\pi = a^3$. Label the exceptional (-2)-curves in Y arising from the resolution of the singularities of X/β by $\{F_1, \ldots, F_8\}$.

There is one more piece of structure on (X, i) that we will need in our discussion. By [**2**, Proposition 3.10], the K3 surface X admits two uniquely defined elliptic fibrations $\Theta_{1,2} \colon X \to \mathbb{P}^1$, the *standard* and *alternate fibrations*. We will be mainly concerned with the alternate fibration Θ_2. This fibration has two sections, one singular fibre of type I_{12}^* and, if $a^3 \neq (b \pm 1)^2$, six singular fibres of type I_1 [**2**, Proposition 4.6].

The alternate fibration Θ_2 is preserved by the Nikulin involution β, so induces an elliptic fibration $\Psi \colon Y \to \mathbb{P}^1$ on the Kummer surface Y. The two sections of Θ_2 are identified to give a section S of Ψ, and Ψ has one singular fibre of type I_6^* and, if $a^3 \neq (b \pm 1)^2$, six I_2's [**2**, Proposition 4.7].

REMARK 2.2. As noted in the introduction, this construction is completely reversible. Clingher and Doran [**3**, Section 1] identify a second distinguished section S' of the fibration Ψ, along with a uniquely defined Nikulin involution β' on Y that preserves Ψ and takes S to S'. The resolved quotient $\widetilde{Y/\beta'}$ is then isomorphic to X, and Ψ induces the alternate fibration Θ_2 on X.

The locations of the I_2 fibres in Ψ are given by [**2**, Proposition 4.7]. They occur at the roots of the polynomials $(P(x) \pm 1)$, where P is the cubic equation

$$P(x) := 4x^3 - 3ax - b, \tag{2.1}$$

(a, b) are the modular parameters associated to X, and x is an affine coordinate on $\mathbb{P}^1$ chosen so that the I_6^* fibre occurs at $x = \infty$.

Finally, using this information we may identify some of the (-2)-curves F_i in Y. By the discussion in [**10**, Section 4.3], $\{F_3, F_4, F_5\}$ (resp.$\{F_6, F_7, F_8\}$) are the (-2)-curves in the I_2 fibres lying over the roots of $(P(x) - 1)$ (resp. $(P(x) + 1)$) that do not meet the section S (this labelling may seem arbitrary, but in fact is chosen to match with that used in [**10**, Section 4.3]).

3. Threefolds fibred by Kummer surfaces

We will now apply this theory to study the K3-fibred threefold $\mathcal{X} \to B$. Via the embedding $M \hookrightarrow N$, we see that a general fibre of $\mathcal{X}$ is an M-polarized K3 surface. Thus, by the discussion in Section 2.1, there is a canonical Shioda-Inose structure on such a fibre, which defines a Nikulin involution on it. This involution extends uniquely to all fibres of $\mathcal{X}_U$ by [**10**, Corollary 2.12].

Let $\mathcal{Y}_U \to U$ denote the family obtained by taking the quotient of $\mathcal{X}_U$ by this involution and resolving the resulting singularities. The discussion from Section 2.1 shows that the fibres of $\mathcal{Y}_U$ are Kummer surfaces $\mathrm{Kum}(E_1 \times E_2)$ associated to products of elliptic curves $E_1 \times E_2$. Furthermore, the alternate fibration Θ_2 on the fibres of $\mathcal{X}_U$ induces a uniquely defined elliptic fibration Ψ on the fibres of $\mathcal{Y}_U$.

REMARK 3.1. As in the K3 surface case, this construction turns out to be reversible. Let β' be the Nikulin involution on a general fibre of $\mathcal{Y}_U$, as described in Remark 2.2. By the description of the action of monodromy in U from [**10**, Section

[1]We note that equivalent results to those attributed to Clingher and Doran [**2**] here were proved independently by Shioda [**23**], using a slightly different characterization of M-polarized K3 surfaces as elliptically fibred K3 surfaces with section and two fibres of type II^*.

4.3] and the description of β' from [**3**, Section 1], it can be shown that the action of β' and the action of monodromy on the Néron-Severi lattice of a general fibre commute. So, by [**10**, Proposition 2.11], β' extends to a involution on $\mathcal{Y}_U$, the resolved quotient by which is isomorphic to $\mathcal{X}_U$.

Our aim is to explicitly construct K3-fibred threefolds $\mathcal{Y}$ over B, so that the restriction of $\mathcal{Y}$ to the open set $U \subset B$ is isomorphic to $\mathcal{Y}_U$, and to study their properties.

3.1. Undoing the Kummer construction. In order to do this, the first step is to *undo the Kummer construction* for $\mathcal{Y}_U$, i.e. to find a family of Abelian surfaces $\mathcal{A}_U \to U$ which gives rise to $\mathcal{Y}_U$ upon fibrewise application of the Kummer construction. To do this, we will use the results from [**10**, Section 4.3]. However, in order to apply these results we need to make the following assumption [**10**, Assumption 4.6]; unless otherwise stated, we will make this assumption throughout the remainder of this section.

ASSUMPTION 3.2. The fibration Ψ on a general fibre Y_p of $\mathcal{Y}_U$ has six singular fibres of type I_2.

REMARK 3.3. Note that each I_2 fibre in the fibration Ψ on Y_p arises as the total transform of an I_1 fibre in the alternate fibration Θ_2 on X_p. Thus to check that Assumption 3.2 is satisfied, it is equivalent to show that the alternate fibration Θ_2 on a general fibre X_p of $\mathcal{X}_U$ has six singular fibres of type I_1

This latter condition is easy to check numerically from the (a, b) parameters associated to X_p. Indeed, the locations of the I_1 fibres in Θ_2 are given by the roots of the polynomials $(P(x) \pm 1)$, where $P(x)$ is defined by Equation (2.1), which are all distinct if and only if $a^3 \neq (b \pm 1)^2$.

Unfortunately, by the discussion in [**10**, Section 4.3], it is not always possible to undo the Kummer construction on $\mathcal{Y}_U$ directly. Instead, we must pull everything back to a cover $f\colon C \to B$.

This cover is constructed by the method described in [**10**, Section 4.3]. Let $p \in U$ be a point and consider the six divisors $\{F_3, \ldots, F_8\}$ in the fibre Y_p of $\mathcal{Y}$ over p. Monodromy in U preserves the fibration Ψ along with its section S (as both are induced from the structure of the alternate fibration Θ_2 on the fibres of $\mathcal{X}_U$), so must act to permute the F_i. We thus have a homomorphism $\rho\colon \pi_1(U, p) \to S_6$; call its image G. Then define an unramified $|G|$-fold cover $f\colon V \to U$ as follows: the preimages of $p \in U$ are labelled by permutations in G and, if γ is a loop in U, monodromy around $f^{-1}(\gamma)$ acts on these labels as composition with $\rho(\gamma)$. This cover extends uniquely to a cover $f\colon C \to B$, with ramification over the points in $B - U$.

Let $\mathcal{Y}'_V$ denote the pull-back of $\mathcal{Y}_U$ to V. Then [**10**, Theorem 4.11] shows that we can undo the Kummer construction for $\mathcal{Y}'_V$, so there exists a family of Abelian surfaces $\mathcal{A}_V \to V$ that gives rise to $\mathcal{Y}'_V$ under fibrewise application of the Kummer construction.

We have the following diagram:

$$\begin{array}{ccccccccc}
\mathcal{A}_V & \overset{\text{Kummer}}{\dashrightarrow} & \mathcal{Y}'_V & \longrightarrow & \mathcal{Y}_U & \overset{\text{Nikulin}}{\dashleftarrow} & \mathcal{X}_U & \hookrightarrow & \mathcal{X} \\
\downarrow & & \downarrow & & \downarrow & & \downarrow & & \downarrow \\
V & = & V & \overset{f}{\longrightarrow} & U & = & U & \hookrightarrow & B
\end{array}$$

3.2. The forward construction. Our next aim is to construct threefolds $\mathcal{A}$, $\mathcal{Y}'$ and $\mathcal{Y}$ that agree with $\mathcal{A}_V$, $\mathcal{Y}'_V$ and $\mathcal{Y}_U$ over V and U respectively. This construction will generalize the *forward construction* of [**4**, Section 7].

We begin by constructing a threefold fibred by Abelian surfaces $\mathcal{A} \to C$ that agrees with $\mathcal{A}_V$ over V. The first step is to identify some special divisors on the fibres of $\mathcal{Y}'_V$.

Recall that a fibre of $\mathcal{Y}'_V$ is isomorphic to $\mathrm{Kum}(E_1 \times E_2)$, where E_1 and E_2 are elliptic curves. There is a special configuration of twenty-four (-2)-curves on $\mathrm{Kum}(E_1 \times E_2)$ arising from the Kummer construction, that we shall now describe (here we note that we use the same notation as [**2**, Definition 3.18], but with the roles of G_i and H_j reversed).

Let $\{x_0, x_1, x_2, x_3\}$ and $\{y_0, y_1, y_2, y_3\}$ denote the two sets of points of order two on E_1 and E_2 respectively. Denote by G_i and H_j $(0 \leq i, j \leq 3)$ the (-2)-curves on $\mathrm{Kum}(E_1 \times E_2)$ obtained as the proper transforms of $E_1 \times \{y_i\}$ and $\{x_j\} \times E_2$ respectively. Let E_{ij} be the exceptional (-2)-curve on $\mathrm{Kum}(E_1 \times E_2)$ associated to the point (x_j, y_i) of $E_1 \times E_2$. This gives 24 curves, which have the following intersection numbers:

$$\begin{aligned} G_i.H_j &= 0, \\ G_k.E_{ij} &= \delta_{ik}, \\ H_k.E_{ij} &= \delta_{jk}. \end{aligned}$$

DEFINITION 3.4. The configuration of twenty-four (-2)-curves

$$\{G_i, H_j, E_{ij} \mid 0 \leq i, j \leq 3\}$$

is called a *double Kummer pencil* on $\mathrm{Kum}(E_1 \times E_2)$.

With this in place, we can prove an analogue of [**4**, Lemma 7.4 and Proposition 7.5].

PROPOSITION 3.5. *$\mathcal{A}_V \to V$ is isomorphic over V to a fibre product $\mathcal{E}_1 \times_C \mathcal{E}_2$ of minimal elliptic surfaces $\mathcal{E}_i \to C$ with section. Furthermore, the j-invariants of the elliptic curves E_1 and E_2 forming the fibres of $\mathcal{E}_1$ and $\mathcal{E}_2$ over a point $p \in C$ are given by the roots of the equation*

$$j^2 - \sigma(p)j + \pi(p) = 0, \tag{3.1}$$

where $\sigma(p)$ and $\pi(p)$ are the σ and π invariants associated to the M-polarized K3 surface fibre $X_{f(p)}$ of $\mathcal{X}_U$ over $f(p)$.

REMARK 3.6. Thus, using the expressions for (σ, π) in terms of (a, b) from Section 2.1, we find that the discriminant of Equation (3.1) is

$$\sigma^2 - 4\pi = a^6 - 2a^3b - 2a^3 + b^4 - 2b^2 + 1 = (a^3 - (b-1)^2)(a^3 - (b+1)^2).$$

But Assumption 3.2 and Remark 3.3 imply that, for generic $p \in C$, this does not vanish, so the roots of Equation (3.1) are generically distinct.

PROOF OF PROPOSITION 3.5. We begin by showing that the fibration $\mathcal{A}_V \to V$ has a section s. Construct a double Kummer pencil $\{G_i, H_j, E_{ij}\}$ on the fibre Y'_p of $\mathcal{Y}'_V$ over $p \in V$ as described in [**10**, Section 4.3]. By [**10**, Theorem 4.11], $\mathcal{Y}'_V$ is $\mathrm{NS}(Y'_p)$-polarized, so the divisors in this pencil are invariant under monodromy around loops in V. In particular, the curve E_{11} is invariant. So E_{11} sweeps out a divisor in $\mathcal{Y}'_V$, which intersects each smooth fibre in a (-2)-curve. Passage to $\mathcal{A}_V$ contracts this divisor to a curve which intersects each smooth fibre in a single point, i.e. a section over V.

Now, let $\mathcal{A}_p$ denote the fibre of $\mathcal{A}_V$ over p, which is isomorphic to a product $E_1 \times E_2$ of elliptic curves. We may identify E_1 (resp. E_2) with the preimages of the curve G_1 (resp. H_1) in the double Kummer pencil on Y'_p. As G_1 and H_1 are invariant under monodromy around loops in V, they sweep out two divisors in Y'_V. Upon passage to $\mathcal{A}_V$ these two divisors become a pair of elliptic surfaces, $\mathcal{E}_{1,V} \to V$ and $\mathcal{E}_{2,V} \to V$, which intersect along the section s. By [**19**, Theorem 2.5], there are unique extensions of $\mathcal{E}_{i,V} \to V$ to minimal elliptic surfaces $\mathcal{E}_i \to C$ over C, for $i = 1, 2$. By construction, we have an isomorphism over V between $\mathcal{A}_V$ and $\mathcal{E}_1 \times_C \mathcal{E}_2$, as required.

Finally, the statement about the j-invariants is an easy consequence of the discussion in Section 2.1. □

Using this, we may construct a threefold $\mathcal{Y}' \to C$ that is isomorphic to $\mathcal{Y}'_V$ over V by applying the Kummer construction to $\mathcal{E}_1 \times_C \mathcal{E}_2$ fibrewise. To further construct a model for $\mathcal{Y}_U$, we need to know how the group G defining the cover f acts on $\mathcal{E}_1 \times_C \mathcal{E}_2$.

LEMMA 3.7. *Let φ denote the action of a permutation in $G \subset S_6$ on C. Then either φ induces automorphisms on $\mathcal{E}_1$ and $\mathcal{E}_2$, or φ induces an isomorphism $\mathcal{E}_1 \to \mathcal{E}_2$.*

PROOF. Note first that φ induces an automorphism $\hat{\varphi}$ of $\mathcal{A}_V$. Furthermore, $\hat{\varphi}$ preserves the section s as, by [**10**, Lemma 4.5], the curve E_{11} in a general fibre Y_p of $\mathcal{Y}_U$ is invariant under monodromy in U.

As in the proof of Proposition 3.5, we identify $\mathcal{E}_{1,V}$ and $\mathcal{E}_{2,V}$ with the elliptic surfaces in $\mathcal{A}_V$ swept out by the preimages of the curves G_1 and H_1. These elliptic surfaces intersect along the section s.

As $\hat{\varphi}$ preserves s, we see that $\hat{\varphi}(\mathcal{E}_{1,V})$ is an elliptic surface in $\mathcal{A}_V$ that contains s as a section. It must therefore either be $\mathcal{E}_{1,V}$ or $\mathcal{E}_{2,V}$. The same holds for $\mathcal{E}_{2,V}$. Thus, we see that φ either induces automorphisms on $\mathcal{E}_{1,V}$ and $\mathcal{E}_{2,V}$, or induces an isomorphism $\mathcal{E}_{1,V} \to \mathcal{E}_{2,V}$.

Thus φ induces either a birational automorphism on $\mathcal{E}_1$ and $\mathcal{E}_2$, or a birational map $\mathcal{E}_1 \to \mathcal{E}_2$. But, by [**17**, Proposition II.1.2], a birational map between minimal elliptic surfaces is an isomorphism. □

It is easy to determine which case of Lemma 3.7 occurs:

LEMMA 3.8. *Let φ denote the action of a permutation in $G \subset S_6$ on C. Then φ induces automorphisms on $\mathcal{E}_1$ and $\mathcal{E}_2$ (resp. φ induces an isomorphism $\mathcal{E}_1 \to \mathcal{E}_2$) if and only if the action of φ preserves (resp. exchanges) the roots of Equation* (3.1) (*which are generically distinct by Remark* 3.6).

PROOF. By Lemma 3.7, we know that either φ induces automorphisms on $\mathcal{E}_1$ and $\mathcal{E}_2$, or φ induces an isomorphism $\mathcal{E}_1 \to \mathcal{E}_2$. To see which occurs, we study the action on a general fibre of $\mathcal{E}_1$.

So let E_i denote the fibre of $\mathcal{E}_i$ over a general point $p \in V$ and let E'_i denote the fibre of $\mathcal{E}_i$ over $\varphi(p)$ (for $i \in \{1, 2\}$). By Proposition 3.5, we see that the j-invariants of $\{E_1, E_2\}$ are equal to those of $\{E'_1, E'_2\}$. Thus, either

- $j(E_1) = j(E_1')$ then $j(E_2) = j(E_2')$, so $E_1 \cong E_1'$ and $E_2 \cong E_2'$ and φ induces automorphisms on $\mathcal{E}_1$ and $\mathcal{E}_2$, or
- $j(E_1) = j(E_2')$, so $E_1 \cong E_2'$ and φ induces an isomorphism $\mathcal{E}_1 \to \mathcal{E}_2$.

But the j-invariants of E_i and E_i' are given by the roots of Equation (3.1), so the first (resp. second) case occurs if and only if the action of φ preserves (resp. exchanges) these roots. □

Let $H \subset G$ denote the subgroup of G that preserves the decomposition of $\{F_3, \ldots, F_8\}$ into subsets $\{F_3, F_4, F_5\}$ and $\{F_6, F_7, F_8\}$. Then we can say more about the action of the subgroup H on $\mathcal{E}_1$ and $\mathcal{E}_2$.

PROPOSITION 3.9. (See [**4**, Lemmas 7.6 and 7.7]) *Let τ be any permutation in $H \subset S_6$ and let φ denote the corresponding map on C. Then*

- *If τ is an odd permutation, then φ induces an isomorphism $\mathcal{E}_1 \to \mathcal{E}_2$.*
- *If τ is an even permutation, then φ induces automorphisms of $\mathcal{E}_1$ and $\mathcal{E}_2$*

PROOF. Suppose first that τ is an odd permutation. Let γ denote a path in V that connects a point $p \in V$ to $\varphi(p)$. We will show that as we move along γ, the j-invariants of E_1 and E_2 are switched. To do this we use f to push everything down to B. The image $f(\gamma)$ is a loop in U starting and ending at $f(p)$. By Lemma 3.8, we therefore need to show that monodromy around $f(\gamma)$ switches the roots of Equation (3.1).

Monodromy around $f(\gamma)$ acts on the set of divisors $\{F_3, \ldots, F_8\}$ in the fibre $Y_{f(p)}$ of $\mathcal{Y}$ over $f(p)$ (see Section 2.1) as the permutation τ. Furthermore, as H is the subgroup of G that preserves the sets $\{F_3, F_4, F_5\}$ and $\{F_6, F_7, F_8\}$, monodromy around $f(\gamma)$ must also preserve these sets.

As these divisors are permuted if and only if the roots of the cubic polynomials $(P(x) \pm 1)$ (see Equation (2.1)) are permuted and as τ is an odd permutation, we see that the product of the discriminants of these cubics must vanish to odd order inside γ. This product is given by

$$\Delta := a^6 - 2a^3b^2 + b^4 - 2a^3 - 2b^2 + 1,$$

where a and b are the (a, b)-parameters associated to the M-polarized fibre $X_{f(p)}$ of $\mathcal{X}$ over $f(p)$.

Now, monodromy around $f(\gamma)$ switches the roots of Equation (3.1) if and only if its discriminant $(\sigma^2 - 4\pi)$ vanishes to odd order inside γ. However, by Remark 3.6, we find that this discriminant is given exactly by Δ. So monodromy around $f(\gamma)$ switches the roots of Equation (3.1) and thus, by Lemma 3.8, it induces an isomorphism $\mathcal{E}_1 \to \mathcal{E}_2$.

This completes the proof in the case when τ is an odd permutation. The proof when τ is an even permutation is similar. □

To construct a model for $\mathcal{Y}_U$, our starting data consists of the cover $f\colon C \to B$ and the two elliptic surfaces $\mathcal{E}_{1,2} \to C$. This data must satisfy the condition that the deck transformation group G of the cover f should act as automorphisms on or isomorphisms between the elliptic surfaces $\mathcal{E}_1$ and $\mathcal{E}_2$, in a way that is compatible with its action on C.

We begin by constructing a model for $\mathcal{Y}_V'$ by performing the Kummer construction fibrewise on $\mathcal{E}_1 \times_C \mathcal{E}_2$ to obtain a new threefold $\mathcal{Y}'$, which is isomorphic to $\mathcal{Y}_V'$ over V. Then, to obtain a model for $\mathcal{Y}_U$, we perform a quotient of $\mathcal{Y}'$ by

G. However, the action of G is not the obvious one induced by the action of G on $\mathcal{E}_1$ and $\mathcal{E}_2$ (if it were, we would be able to undo the Kummer construction on $\mathcal{Y}_U$, which is not possible in general). Instead, we compose this action with the fibrewise automorphism induced by the action of G (as a subset of S_6) on the set of curves $\{F_3, \ldots, F_8\}$.

REMARK 3.10. As the fibrewise Kummer construction defines a natural double Kummer pencil on smooth fibres of $\mathcal{F}$, we can use the results of Kuwata and Shioda [**16**, Section 5.2] to define the elliptic fibration Ψ on the smooth fibres of $\mathcal{F}$. The curves $\{F_3, \ldots, F_8\}$ are then the components of the I_2 fibres that do not meet a chosen section.

Once the curves $\{F_3, \ldots, F_8\}$ are known, the automorphisms permuting them may be computed explicitly as compositions of the relevant symplectic automorphisms from [**14**, Section 4].

Quotienting by this G-action, we obtain a new threefold $\mathcal{Y} \to B$. By construction, $\mathcal{Y}$ is isomorphic to $\mathcal{Y}_U$ over U, as required. We have a diagram:

$$\begin{array}{ccccccc} \mathcal{E}_1 \times_C \mathcal{E}_2 & \overset{\text{Kummer}}{\dashrightarrow} & \mathcal{Y}' & \longrightarrow & \mathcal{Y} & \overset{\text{Nikulin}}{\dashleftarrow} & \mathcal{X} \\ \uparrow & & \uparrow & & \uparrow & & \uparrow \\ \mathcal{A}_V & \longrightarrow & \mathcal{Y}'_V & \longrightarrow & \mathcal{Y}_U & \dashleftarrow & \mathcal{X}_U \end{array}$$

This construction will be illustrated in the next section.

4. Some Calabi-Yau threefolds fibred by Kummer surfaces

In the remainder of this paper, we will illustrate how these methods can be used to construct explicit examples of Calabi-Yau threefolds. We note that, in this paper, a *Calabi-Yau threefold* will always be a smooth projective threefold $\mathcal{X}$ with $\omega_{\mathcal{X}} \cong \mathcal{O}_{\mathcal{X}}$ and $H^1(\mathcal{X}, \mathcal{O}_{\mathcal{X}}) = 0$. We further note that the cohomological condition in this definition implies that any fibration of a Calabi-Yau threefold by K3 surfaces must have base curve $\mathbb{P}^1$, so from this point we restrict our attention to the case where $B \cong \mathbb{P}^1$.

As our starting point, we will take the K3-fibred threefolds $\mathcal{X}_g \to \mathbb{P}^1$ constructed in [**9**]. By construction, the Néron-Severi group of a general fibre in these threefolds is isometric to $M_2 = H \oplus E_8 \oplus E_8 \oplus \langle -4 \rangle$, which admits an obvious embedding $M \hookrightarrow M_2$, and the restriction $\mathcal{X}_U \to U$ of $\mathcal{X}_g$ to the subset $U \subset \mathbb{P}^1$ over which the fibres are smooth is an M_2-polarized family of K3 surfaces. Thus these threefolds satisfy all of the conditions of Section 2.

4.1. A special family. In [**9**], the threefolds $\mathcal{X}_g$ are constructed as resolved pull-backs of a special family $\mathcal{X}_2 \to \mathcal{M}_{M_2}$ over the (compact) 1-dimensional moduli space $\mathcal{M}_{M_2}$ of M_2-polarized K3 surfaces, by a map $g\colon \mathbb{P}^1 \to \mathcal{M}_{M_2}$. To study the threefolds related to the $\mathcal{X}_g$ by the construction detailed above, we will begin by studying $\mathcal{X}_2$.

The family $\mathcal{X}_2 \to \mathcal{M}_{M_2}$ is described in [**10**, Section 5.4.1]. It is given as the minimal resolution of the family of hypersurfaces in $\mathbb{P}^3$ obtained by varying λ in the following expression

$$\lambda w^4 + xyz(x + y + z - w) = 0. \tag{4.1}$$

This family has also been studied extensively by Narumiya and Shiga [**20**], we will use some of their results in the sequel (note, however, that our λ is not the same as the λ used in [**20**], instead, our λ is equal to μ^4 or $\frac{u}{256}$ from [**20**]).

Dolgachev [**8**, Theorem 7.1] proved that $\mathcal{M}_{M_2}$ is isomorphic to the compactification of the modular curve $\Gamma_0(2)^+ \setminus \mathbb{H}$. In [**10**, Section 5.4.1] it is shown that the orbifold points of orders $(2, 4, \infty)$ on this curve occur at $\lambda = (\frac{1}{256}, \infty, 0)$ respectively, and that the K3 fibres of $\mathcal{X}_2$ are smooth away from these three points. Let U_{M_2} denote the open set obtained from $\mathcal{M}_{M_2}$ by removing these three points. Then the restriction $\mathcal{X}_{2,U}$ of $\mathcal{X}_2$ to U_{M_2} is an M_2-polarized family of K3 surfaces, and $\mathcal{X}_2$ satisfies all of the conditions of Section 2.

We next compute the (a, b, d)-parameters of the fibres of $\mathcal{X}_{2,U}$ (considered as M-polarized K3 surfaces, see Section 2.1). To do this, we use the fact that the standard and alternate fibrations on the K3 fibres are torically induced, so their g_2 and g_3 invariants may be computed (in terms of λ) using the toric geometry functionality of the computer software *Sage*. These expressions can then be compared to the corresponding expressions computed for an M-polarized K3 surface in normal form (see, for instance, [**5**, Theorem 3.1]). We thus obtain

$$a = \lambda + \frac{1}{2^4 3^2}, \qquad b = \frac{3}{8}\lambda - \frac{1}{2^6 3^3}, \qquad d = \lambda^3,$$

The σ and π invariants for the family $\mathcal{X}_{2,U}$ are then given in terms of λ by

$$\begin{aligned} \sigma &:= 2 - \frac{23}{2^6 3 \lambda} + \frac{1}{2^6 3^3 \lambda^2}, \\ \pi &:= \left(1 + \frac{1}{2^4 3^2 \lambda}\right)^3. \end{aligned}$$

4.2. Undoing the Kummer construction. Let $\mathcal{Y}_{2,U} \to U_{M_2}$ denote the family of Kummer surfaces obtained from $\mathcal{X}_2$ by quotienting by the Nikulin involution and resolving any resulting singularities. From Remark 3.3 and the (a, b, d) parameters for $\mathcal{X}_2$ computed above, we see that Assumption 3.2 is satisfied by $\mathcal{Y}_{2,U}$. We will explicitly show how to construct a model for $\mathcal{Y}_{2,U}$ from elliptic surfaces, as described in Section 3.2.

Our first step is to undo the Kummer construction for $\mathcal{Y}_{2,U} \to U_{M_2}$. To do this, we proceed to a cover $f\colon C_{M_2} \to \mathcal{M}_{M_2}$, as computed in [**10**, Section 5.3.2]. This cover is calculated in three steps. The first step is to take the cover $f_1\colon \Gamma_0(2) \setminus \mathbb{H} \to \mathcal{M}_{M_2}$. This is a double cover ramified over $\lambda \in \{\frac{1}{256}, \infty\}$, which is given in coordinates by

$$\lambda = -\mu^2 + \frac{1}{256},$$

where μ is a coordinate on $\Gamma_0(2) \setminus \mathbb{H}$ and the orbifold points of orders $(2, \infty, \infty)$ occur at $\mu = (\infty, \frac{1}{16}, -\frac{1}{16})$ respectively.

The second step is to take the cover $f_2\colon \Gamma_0(4) \setminus \mathbb{H} \to \Gamma_0(2) \setminus \mathbb{H}$. This is a double cover ramified over $\mu \in \{\frac{1}{16}, \infty\}$, which is given in coordinates by

$$\mu = -(\mu')^2 + \frac{1}{16},$$

where μ' is a coordinate on $\Gamma_0(4) \setminus \mathbb{H}$ and the three cusps occur at $\mu' = (0, \frac{1}{\sqrt{8}}, -\frac{1}{\sqrt{8}})$.

Finally, the third step is to take the cover $f_3\colon C_{M_2} \to \Gamma_0(4)\setminus\mathbb{H}$. This is a double cover ramified over $\mu' \in \{\frac{1}{\sqrt{8}}, -\frac{1}{\sqrt{8}}\}$, which is given in coordinates by

$$\mu' = \frac{1}{\sqrt{8}}\frac{(1-\nu^2)}{(1+\nu^2)},$$

where ν is a coordinate on the rational curve C_{M_2}, which has four cusps occurring at $\nu \in \{0, 1, -1, \infty\}$.

The composition of these three maps is the map $f\colon C_{M_2} \to \mathcal{M}_{M_2}$ given in coordinates by

$$\lambda = \frac{1}{16}\frac{\nu^2(1-\nu^2)^2}{(1+\nu^2)^4}.$$

This is an 8-fold cover with deck transformation group D_8. It coincides precisely with the change of coordinates computed in [**20**, Equation (7.1)]. Given this, the pulled-back family of Kummer surfaces over C_{M_2} is given in affine coordinates by the expression $K(\nu)$ in [**20**, Section 7]:

$$u^2 = s(s-1)(s-\left(\frac{\nu+1}{\nu-1}\right)^2)t(t-1)(t-\nu^2) \tag{4.2}$$

We pause here to set up a little more notation. Let $V = f^{-1}(U_{M_2})$ and let $\mathcal{Y}'_{2,V}$ denote the pull-back of $\mathcal{Y}_{2,U}$ to V. By [**10**, Propositions 4.1 and 5.6], we may undo the Kummer construction for $\mathcal{Y}'_{2,V}$, so there is a family of Abelian surfaces $\mathcal{A}_{2,V}$ that gives rise to $\mathcal{Y}'_{2,V}$ under fibrewise application of the Kummer construction.

By Proposition 3.5, $\mathcal{A}_{2,V}$ is isomorphic over V to a fibre product $\mathcal{E}_1 \times_{C_{M_2}} \mathcal{E}_2$ of minimal elliptic surfaces $\mathcal{E}_i \to C_{M_2}$ with section. These elliptic surfaces $\mathcal{E}_i \to C_{M_2}$ are given in affine coordinates by the expressions $E_i(\nu)$ in [**20**, Section 7]. We find

$$\begin{aligned} \mathcal{E}_1\colon \quad & z^2 = t(t-1)(t-\nu^2),\\ \mathcal{E}_2\colon \quad & w^2 = s(s-1)(s-\left(\frac{\nu+1}{\nu-1}\right)^2). \end{aligned}$$

From this, we see that $\mathcal{E}_2$ can be obtained from $\mathcal{E}_1$ by applying the involution $\nu \mapsto \frac{\nu+1}{\nu-1}$. Thus, it suffices to study the elliptic surface $\mathcal{E}_1$.

The j-invariant for $\mathcal{E}_1$ is given by the expression (which can be computed directly, or by using Proposition 3.5 and the fibrewise σ and π invariants for $\mathcal{X}_{2,U}$ calculated in Section 4.1)

$$j = \frac{4}{27}\frac{(\nu^4-\nu^2+1)^3}{\nu^4(\nu-1)^2(\nu+1)^2}.$$

The fibres of $\mathcal{E}_1$ with j-invariants in the set $\{0, 1, \infty\}$ are given in Table 4.1, where ν gives the location of the fibre in terms of the parameter ν on C_{M_2}, j gives the corresponding value of the j-invariant, Multiplicity gives the order of vanishing of j and Type gives the type of singular fibre. Finally, in the first column, ω denotes a primitive twelfth root of unity.

From this, we see that $\mathcal{E}_1$ (and thus also $\mathcal{E}_2$) is a rational elliptic surface with all fibres of type I_n (in fact, it is even an *extremal* rational elliptic surface, see [**17**, Section VIII.1]). The fibre product $\mathcal{E}_1 \times_{C_{M_2}} \mathcal{E}_2$ is a singular threefold with isolated singularities in the fibres over $\nu \in \{0, 1, -1, \infty\}$. Such threefolds have been studied by Schoen [**22**, Lemma 3.1], who showed that $\mathcal{E}_1 \times_{C_{M_2}} \mathcal{E}_2$ admits a small projective resolution. Denote such a resolution by $\mathcal{A}_2$.

$\nu \in$	j	Multiplicity	Type
$\{1,-1\}$	∞	2	I_2
$\{0,\infty\}$	∞	4	I_4
$\{i,-i,\sqrt{2},-\sqrt{2},\frac{1}{\sqrt{2}},-\frac{1}{\sqrt{2}}\}$	1	2	I_0
$\{\omega,-\omega,\frac{1}{\omega},-\frac{1}{\omega}\}$	0	3	I_0

TABLE 4.1. Fibres of $\mathcal{E}_1$ for $j \in \{0, 1, \infty\}$.

Our next task is to construct a model for $\mathcal{Y}'_{2,V}$. It follows from [**22**, Lemma 6.1] that the involution defining the fibrewise Kummer construction on $\mathcal{E}_1 \times_{C_{M_2}} \mathcal{E}_2$ lifts to the resolution $\mathcal{A}_2$. We can thus construct a model $\mathcal{Y}'_2 \to C_{M_2}$ for $\mathcal{Y}'_{2,V}$ by performing the fibrewise Kummer construction on $\mathcal{A}_2$.

4.3. Two special Calabi-Yau threefolds. We digress briefly to study the properties of the threefolds $\mathcal{A}_2$ and $\mathcal{Y}'_2$. The threefold $\mathcal{A}_2$ is smooth and projective by construction and is isomorphic over V to $\mathcal{A}_{2,V}$. An easy application of adjunction shows that $\mathcal{A}_2$ has trivial canonical bundle $\omega_{\mathcal{A}_2} \cong \mathcal{O}_{\mathcal{A}_2}$ and, by [**22**, Remark 2.1], we see that it is also simply connected. $\mathcal{A}_2$ is thus a Calabi-Yau threefold.

The invariants of $\mathcal{A}_2$ can be computed using the methods of Schoen [**22**]. By the discussion in [**22**, Section 2], we see that the Euler characteristic $e(\mathcal{A}_2)$ is equal to 64. Furthermore, [**22**, Proposition 7.1] implies that $\mathcal{A}_2$ is a rigid threefold, so $h^{2,1}(\mathcal{A}_2) = 0$. Thus we find that the remaining Hodge number $h^{1,1}(\mathcal{A}_2) = 32$.

The threefold $\mathcal{Y}'_2$ is also smooth and projective, has trivial canonical bundle $\omega_{\mathcal{Y}'_2} \cong \mathcal{O}_{\mathcal{Y}'_2}$ and, by [**22**, Remark 6.5], is simply connected. $\mathcal{Y}'_2$ is therefore also a Calabi-Yau threefold. Note that $\mathcal{Y}'_2$ is a resolution of the threefold given in affine coordinates by Equation (4.2).

The Euler characteristic of $\mathcal{Y}'_2$ can be computed using [**22**, Lemma 6.3], to obtain $e(\mathcal{Y}'_2) = 80$. However, the Hodge numbers of $\mathcal{Y}'_2$ are not quite so simple to compute.

PROPOSITION 4.1. *$\mathcal{Y}'_2$ has Hodge numbers $h^{2,1}(\mathcal{Y}'_2) = 0$ and $h^{1,1}(\mathcal{Y}'_2) = 40$.*

PROOF. We compute $h^{2,1}(\mathcal{Y}'_2)$ using the method of [**13**, Section 2.1]. From the affine form (4.2), we see that $\mathcal{Y}'_2$ is birational to a double cover of the weighted projective space $\mathbb{WP}(1,1,2,2)$ ramified over the pair of weighted cones given by

$$s(s-(\mu+\nu)^2)(s-(\mu-\nu)^2)t(t-\mu^2)(t-\nu^2) = 0, \tag{4.3}$$

where (μ,ν,s,t) are homogeneous coordinates on $\mathbb{WP}(1,1,2,2)$ of weights $(1,1,2,2)$ respectively. We may therefore compute the space of deformations of $\mathcal{Y}'_2$ using the methods of [**7**].

However, in order to apply these methods we first need to ensure that our base space is smooth. Let Z denote the blow-up of $\mathbb{WP}(1,1,2,2)$ along $\mu = \nu = 0$. Then Z is a smooth variety. Let $B \subset Z$ denote the pull-back of the branch locus (4.3) and let $(\hat{Z},\hat{B}) \to (Z,B)$ be a resolution of the singularities of B. Then $\mathcal{Y}'_2$ is birational to a double cover of $\hat{Z}$ ramified over $\hat{B}$.

By [**7**, Proposition 2.1], the space of deformations of $\mathcal{Y}'_2$ is isomorphic to

$$H^1(\hat{Z},\Theta_{\hat{Z}}(\log \hat{B})) \oplus H^1(\hat{Z},\Theta_{\hat{Z}} \otimes \mathcal{O}_{\hat{Z}}(-6)),$$

where $\Theta_{\hat{Z}}$ denotes the tangent bundle of $\hat{Z}$ and $\Theta_{\hat{Z}}(\log \hat{B})$ is the kernel of the natural restriction map $\Theta_{\hat{Z}} \to \mathcal{N}_{\hat{B}|\hat{Z}}$. We will show that each of these direct summands vanishes, so that the space of deformations of $\mathcal{Y}_2'$ is trivial.

Since Z is rigid (see [**13**, Remark 2.21]), applying [**7**, Corollary 2] we find that $H^1(\hat{Z}, \Theta_{\hat{Z}}(\log \hat{B}))$ is isomorphic to the space of equisingular deformations of B in Z. Furthermore, by [**13**, Proposition 2.19], any equisingular deformation of B in Z induces a deformation of $\mathcal{A}_2$. But we saw above that $\mathcal{A}_2$ is rigid, so we must have $H^1(\hat{Z}, \Theta_{\hat{Z}}(\log \hat{B})) = 0$.

To compute $H^1(\hat{Z}, \Theta_{\hat{Z}} \otimes \mathcal{O}_{\hat{Z}}(-6))$ we use [**7**, Proposition 5.1]. Suppose that the resolution $\hat{Z} \to Z$ is a composition of blow-ups along subvarieties C_i. Then [**7**, Proposition 5.1] gives an equation

$$h^1(\hat{Z}, \Theta_{\hat{Z}} \otimes \mathcal{O}_{\hat{Z}}(-6)) = h^1(Z, \Theta_Z \otimes \mathcal{O}_Z(-6)) + \sum_{\mathrm{codim} C_i = 2} h^0(C_i, K_{C_i}).$$

Now, as $K_Z = \mathcal{O}_Z(-6)$, we have

$$H^1(Z, \Theta_Z \otimes \mathcal{O}_Z(-6)) \cong H^1(Z, \Omega_Z^2 \otimes K_Z^\vee \otimes \mathcal{O}_Z(-6)) = H^1(Z, \Omega_Z^2),$$

which vanishes by [**1**, (2.3)].

Thus to show that $H^1(\hat{Z}, \Theta_{\hat{Z}} \otimes \mathcal{O}_{\hat{Z}}(-6)) = 0$ it suffices to show that the curves C_i are all rational. These curves arise from the multiple curves in the divisor B, which can be divided into three classes:

(1) multiple curves arising from the preimages of the two points $(0,0,1,0)$ and $(0,0,0,1)$ under the blow-up $Z \to \mathbb{WP}(1,1,2,2)$,
(2) the strict transforms in Z of the double curves lying in the weighted cones $s(s-(\mu+\nu)^2)(s-(\mu-\nu)^2) = 0$ and $t(t-\mu^2)(t-\nu^2) = 0$, and
(3) the strict transforms in Z of the double curves arising from the intersection between these two weighted cones.

The preimages in Z of the two points $(0,0,1,0)$ and $(0,0,0,1)$ are copies of $\mathbb{P}^1$ and appear with multiplicity 3 in B. To resolve the singularities of B, we blow up once along each copy of $\mathbb{P}^1$, then again along the six further $\mathbb{P}^1$'s arising as the intersection between the exceptional loci and the strict transform of B. As all curves blown up by this procedure are $\mathbb{P}^1$'s, they do not contribute to $h^1(\hat{Z}, \Theta_{\hat{Z}} \otimes \mathcal{O}_{\hat{Z}}(-6))$.

The curves of the second class are also all isomorphic to $\mathbb{P}^1$ and each gets blown up once to resolve B. Therefore they also do not contribute to $h^1(\hat{Z}, \Theta_{\hat{Z}} \otimes \mathcal{O}_{\hat{Z}}(-6))$.

The curves of the third class are the most difficult. There are nine of them and each gets blown up once to resolve B. To see that they are all rational, we note that each is a section of the fibration $Z \to \mathbb{P}^1$ given by projecting onto (μ, ν). So they do not contribute to $h^1(\hat{Z}, \Theta_{\hat{Z}} \otimes \mathcal{O}_{\hat{Z}}(-6))$, which is therefore trivial.

Thus we find that the space of deformations of $\mathcal{Y}_2'$ is trivial. But this implies that $h^{2,1}(\mathcal{Y}_2') = 0$. Finally, the statement that $h^{1,1}(\mathcal{Y}_2') = 40$ follows immediately from the fact that $e(\mathcal{Y}_2') = 80$. □

REMARK 4.2. We note that we can identify a lot of data about the Kummer fibration on $\mathcal{Y}_2'$ explicitly from $\hat{Z}$. This fibration is induced by the fibration $Z \to \mathbb{P}^1$ given by projecting onto (μ, ν). The divisors G_i in the double Kummer pencil on $\mathcal{Y}_2'$ are the lifts to $\mathcal{Y}_2'$ of the strict transforms of the three components of the cone $t(t-\mu^2)(t-\nu^2) = 0$ and the first exceptional divisor arising from the blow up of the curve in Z over $(0,0,1,0)$, and the divisors H_j are given by the same procedure

λ	Monodromy
0	(14)(25)(36)
$\frac{1}{256}$	(12)
∞	(1524)(36)

TABLE 4.2. Action of monodromy around punctures in U_{M_2} on $\{F_3, \ldots, F_8\}$.

applied to the cone $s(s-(\mu+\nu)^2)(s-(\mu-\nu)^2) = 0$. Finally, the sixteen divisors E_{ij} are given by the lifts to $\mathcal{Y}_2'$ of the nine exceptional divisors arising from blow-ups of curves of class (3), the six exceptional divisors arising from the second blow-up of curves of class (1), and the divisor over $\mu = \nu = 0$.

4.4. The action of D_8. To get from $\mathcal{Y}_2'$ to a model for $\mathcal{Y}_{2,U}$ we need to understand the action of the group D_8 on $\mathcal{Y}_2'$. The action of D_8 on the base curve C_{M_2} is generated by the automorphisms

$$\alpha\colon \nu \longmapsto \frac{\nu-1}{\nu+1},$$
$$\beta\colon \nu \longmapsto -\nu,$$

satisfying $\alpha^4 = \beta^2 = \mathrm{Id}$ and $\beta \circ \alpha \circ \beta = \alpha^{-1}$. Note that the automorphism α interchanges $\mathcal{E}_1$ and $\mathcal{E}_2$, whilst β preserves them.

We next compute the action of α and β on the threefold $\mathcal{Y}_2'$. To do this, we need to understand how these automorphisms act on the fibres of $\mathcal{Y}_2'$. But this action is identical to that induced on the fibres of $\mathcal{Y}_{2,U}$ by monodromy around appropriately chosen loops in U_{M_2}.

By the results of [**10**, Section 4.3], the action of monodromy around loops in U_{M_2} on the fibres of $\mathcal{Y}_{2,U}$ is completely encoded by its action on the divisors $\{F_3, \ldots, F_8\}$. This action may be calculated using the fact (see Section 2.1) that these divisors lie in the six I_2 fibres of the elliptic fibration Ψ on the K3 fibres of $\mathcal{Y}_{2,U}$, the locations of which are given by the roots of the polynomials $(P(x) \pm 1)$, where $P(x)$ is as defined in Equation (2.1) (and (a, b) in this equation are the (a, b) values associated to the M-polarized fibres of $\mathcal{X}_{2,U}$, calculated in Section 4.1, normalized so that $d = 1$).

Thus, to compute the action of monodromy on the divisors $\{F_3, \ldots, F_8\}$, it suffices to compute its action on the roots of $(\Gamma(x) \perp 1)$. This action may be calculated explicitly using the `monodromy` command in *Maple*'s `algcurves` package, with the base point $\lambda = -\frac{257}{256}$ (which was chosen for ease of lifting to the covers of U_{M_2}). We obtain Table 4.2, where λ gives the λ-value at the puncture in U_{M_2} around which monodromy occurs and Monodromy gives the action of anticlockwise monodromy around that point on the divisors $\{F_3, \ldots, F_8\}$ in the fibre over $\lambda = -\frac{257}{256}$, expressed as a permutation in S_6, where we assign the labels $1, \ldots, 6$ to $F_3, \ldots, F_8$ respectively.

Now we lift this information to $\mathcal{Y}_2'$. There is a natural double Kummer pencil on the fibres of $\mathcal{Y}_{2,V}'$, defined by the fibrewise Kummer construction, so we may use the method of Remark 3.10 to define the elliptic fibration Ψ and the divisors $\{F_3, \ldots, F_8\}$ on the K3 fibres of $\mathcal{Y}_{2,V}'$. We can then compute the action of the automorphisms α and β on the divisors $\{F_3, \ldots, F_8\}$ (calculated using the base point

$\nu = -\frac{\sqrt{17}}{4}$). We find that α acts on these divisors as the permutation (1524)(36) and β acts as the permutation (14)(25)(36).

With this in place, we are now ready to compute the action of α and β on the threefold $\mathcal{Y}_2'$. The expressions we compute will use the coordinates (ν, s, t, u) from the affine description given in Equation (4.2).

We begin with the involution β. From the calculations above, β should be induced by $\nu \mapsto -\nu$ on the base C_{M_2}, so fixes the points $\nu \in \{0, \infty\}$. The monodromy calculations above show that β should exchange the divisors $F_i \leftrightarrow F_{i+3}$ (for $i \in \{3,4,5\}$) in the fibres of $\mathcal{Y}_2'$. It is easy to show that this action is realised by the involution β given by

$$\beta\colon (\nu, s, t, u) \longmapsto \left(-\nu, \left(\frac{\nu-1}{\nu+1}\right)^2 s, t, \left(\frac{\nu-1}{\nu+1}\right)^3 u\right).$$

Instead of computing the action of α directly, we will instead compute the action of the involution $\alpha \circ \beta$. This involution should be induced by $\nu \mapsto \frac{\nu+1}{\nu-1}$ on the base C_{M_2}, so fixes the points $\nu = 1 \pm \sqrt{2}$. The monodromy calculations above show that $\alpha \circ \beta$ should fix the divisors $\{F_3, F_4, F_5, F_8\}$ and exchange $F_6 \leftrightarrow F_7$ in the fibres of $\mathcal{Y}_2'$.

If we compute the involution induced on $\mathcal{Y}_2'$ by $\nu \mapsto \frac{\nu+1}{\nu-1}$, we obtain

$$\iota\colon (\nu, s, t, u) \longmapsto \left(\frac{\nu+1}{\nu-1}, t, s, u\right).$$

However, ι cannot be equal to $\alpha \circ \beta$: it does not preserve the fibration Ψ on the fibres of $\mathcal{Y}_2'$ (as it switches the divisors G_i and H_i from the double Kummer pencil, which has the effect of permuting the sections of Ψ) and it does not exchange the divisors $F_6 \leftrightarrow F_7$. Both of these problems can be rectified by composing with the unique fibrewise symplectic automorphism φ from [**14**, Section 4.1] that exchanges $G_i \leftrightarrow H_i$ and $F_6 \leftrightarrow F_7$ (which may be computed by the method of [**4**, Example 7.9]).

Thus, we find that $\alpha \circ \beta = \varphi \circ \iota$. Composing with β, we find that α is given as the composition $\varphi \circ \iota'$, where

$$\iota'\colon (\nu, s, t, u) \longmapsto \left(\frac{\nu-1}{\nu+1}, \frac{t}{\nu^2}, s, \frac{u}{\nu^3}\right)$$

and φ is as before.

Define $\mathcal{Y}_2$ to be the threefold obtained by quotienting $\mathcal{Y}_2'$ by the action of D_8 described above and resolving any resulting singularities. Then $\mathcal{Y}_2$ is isomorphic to $\mathcal{Y}_{2,U}$ over U_{M_2} by construction, so gives the required model for $\mathcal{Y}_{2,U}$. These threefolds fit together in a diagram:

$$\begin{array}{ccccccccc}
\mathcal{E}_1 \times_{C_{M_2}} \mathcal{E}_2 & \xleftarrow{\text{resolve}} & \mathcal{A}_2 & \xrightarrow{\text{Kummer}} & \mathcal{Y}_2' & \dashrightarrow & \mathcal{Y}_2 & \xleftarrow{\text{Nikulin}} & \mathcal{X}_2 \\
\downarrow & & \downarrow & & \downarrow & & \downarrow & & \downarrow \\
C_{M_2} & = & C_{M_2} & = & C_{M_2} & \xrightarrow{f} & \mathcal{M}_{M_2} & = & \mathcal{M}_{M_2}
\end{array}$$

4.5. Singular fibres. Our next task is to study the forms of the singular fibres in the threefolds from the above diagram. These singular fibres come in three flavours, lying over $\lambda \in \{0, \frac{1}{256}, \infty\} \subset \mathcal{M}_{M_2}$; we will study each in turn.

4.5.1. *Fibres over* $\lambda = 0$. The first singular fibres we will study lie over the point $\lambda = 0$, which is the cusp in the orbifold base curve. Let Δ_0 denote a small disc around $\lambda = 0$ and let Δ'_0 denote one of the connected components of the preimage of Δ under f. The map $\Delta'_0 \to \Delta_0$ is a double cover ramified over $\lambda = 0 \in \Delta_0$. The preimages of $\lambda = 0$ are $\nu \in \{0, 1, -1, \infty\}$; without loss of generality we will assume that Δ'_0 is the connected component containing $\nu = 0$, the other choices give identical results.

The singular fibre of $\mathcal{E}_1 \times_{C_{M_2}} \mathcal{E}_2$ over $\nu = 0$ is a product $I_2 \times I_4$ of singular elliptic curves and has 8 components, each of which is isomorphic to $\mathbb{P}^1 \times \mathbb{P}^1$. The threefold $\mathcal{E}_1 \times_{C_{M_2}} \mathcal{E}_2$ has eight nodes occurring at the points where four components of the central fibre intersect; these may be resolved by a small projective resolution [**22**, Lemma 3.1] to give the threefold $\mathcal{A}_2$, which is smooth. The singular fibre of $\mathcal{A}_2$ over $\nu = 0$ again has 8 components, which are rational surfaces.

The involution defining the Kummer construction fixes four of the components of the $\nu = 0$ singular fibre of $\mathcal{A}_2$ and acts to exchange the other four as two pairs. The resultant singular fibre of $\mathcal{Y}'_2$ has six components, each of which is a rational surface, arranged in a cube and the threefold $\mathcal{Y}'_2$ is again smooth.

Finally, the involution defining the quotient $\mathcal{Y}'_2 \to \mathcal{Y}_2$ is given by β from Section 4.4. This involution acts trivially on two of the components of the $\nu = 0$ fibre in $\mathcal{Y}'_2$ (the "top" and "bottom" of the cube) and acts as an involution on each of the remaining four (the "sides" of the cube). Upon performing the quotient, these four components become exceptional and may be contracted, giving a singular fibre consisting of two rational components in the threefold $\mathcal{Y}_2$ (which is smooth over Δ_0). These rational components meet along four rational curves, which form a cycle in each component.

4.5.2. *Fibres over* $\lambda = \frac{1}{256}$. The next type of singular fibres lie over the point $\lambda = \frac{1}{256}$, which is an orbifold point of order 2 in the base curve. As before, let $\Delta_{\frac{1}{256}}$ denote a small disc around $\lambda = \frac{1}{256}$ and let $\Delta'_{\frac{1}{256}}$ denote one of the connected components of the preimage of $\Delta_{\frac{1}{256}}$ under f. The map $\Delta'_{\frac{1}{256}} \to \Delta_{\frac{1}{256}}$ is a double cover ramified over $\lambda = \frac{1}{256} \in \Delta$. The preimages of $\lambda = \frac{1}{256}$ are $\nu \in \{1 \pm \sqrt{2}, -1 \pm \sqrt{2}\}$; without loss of generality we will assume that $\Delta'_{\frac{1}{256}}$ is the connected component containing $\nu = 1 + \sqrt{2}$, the other choices give identical results.

The fibre of $\mathcal{E}_1 \times_{C_{M_2}} \mathcal{E}_2$ over $\nu = 1 + \sqrt{2}$ is a product $E \times E$ of a smooth elliptic curve E with $j(E) = \frac{125}{27}$ with itself, so the threefold $\mathcal{E}_1 \times_{C_{M_2}} \mathcal{E}_2$ is smooth and isomorphic to $\mathcal{A}_2$ over $\Delta'_{\frac{1}{256}}$. The fibre of $\mathcal{Y}'_2$ over $\nu = 1 + \sqrt{2}$ is thus just the Kummer surface associated to a product of an elliptic curve with itself and the threefold $\mathcal{Y}'_2$ is smooth.

The involution defining the quotient $\mathcal{Y}'_2 \to \mathcal{Y}_2$ is given by $\beta\alpha = \varphi \circ \iota$ from Section 4.4. Its action is calculated in the same way as [**4**, Example 7.9] and yields the same result: for the action of the fibrewise involution φ to be well-defined we must first contract two curves lying in the fibre of $\mathcal{Y}'_2$ over $\nu = 1 + \sqrt{2}$, giving two nodes in the threefold total space. After this contraction has been performed the involution $\beta\alpha$ is well-defined and acts trivially on the fibre over $\nu = 1 + \sqrt{2}$.

We find that the threefold $\mathcal{Y}_2$ is smooth over $\Delta_{\frac{1}{256}}$ and its fibre over $\lambda = \frac{1}{256}$ is a singular K3 surface containing two A_1 singularities. Under the double cover ramified over the fibre $\lambda = \frac{1}{256}$ these become two nodes in the threefold total space, which may be crepantly resolved to give the threefold $\mathcal{Y}_2'$.

4.5.3. *Fibres over* $\lambda = \infty$. The final, and most difficult, type of singular fibres lie over the point $\lambda = \infty$, which is an orbifold point of order 4 in the base curve. As before, let Δ_∞ denote a small disc around $\lambda = \infty$ and let Δ_∞' denote one of the connected components of the preimage of Δ_∞ under f. The map $\Delta_\infty' \to \Delta_\infty$ is a cyclic four-fold cover ramified over $\lambda = \infty \in \Delta_\infty$. The preimages of $\lambda = \infty$ are $\nu = \pm i$; without loss of generality we will assume that Δ_∞' is the connected component containing $\nu = i$, the other choice gives the same result.

The fibre of $\mathcal{E}_1 \times_{C_{M_2}} \mathcal{E}_2$ over $\nu = i$ is a product $E \times E$ of a smooth elliptic curve E with $j(E) = 1$ with itself, so the threefold $\mathcal{E}_1 \times_{C_{M_2}} \mathcal{E}_2$ is smooth and isomorphic to $\mathcal{A}_2$ over Δ_∞'. The fibre of $\mathcal{Y}_2'$ over $\nu = i$ is thus just the Kummer surface associated to a product of an elliptic curve with itself and the threefold $\mathcal{Y}_2'$ is smooth.

We compute the fibre of $\mathcal{Y}_2$ over $\lambda = \infty$ in two stages. First, we compute the quotient of $\mathcal{Y}_2'$ by the involution α^2. This involution acts on the fibre over $\nu = i$ as $u \mapsto -u$, which fixes the four curves G_i and H_j. The quotient by α^2 therefore has a fibre of multiplicity 2 and eight disjoint curves of cA_1 singularities, given by the images of the G_i and H_j. These curves may be crepantly resolved to give 8 exceptional divisors. The resulting threefold is smooth and has a singular fibre with 9 components: a rational component of multiplicity 2, isomorphic to $\mathbb{P}^1 \times \mathbb{P}^1$ blown up at sixteen points (the sixteen (-1)-curves are the images of the E_{ij}), meeting eight disjoint exceptional components of multiplicity 1, each of which is isomorphic to $\mathbb{F}_2$. It is an example of a "flowerpot degeneration" [**6**] with flowers of type 4α (see [**6**, Table 3.3]).

Next, we compute the quotient of this threefold by α, which now acts as an involution. This involution acts on the images of the G_i and H_j as follows: α fixes H_0 and H_1 pointwise, exchanges H_2 and H_3, and acts on each G_i as an involution that fixes the intersections between G_i and the images of E_{i0} and E_{i1}.

Thus, on the degenerate fibre, α acts as an involution on each of the components over the curves G_i, H_0 and H_1, and exchanges the components over H_2 and H_3. This action has twelve disjoint fixed curves: the curves H_0 and H_1, the eight fibres in the $\mathbb{F}_2$-components over the G_i that lie above the intersections $G_i \cap E_{i0}$ and $G_i \cap E_{i1}$ (two curves in each of four components), and the (-2)-sections in the $\mathbb{F}_2$-components over H_0 and H_1.

The degenerate fibre in the quotient by α therefore has eight components: one of multiplicity 4, coming from the quotient of the component of multiplicity 2, and seven of multiplicity 2, coming from the quotients of the eight components of multiplicity 1 (recall that the components over H_2 and H_3 are identified). Furthermore, the 12 fixed curves give rise to 12 disjoint curves of cA_1 singularities in the threefold total space, which may be crepantly resolved to obtain a further 12 exceptional components.

After performing this resolution, we find that the threefold $\mathcal{Y}_2$ is smooth over Δ_∞. Its fibre over $\lambda = \infty$ has 20 components: one of multiplicity 4, two of multiplicity 3, seven of multiplicity 2 and ten of multiplicity 1.

From this calculation and the adjunction formula for multiple covers, it is easy to see that:

THEOREM 4.3. *$\mathcal{Y}_2 \to \mathcal{M}_{M_2}$ is a smooth threefold fibred by Kummer surfaces with canonical bundle $\omega_{\mathcal{Y}_2} \cong \mathcal{O}_{\mathcal{Y}_2}(-F)$, where F is the class of a K3 surface fibre.*

Moreover, the restriction $\mathcal{Y}_{2,U}$ of $\mathcal{Y}_2$ to the open set U_{M_2} is isomorphic to the resolved quotient of $\mathcal{X}_{2,U}$ by the fibrewise Nikulin involution.

4.6. Constructing Calabi-Yau threefolds. Recall that, in [**9**], the threefolds $\mathcal{X}_g$ were constructed by pulling-back the family $\mathcal{X}_2 \to \mathcal{M}_{M_2}$ by a map $g\colon \mathbb{P}^1 \to \mathcal{M}_{M_2}$ and resolving singularities. The aim of this section is to perform the same construction with the threefold $\mathcal{Y}_2$, then to see how the resulting threefolds are related to the $\mathcal{X}_g$.

As in [**9**], let $g\colon \mathbb{P}^1 \to \mathcal{M}_{M_2}$ be an n-fold cover and let $[x_1, \dots, x_k]$, $[y_1, \dots, y_l]$ and $[z_1, \dots, z_m]$ be partitions of n encoding the ramification profiles of g over $\lambda = 0$, $\lambda = \infty$ and $\lambda = \frac{1}{256}$ respectively. Let r denote the degree of ramification of g away from $\lambda \in \{0, \frac{1}{256}, \infty\}$, defined to be

$$r := \sum_{\substack{p \in \mathbb{P}^1 \\ g(p) \notin \{0, \frac{1}{256}, \infty\}}} (e_p - 1),$$

where e_p denotes the ramification index of g at the point $p \in \mathbb{P}^1$.

Now let $\bar{\mathcal{Y}}_2$ denote the threefold obtained from $\mathcal{Y}_2$ by contracting all of the components in the fibre over $\lambda = \infty$ that have multiplicity less than 4 (in a neighbourhood of $\lambda = \infty$, the threefold $\bar{\mathcal{Y}}_2$ is isomorphic to the quotient of $\mathcal{Y}_2'$ by the action of D_8). Let $\bar{\psi}_g\colon \bar{\mathcal{Y}}_g \to \mathbb{P}^1$ denote the normalization of the pull-back $g^*(\bar{\mathcal{Y}}_2)$. Then we have the following analogue of [**9**, Proposition 2.4].

PROPOSITION 4.4. *The threefold $\bar{\mathcal{Y}}_g$ has trivial canonical sheaf if and only if $k + l + m - n - r = 2$ and either $l = 2$ with $y_1, y_2 \in \{1, 2, 4\}$, or $l = 1$ with $y_1 = 8$.*

PROOF. This is proved in exactly the same way as [**9**, Proposition 2.4]. □

Next we prove an analogue of [**9**, Proposition 2.5].

PROPOSITION 4.5. *If Proposition 4.4 holds, then there exists a projective birational morphism $\mathcal{Y}_g \to \bar{\mathcal{Y}}_g$, where $\mathcal{Y}_g$ is a normal threefold with trivial canonical sheaf and at worst $\mathbb{Q}$-factorial terminal singularities. Furthermore, any singularities of $\mathcal{Y}_g$ occur in its fibres over $g^{-1}(\frac{1}{256})$, and $\mathcal{Y}_g$ is smooth if g is unramified over $\lambda = \frac{1}{256}$ (which happens if and only if $m = n$).*

PROOF. We follow the same method that was used to prove [**9**, Proposition 2.5] and show that the singularities of $\bar{\mathcal{Y}}_g$ may all be crepantly resolved, with the possible exception of some $\mathbb{Q}$-factorial terminal singularities lying in fibres over $g^{-1}(\frac{1}{256})$.

First note that the threefold $\bar{\mathcal{Y}}_g$ is smooth away from the fibres lying over $g^{-1}\{0, \frac{1}{256}, \infty\}$, so it suffices to compute crepant resolutions in a neighbourhood of each of these fibres.

First let Δ_∞ denote a disc in $\mathcal{M}_{M_2}$ around $\lambda = \infty$ and let Δ'_∞ denote one of its preimages under g. Then $g\colon \Delta'_\infty \to \Delta_\infty$ is a y_i-fold cover ramified totally over $\lambda = \infty$, for some $y_i \in \{1, 2, 4, 8\}$.

However, in the three cases $y_i \in \{1, 2, 4\}$, over Δ_∞ we have that $\bar{\mathcal{Y}}_g$ is isomorphic to a quotient of $\mathcal{Y}_2'$. Crepant resolutions of such quotients were computed in Subsection 4.5.3: these resolutions have 20, 9 and 1 components in the cases $y_i = 1$, 2 and 4 respectively. The case $y_i = 8$ is a double cover of the case $y_i = 4$, in this case $\bar{\mathcal{Y}}_g$ is smooth with one component.

Next let Δ_0 denote a disc in $\mathcal{M}_{M_2}$ around $\lambda = 0$ and let Δ_0' denote one of its preimages under g. Then $g\colon \Delta_0' \to \Delta_0$ is an x_i-fold cover ramified totally over $\lambda = 0$, for some x_i. The fibre of $\bar{\mathcal{Y}}_2$ over $\lambda = 0$ was computed in Subsection 4.5.1: it consists of two rational surfaces meeting along four rational curves $D_1, \ldots, D_4$, which are arranged in a cycle in each component.

Upon proceeding to the x_i-fold cover, we find that the threefold $\bar{\mathcal{Y}}_g$ contains four curves of cA_{x_i-1} singularities in its fibre over $g^{-1}(0)$, given by the pull-backs of the curves D_i. This has a crepant resolution which contains:

- 2 components that are strict transforms of the original 2,
- $4(x_i-1)$ components arising from the blow-ups of the four curves of cA_{x_i-1} singularities lying over the D_i, and
- $(x_i-2)^2$ (if x_i is even) or $(x_i-2)^2-1$ (if x_i is odd) components arising from the blow-ups of the intersections between these four curves.

To see this, assume first that x_i is even. Then we may factorize the map $g\colon \Delta_0' \to \Delta_0$ into an $\frac{x_i}{2}$-fold cover followed by a double cover. We compute a crepant resolution of $\bar{\mathcal{Y}}_g$ over Δ_0' as follows. First pull-back $\bar{\mathcal{Y}}_2$ to a double cover of Δ_0 ramified over $\lambda = 0$. The resulting threefold has a crepant resolution, which is locally isomorphic to $\mathcal{Y}_2'$ (in a neighbourhood of $\nu = 0$). Then pull-back again to an $\frac{x_i}{2}$-fold cover ramified over the preimage of $\lambda = 0$. As the fibre of $\mathcal{Y}_2'$ over $\nu = 0$ is semistable, consisting of six rational components arranged in a cube (by Subsection 4.5.1), we may compute a crepant resolution of this $\frac{x_i}{2}$-fold cover using results of Friedman [**11**, Section 1]. This gives a crepant resolution of $\bar{\mathcal{Y}}_g$ with the required properties.

Next, assume that x_i is odd. Consider the $2x_i$-fold cover $g'\colon \Delta_0'' \to \Delta_0$ that is ramified totally over $\lambda = 0$. Then a crepant resolution $\mathcal{Y}_{g'}$ of the pull-back of $\bar{\mathcal{Y}}_2$ by g' can be computed as above. Furthermore, there is an involution on this resolution, the quotient by which gives a threefold birational to $\bar{\mathcal{Y}}_g$.

This involution preserves every component of the fibre of $\mathcal{Y}_{g'}$ over $(g')^{-1}(0)$. Its fixed locus consists of the strict transforms of the two components of $\bar{\mathcal{Y}}_2$, along with (x_i-1) of the exceptional components arising from the blow-up of each curve of cA_{x_i-1} singularities and $(\frac{x_i}{2}-1)^2-\frac{1}{4}$ of the exceptional components arising from the blow-up of each of their intersections. The components appearing in this fixed locus are uniquely determined by the properties that no two of them meet in a double curve and that every non-fixed component meets precisely two fixed ones.

Under the quotient, the non-fixed components become exceptional and may be contracted, resulting in a smooth threefold that resolves $\bar{\mathcal{Y}}_g$. This resolution is crepant by the adjunction formula for double covers.

Finally, let $\Delta_{\frac{1}{256}}$ be a disc in $\mathcal{M}_{M_2}$ around $\lambda = \frac{1}{256}$ and let $\Delta'_{\frac{1}{256}}$ be one of the connected components of its preimage under g. Then $g\colon \Delta'_{\frac{1}{256}} \to \Delta_{\frac{1}{256}}$ is a z_i-fold cover ramified totally over $\lambda = \frac{1}{256}$, for some z_i.

The threefold $\bar{\mathcal{Y}}_2$ is smooth over $\Delta_{\frac{1}{256}}$, but its fibre over $\lambda = \frac{1}{256}$ has two isolated A_1 singularities. Upon proceeding to the z_i-fold cover $\Delta'_{\frac{1}{256}} \to \Delta_{\frac{1}{256}}$, these become a pair of isolated terminal singularities of type cA_{z_i-1} in $\bar{\mathcal{Y}}_g$.

Thus $\mathcal{Y}_g$ is smooth away from its fibres over $g^{-1}(\frac{1}{256})$, where it can have isolated terminal singularities. By [**15**, Theorem 6.25], we may further assume that $\mathcal{Y}_g$ is $\mathbb{Q}$-factorial. To complete the proof, we note that if g is a local isomorphism over $\Delta_{\frac{1}{256}}$, then $\mathcal{Y}_g$ is also smooth over $g^{-1}(\frac{1}{256})$ and thus smooth everywhere. $\square$

Let $\psi_g\colon \mathcal{Y}_g \to \mathbb{P}^1$ denote the fibration induced on $\mathcal{Y}_g$ by the map $\bar{\psi}_g\colon \bar{\mathcal{Y}}_g \to \mathbb{P}^1$. Then $\mathcal{Y}_g$ is a threefold fibred by Kummer surfaces. It follows that:

PROPOSITION 4.6. *Let $\mathcal{Y}_g$ be a threefold as in Proposition 4.5. If $\mathcal{Y}_g$ is smooth, then $\mathcal{Y}_g$ is a Calabi-Yau threefold.*

PROOF. By Proposition 4.5, we see that $\mathcal{Y}_g$ has trivial canonical bundle. The condition on the vanishing of the first cohomology $H^1(\mathcal{Y}_g, \mathcal{O}_{\mathcal{Y}_g})$ is proved in exactly the same way as [**9**, Proposition 2.7]. □

This proposition enables the construction of many Kummer surface fibred Calabi-Yau threefolds $\psi_g\colon \mathcal{Y}_g \to \mathbb{P}^1$. These Calabi-Yau threefolds are related to the Shioda-Inose fibred threefolds $\pi_g\colon \mathcal{X}_g \to \mathbb{P}^1$ constructed in [**9**] as follows. Let $U := g^{-1}(U_{M_2})$. Then, by construction, the restriction $\psi_g\colon \mathcal{Y}_g|_U \to U$ of $\mathcal{Y}_g$ to U is isomorphic to the threefold obtained from the restriction $\pi_g\colon \mathcal{X}_g|_U \to U$ by quotienting by the fibrewise Nikulin involution and resolving singularities (note that the maps $g\colon \mathbb{P}^1 \to \mathcal{M}_{M_2}$ defining $\mathcal{X}_g$ and $\mathcal{Y}_g$ here are the same). So we may think of the $\mathcal{Y}_g$ as arising from the $\mathcal{X}_g$ through the process described in Section 3.

REMARK 4.7. In fact, there is a kind of duality between the Shioda-Inose fibred threefolds $\pi_g\colon \mathcal{X}_g \to \mathbb{P}^1$ and the Kummer fibred threefolds $\Psi_g\colon \mathcal{Y}_g \to \mathbb{P}^1$. As noted above, for $U := g^{-1}(U_{M_2})$, the restriction $\mathcal{Y}_g|_U$ is isomorphic to the resolved quotient of $\mathcal{X}_g|_U$ by the fibrewise Nikulin involution. However, $\mathcal{X}_g|_U$ is *also* isomorphic to the resolved quotient of $\mathcal{Y}_g|_U$ by the fibrewise Nikulin involution given in Remark 3.1. So we can move back and forth between $\mathcal{X}_g|_U$ and $\mathcal{Y}_g|_U$ by quotienting by Nikulin involutions and resolving singularities.

4.7. Properties of the constructed threefolds. The properties of the threefolds $\mathcal{Y}_g$ are closely linked to those of the related threefolds $\mathcal{X}_g$. Of particular importance to these calculations is the curve $C_g \subset \mathcal{X}_g$, defined as the closure of the fixed locus of the fibrewise Nikulin involution on $\mathcal{X}_g|_U$. As the Nikulin involution has 8 fixed points in a general fibre of $\mathcal{X}_g$, the curve C_g is an 8-fold cover of $\mathbb{P}^1$.

The curve C_g is easily calculated as the pull-back of the curve $C_2 \subset \mathcal{X}_2$ (defined in the same way as $C_g \subset \mathcal{X}_g$) by the map g. The properties of C_2 are as follows:

LEMMA 4.8. *The curve $C_2 \subset \mathcal{X}_2$ has three irreducible components, all of which have genus 0. Two of these components are double covers of $\mathcal{M}_{M_2}$ ramified over $\lambda \in \{0, \infty\}$. The third component is a 4-fold cover of $\mathcal{M}_{M_2}$ that has ramification profile $[2,1,1]$ over $\lambda = \frac{1}{256}$, ramification profile $[2,2]$ over $\lambda = 0$, and ramification profile $[4]$ over $\lambda = \infty$.*

PROOF. Let $p \in U_{M_2}$ be a general point and let X_p (resp. Y_p) be the fibre of $\mathcal{X}_2$ (resp. $\mathcal{Y}_2$) over p. Define the divisors $\{F_1, \ldots, F_8\}$ in Y_p as in Section 2.1. The F_i arise as the exceptional curves in the resolution of X_p/β, where β is the Nikulin involution on X_p. Thus, the action of monodromy in $\pi_1(U_{M_2}, p)$ on the 8 fixed points of β in X_p, which determines the curve C_2, is the same as the action of monodromy on the divisors F_i in Y_p.

The action of monodromy on the divisors $\{F_3, \ldots, F_8\}$ was computed explicitly in Table 4.2. Using this, the action of monodromy on $\{F_1, F_2\}$ may be computed from [**10**, Proposition 4.8]. From this it is easy to compute the description of the ramification profiles of the components of C_2, their genera may be computed by Hurwitz's theorem. □

It turns out that many of the properties of the Calabi-Yau threefolds $\mathcal{Y}_g$ can be calculated from knowledge of the curve C_g and the ramification behaviour of the map g. At this point we restrict ourselves to the case $l = 2$, to avoid pathologies occurring when $l = 1$ (see [**9**, Remark 3.1]).

PROPOSITION 4.9. *Let $\mathcal{Y}_g$ be a Calabi-Yau threefold as in Proposition 4.6 and suppose that $g^{-1}(\infty)$ consists of two points (so that $l = 2$). Then*

$$h^{1,1}(\mathcal{Y}_g) = 12 + \sum_{x_i \text{ odd}} x_i^2 + \sum_{x_i \text{ even}} (x_i^2 + 1) + s + c_1 + c_2,$$

where $[x_1, \ldots, x_k]$ is the partition of n encoding the ramification profile of g over $\lambda = 0$, s is the number of irreducible components of C_g, and c_1, c_2 are given in terms of the partition $[y_1, y_2]$ of n encoding the ramification profile of g over $\lambda = \infty$ by $c_j = 19$ (resp. 8, 0) if and only if $y_j = 1$ (resp. 2, 4).

PROOF. We follow the same method used to prove [**9**, Proposition 3.5], by noting that $h^{1,1}(\mathcal{Y}_g)$ is equal to the sum of the ranks of the groups of horizontal divisors $\mathrm{Pic}^h(\mathcal{Y}_g)$ and vertical divisors $\mathrm{Pic}^v(\mathcal{Y}_g)$.

We begin with the subspace of horizontal divisors. As before, we have an embedding $\mathrm{Pic}^h(\mathcal{Y}_g) \hookrightarrow \mathrm{Pic}(Y)$, where Y denotes a general fibre of $\mathcal{Y}_g$, given by restriction. Furthermore, by [**10**, Corollary 3.2], monodromy around singular fibres can only act non-trivially on the 8-dimensional sublattice of $\mathrm{Pic}(Y)$ generated by the eight curves $\{F_1, \ldots, F_8\}$ (defined as in Section 2.1). Thus, every divisor in the 11-dimensional orthogonal complement to this set is preserved under monodromy, so sweeps out a unique divisor in $\mathrm{Pic}^h(\mathcal{Y}_g)$. This contributes 11 to the rank of $\mathrm{Pic}^h(\mathcal{Y}_g)$.

To finish computing the rank of $\mathrm{Pic}^h(\mathcal{Y}_g)$, we thus have to compute how many distinct divisors in $\mathcal{Y}_g$ are swept out by the F_i's. However, as the divisors F_i occur as the blow-ups of the singularities arising from the fixed points of the fibrewise Nikulin involution on $\mathcal{X}_g$, each distinct divisor swept out by the F_i's corresponds to an irreducible component of C_g. There are s such components, so the rank of $\mathrm{Pic}^h(\mathcal{Y}_g)$ is equal to $11 + s$.

Next we consider the vertical divisors. As before, the class of a generic fibre contributes one divisor class to $\mathrm{Pic}^v(\mathcal{Y}_g)$; the remaining divisor classes arise from singular fibres. However, we computed the singular fibres in $\mathcal{Y}_g$ explicitly in the proof of Proposition 4.5. There we found that the fibre over a point p with $g(p) = 0$ and ramification order x at p has $2 + 4(x-1) + (x-2)^2 = x^2 + 2$ components (if x is even) or $2 + 4(x-1) + (x-2)^2 - 1 = x^2 + 1$ components (if x is odd), so each fibre of this kind contributes $x^2 + 1$ (resp. x^2) new classes to $\mathrm{Pic}^v(\mathcal{Y}_g)$ when x is even (resp. odd).

Furthermore, the fibre of $\mathcal{Y}_g$ over a point p with $g(p) = \infty$ and ramification order y at p has 20 (resp. 9, 1) components when $y = 1$ (resp. 2, 4). Thus, such fibres contribute 19 (resp. 8, 0) new classes when $y = 1$ (resp. 2, 4). Summing over all singular fibres of $\mathcal{Y}_g$, we find that

$$\mathrm{rank}(\mathrm{Pic}^v(\mathcal{Y}_g)) = 1 + \sum_{x_i \text{ odd}} x_i^2 + \sum_{x_i \text{ even}} (x_i^2 + 1) + c_1 + c_2,$$

where x_i and c_j are as in the statement of the proposition. Adding in the $11 + s$ horizontal divisor classes, we obtain the result. □

PROPOSITION 4.10. *Let $\mathcal{Y}_g$ be a Calabi-Yau threefold as in Proposition 4.6 and suppose that $g^{-1}(\infty)$ consists of two points (so that $l = 2$). Then*

$$h^{2,1}(\mathcal{Y}_g) = k + \left(\frac{m_{\text{odd}} - n}{2}\right) + p_g(C_g),$$

where k denotes the number of ramification points of g over $\lambda = 0$, m_{odd} denotes the number of ramification points of odd order of g over $\lambda = \frac{1}{256}$, n is the degree of g, and $p_g(C_g)$ denotes the geometric genus of the curve C_g (if C_g is singular, this is equal to the sum of the genera of the components in the normalization of C_g).

Moreover, if g is unramified over $\lambda = \frac{1}{256}$, then

$$h^{2,1}(\mathcal{Y}_g) = k + p_g(C_g) = r + p_g(C_g),$$

where r is the degree of ramification of the map g away from $\lambda \in \{0, \frac{1}{256}, \infty\}$.

PROOF. To compute $h^{2,1}(\mathcal{Y}_g)$, we first find the third Betti number $b_3(\mathcal{Y}_g)$, then use the fact that $\mathcal{Y}_g$ is a Calabi-Yau threefold, so that $b_3(\mathcal{Y}_g) = 2h^{2,1}(\mathcal{Y}_g) + 2$.

As above, let $U := g^{-1}(U_{M_2})$. Then let $j\colon U \to \mathbb{P}^1$ denote the inclusion map and let $\psi_U\colon \mathcal{Y}_g|_U \to U$ denote the restriction of the fibration $\psi_g\colon \mathcal{Y}_g \to \mathbb{P}^1$ to U. Applying [**9**, Lemma 3.4], noting that the condition on the singular fibres of $\mathcal{Y}_g$ is satisfied by the description of these fibres given in the proof of Proposition 4.5, we see that $H^3(\mathcal{Y}_g, \mathbb{Q}) \cong H^1(\mathbb{P}^1, j_*R^2(\psi_U)_*\mathbb{Q})$. It therefore suffices to compute the rank of this latter group.

By the discussion in [**10**, Section 2.1], there is a splitting of $R^2(\psi_U)_*\mathbb{Q}$ as a direct sum of two irreducible $\mathbb{Q}$-local systems

$$R^2(\psi_U)_*\mathbb{Q} = \mathcal{NS}(\mathcal{Y}_g) \oplus \mathcal{T}(\mathcal{Y}_g),$$

where $\mathcal{NS}(\mathcal{Y}_g)$ consists of those classes which are in $\text{NS}(Y_p) \otimes \mathbb{Q}$ for every smooth fibre Y_p of $\mathcal{Y}_g$, and $\mathcal{T}(\mathcal{Y}_p)$ is the orthogonal complement of $\mathcal{NS}(\mathcal{Y}_g)$. We may therefore split

$$H^3(\mathcal{Y}_g, \mathbb{Q}) = H^1(\mathbb{P}^1, j_*R^2(\psi_U)_*\mathbb{Q}) = H^1(\mathbb{P}^1, j_*\mathcal{NS}(\mathcal{Y}_g)) \oplus H^1(\mathbb{P}^1, j_*\mathcal{T}(\mathcal{Y}_g)).$$

Now let $\pi_g\colon \mathcal{X}_g \to \mathbb{P}^1$ denote the threefold fibred by M_2-polarized K3 surfaces related to $\psi_g\colon \mathcal{Y}_g \to \mathbb{P}^1$. Then, by [**10**, Proposition 3.1], the transcendental variations of Hodge structure $\mathcal{T}(\mathcal{Y}_g)$ and $\mathcal{T}(\mathcal{X}_g)$ are isomorphic over $\mathbb{R}$. So, by [**9**, Proposition 3.8], we see that

$$h^1(\mathbb{P}^1, j_*\mathcal{T}(\mathcal{Y}_g)) = h^1(\mathbb{P}^1, j_*\mathcal{T}(\mathcal{X}_g)) = 2 + 2k + (m_{\text{odd}} - n).$$

Next we consider the $\mathbb{Q}$-local system $\mathcal{NS}(\mathcal{Y}_g)$. Let $\mathbb{L}_{\text{Nik}}$ denote the sub-$\mathbb{Q}$-local system of $\mathcal{NS}(\mathcal{Y}_g)$ generated by the classes of the divisors $\{F_1, \ldots, F_8\}$. Then it follows from [**10**, Corollary 3.2] that there is a decomposition $\mathcal{NS}(\mathcal{Y}_g) \cong \mathbb{Q}^{12} \oplus \mathbb{L}_{\text{Nik}}$

We compute $\mathbb{L}_{\text{Nik}}$ as follows. Let $\pi_U\colon \mathcal{X}_g|_U \to U$ (resp. C_U) denote the restriction of the fibration $\pi_g\colon \mathcal{X}_g \to \mathbb{P}^1$ (resp. the curve $C_g \subset \mathcal{X}_g$) to the open set $g^{-1}(U)$. Recall that the fixed locus of the fibrewise Nikulin involution on $\mathcal{X}_g|_U$ is given by the curve C_U, and that the divisors $F_i \subset \mathcal{Y}_g$ arise from the resolution of the singularities in the quotient by this resolution. It therefore follows that there is an isomorphism of $\mathbb{Q}$-local systems

$$\mathbb{L}_{\text{Nik}} \cong (\pi_U|_{C_U})_*\mathbb{Q}_{C_U},$$

where $\mathbb{Q}_{C_U}$ denotes the constant sheaf with stalk $\mathbb{Q}$ on C_U.

Thus, we have

$$h^1(\mathbb{P}^1, j_*\mathcal{NS}(\mathcal{Y}_g)) = h^1(\mathbb{P}^1, j_*\mathbb{L}_{\mathrm{Nik}}) = h^1(\mathbb{P}^1, j_*(\pi_U|_{C_U})_*\mathbb{Q}_{C_U}).$$

Moreover, by the Leray spectral sequence, we have an isomorphism

$$H^i(\mathbb{P}^1, j_*(\pi_U|_{C_U})_*\mathbb{Q}_{C_U}) \cong H^i(\hat{C}_g, \mathbb{Q}),$$

for every $i \geq 0$, where $\hat{C}_g$ denotes the normalization of C_g. We therefore find that

$$h^1(\mathbb{P}^1, j_*\mathcal{NS}(\mathcal{Y}_g)) = h^1(\hat{C}_g, \mathbb{Q}) = 2p_g(C_g).$$

Putting everything together, we obtain

$$b_3(\mathcal{Y}_g) = 2 + 2k + (m_{\mathrm{odd}} - n) + 2p_g(C_g),$$

and the result follows. □

REMARK 4.11. We note that the analogue of [**9**, Proposition 4.1] is not true for $\mathcal{Y}_g$: even if we assume that g is unramified over $\lambda = \frac{1}{256}$, in general we cannot realize every small deformation of $\mathcal{Y}_g$ by simply deforming the map g.

The reason for this is as follows. In [**9**], the Néron-Severi group of a general fibre in the K3 fibration on $\mathcal{X}_g$ is isometric to M_2 and the smooth fibres form an M_2-polarized family. Consequently, the Néron-Severi group of a general fibre is preserved under monodromy. It follows that any small deformation of $\mathcal{X}_g$ is also fibred by M_2-polarized K3 surfaces, so can be realized as a pull-back of the family $\mathcal{X}_2$, as seen in [**9**, Proposition 4.1].

However, in the setting presented here, we have seen that monodromy acts non-trivially on the Néron-Severi group of a general fibre of $\mathcal{Y}_g$. Due to this, deformations need not preserve the entire Néron-Severi group; they only need to preserve the part that is fixed under monodromy. In other words, any deformation of $\mathcal{Y}_g$ must be fibred by K3 surfaces, but the rank of the Néron-Severi group of a general fibre in such fibrations may drop. Deformations of this type can obviously no longer be pull-backs of $\mathcal{Y}_2$.

EXAMPLE 4.12. As a very simple example, we compute the Hodge numbers of the Calabi-Yau threefold $\mathcal{Y}_2'$. In this case g is the map with $(k,l,m,n,r) = (4,2,4,8,0)$ and $[x_1,x_2,x_3,x_4] = [2,2,2,2]$, $[y_1,y_2] = [4,4]$, and $[z_1,z_2,z_3,z_4] = [2,2,2,2]$.

The action of monodromy around singular fibres of $\mathcal{Y}_2'$ fixes the Néron-Severi lattice of a general fibre (by construction), so in particular it fixes the eight divisors F_i. The curve C_g thus has eight components, all of which are rational curves. From Propositions 4.9 and 4.10 we obtain

$$\begin{aligned} h^{1,1}(\mathcal{Y}_2') &= 12 + \sum_{x_i \text{ odd}} x_i^2 + \sum_{x_i \text{ even}} (x_i^2 + 1) + s + c_1 + c_2 \\ &= 12 + 0 + 4(5) + 8 + 0 + 0 \\ &= 40 \end{aligned}$$

and $h^{2,1}(\mathcal{Y}_2') = k + \frac{1}{2}(m_{\mathrm{odd}} - n) + p_g(C_g) = 4 + \frac{1}{2}(0-8) + 0 = 0$, as expected from Proposition 4.1.

EXAMPLE 4.13. As a harder example, we consider the map $g\colon \mathbb{P}^1 \to \mathcal{M}_{M_2}$ defined by $(k,l,m,n,r) = (1,2,5,5,1)$, $[x_1] = [5]$, $[y_1,y_2] = [1,4]$, and $[z_1,\ldots,z_5] = [1,1,1,1,1]$. By [**10**, Theorem 5.10], the Shioda-Inose fibred threefold $\mathcal{X}_g$ corresponding to this map g is birational to the mirror to the quintic threefold. We will compute the Hodge numbers of the corresponding Kummer fibred threefold $\mathcal{Y}_g$, which is Calabi-Yau by Proposition 4.6.

To do this, we begin by computing the curve C_g. This is given by the pull-back of C_2 by g. It has three irreducible components, given by the pull-backs of the irreducible components of C_2. Let $a_1 \in \mathbb{P}^1$ denote the unique point lying over $\lambda = 0$, $b_1 \in \mathbb{P}^1$ (resp. $b_2 \in \mathbb{P}^1$) denote the point over $\lambda = \infty$ where the ramification order of g is 1 (resp. 4), and $d_1, \ldots, d_5$ denote the five points over $\lambda = \frac{1}{256}$.

Then two of the irreducible components of C_g (the pull-backs of the components of C_2 that are double covers of $\mathcal{M}_{M_2}$) are double covers of $\mathbb{P}^1$ having a singularity of type A_4 at a_1, a simple ramification over b_1, and a singularity of type A_3 over b_2. The normalizations of these components are simply ramified over a_1 and b_1, so both have genus 0.

The remaining irreducible component of C_g (the pull-back of the component of C_2 that is a 4-fold cover of $\mathcal{M}_{M_2}$) is a 4-fold cover of $\mathbb{P}^1$ having a pair of singularities of type A_4 over a_1, a 4-fold ramification over b_1, a simple quadruple point over b_2, and ramification profile $[2,1,1]$ over each d_i. Its normalization has ramification profiles $[2,2]$ over a_1, $[4]$ over b_1, $[1,1,1,1]$ over b_2, and $[2,1,1]$ over each d_i. By Hurwitz's theorem, this normalization has genus 2.

From this we can calculate the Hodge numbers of $\mathcal{Y}_g$. From Propositions 4.9 and 4.10 we obtain

$$\begin{aligned} h^{1,1}(\mathcal{Y}_g) &= 12 + \sum_{x_i \text{ odd}} x_i^2 + \sum_{x_i \text{ even}} (x_i^2+1) + s + c_1 + c_2 \\ &= 12 + 25 + 0 + 3 + 0 + 19 \\ &= 59 \end{aligned}$$

and $h^{2,1}(\mathcal{Y}_g) = r + p_g(C_g) = 1 + 2 = 3$.

References

[1] K. Altmann and D. van Straten, *The polyhedral Hodge number $h^{2,1}$ and vanishing of obstructions*, Tohoku Math. J. (2) **52** (2000), no. 4, 579–602, DOI 10.2748/tmj/1178207756. MR1793937 (2001k:14096)

[2] A. Clingher and C. F. Doran, *Modular invariants for lattice polarized K3 surfaces*, Michigan Math. J. **55** (2007), no. 2, 355–393, DOI 10.1307/mmj/1187646999. MR2369941 (2009a:14049)

[3] A. Clingher and C. F. Doran, *Note on a geometric isogeny of K3 surfaces*, Int. Math. Res. Not. IMRN **16** (2011), 3657–3687. MR2824841 (2012f:14072)

[4] A. Clingher, C. F. Doran, J. Lewis, A. Y. Novoseltsev, and A. Thompson, *The 14th case VHS via K3 fibrations*, Preprint, December 2013, `arXiv:1312.6433`.

[5] A. Clingher, C. F. Doran, J. Lewis, and U. Whitcher, *Normal forms, K3 surface moduli, and modular parametrizations*, Groups and symmetries, CRM Proc. Lecture Notes, vol. 47, Amer. Math. Soc., Providence, RI, 2009, pp. 81–98. MR2500555 (2011c:14104)

[6] B. Crauder and D. R. Morrison, *Triple-point-free degenerations of surfaces with Kodaira number zero*, The birational geometry of degenerations (Cambridge, Mass., 1981), Progr. Math., vol. 29, Birkhäuser Boston, Boston, MA, 1983, pp. 353–386. MR690270 (85g:14040)

[7] S. Cynk and D. van Straten, *Infinitesimal deformations of double covers of smooth algebraic varieties*, Math. Nachr. **279** (2006), no. 7, 716–726, DOI 10.1002/mana.200310388. MR2226407 (2007d:14023)

[8] I. V. Dolgachev, *Mirror symmetry for lattice polarized K3 surfaces*, J. Math. Sci. **81** (1996), no. 3, 2599–2630, DOI 10.1007/BF02362332. Algebraic geometry, 4. MR1420220 (97i:14024)

[9] C. F. Doran, A. Harder, A. Y. Novoseltsev, and A. Thompson, *Calabi-Yau threefolds fibred by mirror quartic K3 surfaces*, Preprint, January 2015, `arXiv:1501.04019`.

[10] C. F. Doran, A. Harder, A. Y. Novoseltsev, and A. Thompson, *Families of lattice polarized K3 surfaces with monodromy*, published online by Int. Math. Res. Not., 2015, `http://dx.doi.org/10.1093/imrn/rnv071`.

[11] R. Friedman, *Base change, automorphisms, and stable reduction for type* III $K3$ *surfaces*, The birational geometry of degenerations (Cambridge, Mass., 1981), Progr. Math., vol. 29, Birkhäuser, Boston, Mass., 1983, pp. 277–298. MR690268 (84m:14039)

[12] H. Inose, *Defining equations of singular* $K3$ *surfaces and a notion of isogeny*, Proceedings of the International Symposium on Algebraic Geometry (Kyoto Univ., Kyoto, 1977), Kinokuniya Book Store, Tokyo, 1978, pp. 495–502. MR578868 (81h:14021)

[13] G. Kapustka and M. Kapustka, *Fiber products of elliptic surfaces with section and associated Kummer fibrations*, Internat. J. Math. **20** (2009), no. 4, 401–426, DOI 10.1142/S0129167X09005339. MR2515047 (2010c:14042)

[14] J. Keum and S. Kondō, *The automorphism groups of Kummer surfaces associated with the product of two elliptic curves*, Trans. Amer. Math. Soc. **353** (2001), no. 4, 1469–1487, DOI 10.1090/S0002-9947-00-02631-3. MR1806732 (2001k:14075)

[15] J. Kollár and S. Mori, *Birational geometry of algebraic varieties*, Cambridge Tracts in Mathematics, vol. 134, Cambridge University Press, Cambridge, 1998. With the collaboration of C. H. Clemens and A. Corti; Translated from the 1998 Japanese original. MR1658959 (2000b:14018)

[16] M. Kuwata and T. Shioda, *Elliptic parameters and defining equations for elliptic fibrations on a Kummer surface*, Algebraic geometry in East Asia—Hanoi 2005, Adv. Stud. Pure Math., vol. 50, Math. Soc. Japan, Tokyo, 2008, pp. 177–215. MR2409557 (2009g:14039)

[17] R. Miranda, *The basic theory of elliptic surfaces*, Dottorato di Ricerca in Matematica. [Doctorate in Mathematical Research], ETS Editrice, Pisa, 1989. MR1078016 (92e:14032)

[18] D. R. Morrison, *On* $K3$ *surfaces with large Picard number*, Invent. Math. **75** (1984), no. 1, 105–121, DOI 10.1007/BF01403093. MR728142 (85j:14071)

[19] N. Nakayama, *On Weierstrass models*, Algebraic geometry and commutative algebra, Vol. II, Kinokuniya, Tokyo, 1988, pp. 405–431. MR977771 (90m:14030)

[20] N. Narumiya and H. Shiga, *The mirror map for a family of* $K3$ *surfaces induced from the simplest 3-dimensional reflexive polytope*, Proceedings on Moonshine and related topics (Montréal, QC, 1999), CRM Proc. Lecture Notes, vol. 30, Amer. Math. Soc., Providence, RI, 2001, pp. 139–161. MR1877764 (2002m:14030)

[21] V. V. Nikulin, *Integer symmetric bilinear forms and some of their geometric applications* (Russian), Izv. Akad. Nauk SSSR Ser. Mat. **43** (1979), no. 1, 111–177, 238. MR525944 (80j:10031)

[22] C. Schoen, *On fiber products of rational elliptic surfaces with section*, Math. Z. **197** (1988), no. 2, 177–199, DOI 10.1007/BF01215188. MR923487 (89c:14062)

[23] T. Shioda, *Kummer sandwich theorem of certain elliptic* $K3$ *surfaces*, Proc. Japan Acad. Ser. A Math. Sci. **82** (2006), no. 8, 137–140. MR2279280 (2008b:14064)

[24] T. Shioda and H. Inose, *On singular* $K3$ *surfaces*, Complex analysis and algebraic geometry, Iwanami Shoten, Tokyo, 1977, pp. 119–136. MR0441982 (56 #371)

Department of Mathematical and Statistical Sciences, 632 CAB, University of Alberta, Edmonton, AB, T6G 2G1, Canada
E-mail address: `charles.doran@ualberta.ca`

Department of Mathematical and Statistical Sciences, 632 CAB, University of Alberta, Edmonton, AB, T6G 2G1, Canada
E-mail address: `aharder@ualberta.ca`

Department of Mathematical and Statistical Sciences, 632 CAB, University of Alberta, Edmonton, AB, T6G 2G1, Canada
E-mail address: `novoselt@ualberta.ca`

Department of Pure Mathematics, University of Waterloo, 200 University Ave West, Waterloo, ON, N2L 3G1, Canada
E-mail address: `am6thomp@uwaterloo.ca`

Proceedings of Symposia in Pure Mathematics
Volume **93**, 2016
http://dx.doi.org/10.1090/pspum/093/01610

Weighted Hurwitz numbers and hypergeometric τ-functions: an overview

J. Harnad

ABSTRACT. This is a survey of recent developments in the use of 2D Toda τ-functions of hypergeometric type as generating functions for multiparametric families of weighted Hurwitz numbers. Such τ-functions are obtained by using diagonal group elements in their expression as fermionic vacuum expectation values, implying that their expansion in a basis of tensor products of Schur functions is diagonal. A corresponding abelian group action on the center of the S_n group algebra is defined by evaluating symmetric functions formed multiplicatively from a weight generating function $G(z)$ on the Jucys-Murphy elements of the group algebra. The resulting central elements act diagonally in the basis of orthogonal idempotents, with eigenvalues $r_\lambda^{G(z)}$ that coincide with the coefficients in the double Schur function expansion. The group action is represented in the basis of cycle sums by matrices whose elements, expanded as power series in z, are the weighted double Hurwitz numbers. Both their geometrical meaning, as weighted sums over n-sheeted branched coverings, and combinatorial one, as weighted enumeration of paths in the Cayley graph of S_n generated by transpositions, follow from expanding the Cauchy-Littlewood generating functions over dual pairs of bases of the algebra of symmetric functions and evaluating on the Jucys-Murphy elements. It follows that the coefficients in the expansion of the τ-function in the basis of products of power sum symmetric functions are the weighted Hurwitz numbers. All previously studied cases are obtained by making suitable choices for $G(z)$. Expansion in powers of some of the parameters determining the weighting provide generating series for multispecies weighted Hurwitz numbers. Replacement of the Cauchy-Littlewood generating function by the one for Macdonald polynomials provides (q,t)-deformations that are generating functions for quantum weighted Hurwitz numbers.

CONTENTS

2010 *Mathematics Subject Classification.* Primary 05Axx, 05E05, 20C30, 14N10.

Key words and phrases. Weighted Hurwitz numbers, tau-functions, generating functions, symmetric functions.

Work supported by the Natural Sciences and Engineering Research Council of Canada (NSERC) and the Fonds Québecois de la recherche sur la nature et les technologies (FQRNT).

1. Hurwitz numbers

The study of Hurwitz numbers, which enumerate branched covers of the Riemann sphere with specified ramification profiles, began with the pioneering work of Hurwitz [**27**, **28**]. Their relation to enumerative factorization problems in the symmetric group and irreducible characters was developed by Frobenius [**19**, **20**] and Schur [**46**]. In recent years, following the discovery by Pandharipande [**44**] and Okounkov [**42**] that certain KP and 2D Toda τ-functions [**49**–**51**], fundamental to the modern theory of integrable systems [**45**, **47**], could serve as generating functions for weighted Hurwitz numbers, there has been a flurry of activity [**1**–**4**, **12**, **13**, **17**, **18**, **22**, **23**, **26**, **33**, **40**, **41**, **52**] concerned with finding new classes of τ-functions that can similarly serve as generating functions for various types of weighted Hurwitz numbers.

Two closely related interpretations of these weighted Hurwitz numbers exist. The enumerative geometrical one consists of weighted sums of Hurwitz numbers for n-sheeted branched coverings of the Riemann sphere. The other consists of weighted enumeration of factorizations of elements of the symmetric group S_n in which the factors are in specified conjugacy classes. This may equivalently be interpreted as a weighted counting of paths in the Cayley graph generated by transpositions, starting and ending in specified classes. The two approaches are related by the monodromy representation of the fundamental group of the sphere punctured at the branch points obtained by lifting closed paths to the covering surface. Variants of this also exist for branched coverings of higher genus surfaces [**35**] and other groups.

Some generating functions of enumerative invariants are known to also have representations as matrix integrals [**2**–**4**, **12**, **13**, **17**, **40**, **41**]. These include, in particular, the well-known Harish-Chandra-Itzykson-Zuber (HCIZ) integral [**21**, **29**], which plays a fundamental rôle both in representation theory and in coupled matrix models. In [**12**, **13**, **17**], it was shown that when the Toda flow parameters are equated to the trace invariants of a pair of $N \times N$ hermitian matrices, and the expansion parameter is equated to $-1/N$, this gives the generating function for the enumeration of weakly monotonic paths in the Cayley graph with a fixed number of steps while, geometrically, it coincides with signed enumeration of branched coverings of fixed genus and variable numbers of branch points [**18**, **26**]. Other matrix integrals give “hybrid” paths consisting of both weakly and strongly monotonic segments or, equivalently, enumeration of coverings with multispecies “coloured” branch points [**17**, **26**]. Certain of these may also be shown to satisfy differential constraints, the so-called *Virasoro constraints* [**33**, **37**, **52**], due to reparametrization invariance, and *loop equations* [**2**–**4**, **6**] following from the structure of the underlying matrix integrals.

These, and other generating functions for various enumerative, topological, combinatorial and geometrical invariants related to Riemann surfaces, such as intersection numbers [**34**], higher Gromov-Witten invariants, Hodge numbers [**31**, **32**], knot invariants [**5**, **38**], and a growing number of other cases, can be placed into the *topological recursion* scheme [**8**–**10**, **15**], which aims at determining the generating functions through algorithmic recursion sequences stemming from an underlying *spectral curve* [**4**, **33**]. This has turned out to be a very effective approach to a broad class of examples.

However not all such generating functions are known to be τ-functions in the usual sense of integrable systems, nor partition functions or correlaters for matrix models. It remains something of a mystery exactly which class of invariants is amenable to such a representation. A further remarkable fact is that, in some cases, different generating functions corresponding to distinct enumerative problems, such as Hurwitz numbers and Hodge integrals, may be τ-functions that are related through algebraic transformations that themselves involve the spectral curve [**31, 32**].

The present work is concerned solely with the case of Hurwitz numbers, but in the generalized sense, allowing infinite parametric families of weightings. It provides a unified approach encompassing all cases of weighted Hurwitz numbers that have appeared to date, interpreting these as special cases of an infinite parametric family of weighting functions determining mKP or 2D-Toda τ-functions of generalized hypergeometric type. The parameters serve to specify the particular weighting used when summing over the various configurations. Their values are determined by a "weight generating" function $G(z)$, and define the weighting by evaluation of the standard bases $(e_\lambda, h_\lambda, m_\lambda, f_\lambda)$, for the space Λ of symmetric functions [**36**] in an infinite number of indeterminates consisting of *elementary*, *complete*, *monomial* and *forgotten* symmetric functions, respectively, at the given set of parameters $(c_1, c_2, \dots)$ determined by $G(z)$, viewed as an infinite product. The other two standard bases, the Schur functions $\{s_\lambda\}$ and the power sum symmetric functions $\{p_\mu\}$, serve as bases for expansions of the τ-function, in which the coefficients in the first are diagonal and of *content product* form, guaranteeing that the Hirota bilinear equations of the integrable hierarchy are satisfied, while those in the second provide the weighted Hurwitz numbers as coefficients.

Besides the various "classically weighted" cases, arising through different choices of the parameters $(c_1, c_2, \dots)$, there are also "quantum deformations", depending on an additional pair (q, t) that are closely linked to the Macdonald symmetric functions [**36**]. This leads to the notion of "quantum weighted" Hurwitz numbers, of various types [**18, 23**], which may depend both on the infinite set of classical weighting parameters $(c_1, c_2, \dots)$, and the further pair (q, t) in a specific way, involving q-deformations. Another generalization involves multiple expansion parameters $(z_1, z_2, \dots)$, leading to generating functions for "multispecies" weighted Hurwitz numbers [**22**], which are counted with different weighting factors, depending on the species type, or "colour".

In section 2, a quick review is given of the fermionic approach to τ-functions for the KP hierarchy and the modified KP sequence of τ-functions as introduced by Sato [**45, 47**], as well as the 2D Toda case introduced in [**49–51**]. Section 3 recalls basic notions regarding the S_n group algebra, including the commuting *Jucys-Murphy elements* [**30, 39**], Frobenius' characteristic map from the center $\mathbf{Z}(\mathbf{C}[S_n])$ to the algebra Λ of symmetric functions, and the abelian group within $\mathbf{Z}(\mathbf{C}[S_n])$ that is generated by combining these. Section 4 gives a summary of the new approach to the construction of τ-functions of hypergeometric type interpretable as generating functions for infinite parametric families of weighted Hurwitz numbers developed in [**17, 18, 22, 23, 26**]. The weightings are interpreted both geometrically, as weighted enumeration of n-sheeted branched covers of the Riemann sphere, and combinatorially, as weighted enumeration of paths in the Cayley graph of S_n generated by transpositions. The relation between these is easily seen algebraically through the Cauchy-Littlewood generating functions for dual pairs of bases for Λ.

Section 5 is devoted to various examples that have so far been considered in the literature. These include: the original case of single and double Hurwitz numbers, generated by the special KP and 2D Toda τ-functions, studied by Pandharipande and Okounkov [**42, 44**]; the case of the HCIZ integral [**12, 13, 17, 21, 29**], which is known to have a combinatorial interpretation as counting weakly monotonically increasing paths of transpositions in the Cayley graph, to which is added the geometrical one of signed enumeration of branched coverings with an arbitrary number of branch points and arbitrary ramification profiles, but fixed genus. Another case [**17**], which counts strongly monotonic paths can be related to the special case of counting Belyi curves [**2, 3, 33, 52**] (with three branch points) or "Dessins d'enfants". A hybrid case [**17**] combines the two, and counts branching configurations of multiple "colour" type and, moreover also has a matrix model representation. More general "multispecies" branched coverings, with their associated combinatorial equivalents [**22, 26**], and other, more general parametric families of weighted Hurwitz numbers are considered in section 6.

Already within the "classical" setting, it is possible to select the parameters $\{c_1, c_2, \dots |$ appearing in the associated weight generating functions as powers $\{q^i\}$ of an auxiliary "quantum" parameter that may be interpreted in terms of Planck's constant $\hbar$ and temperature, and the resulting weights related to the energy distribution law for a Bose gas with linear energy spectrum.

In section 7, we extend the family of weight generating functions by introducing a further pair (q, t) of deformation parameters that play the same rôle as those appearing in the Macdonald symmetric functions [**36**], with the Cauchy-Littlewood generating functiont replaced by the corresponding one for Macdonald functions [**23**]. The resulting weighted Hurwitz numbers are interpretable as multispecies quantum Hurwitz numbers, whose distributions are again related to those for a Bosonic gas, but also depend more generally upon the infinite parameter family of classical weight generating parameters $\{c_1, c_2, \dots\}$. Various specializations are obtained by choosing specific values for the parameters q and t, or relations between them, or various limits. Besides recovering the purely classical weights for $q = t$, this leads to various other specialized cases, such as weights involving the quantum analog of the elementary and complete symmetric functions, the Hall-Littlewood polynomials and the Jack polynomials.

1.1. Enumerative geometrical Hurwitz numbers

For any set of partitions $\{\mu^{(1)}, \dots, \mu^{(k)}\}$ of $n \in \mathbf{N}^+$, the geometrically defined Hurwitz number $H(\mu^{(1)}, \dots, \mu^{(k)})$ is equal to the number of n-sheeted branched coverings of the Riemann sphere having no more than k branch points $\{q_1, \dots, q_k\}_{i=1,\dots,k}$, with ramification profiles of type $\{\mu^{(i)}\}_{i=1,\dots,k}$, weighted by the inverse of the order of their automorphism groups. The Frobenius-Schur formula [**19, 20, 35, 46**] expresses these in terms of the irreducible characters $\chi_\lambda(\mu^{(i)})$ of the symmetric group S_n

$$H(\mu^{(1)}, \dots, \mu^{(k)}) = \sum_{\lambda, |\lambda|=n} h_\lambda^{k-2} \prod_{i=1}^{k} z_{\mu^{(i)}}^{-1} \chi_\lambda(\mu^{(i)}), \tag{1.1}$$

where λ is the partition corresponding to the irreducible representation with Young symmetrizer of type λ, and the parts of the partitions $\{\mu^{(i)}\}$ are the cycle lengths defining the ramifications profiles that determine the conjugacy classes $\mathrm{cyc}(\mu^{(i)})$ on which χ_λ is evaluated. Here

$$z_\mu = \prod_{i=1}^{\ell(\mu)} i^{m_i(\mu)}(m_i(\mu))! \tag{1.2}$$

is the order of the stabilizer of any element in $\mathrm{cyc}(\mu)$ under conjugation, where $m_i(\mu)$ is the number of parts of μ equal to i and

$$h_\lambda := \det\left(\frac{1}{(\lambda_i - i + j)!}\right)^{-1} \tag{1.3}$$

is the product of the hook lengths of the partition λ.

1.2. Combinatorial Hurwitz numbers

An equivalent definition of the Hurwitz number, which we denote $F(\mu^{(1)}, \dots, \mu^{(k)})$, is the following: $n!F(\mu^{(1)}, \dots, \mu^{(k)})$ is the number of ways the identity element $\mathbf{I} \in S_n$ may be factorized into a product

$$\mathbf{I} = h_1 \cdots h_k, \tag{1.4}$$

in which the ith factor $h_i \in S_n$ is in the conjugacy class $\mathrm{cyc}(\mu^{(i)})$. The equality of these two quantities

$$F(\mu^{(1)}, \dots, \mu^{(k)}) = H(\mu^{(1)}, \dots, \mu^{(k)}) \tag{1.5}$$

follows from the monodromy representation of the fundamental group $\pi_1(\mathbf{CP}^1 \backslash \{q_1, \dots, q_k\})$ of the punctured sphere with the branch points removed [**35**, Appendix A].

As shown in subsection 3.3, relation (1.1) follows from (1.5) and the Frobenius character formula. Avatars of this equality will be seen to recur repeatedly in the various versions of weighted Hurwitz numbers studied below.

2. mKP and 2D Toda τ-functions

2.1. Fermionic Fock space

The fermionic Fock space $\mathcal{F}$ is defined [**47**] as the semi-infinite wedge product space

$$\mathcal{F} := \Lambda^{\infty/2}\mathcal{H} \tag{2.1}$$

constructed from a separable Hilbert space $\mathcal{H}$ with orthonormal basis $\{e_i\}_{i\in\mathbf{Z}}$, that is split into an orthogonal direct sum of two subspaces

$$\mathcal{H} = \overline{\mathcal{H}_+ \oplus \mathcal{H}_-}, \tag{2.2}$$

where

$$\mathcal{H}_- = \mathrm{span}\{e_i\}_{i\in\mathbf{N}}, \quad \mathcal{H}_+ = \mathrm{span}\{e_{-i}\}_{i\in\mathbf{N}^+}. \tag{2.3}$$

and $\{e_i\}_{i\in\mathbf{Z}}$ is an orthonormal basis.

2.1 REMARK. The curious convention of using negative i's to label the basis for $\mathcal{H}_+$ and positive ones for $\mathcal{H}_-$ stems from the notion of the "Dirac sea", in which all negative energy levels are filled and all positive ones empty, where the integer lattice is identified with the energies. If we take Segal and Wilson's [**48**] model for $\mathcal{H}$

$$\mathcal{H} := L^2(S^1) = \overline{\operatorname{span}\{z^i\}_{i\in\mathbf{Z}}} \text{ with } e_i := z^{-i-1}, \tag{2.4}$$

we may view $\mathcal{H}_+$ and $\mathcal{H}_-$ either as the subspaces of positive and negative Fourier series on the circle S^1 or, equivalently, the Hardy spaces of square integrable functions admitting a holographic extension to inside and outside the unit circle, with the latter vanishing at $z = \infty$.

$\mathcal{F}$ is the graded sum

$$\mathcal{F} = \oplus_{N\in\mathbf{Z}}\mathcal{F}_N \tag{2.5}$$

of the subspaces $\mathcal{F}_N$ with fermionic charge $N \in \mathbf{Z}$. An orthonormal basis $\{|\lambda; N\rangle\}$ for these is provided by the semi-infinite wedge product states

$$|\lambda; N\rangle := e_{\ell_1} \wedge e_{\ell_2} \wedge \cdots \tag{2.6}$$

labeled by pairs of partitions λ and integers $N \in \mathbf{Z}$, where

$$\{\ell_i := \lambda_i - i + N\} \tag{2.7}$$

are the "particle coordinates", indicating the occupied points on the integer lattice, corresponding to the parts of the partition λ, with the usual convention that $\lambda_i := 0$ for i greater than the length $\ell(\lambda)$ of the partition. The vacuum state in the charge N sector $\mathcal{F}_N$ of the Fock space is denoted

$$|N\rangle := |0; N\rangle. \tag{2.8}$$

In Segal and Wilson's [**48**] sense, the image in the projectivization $\mathbf{P}(\mathcal{F})$ of an element $W \in Gr_{\mathcal{H}_+}(\mathcal{H})$ of the infinite Grassmannian modeled on $\mathcal{H}_+ \subset \mathcal{H}$, having *virtual dimension* N (i.e., such that the Fredholm index of the orthogonal projection map $\pi^\perp : W \to \mathcal{H}_+$ is N) under the Plücker map

$$\begin{aligned}\mathcal{P} : Gr_{\mathcal{H}_+}(\mathcal{H}) &\to \mathbf{P}(\mathcal{F})\\ \mathcal{P} : W &\mapsto \mathcal{P}(W)\\ \mathcal{P} : \operatorname{span}\{w_i \in \mathcal{H}\}_{i\in\mathbf{N}^+} &\mapsto [w_1 \wedge w_2 \wedge \cdots],\end{aligned} \tag{2.9}$$

is in the charge N sector $\mathcal{P}(W) \in \mathcal{F}_N \subset \mathcal{F}$, and the entire image consists of all decomposable elements of $\mathcal{F}$. In particular, $\mathcal{H}_+$ is mapped to the projectivization of the vacuum element

$$\mathcal{P} : \mathcal{H}_+ \mapsto [|0\rangle] := [|0; 0\rangle] = [e_{-1} \wedge e_{-2} \wedge \cdots]. \tag{2.10}$$

The Fermi creation and annihilation operators ψ_i, $\psi_i^\dagger$ are defined as exterior multiplication by the basis element e_i and interior multiplication by the dual basis element $\tilde{e}^i$, respectively.

$$\psi := e_i \wedge \quad \psi^\dagger := i(\tilde{e}^i). \tag{2.11}$$

These satisfy the usual anticommutation relations

$$[\psi_i, \psi_j^\dagger]_+ = \delta_{ij} \tag{2.12}$$

defining the corresponding Clifford algebra on $\mathcal{H} + \mathcal{H}^*$ with respect to the natural quadratic form in which both $\mathcal{H}$ and $\mathcal{H}^*$ are totally isotropic.

The infinite general linear algebra $\mathfrak{gl}(\mathcal{H}) \subset \Lambda^2(\mathcal{H}+\mathcal{H}^*)$, in the standard Clifford representation, is spanned by the elements $:\psi_i\psi_j^\dagger:$, with the usual convention for normal ordering

$$:\psi_i\psi_j^\dagger: =: \psi_i\psi_j^\dagger: - \langle\psi_i\psi_j^\dagger\rangle, \tag{2.13}$$

where $\langle\mathcal{O}\rangle$ denotes the vacuum expectation value

$$\langle\mathcal{O}\rangle := \langle 0|\mathcal{O}|0\rangle. \tag{2.14}$$

The corresponding group $\mathfrak{Gl}(\mathcal{H})$ consists of invertible endomorphisms, having well defined determinants. (See [**45**, **47**, **48**] for more detailed definitions.)

A typical exponentiated element in the Clifford representation is of the form

$$\hat{g} = e^{\sum_{ij\in\mathbf{Z}} A_{ij}:\psi_i\psi_j^\dagger:}, \tag{2.15}$$

where the doubly infinite square matrix with elements A_{ij} satisfies suitable convergence conditions [**45**, **45**, **47**, **48**] that will not be detailed here.

2.2. Abelian group actions, mKP and 2D Toda lattice τ-functions and Hirota relations

The KP and 2D-Toda flows are generated by the multiplicative action on $\mathcal{H}$ of the two infinite abelian subgroups $\Gamma_\pm \subset \mathfrak{Gl}_0(\mathcal{H})$ of the identity component $\mathfrak{Gl}_0(\mathcal{H})$ of the general linear group $\mathfrak{Gl}(\mathcal{H})$, defined by:

$$\Gamma_+ := \{\gamma_+(\mathbf{t}) := e^{\sum_{i=1}^\infty t_i z^i}\}, \quad \text{and} \quad \Gamma_- := \{\gamma_-(\mathbf{s}) := e^{\sum_{i=1}^\infty s_i z^{-i}}\}, \tag{2.16}$$

where $\mathbf{t} = (t_1, t_2, \dots)$ is an infinite sequence of (complex) flow parameters corresponding to one-parameter subgroups, and $\mathbf{s} = (s_1, s_2, \dots)$ is a second such sequence. These in turn have the following Clifford group representations on $\mathcal{F}$

$$\hat{\Gamma}_+ := \{\hat{\gamma}_+(\mathbf{t}) := e^{\sum_{i=1}^\infty t_i J_i}\}, \quad \text{and} \quad \hat{\Gamma}_- := \{\hat{\gamma}_-(\mathbf{s}) := e^{\sum_{i=1}^\infty s_i J_{-i}}\}, \tag{2.17}$$

where

$$J_i := \sum_{k\in\mathbf{Z}} :\psi_k\psi_{i+k}^\dagger:, \quad \pm i \in \mathbf{N}^+ \tag{2.18}$$

are referred to as the "current components". In this infinite dimensional setting, whereas the abelian groups $\Gamma_\pm$ commute, their Clifford representations $\hat{\Gamma}_\pm$ involve a central extension, so that

$$\hat{\gamma}_+(\mathbf{t})\hat{\gamma}_-(\mathbf{s}) = \hat{\gamma}_-(\mathbf{s})\hat{\gamma}_+(\mathbf{t})e^{\sum_{i\in\mathbf{Z}} it_is_i}. \tag{2.19}$$

The mKP-chain and 2D-Toda τ-functions corresponding to the element $g \in \mathfrak{Gl}_0(\mathcal{H}$ are given, within a nonzero multiplicative constant, by the vacuum expectations values (VEV's)

$$\tau_g^{mKP}(N, \mathbf{t}) := \langle N|\hat{\gamma}_+(\mathbf{t})\hat{g}|N\rangle, \tag{2.20}$$

$$\tau_g^{(2Toda)}(N, \mathbf{t}, \mathbf{s}) := \langle N|\hat{\gamma}_+(\mathbf{t})\hat{g}\hat{\gamma}_-(\mathbf{s})|N\rangle. \tag{2.21}$$

If the group element $g \in \mathfrak{Gl}_0(\mathcal{H}$ is interpreted, relative to the $\{e_i\}_{i\in\mathbb{Z}}$ basis, as a matrix exponential $g = e^A$, where the algebra element $A \in \mathfrak{gl}(\mathcal{H})$ is represented by the infinite matrix with elements $\{A_{ij}\}_{i,j\in\mathbb{Z}}$, then the corresponding representation of $GL(\mathcal{H})$ on $\mathcal{F}$ is given by

$$\hat{g} := e^{\sum_{i,j\in\mathbb{Z}} A_{ij}:\psi_i\psi_j^\dagger:}, \tag{2.22}$$

These satisfy the Hirota bilinear relations

$$\oint_{z=\infty} z^{N'-N} e^{-\xi(\delta \mathbf{t}, z)} \tau_g^{mKP}(N, \mathbf{t} + \delta \mathbf{t} + [z^{-1}]) \tau_g^{mKP}(N', \mathbf{t} - [z^{-1}])\, dz = 0 \tag{2.23}$$

(2.24)

$$\oint_{z=\infty} z^{N'-N} e^{-\xi(\delta \mathbf{t}, z)} \tau^{2DT}(N, \mathbf{t} + [z^{-1}], \mathbf{s}) \tau^{2DT}(N', \mathbf{t} + \delta\mathbf{t} - [z^{-1}], \mathbf{s} + \delta \mathbf{s})\, dz =$$

$$\oint_{z=0} z^{N'-N} e^{-\xi(\delta \mathbf{s}, z^{-1})} \tau^{2DT}(N-1, \mathbf{t}, \mathbf{s} + [z]) \tau^{2DT}(N'+1, \mathbf{t} + \delta \mathbf{t}, \mathbf{s} + \delta \mathbf{s} - [z])\, dz$$

understood to hold identically in $\delta \mathbf{t} = (\delta t_1, \delta t_2, \dots)$, $\delta \mathbf{s} := (\delta s_1, \delta s_2, \dots)$, where

$$[z]_i := \frac{1}{i} z^i. \tag{2.25}$$

2.3. Bose-Fermi equivalence and Schur function expansions

It follows from the identities [**45, 47**]

$$\langle N | \hat{\gamma}_+(\mathbf{t}) | \lambda; N \rangle = \langle \lambda; N | \hat{\gamma}_-(\mathbf{t}) | N \rangle = s_\lambda(\mathbf{t}), \tag{2.26}$$

where s_λ is the Schur function corresponding to partition λ, viewed as function of the parameters

$$t_i := \frac{p_i}{i}, \tag{2.27}$$

where the p_i's are the power sums, that the τ-functions may be expressed, at least formally, as single and double Schur functions expansions

$$\tau_g^{mKP}(N, \mathbf{t}) = \sum_\lambda \pi_\lambda(N, g) s_\lambda(\mathbf{t}) \tag{2.28}$$

$$\tau_g^{2Toda}(N, \mathbf{t}, \mathbf{s})) = \sum_\lambda \sum_\mu B_{\lambda\mu}(N, g) s_\lambda(\mathbf{t}) s_\mu(\mathbf{t}) \tag{2.29}$$

where

$$\pi_\lambda(N, g) := \langle \lambda; N | \hat{g} | N \rangle, \quad B_{\lambda\mu}(N, g) := \langle \lambda; N | \hat{g} | \mu; N \rangle, \tag{2.30}$$

are the Plücker coordinates of the elements $\hat{g}|N\rangle$ and $\hat{g}|\mu; N\rangle$ when $g \in \mathfrak{Gl}_0(\mathcal{H})$ is in the identity component of $\mathfrak{Gl}(\mathcal{H})$. The Hirota bilinear relations (2.23), (2.24) are then equivalent to the Plücker relations satisfied by these coefficients.

The "Bose-Fermi equivalence" gives an isomorphism between a completion $\mathcal{B}_0$ of the space of symmetric functions Λ of an infinite number of "bosonic" variables $\{x_i\}_{i \in \mathbf{N}^+}$, labelled by the natural numbers and the $N = 0$ (zero charge) sector of the Fermionic Fock space $\mathcal{F}_0 \subset \mathcal{F}$. It identifies the basis states $\{|\lambda; 0\rangle\}$ with the basis of Schur functions $\{s_\lambda \in \Lambda\}$ through the "bosonization" map:

$$\begin{aligned} \mathbf{B} : \mathcal{F}_0 &\to \mathcal{B}_0 \\ \mathbf{B} : |v\rangle &\mapsto \langle 0 | \hat{\gamma}_+ | v \rangle \\ \mathbf{B} : |\lambda; 0\rangle &\mapsto s_\lambda. \end{aligned} \tag{2.31}$$

More generally, this can be extended to the full (graded) fermonic Fock space $\mathcal{F} = \oplus_{N \in \mathbf{Z}} \mathcal{F}_N$ by adding a parameter ζ to the Bosonic Fock space, taking formal Laurent expansions in this

$$\mathcal{B} := \mathcal{B}_0[[\zeta]], \tag{2.32}$$

and defining

$$\begin{aligned} &\mathbf{B}: \mathcal{F}_N \to \mathcal{B}_N \\ &\mathbf{B}: |v\rangle \mapsto \langle N|\hat{\gamma}_+|v\rangle\zeta^N. \end{aligned} \tag{2.33}$$

Using $\mathbf{B}$ as an intertwining map, this defines identifications between operators in $\mathrm{End}(\mathcal{F})$ and those in $\mathrm{End}(\mathcal{B})$. However, what appears in the Fermonic representation as a "locally" defined element of the Clifford algebra or group is in general a nonlocal operator in the Bosonic representation (involving exponentiated differential operators in terms of the $\mathbf{t}$ coordinates), as is the case, e.g. , for the Bosonic representations of the operators ψ, $\psi_i^\dagger$, which are special types of "vertex operators". In particular, the Bosonization of fermionic states of the type $\hat{g}\hat{\gamma}_-(\mathbf{s})|0\rangle$ is given by application of such nonlocal operators to the trivial (linear exponential) τ-functions corresponding to gauge transforms of the vacuum state, defined by Bosonization of $\hat{\gamma}_-(\mathbf{s})|0\rangle$.

2.4. Hypergeometric τ-functions and convolution symmetries

A special subfamily of the above consists of those τ-functions for which the group element $\hat{g}$ is diagonal

$$\hat{g} = e^{\sum_{i\in\mathbf{Z}} T_i :\psi_i\psi_i^\dagger}, \quad A_{ij} = T_i\delta_{ij} \tag{2.34}$$

in the basis $|\lambda; N\rangle$. These were called *convolution symmetries* in [**25**], since in the Segal-Wilson representations of $\mathfrak{Gl}(\mathcal{H})$ they may be interpreted as (generalized) convolution products on $\mathcal{H} \sim L^2(S_1)$. Their eigenvalues $r_\lambda(N, g)$ in the basis $|\lambda; N\rangle$

$$e^{\sum_{i\in\mathbf{Z}} T_i :\psi_i\psi_i^\dagger}|\lambda; N\rangle = r_\lambda(N, g)|\lambda; N\rangle \tag{2.35}$$

can be written in the form of a *content product* [**25, 43**]:

$$r_\lambda(N, g) := r_0(N, g) \prod_{(i,j)\in\lambda} r_{N+j-i}(g), \quad r_i(g) := e^{T_i - T_{i-1}} \tag{2.36}$$

where

$$r_0(N, g) := \begin{cases} \prod_{i=0}^{N-1} e^{T_i} & \text{if} \quad N > 0 \\ 1 & \text{if} \quad N = 0 \\ \prod_{i=N}^{-1} e^{-T_i} & \text{if} \quad N < 0. \end{cases} \tag{2.37}$$

The double Schur function expansion (2.21) in this case reduces to the diagonal form

$$\tau_g^{2Toda}(N, \mathbf{t}, \mathbf{s})) = \sum_\lambda r_\lambda(N, g) s_\lambda(\mathbf{t}) s_\lambda(\mathbf{s}). \tag{2.38}$$

If we view the second set of parameters $(s_1, s_2, \dots)$ as fixed, and consider only the first set $(t_1, t_2, \dots)$ as KP flow parameters, we may interpret (2.38) as defining a chain of mKP τ functions. A specific value of special interest is $(s_1, s_2, \dots) = (1, 0, 0 \dots)$, for which the Schur function evaluates to

$$s_\lambda(1, 0, \dots) = h_\lambda^{-1} \tag{2.39}$$

and (2.38) reduces to

$$\tau_g^{mKP}(N, \mathbf{t}, \mathbf{s})) = \sum_\lambda r_\lambda(N, g) h_\lambda^{-1} s_\lambda(\mathbf{t}). \tag{2.40}$$

In the following, only such *hypergeometric* τ-functions will be needed. By defining suitable parametric families of the latter, and expanding these in powers of some auxiliary parameters, while leaving the others to define the weightings, it will be seen that we can interpret such τ-functions as generating functions for finite or infinite parametric families of weighted Hurwitz numbers, both classical and quantum, obtaining both a natural enumerative geometric and combinatorial interpretation in all cases.

3. The center $\mathbf{Z}(\mathbf{C}[S_n])$ of the S_n group algebra and symmetric functions

3.1. The $\{C_\mu\}$ and $\{F_\lambda\}$ bases

There are two natural bases for the center $\mathbf{Z}(\mathbf{C}[S_n])$ of the group algebra of the symmetric group S_n, both labelled by partitions of n. The first is the basis of cycle sums $\{C_\mu\}_{|\mu|=n}$, defined by

$$C_\mu := \sum_{h\in \mathrm{cyc}(\mu)} h. \tag{3.1}$$

The second is the basis of orthogonal idempotents $\{F_\lambda\}_{|\lambda|=n}$, which project onto the irreducible representations of type λ and satisfy

$$F_\lambda F_\mu = F_\lambda \delta_{\lambda\mu}. \tag{3.2}$$

These are related by

$$F_\lambda = h_\lambda^{-1} \sum_{\mu,\,|\mu|=|\lambda|=n} \chi_\lambda(\mu) C_\mu, \tag{3.3}$$

$$C_\mu = z_\mu^{-1} \sum_{\lambda,\,|\lambda|=|\mu|=n} h_\lambda \chi_\lambda(\mu) F_\lambda \tag{3.4}$$

which is equivalent to the Frobenius character formula (see below). The main property of the $\{F_\lambda\}$ basis is that multiplication by any element of the center $\mathbf{Z}(\mathbf{C}[S_n])$ is diagonal in this basis (as follows immediately from (3.2)).

3.2. The characteristic map

Frobenius' characteristic map defines a linear isomorphism between the characters of S_n and the characters of tensor representations of $GL(k)$, of total tensor weight n, for k sufficiently large. It maps the irreducible character χ_λ to the Schur function s_λ, viewed as the corresponding $GL(k)$ character through the Weyl character formula for any $k \geq \ell(\lambda)$. Equivalently, it defines a linear endomorphism

$$\begin{aligned} \mathrm{ch} : \mathbf{Z}(\mathbf{C}[S_n]) &\to \Lambda \\ \mathrm{ch} : F_\lambda &\mapsto \frac{s_\lambda}{h_\lambda} \end{aligned} \tag{3.5}$$

from the center $\mathbf{Z}(\mathbf{C}[S_n])$ of the group algebra to the algebra Λ of symmetric functions [**36**]. The change of basis formulae (3.3), (3.4), together with the Frobenius character formula

$$s_\lambda = \sum_{\mu,\,|\mu|=|\lambda|=n} z_\mu^{-1} \chi_\lambda(\mu) p_\mu, \tag{3.6}$$

where

$$p_\mu := \prod_{i=1}^{\ell(\mu)} p_{\mu_i} \tag{3.7}$$

is the power sum symmetric function, then imply that the characteristic map takes the cycle sum basis into the $\{p_\mu\}$ basis for Λ

$$\mathrm{ch} : C_\mu \mapsto \frac{p_\mu}{z_\mu}. \tag{3.8}$$

3.3. Combinatorics of Hurwitz numbers and the Frobenius-Schur formula

The two bases $\{C_\mu\}$, $\{F_\lambda\}$ can be used to deduce the Frobenius-Schur formula (1.1), expressing $H(\mu^{(1)}, \dots, \mu^{(k)})$ in terms of the irreducible group characters $\chi_\lambda(\mu)$. The product $\prod_{i=1}^k C_{\mu^{(i)}}$ of elements of the cycle sum basis is central and hence can be expressed relative to the same basis:

$$\prod_{i=1}^{k} C_{\mu^{(i)}} = \sum_{\nu, |nu|=n} H(\mu^{(1)}, \dots, \mu^{(i)}, \nu) z_\nu C_\nu, \tag{3.9}$$

and, in particular, the coefficient of the identity class, for which $\mu = (1)^n$ is $n!$ times the Hurwitz number

$$[\mathbf{I} = C_{(1)^n}] \prod_{i=1}^{k} C_{\mu^{(i)}} = n! H(\mu^{(1)}, \dots, \mu^{(k)}), \tag{3.10}$$

giving the number of factorizations of the identity element into a product of k elements within the conjugacy classes $\{\mathrm{cyc}(\mu^{(i)}\}_{i=1,\dots,k}$.

Substituting the change of basis formula (3.4) into (3.10), applying both sides to the basis element $\{F_\lambda\}$ and equating the eigenvalues that result gives the Frobenius-Schur formula:

$$H(\mu^{(1)}, \dots, \mu^{(k)}) = \sum_{\lambda,\, |\lambda|=|\mu|=n} h_\lambda^{k-2} \prod_{i=1}^{k} \frac{\chi_\lambda(\mu^{(i)})}{z_\mu^{(i)}}. \tag{3.11}$$

3.4. Jucys-Murphy elements, central elements and weight generating functions

We now recall the special commuting elements $(\mathcal{J}_1, \dots, \mathcal{J}_n\}$ of the group algebra $\mathbf{C}[S_n]$ introduced by Jucys [**30**] and Murphy [**39**]. (See also [**7**]). These are defined by

$$\mathcal{J}_b := \sum_{a=1}^{n-1} (ab) \quad \text{for } b > 1, \text{ and } J_1 := 0. \tag{3.12}$$

where $(ab) \in S_n$ is the transposition that interchanges a with b. Although these are not central elements, they have two remarkable properties: Any symmetric function $f(\mathcal{J}_1, \dots, \mathcal{J}_n)$, $f \in \Lambda_n$ formed from them is central, and this central element has eigenvalues in the F_λ basis that are equal to the evaluation on the *content* of the partition λ; i.e. the set of number $j - i$, where $\{(i,j) \in \lambda\}$ are the set of positions (in the English convention) in the Young diagram of λ:

$$f(\mathcal{J}_1, \dots, \mathcal{J}_n) F_\lambda = f(\{j-i\}_{(ij)\in\lambda}) F_\lambda. \tag{3.13}$$

A particular class of symmetric functions of n variables is obtained by taking a single generating function $G(z)$, expressed formally either as an infinite product

$$G(z) = \prod_{i=1}^{\infty}(1 + c_i z), \tag{3.14}$$

or some limit thereof, or an infinite sum

$$G(z) = 1 + \sum_{i=1}^{\infty} G_i z^i, \tag{3.15}$$

and defining the central element as a product

$$G_n(z, \mathcal{J}) := \prod_{b=1}^{n} G(z\mathcal{J}_a). \tag{3.16}$$

(For the present, we are not concerned with whether $G(z)$ is polynomial, rational, a convergent series, in some field extension or just a formal infinite series or infinite product; the considerations that follow are mainly algebraic, but are easily extended to include either convergent series, through suitable completions, or formal series and products, as in the generating functions for symmetric functions.) When applied multiplicatively to the $\{F_\lambda\}$ basis, the central element $G_n(z, \mathcal{J})$ has eigenvalues that are expressible as content products

$$G_n(z, \mathcal{J})F_\lambda = \prod_{(ij)\in\lambda} G(z(j-i))F_\lambda, \quad |\lambda| = n. \tag{3.17}$$

We also consider the "dual" generating function:

$$\tilde{G}(z) := \frac{1}{G(-z)} = \prod_{i=1}^{\infty}(1 - c_i z)^{-1} \tag{3.18}$$

and associated central element

$$\tilde{G}_n(z, \mathcal{J}) := \prod_{b=1}^{n} \tilde{G}(z\mathcal{J}_a), \tag{3.19}$$

which similarly satisfies

$$\tilde{G}_n(z, \mathcal{J})F_\lambda = \prod_{(ij)\in\lambda} \tilde{G}(z(j-i))F_\lambda, \quad |\lambda| = n. \tag{3.20}$$

This suggests comparison with the "convolution symmetry" elements in the fermionic representation of the group $\mathfrak{Gl}(\mathcal{H})$ and an extension of the Bose-Fermi equivalence, using the characteristic map, to a correspondence between the direct sum $\oplus_{n\in\mathbf{N}}\mathbf{Z}(\mathbf{C}[S_n])$ and the $N = 0$ sector $\mathcal{F}_0 \subset \mathcal{F}$ of the fermonic Fock space.

3.5. Bose-Fermi equivalence and $\oplus_{n\in\mathbf{N}}\mathbf{Z}(\mathbf{C}[S_n])$

Composing the characteristic map with the Bose-Fermi equivalence, we obtain an endomorphism $\mathcal{E}$ from the direct sum $\oplus_{n\in\mathbf{N}}\mathbf{Z}(\mathbf{C}[S_n])$ of the centers of the group algebras to the zero charge sector $\mathcal{F}_0$ in the Fermionic Fock space

$$\begin{aligned} \mathcal{E} : \oplus_{n\in\mathbf{N}}\mathbf{Z}(\mathbf{C}[S_n]) &\to \mathcal{F}_0 \\ \mathcal{E} : F_\lambda &\mapsto h_\lambda^{-1}|\lambda; 0\rangle. \end{aligned} \tag{3.21}$$

This provides an intertwining map between the central elements in the completion of the group algebra formed from products of functions of a single variable, acting by multiplications, and the convolution symmetries discussed in subsection 2.4.

Choosing the parameters T_j in (2.34) as

(3.22)
$$T_j^{G(z)} = \sum_{k=1}^{j} \ln G(zk), \quad T_0^{G(z)}(z) = 0, \quad T_{-j}^{G(z)}(z) = -\sum_{k=0}^{j-1} \ln G(-zk) \quad \text{for } j > 0,$$

so that

$$\hat{g} = \hat{C}_G := e^{\sum_{i\in\mathbf{Z}} T_i^G(z) :\psi_i\psi_i^\dagger:}, \tag{3.23}$$

it follows that

$$r_j(g) := r_j^{G(z)} = G(jz) \tag{3.24}$$

and

$$\hat{C}_G|\lambda; N\rangle = r_\lambda^{G(z)}(N)|\lambda; N\rangle \tag{3.25}$$

with eigenvalues

$$r_\lambda^{G(z)}(N) := r_0^{G(z)}(N) \prod_{(i,j)\in\lambda} G(z(N+j-i)), \tag{3.26}$$

where

(3.27)
$$r_0^{G(z)}(N) = \prod_{j=1}^{N-1} G((N-j)z)^j, \quad r_0(0) = 1, \quad r_0^{G(z)}(-N) = \prod_{j=1}^{N} G((j-N)z)^{-j}, \quad N > 1$$

The map $\mathcal{E}$ defined in (3.21) therefore intertwines the action of $\oplus_{n\in\mathbf{N}} G_n(z, \mathcal{J})$ on $\oplus_{n\in\mathbf{N}} \mathbf{Z}(\mathbf{C}[S_n])$ with that of $\hat{C}_G$ on $\mathcal{F}_0$. The same applies to the dual generating functions $\tilde{G}(z)$, for which we obtain the corresponding content product formula expression

$$r_\lambda^{\tilde{G}(z)}(N) := r_0^{\tilde{G}(z)}(N) \prod_{(i,j)\in\lambda} \tilde{G}(z(N+j-i)). \tag{3.28}$$

For the following, we only have need of the $N = 0$ case, for which we simplify the notation for the content product coefficients to

$$r_\lambda^{G(z)} := r_\lambda^{G(z)}(0) = \prod_{(i,j)\in\lambda} G(z(j-i)), \tag{3.29}$$

$$r_\lambda^{\tilde{G}(z)} := r_\lambda^{\tilde{G}(z)}(0) = \prod_{(i,j)\in\lambda} \tilde{G}(z(j-i)), \tag{3.30}$$

4. Hypergeometric τ-functions as generating functions for weighted Hurwitz numbers

We are now ready to state the main results, which show that the KP and 2D Toda τ-functions of hypergeometric type

$$\tau^{G(z)}(\mathbf{t}) = \sum_\lambda r_\lambda^{G(z)} h_\lambda^{-1} s_\lambda(\mathbf{t}), \tag{4.1}$$

$$\tau^{G(z)}(\mathbf{t}, \mathbf{s})) = \sum_\lambda r_\lambda^{G(z)} s_\lambda(\mathbf{t}) s_\lambda(\mathbf{s}), \tag{4.2}$$

when expanded in bases of (products of) the power sum symmetric functions $\{p_\mu\}$, are interpretable as generating functions for suitably defined infinite parametric weighted Hurwitz numbers, both in the enumerative geometric and the combinatorial sense. The details and proofs may be found in [**17, 18, 22, 23, 26**].

4.1. The Cauchy-Littlewood formula and dual bases for Λ

We have already encountered the two bases for the ring Λ of symmetric functions in an arbitrary number of indeterminates [**36**] consisting of Schur functions $\{s_\lambda\}$ and power sum symmetric functions $\{p_\lambda\}$. In addition to these, there are four other useful bases, consisting of the products of the elementary symmetric functions

$$e_\lambda(\mathbf{x}) := \prod_{i=1}^{\ell(\lambda} e_{\lambda_i} \tag{4.3}$$

the complete symmetric functions

$$h_\lambda(\mathbf{x}) := \prod_{i=1}^{\ell(\lambda} h_{\lambda_i}, \tag{4.4}$$

with generating functions

$$E(z) = \prod_{ij}(1 + zx_i) = \sum_{i=0}^{\infty} e_i z^i, \quad H(z) = \prod_{ij}(1 - zx_i)^{-1} = \sum_{i=0}^{\infty} h_i z^i, \tag{4.5}$$

the monomial sum symmetric functions

$$m_\lambda(\mathbf{x}) := \frac{1}{|\mathrm{aut}(\lambda)|} \sum_{\sigma\in S_k} \sum_{1\le i_1<\cdots<i_k} x_{i_{\sigma(1)}}^{\lambda_1} \cdots x_{i_\sigma(k)}^{\lambda_k}, \tag{4.6}$$

and the "forgotten" symmetric functions

$$f_\lambda(\mathbf{x}) := \frac{(-1)^{\ell^*(\lambda)}}{|\mathrm{aut}(\lambda)|} \sum_{\sigma\in S_k} \sum_{1\le i_1\le\cdots\le i_k} x_{i_\sigma(1)}^{\lambda_1}, \cdots x_{i_\sigma(k)}^{\lambda_k}, \tag{4.7}$$

where

$$\mathbf{x} := (x_1, x_2, \dots) \tag{4.8}$$

is an infinite sequence of indeterminates, and

$$|\mathrm{aut}(\lambda)| := \prod_{i=1}^{\ell(\lambda)} (m_i(\lambda))! \tag{4.9}$$

is the order of the automorphism group of the conjugacy class of type λ under conjugation.

These bases have the following duality and orthogonality properties with respect to the standard scalar product $(\, ,\,)$ in which the Schur functions are orthonormal [**36**]:

$$(s_\lambda, s_\mu) = \delta_{\mu\nu}, \quad (p_\lambda, p_\mu) = z_\mu \delta_{\mu\nu}, \quad (e_\lambda, m_\mu) = \delta_{\mu\nu}, \qquad (f_\lambda, h_\mu) = \delta_{\mu\nu}. \tag{4.10}$$

It follows [**36**] that the Cauchy-Littlewood formula is expressible bilinearly in terms of these dually paired bases

$$\prod_{i=1}^{\infty}\prod_{j=1}^{\infty}(1-x_iy_j)^{-1} = \sum_\lambda s_\lambda(\mathbf{x})s_\lambda(\mathbf{y}) \tag{4.11}$$

$$= \sum_\lambda z_\mu^{-1} p_\lambda(\mathbf{x})p_\lambda(\mathbf{y}) \tag{4.12}$$

$$= \sum_\lambda e_\lambda(\mathbf{x})m_\lambda(\mathbf{y}) = \sum_\lambda e_\lambda(\mathbf{y})m_\lambda(\mathbf{x}) \tag{4.13}$$

$$= \sum_\lambda f_\lambda(\mathbf{x})h_\lambda(\mathbf{y}) = \sum_\lambda f_\lambda(\mathbf{y})h_\lambda(\mathbf{x}). \tag{4.14}$$

The dual Cauchy-Littlewood generating function is similarly expressed in terms these in a dual way [**36**]:

$$\prod_{i=1}^{\infty}\prod_{j=1}^{\infty}(1+x_iy_j) = \sum_\lambda s_\lambda(\mathbf{x})s_{\lambda'}(\mathbf{y}) \tag{4.15}$$

$$= \sum_\lambda (-1)^{\ell^*(\lambda)} z_\mu^{-1} p_\lambda(\mathbf{x})p_\lambda(\mathbf{y}) \tag{4.16}$$

$$= \sum_\lambda e_\lambda(\mathbf{x})f_\lambda(\mathbf{y}) = \sum_\lambda e_\lambda(\mathbf{y})f_\lambda(\mathbf{x}) \tag{4.17}$$

$$= \sum_\lambda h_\lambda(\mathbf{x})m_\lambda(\mathbf{y}) = \sum_\lambda h_\lambda(\mathbf{y})m_\lambda(\mathbf{x}), \tag{4.18}$$

where λ' is the partition whose Young diagram is the transpose of that for λ and

$$\ell^*(\lambda) = |\lambda| - \ell(\lambda) \tag{4.19}$$

is the *colength* of λ (i.e., the complement of its length).

We make use of these formulae in the following way: for the indeterminates $(x_1, x_2, \dots)$ we substitute the parameters $(c_1, c_2, \dots)$ defining the weight generating function $G(z)$ as an infinite product, or its dual $\tilde{G}(z)$, while for the indeterminates $(y_1, y_2, \dots)$, we substitute the z times the Jucys-Murphy elements $(z\mathcal{J}_1, z\mathcal{J}_2, \dots)$, up to a finite number n of these, and 0 for the rest, to obtain a finite product over j in (4.11) and (4.15).

This gives the following central elements, expressed as sums of products of these bases, either evaluated on the contents $(c_1, c_2, \dots)$ or the commuting elements $(\mathcal{J}_1, \mathcal{J}_2, \dots)$:

$$G_n(z, \mathcal{J}) = \prod_{i=1}^{n}\prod_{a=1}^{n}(1 + zc_i\mathcal{J}_a) \tag{4.20}$$

$$= \sum_{d=0}^{\infty} z^d \sum_{\lambda,\, |\lambda|=d} e_\lambda(\mathbf{c})m_\lambda(\mathcal{J}) \tag{4.21}$$

$$= \sum_{d=0}^{\infty} z^d \sum_{\lambda,\, |\lambda|=d} m_\lambda(\mathbf{c})e_\lambda(\mathcal{J}) \tag{4.22}$$

and

$$\tilde{G}_n(z,\mathcal{J}) = \prod_{i=1}^{n}\prod_{a=1}^{n}(1 - zc_i\mathcal{J}_a)^{-1} \tag{4.23}$$

$$= \sum_{d=0}^{\infty} z^d \sum_{\lambda,\, |\lambda|=d} h_\lambda(\mathbf{c}) m_\lambda(\mathcal{J}) \tag{4.24}$$

$$= \sum_{d=0}^{\infty} z^d \sum_{\lambda,\, |\lambda|=d} f_\lambda(\mathbf{c}) e_\lambda(\mathcal{J}). \tag{4.25}$$

Recall that the elements $G_n(z,\mathcal{J}), \tilde{G}_n(z,\mathcal{J}) \in \mathbf{Z}(\mathbf{C}[S_n])$ are diagonal in the basis of orthogonal idempotents, with the content product coefficients as eigenvalues

$$G_n(z,\mathcal{J})F_\lambda = r_\lambda^{G(z)} F_\lambda, \quad \tilde{G}_n(z,\mathcal{J})F_\lambda = r_\lambda^{\tilde{G}}(z) F_\lambda \tag{4.26}$$

where

$$r_\lambda^{G(z)} = \prod_{(ij)\in\lambda}\prod_{k=1}^{\infty}(1 + zc_k(j-i)) \tag{4.27}$$

$$r_\lambda^{\tilde{G}(z)} = \prod_{(ij)\in\lambda}\prod_{k=1}^{\infty}(1 - zc_k(j-i))^{-1}. \tag{4.28}$$

4.2. Multiplication by $m_\lambda(\mathcal{J})$ and $e_\lambda(\mathcal{J})$ in the C_μ basis

To proceed further, we need to compute the effect of multiplication by $G_n(z,\mathcal{J})$ and $\tilde{G}_n(z,\mathcal{J})$ on the basis $\{C_\mu\}$ of cycle sums. Combinatorially, this requires the notion of the *signature* of a path in the Cayley graph of S_n generated by the transpositions.

DEFINITION 4.1. Given a d-step path in the Cayley graph of S_n generated by transpositions (ab), $a, b \in \{1,\dots,n\}$, $a < b$, consisting of the sequence:

$$h, \to (a_1b_1)h \to (a_2b_2)(a_1b_1)h \to \dots \to (a_db_d)\cdots(a_1b_1)h, \tag{4.29}$$

its *signature* λ is the partition of weight $|\lambda| = d$ whose length $\ell(\lambda)$ equals the number of distinct second elements appearing in the sequence, and whose parts $\{\lambda_i\}_{i=1,\dots,d}$ are the number of times each positive integer appears as a second element b_j.

The effect of multiplication of C_μ by the central element $m_\lambda(\mathcal{J}) \in \mathbf{Z}(\mathbf{C}[S_n])$ is given by the following easily proved lemma [**18**]:

LEMMA 4.1. *Multiplication by $m_\lambda(\mathcal{J})$ defines an endomorphism of $\mathcal{Z}(\mathbf{C}[S_n])$ which, expressed in the $\{C_\mu\}$ basis, is given by*

$$m_\lambda(\mathcal{J})C_\mu = \frac{1}{|\mu|!}\sum_{\nu,\, |\nu|=|\mu|} m_{\mu\nu}^{\lambda} z_\nu C_\nu, \tag{4.30}$$

where $m_{\mu\nu}^{\lambda}$ is the number of monotonic $|\lambda|$-step paths in the Cayley graph of S_n generated by all transpositions, starting from an element h in the conjugacy class $\mathrm{cyc}(\nu)$ *and ending in* $\mathrm{cyc}(\mu)$, *with signature λ. Equivalently,*

$$m_{\mu\nu}^{\lambda} := \frac{\prod_{i=1}^{\ell(\lambda)}\lambda_i!}{|\lambda|!}\tilde{m}_{\mu\nu}^{\lambda} \tag{4.31}$$

where $\tilde{m}^{\lambda}_{\mu\nu}$ *is the number of* $|\lambda|$ *step paths of signature* λ *in the Cayley graph of* S_n *generated by transpositions, starting at the conjugacy class* $\mathrm{cyc}\,\mu$ *and ending in the class* $\mathrm{cyc}(\nu)$

On the other hand, the effect of multiplication of C_μ by the central element $e_\lambda(\mathcal{J}) \in \mathbf{Z}(\mathbf{C}[S_n])$ is given by the following [**18**]:

LEMMA 4.2. *Multiplication by* $e_\lambda(\mathcal{J})$ *defines an endomorphism of* $\mathcal{Z}(\mathbf{C}[S_n])$ *which, expressed in the* $\{C_\mu\}$ *basis, is given by*

$$e_\lambda(\mathcal{J})C_\mu = \sum_{\substack{\mu^{(1)},\dots,\mu^{(k)},\\ \{\ell^*(\mu^{(i)})=\lambda_i\}}} \left(\prod_{i=1}^{k} C_{\mu^{(i)}}\right) C_\mu \tag{4.32}$$

$$= \sum_{\substack{\mu^{(1)},\dots,\mu^{(k)},\\ \{\ell^*(\mu^{(i)})=\lambda_i\}}} H(\mu^{(1)},\dots,\mu^{(k)},\mu,\nu) z_\nu C_\nu, \tag{4.33}$$

*(where the identity (*3.9*) has been used in the second line).*

4.3. Weighted double Hurwitz numbers: enumerative geometric and combinatorial

We now proceed to the enumerative geometrical definition of weighted Hurwitz numbers. For a fixed pair of branch points, say at $(0,\infty)$, with ramification profiles (μ,ν) and an additional set of k branch points $(q_1,\dots,q_k)$ with ramification profiles $(\mu^{(1)},\dots,\mu^{(k)})$, we define the weights to be given by the evaluation of the monomial sum and "forgotten" symmetric functions at the parameter values $\mathbf{c}=(c_1,c_2,\dots)$ for the two cases corresponding to the dual weight generating functions $G(z)$ and $\tilde{G}(z)$.

$$W_G(\mu^{(1)},\dots,\mu^{(k)}) := m_\lambda(\mathbf{c}) \tag{4.34}$$

$$W_{\tilde{G}}(\mu^{(1)},\dots,\mu^{(k)}) := f_\lambda(\mathbf{c}). \tag{4.35}$$

The weighted geometrical Hurwitz numbers $H^d_G(\mu,\nu)$, giving the weighted count of such n-sheeted branched coverings of the Riemann sphere, having a pair of specified branch points with ramification profiles μ and ν and any number k of further branch points, with arbitrary ramification profiles $(\mu^{(1)},\dots,\mu^{(k)})$, but fixed genus, are defined to be the weighted sum

$$H^d_G(\mu,\nu) := \sum_{k=0}^{\infty} \sideset{}{'}\sum_{\substack{\mu^{(1)},\dots\mu^{(k)}\\ \sum_{i=1}^{k}\ell^*(\mu^{(i)})=d}} W_G(\mu^{(1)},\dots,\mu^{(k)}) H(\mu^{(1)},\dots,\mu^{(k)},\mu,\nu) \tag{4.36}$$

$$H^d_{\tilde{G}}(\mu,\nu) := \sum_{k=0}^{\infty} \sideset{}{'}\sum_{\substack{\mu^{(1)},\dots\mu^{(k)}\\ \sum_{i=1}^{d}\ell^*(\mu^{(i)})=d}} W_{\tilde{G}}(\mu^{(1)},\dots,\mu^{(k)}) H(\mu^{(1)},\dots,\mu^{(k)},\mu,\nu), \tag{4.37}$$

where $\sum'$ denotes the sum over all partitions other than the cycle type of the identity element. The genus g of the covering cover is given by the Riemann-Hurwitz formula [**35**],

$$2-2g = \ell(\mu)+\ell(\nu)-d. \tag{4.38}$$

where

$$d := \sum_{i=1}^{k} \ell^*(\mu^{(i)}). \tag{4.39}$$

The weighted combinatorial Hurwitz numbers $F_G^d(\mu, \nu)$ give weighted enumerations of the paths in the Cayley graph of S_n generated by transpositions. Expanding the weight generating functions $G(z)$ as a Taylor series

$$G(z) = 1 + \sum_{i=1}^{\infty} G_i z^i, \quad \tilde{G}(z) = 1 + \sum_{i=1}^{\infty} \tilde{G}_i z^i, \tag{4.40}$$

the weight for a given path depends only upon the signature λ, and is chosen to be the product of the coefficients of the Taylor series of the generating functions $G(z)$ (or $\tilde{G}(z)$) corresponding to the parts λ_i

$$G_\lambda := \prod_{i=1}^{\ell(\lambda)} G_{\lambda_i} = e_\lambda(\mathbf{c}) \tag{4.41}$$

$$\tilde{G}_\lambda := \prod_{i=1}^{\ell(\lambda)} \tilde{G}_{\lambda_i} = h_\lambda(\mathbf{c}). \tag{4.42}$$

The path weights for signature λ, are thus chosen to be either the products $e_\lambda(\mathbf{c})$ of the elementary symmetric functions evaluated at the weighting parameters $\mathbf{c} = (c_1, c_2, \dots)$ entering in the infinite product representation of $G(z)$ or, in the dual case $\tilde{G}(z)$, the products $h_\lambda(\mathbf{c})$ of the complete symmetric functions.

The weighted combinatorial Hurwitz numbers $F_G^d(z)(\mu, \nu)$ and $F_{\tilde{G}}^d(\mu, \nu)$ are defined to be the weighted number of d-step paths, starting in the conjugacy class $\mathrm{cyc}(\mu)$ and ending in $\mathrm{cyc}(\nu)$

$$F_G^d(\mu, \nu) := \frac{1}{n!} \sum_{\lambda,\ |\lambda|=d} e_\lambda(\mathbf{c}) m_{\mu\nu}^\lambda \tag{4.43}$$

$$F_{\tilde{G}}^d(\mu, \nu) := \frac{1}{n!} \sum_{\lambda,\ |\lambda|=d} h_\lambda(\mathbf{c}) m_{\mu\nu}^\lambda. \tag{4.44}$$

4.4. Hypergeometric 2D Toda τ-functions as generating functions for weighted Hurwitz number

Applying the central elements $G_n(z, \mathcal{J})$, $\tilde{G}_n(z, \mathcal{J})$ to the cycle sums C_μ, $|\mu| = n$, and using (4.21), (4.22), (4.24), (4.25), (4.32) and (4.33), gives [**18**]

PROPOSITION 4.3.

$$G_n(z, \mathcal{J}) C_\mu = \sum_{d=1}^{\infty} z^d \sum_{\substack{\nu \\ |\nu|=|\mu|=n}} F_G^d(\mu, \nu)\, C_\nu = \sum_{d=1}^{\infty} z^d \sum_{\substack{\nu \\ |\nu|=|\mu|=n}} H_G^d(\mu, \nu)\, C_\nu \tag{4.45}$$

$$\tilde{G}_n(z, \mathcal{J}) C_\mu = \sum_{d=1}^{\infty} z^d \sum_{\substack{\nu \\ |\nu|=|\mu|=n}} F_{\tilde{G}}^d(\mu, \nu)\, C_\nu = \sum_{d=1}^{\infty} z^d \sum_{\substack{\nu \\ |\nu|=|\mu|=n}} H_{\tilde{G}}^d(\mu, \nu)\, C_\nu \tag{4.46}$$

This implies, in particular, that the two definitions of weighted Hurwitz numbers coincide:

COROLLARY 4.4.

$$F^d_G(\mu,\nu) = H^d_G(\mu,\nu). \tag{4.47}$$

Since

$$G_n(z,\mathcal{J})F_\lambda = r_\lambda^{G(z)} F_\lambda, \quad |\lambda| = n, \tag{4.48}$$

$$\tilde{G}_n(z,\mathcal{J})F_\lambda = r_\lambda^{\tilde{G}(z)} F_\lambda, \tag{4.49}$$

the change of basis formulae (3.3), (3.4) together with Proposition 4.3 imply that

$$\sum_{d=0}^\infty z^d H^d_G(\mu,\nu) = \sum_{d=0}^\infty z^d F^d_G(\mu,\nu) = \sum_{\substack{\lambda \\ |\lambda|=|\mu|=|\nu|}} r_\lambda^{G(z)} z_\mu^{-1} z_\nu^{-1} \chi_\lambda(\mu)\chi_\lambda(\nu) \tag{4.50}$$

This leads to our main result [**18**]

THEOREM 4.5. *The 2D Toda τ-functions $\tau^{G(z)}(\mathbf{t},\mathbf{s})$, $\tau^{\tilde{G}(z)}(\mathbf{t},\mathbf{s})$ can be expressed as*

$$\begin{aligned}\tau^{G(z)}(\mathbf{t},\,\mathbf{s}) &= \sum_{d=0}^\infty z^d \sum_{\substack{\mu,\nu \\ |\mu|=|\nu|}} H^d_G(\mu,\nu) p_\mu(\mathbf{t}) p_\nu(\mathbf{s}) \\ &= \sum_{d=0}^\infty z^d \sum_{\substack{\mu,\nu \\ |\mu|=|\nu|}} F^d_G(\mu,\nu) p_\mu(\mathbf{t}) p_\nu(\mathbf{s})\end{aligned} \tag{4.51}$$

$$\begin{aligned}\tau^{\tilde{G}(z)}(\mathbf{t},\mathbf{s}) &= \sum_{d=0}^\infty z^d \sum_{\substack{\mu,\nu \\ |\mu|=|\nu|}} H^d_{\tilde{G}}(\mu,\nu) p_\mu(\mathbf{t}) p_\nu(\mathbf{s}) \\ &= \sum_{d=0}^\infty z^d \sum_{\substack{\mu,\nu \\ |\mu|=|\nu|}} F^d_{\tilde{G}}(\mu,\nu) p_\mu(\mathbf{t}) p_\nu(\mathbf{s}),\end{aligned} \tag{4.52}$$

and hence are generating functions for the weighted Hurwitz numbers $H^d_G(\mu,\nu)$, $F^d_G(\mu,\nu)$, $H^d_{\tilde{G}}(\mu,\nu)$, $F^d_{\tilde{G}}(\mu,\nu)$.

5. Examples of weighted double Hurwitz numbers

We now consider several examples of different types of weighted Hurwitz numbers that are special cases of the above. All these have appeared in the recent literature on the subject [**1–3**, **12**, **13**, **17**, **18**, **22**, **23**, **26**, **33**, **42**, **44**, **52**], and new examples are easily constructed. Further details for all cases may be found in [**17**, **18**, **22**, **23**, **26**]

5.1. Double Hurwitz numbers for simple branchings; enumeration of d-step paths in the Cayley graph with uniform weight [42, 44]

This was the original case studied by Okounkov [**42**], extending an earlier result of Pandharipande [**44**]. The weight generating function in this case is just the exponential $G = \exp$

$$G(z) = \tilde{G}(z) = e^z. \tag{5.1}$$

The central element $G_n(z,\mathcal{J}) \in \mathbf{Z}(\mathbf{C}[S_n])$ is therefore

$$exp_n(z,\mathcal{J}) = e^{z\sum_{b=1}^n \mathcal{J}_b} \tag{5.2}$$

and the content product formula and fermionic exponent coefficients are given by

$$r_j^{\exp(z)} = e^{jz}, \quad r_\lambda^{\exp}(z) = e^{\frac{z}{2}\sum_{i=1}^{\ell(\lambda)} \lambda_i(\lambda_i - 2i+1)} \tag{5.3}$$

$$T_j^{\exp(z)} = \frac{1}{2} j(j+1)z. \tag{5.4}$$

The generating hypergeometric 2D Toda τ-function is thus

$$\tau^{\exp(z)}(\mathbf{t}) = \sum_\lambda e^{\frac{z}{2}\sum_{i=1}^{\ell(\lambda)} \lambda_i(\lambda_i - 2i+1)} s_\lambda(\mathbf{t}) s_\lambda(\mathbf{s}), \tag{5.5}$$

For this case, the infinite product form (3.14) of the generating function must be interpreted as a limit

$$e^z = \lim_{m\to\infty} \left(1 + \frac{z}{m}\right)^m, \tag{5.6}$$

and the expression (4.34) for the geometrical weighting becomes:

$$W_{\exp}(\mu^{(1)}, \dots, \mu^{(k)}) = \prod_{i=1}^k \delta_{(\ell^*(\mu^{(i)}),1)} \tag{5.7}$$

(since we require $\ell^*(\mu^{(i)}) \geq 1$, $\forall i$). Theorem 4.5 therefore gives the generating function

$$\tau^{\exp(z)}(\mathbf{t}) = \sum_{d=0}^\infty z^d \sum_{\substack{\mu,\nu \\ |\mu|=|\nu|}} H^d_{\exp}(\mu,\nu) p_\mu(\mathbf{t}) p_\nu(\mathbf{s}) \tag{5.8}$$

for the (weighted) number

$$H^d_{\exp}(\mu,\nu) := H(\underbrace{(2,(1)^{n-2}), \dots, (2,(1)^{n-2})}_{d \text{ times}}, \mu, \nu) \tag{5.9}$$

of n-sheeted branched coverings of the Riemann sphere having d branch points with simple ramification (i.e. , profile $(2,(1)^{n-2})$ and two more (say, at 0 and ∞) with profiles μ and ν (weighted, as usual, by the inverse of the automorphism group).

The combinatorial definition of the weighted Hurwitz number (4.43) gives

$$\begin{aligned} F^d_{\exp}(\mu,\nu) &= \frac{1}{n!} \sum_{\lambda,\ |\lambda|=d} \frac{1}{\prod_{i=1}^{\ell(\lambda)} \lambda_i!} m^\lambda_{\mu\nu} \\ &= \frac{1}{d!n!} \times (\#\ d\text{-step paths from an element } h \in \mathrm{cyc}(\mu) \text{ to } \mathrm{cyc}(\nu)). \end{aligned} \tag{5.10}$$

5.2. Coverings with three branch points (Belyi curves): strongly monotonic paths [2, 18, 26, 33, 52]

In this case the weight generating function is

$$G(z) = E(z) := 1 + z, \tag{5.11}$$

so

$$c_1 = 1, \quad c_i = 0, \quad i > 1. \tag{5.12}$$

Therefore the central element $E_n(z,\mathcal{J}) \in \mathbf{Z}(\mathbf{C}[S_n])$ is

$$E_n(z,\mathcal{J}) = \prod_{a=1}^{n}(1+z\mathcal{J}_a), \tag{5.13}$$

$$r_j^{E(z)} = 1+zj, \quad r_\lambda^{E(z)} = \prod_{(i,j)\in\lambda}(1+z(j-i)) = z^{|\lambda|}\,(1/z)_\lambda, \tag{5.14}$$

$$T_j^{E(z)} = \sum_{i=1}^{j}\ln(1+iz), \quad T_{-j}^{E(z)} = -\sum_{i=1}^{j-1}\ln(1-iz), \quad j>0, \tag{5.15}$$

where

$$(u)_\lambda := \prod_{i=1}^{\ell(\lambda)}(u-i+1)_{\lambda_i} \tag{5.16}$$

is the multiple Pochhammer symbol corresponding to the partition λ.

The generating τ-function is thus [**17**, **26**]

$$\begin{aligned}\tau^{E(z)}(\mathbf{t},\mathbf{s}) &= \sum_\lambda z^{|\lambda|}(1/z)_\lambda s_\lambda(\mathbf{t})s_\mu(\mathbf{s})\\ &= \sum_{d=0}^{\infty} z^d \sum_{\mu,\nu,\ |\mu|=|\nu|} H_E^d(\mu,\nu)p_\mu(\mathbf{t})p_\nu(\mathbf{s}),\end{aligned} \tag{5.17}$$

where

$$H_E^d(\mu,\nu) = {\sum_{\mu^{(1)},\ \ell^*(\mu_1)=d}}' H(\mu^{(1)},\mu,\nu) \tag{5.18}$$

is the number of $n=|\mu|=|\nu|=|\mu^{(1)}|$ sheeted branched covers with branch points of ramification type (μ,ν) at $(0,\infty)$, and one further branch point, with colength $\ell^*(\mu^{(1)})=d$;

These are the double Hurwitz numbers for Belyi curves [**2**, **18**, **26**, **33**, **52**], which enumerate n-sheeted branched coverings of the Riemann sphere having three ramification points, with ramification profile type μ and ν at 0 and ∞, and a single additional branch point, with ramification profile $\mu^{(1)}$ of colength

$$\ell^*(\mu^{(1)}) := n-\ell(\mu^{(1)}) = d, \tag{5.19}$$

i.e., with $n-d$ preimages. The genus is again given by the Riemann-Hurwitz formula (4.38).

Combinatorially, we have the weight

$$e_\lambda(\mathbf{c}) = \delta_{\lambda,(1)^{|\lambda|}} \tag{5.20}$$

and therefore

$$\sum_{\lambda,|\lambda|=d} e_\lambda(\mathbf{c})m_\lambda(\mathcal{J}) = \sum_{b_1<\cdots<b_d}\mathcal{J}_{b_1}\cdots\mathcal{J}_{b_d}. \tag{5.21}$$

The coefficient $F_E^d(\mu,\nu)$ is thus

$$F_E^d(\mu,\nu) = m_{\mu\nu}^{(1)^d}, \tag{5.22}$$

which enumerates all d-step paths in the Cayley graph of S_n starting at an element in the conjugacy class of cycle type ν and ending in the class of type μ, that are strictly monotonically increasing in their second elements [**18**, **26**].

5.3. Fixed number of branch points and genus: multimonotonic paths [26]

In this case the weight generating function is :

$$G(z) = E^k(z) := (1+z)^k, \tag{5.23}$$

and hence

$$c_i = 1,\ 1 \le i \le k, \quad c_i = 0,\ \forall\, i > k. \tag{5.24}$$

Therefore the central element is

$$E_n(z, \mathcal{J})^k = \prod_{a=1}^{n} (1 + z\mathcal{J}_a)^k, \tag{5.25}$$

and

$$r_j^{E^k(z)} = (1+zj)^k, \quad r_\lambda^{E^k(z)} = \prod_{(i,j)\in\lambda} (1 + z(j-i))^k = z^{k|\lambda|}((1/z)_\lambda)^k, \tag{5.26}$$

$$T_j^{E^k(z)} = k\sum_{i=1}^{j} \ln(1+iz), \quad T_{-j}^{E^k(z)} = -k\sum_{i=1}^{j-1} \ln(1-iz), \quad j > 0. \tag{5.27}$$

The generating τ-function is

$$\begin{aligned} \tau^{E^k(z)}(\mathbf{t}, \mathbf{s}) &= \sum_{\lambda} z^{k|\lambda|}((1/z)_\lambda)^k s_\lambda(\mathbf{t}) s_\lambda(\mathbf{s}) \\ &= \sum_{d=0}^{\infty} z^d \sum_{\mu,\nu,\ |\mu|=|\nu|} H^d_{E^k}(\mu,\nu) p_\mu(\mathbf{t}) p_\nu(\mathbf{s}), \end{aligned} \tag{5.28}$$

where

$$H^d_{E^k}(\mu, \nu) = \sum_{\substack{\mu^{(1)},\dots,\mu^{(k)} \\ \sum_{i=1}^k \ell^*(\mu_i) = d}}{}' H(\mu^{(1)}, \dots \mu^{(k)}, \mu, \nu) \tag{5.29}$$

is the number of $n = |\mu| = |\nu| = |\mu^{(i)}|$ sheeted branched covers with branch points of ramification type (μ, ν) at $(0, \infty)$, and k further branch points, such that the sum of the colengths of their ramification profile type (i.e., the "defect" in the Riemann Hurwitz formula (4.38)) is equal to d:

$$\sum_{i=1}^{k} \ell^*(\mu^{(i)}) = kn - \sum_{i=1}^{k} \ell(\mu^{(i)}) = d. \tag{5.30}$$

This amounts to counting covers with the genus fixed by (4.38) and the number of additional branch points fixed at k, but no restriction on their simplicity.

The combinatorial weighting for paths of signature λ is

$$e_\lambda(\mathbf{c}) = \prod_{i=1}^{\ell(\lambda)} \binom{k}{\lambda_i} \tag{5.31}$$

and hence,

$$\sum_{\lambda, |\lambda|=d} \left(\prod_{i=1}^{\ell(\lambda)} \binom{k}{\lambda_i} \right) m_\lambda(\mathcal{J}) = [z^d] \prod_{a=1}^{n} (1 + zJ_a)^k \tag{5.32}$$

where $[z^d]$ means the coefficient of z^d in the polynomial.

The weighted combinatorial Hurwitz number

$$F^d_{E^k}(\mu,\nu) = \sum_{\lambda, |\lambda|=k} \left(\prod_{i=1}^{\ell(\lambda)} \binom{k}{\lambda_i}\right) m^\lambda_{\mu\nu} \tag{5.33}$$

is thus the number of $(d+1)$-term products $(a_1\, b_1)\cdots(a_d\, b_d)h$ such that $h \in \mathrm{cyc}(\mu)$, while $(a_1\, b_1)\cdots(a_d\, b_d)h \in \mathrm{cyc}(\nu)$, which consist of a product of k consecutive subsequences, each of which is strictly monotonically increasing in their second elements [**17**, **26**].

5.4. Signed Hurwitz numbers at fixed genus: weakly monotonic paths [12, 13, 17]

This case was studied from the combinatorial viewpoint, and related to the HCIZ integral in [**12**, **13**, **17**]. It is the dual $\tilde{E}$ of the weight generating function of subsection 5.2.

$$\tilde{E}(z) := H(z) := \frac{1}{1-z} \tag{5.34}$$

and hence we have

$$c_1 = 1,\ c_i = 0,\ \forall\, i > 1 \tag{5.35}$$

as before, but the relevant combinatorial weighting factor is

$$h_\lambda(\mathbf{c}) = 1 \quad \forall \lambda. \tag{5.36}$$

The corresponding central element is

$$H_n(z, \mathcal{J}) = \prod_{a=1}^{n} (1 - z\mathcal{J}_a)^{-1}, \tag{5.37}$$

and therefore

$$\begin{aligned} r_j^{H(z)} &= (1-zj)^{-1}, \quad r_\lambda^{H(z)}(z) = \prod_{(i,j)\in\lambda} (1 - z(j-i))^{-1} \\ &= (-z)^{-|\lambda|}((-1/z)_\lambda)^{-1}, \end{aligned} \tag{5.38}$$

$$T_j^{H(z)} = -\sum_{i=1}^{j} \ln(1-iz), \quad T_{-j}^{H(z)} = \sum_{i=1}^{j-1} \ln(1+iz), \quad j > 0. \tag{5.39}$$

The generating τ-function for this case is [**17**, **26**]

$$\begin{aligned} \tau^{H(z)}(\mathbf{t},\mathbf{s}) &= \sum_\lambda (-z)^{-|\lambda|}\, (-1/z)_\lambda^{-1}\, s_\lambda(\mathbf{t}) s_\mu(\mathbf{s}) \\ &= \sum_{d=0}^{\infty} z^d \sum_{\mu,\nu,\ |\mu|=|\nu|} H^d_H(\mu,\nu) p_\mu(\mathbf{t}) p_\nu(\mathbf{s}) \end{aligned} \tag{5.40}$$

where

$$H^d_H(\mu,\nu) = (-1)^{n+d} \sum_{k=1}^{\infty} (-1)^k \sideset{}{'}\sum_{\substack{\mu^{(1)},\dots,\mu^{(k)} \\ \sum_{i=1}^k \ell^*(\mu_i) = d}} H(\mu^{(1)}, \dots \mu^{(k)}, \mu, \nu) \tag{5.41}$$

is the signed enumeration of $n = |\mu| = |\nu|$ sheeted branched covers with branch points of ramification type (μ,ν) at $(0,\infty)$, and any number further branch points, the sum of whose colengths is d, with sign determined by the parity of the number of branch points [**26**].

These are thus signed double Hurwitz numbers for n-sheeted branched coverings of the Riemann sphere with branch points having ramification profile type (μ, ν) at $(0, \infty)$ and an arbitrary number of further branch points, such that the sum of the colengths of their ramification profile lengths is again equal to d

$$\sum_{i=1}^{k} \ell^*(\mu^{(i)}) = kn - \sum_{i=1}^{k} \ell(\mu^{(i)}) = d. \tag{5.42}$$

The latter are counted with a sign, which is $(-1)^{n+d}$ times the parity of the number of branch points [**26**]. The genus is again given by (4.38).

The combinatorial Hurwitz number $F_H^d(\mu, \nu)$, derived from

$$\sum_{\lambda, |\lambda|=d} h_\lambda(\mathbf{c}) m_\lambda(\mathcal{J}) = \sum_{b_1 \leq \cdots \leq b_d} \mathcal{J}_{b_1} \cdots \mathcal{J}_{b_d}. \tag{5.43}$$

is therefore given by

$$F_H^d(\mu, \nu) = \sum_{\lambda, \ |\lambda|=d} m_{\mu\nu}^\lambda, \tag{5.44}$$

which is the number of products of the form $(a_1\, b_1) \cdots (a_d\, b_d) h$ for $g \in \mathrm{cyc}(\mu)$ that are weakly monotonically increasing, such that $(a_1\, b_1) \cdots (a_d\, b_d) h \in \mathrm{cyc}(\nu)$. These thus enumerate the d-step paths in the Cayley graph of S_n from an element in the conjugacy class of cycle type μ to the class cycle type ν, that are weakly monotonically increasing in their second elements [**17**].

This case is known to have a matrix model representation [**12, 13**] when the flow parameters $\mathbf{t}$ and $\mathbf{s}$ are restricted to be trace invariants of a pair of $N \times N$ normal matrices A, B:

$$t_i = \frac{1}{i} \mathrm{tr}(A^i), \quad s_i = \frac{1}{i} \mathrm{tr}(B^i), \tag{5.45}$$

Within a normalization, setting

$$z = -\frac{1}{N} \tag{5.46}$$

as the expansion parameter, we have equality with the HCIZ double matrix integral (5.47)

$$\tau^{H(-\frac{1}{N})}(\mathbf{t}, \mathbf{s}) = \mathcal{I}_N(A, B) := \int_{U \in U(N)} e^{\mathrm{tr}(UAU^\dagger B)} d\mu(U) = \left(\prod_{k=0}^{N-1} k!\right) \frac{\det\left(e^{a_i b_j}\right)_{1 \leq i,j \leq N}}{\Delta(\mathbf{a}) \Delta(\mathbf{b})},$$

where $d\mu(U)$ is the Haar measure on $U(N)$, $\mathbf{a} = (a_1, \ldots, a_N)$, $\mathbf{b} = (b_1, \ldots, b_N)$ are the eigenvalues of A and B respectively, and $\Delta(\mathbf{a})$, $\Delta(\mathbf{b})$ are the Vandermonde determinants.

The identification (5.46), however, gives rise to a cutoff in the expansion (5.40), giving a sum only over partitions λ of length $\ell(\lambda) \leq N$:

$$\tau^{H(-\frac{1}{N})}(\mathbf{t}, \mathbf{s}) = \sum_{\lambda, \ \ell(\lambda) \leq N} \frac{N^{|\lambda|}}{(N)_\lambda} s_\lambda(\mathbf{t}) s_\mu(\mathbf{s}). \tag{5.48}$$

5.5. Quantum weighted branched coverings and paths [18]

In [**18**] three variants of quantum Hurwitz numbers were studied, with weight generating functions denoted $E(q,z)$, $H(q,z)$ and $E'(q,z)$. We only consider the case $E'(q,z)$, which has the most interesting interpretation in relation to Bosonic gases. The other two are developed in detail in [**18**] and may also be obtained as special cases of the Macdonald polynomial approach to quantum Hurwitz numbers developed in [**23**], which is summarized in section 7 below.

The weight generating function is

$$E'(q,z) := \prod_{i=1}^{\infty}(1+q^i z) = 1 + \sum_{i=0}^{\infty} E'_i(q) z^i, \tag{5.49}$$

$$E'_i(q) := \frac{q^{\frac{1}{2}i(i+1)}}{\prod_{j=1}^{i}(1-q^j)}, \quad i \geq 1, \tag{5.50}$$

where q is viewed as a quantum deformation parameter that may interpreted (see below) in terms of the energy distribution of Bosonic gases with a linear energy spectrum. This is related to the quantum dilogarithm function by

$$(1+z)E'(q,z) = e^{-\operatorname{Li}_2(q,-z)}, \quad \operatorname{Li}_2(q,z) := \sum_{k=1}^{\infty} \frac{z^k}{k(1-q^k)}. \tag{5.51}$$

We thus have

$$c_i = q^i, \quad i \geq 1, \quad e_\lambda(\mathbf{c}) =: E'_\lambda(q) = \prod_{i=1}^{\ell(\lambda)} \frac{q^{\frac{1}{2}\lambda_i(\lambda_i+1)}}{\prod_{j=1}^{\lambda_i}(1-q^j)}. \tag{5.52}$$

The central element $E'_n(q, x\mathcal{J}) \in \mathbf{Z}(\mathbf{C}[S_n])$ is given by

$$E'_n(q, z\mathcal{J}) = \prod_{a=1}^{n}\prod_{k=1}^{\infty}(1+q^k z\mathcal{J}_a), \tag{5.53}$$

and hence the content product coefficient is

$$r_j^{E'(q,z)} = \prod_{k=1}^{\infty}(1+q^k zj), \tag{5.54}$$

$$r_\lambda^{E'(q,z)}(z) = \prod_{k=1}^{\infty}\prod_{(i,j)\in\lambda}(1+q^k z(j-i)) = \prod_{k=1}^{\infty}(zq^k)^{|\lambda|}(1/(zq^k))_\lambda. \tag{5.55}$$

The generating τ-function is therefore [**18**]

$$\begin{aligned} \tau^{E'(q,z)}(\mathbf{t},\mathbf{s}) &= \sum_\lambda \left(\prod_{k=1}^{\infty}(zq^k)^{|\lambda|}(1/(zq^k))_\lambda\right) s_\lambda(\mathbf{t}) s_\mu(\mathbf{s}) \\ &= \sum_{d=0}^{\infty} z^d \sum_{\mu,\nu,\ |\mu|=|\nu|} H^d_{E'(q)}(\mu,\nu) p_\mu(\mathbf{t}) p_\nu(\mathbf{s}), \end{aligned} \tag{5.56}$$

where

(5.57)

$$H^d_{E'(q)}(\mu,\nu) := \sum_{k=0}^{\infty} \sideset{}{'}\sum_{\substack{\mu^{(1)},\dots\mu^{(k)} \\ \sum_{i=1}^k \ell^*(\mu^{(i)})=d}} W_{E(q)}(\mu^{(1)},\dots,\mu^{(k)}) H(\mu^{(1)},\dots,\mu^{(k)},\mu,\nu)$$

is the quantum weighted enumeration of $n = |\mu| = |\nu|$ sheeted branched coverings with genus g given by (4.38) and weight $W_{E'(q)}(\mu^{(1)}, \dots, \mu^{(k)})$ for branched coverings of type $(\mu^{(1)}, \dots, \mu^{(k)}, \mu, \nu)$ given by

$$
\begin{aligned}
W_{E'(q)}(\mu^{(1)}, \dots, \mu^{(k)}) &:= \frac{1}{|\mathrm{aut}(\lambda)|} \sum_{\sigma \in S_k} \sum_{1 \le i_1 < \cdots < i_k}^{\infty} q^{i_1 \ell^*(\mu^{(\sigma(1))})} \cdots q^{i_k \ell^*(\mu^{(\sigma(k))})} \\
&= \frac{1}{|\mathrm{aut}(\lambda)|} \sum_{\sigma \in S_k} \frac{q^{k\ell^*(\mu^{(\sigma(1))})} \cdots q^{\ell^*(\mu^{(\sigma(k))})}}{(1 - q^{\ell^*(\mu^{(\sigma(1))})}) \cdots (1 - q^{\ell^*(\mu^{(\sigma(1))})} \cdots q^{\ell^*(\mu^{(\sigma(k))})})} \\
&= \frac{1}{|\mathrm{aut}(\lambda)|} \sum_{\sigma \in S_k} \frac{1}{(q^{-\ell^*(\mu^{(\sigma(1))})} - 1) \cdots (q^{-\ell^*(\mu^{(\sigma(1))})} \cdots q^{-\ell^*(\mu^{(\sigma(k))})} - 1)},
\end{aligned}
\tag{5.58}
$$

where λ is the partition with parts $\{\ell^*(\mu^{(i)})\}_{i=1,\dots,k}$

The combinatorial Hurwitz number $F^d_{E'(q)}(\mu, \nu)$ giving the weighted enumeration of paths is

$$
F^d_{E'(q)}(\mu, \nu) = \sum_{\lambda,\ |\lambda| = d} \frac{q^{\frac{1}{2}i(i+1)}}{\prod_{j=1}^{i}(1 - q^j)} m^{\lambda}_{\mu\nu} \tag{5.59}
$$

and we have the usual equality

$$
H^d_{E'(q)}(\mu, \nu) = F^d_{E'(q)}(\mu, \nu). \tag{5.60}
$$

5.1 REMARK. **Relation to Bosonic gas distribution.** If we identify

$$
q := e^{-\beta \hbar \omega_0}, \quad \beta = k_B T, \tag{5.61}
$$

where $\hbar\omega_0$ is the lowest energy state in a gas of identical Bosonic particles, assume the energy spectrum to consist of integer multiples of $\hbar\omega_0$

$$
\epsilon_k = k\hbar\omega_0, \tag{5.62}
$$

and assign the energy

$$
\epsilon(\mu^{(1)}, \dots, \mu^{(k)}) = \sum_{i=1}^{k} \epsilon_{\ell^*(\mu^{(i)})} \tag{5.63}
$$

to a configuration with ramification profiles $(\mu^{(1)}, \dots, \mu^{(k)}, \mu, \nu)$, the distribution function for Bosonic gases gives the weight

$$
W(\mu^{(1)}, \dots, \mu^{(k)}) = \frac{1}{e^{\beta\epsilon(\mu^{(1)}, \dots, \mu^{(k)})} - 1}. \tag{5.64}
$$

The weighting factor $W_{E'(q)}(\mu^{(1)}, \dots, \mu^{(k)})$ in eq. (5.58) is thus the symmetrized product

$$
W_{E'(q)}(\mu^{(1)}, \dots, \mu^{(k)}) = \frac{1}{|\mathrm{aut}(\lambda)|} \sum_{\sigma \in S_k} W(\mu^{(\sigma(1)}) \cdots W(\mu^{\sigma(1)}, \dots, \mu^{\sigma(k)}) \tag{5.65}
$$

of that for each subconfiguration.

In [**18**], a dual pair of similar weight generating functions $E(q, z)$, $H(q, z)$ was introduced, which correspond to two slightly different definitions of quantum Hurwitz numbers. These are the q-analogs of what, when extended to the Cauchy-Littlewood formula, become the generating functions of the elementary and the complete symmetric functions:

$$E(q,z) := \prod_{k=0}^{\infty}(1+zq^k) \tag{5.66}$$

$$H(q,z) := \prod_{k=0}^{\infty}(1-zq^k)^{-1} \tag{5.67}$$

The corresponding weights for branched covers with ramification profiles $(\mu^{(1)}, \dots, \mu^{(k)})$ and $(\nu^{(1)}, \dots, \nu^{(\tilde{k})})$ at the branch points are:

$$\begin{aligned} W_{E(q)}(\mu^{(1)}, \dots, \mu^{(k)}) & \frac{1}{|\mathrm{aut}(\lambda)|} \sum_{\sigma\in S_k} \sum_{0\le i_1<\cdots<i_k}^{\infty} q^{i_1\ell^*(\mu^{(\sigma(1))})} \cdots q^{i_k\ell^*(\mu^{(\sigma(k))})} \\ &= \frac{1}{|\mathrm{aut}(\lambda)|} \sum_{\sigma\in S_k} \frac{q^{(k-1)\ell^*(\mu^{(\sigma(1))})} \cdots q^{\ell^*(\mu^{(\sigma(k-1))})}}{(1-q^{\ell^*(\mu^{(\sigma(1))})})\cdots(1-q^{\ell^*(\mu^{(\sigma(1))})}\cdots q^{\ell^*(\mu^{(\sigma(k))})})}, \end{aligned} \tag{5.68}$$

where λ is the partition with parts $(\ell^*(\mu^{(1)}), \dots, \ell^*(\mu^{(\tilde{k})}))$, and

$$\begin{aligned} W_{H(q)}(\nu^{(1)}, \dots, \nu^{(\tilde{k})}) &:= \frac{(-1)^{\ell^*(\lambda)}}{|\mathrm{aut}(\lambda)|} \sum_{\sigma\in S_{\tilde{k}}} \sum_{0\le i_1\le\cdots\le i_{\tilde{k}}}^{\infty} q^{i_1\ell^*(\nu^{(\sigma(1))})} \cdots q^{i_k\ell^*(\nu^{(\sigma(\tilde{k}))})} \\ &= \frac{(-1)^{\ell^*(\lambda)}}{|\mathrm{aut}(\lambda)|} \sum_{\sigma\in S_{\tilde{k}}} \frac{1}{(1-q^{\ell^*(\nu^{(\sigma(1))})})\cdots(1-q^{\ell^*(\nu^{(\sigma(1))})}\cdots q^{\ell^*(\nu^{(\sigma(\tilde{k}))})})}, \end{aligned} \tag{5.69}$$

where λ is the partition with parts $(\ell^*(\nu^{(1)}), \dots, \ell^*(\nu^{(\tilde{k})}))$.

The associated hypergeometric τ-functions $\tau^{E(q,z))}(\mathbf{t}, \mathbf{s})$, and $\tau^{H(q,z))}(\mathbf{t}, \mathbf{s})$ are defined similarly to $\tau^{E'(q,z))}(\mathbf{t}, \mathbf{s})$ and are generating functions for the correspondingly modified Hurwitz numbers $F^d_{E(q)}(\mu, \nu) = H^d_{E(q)}(\mu, \nu)$ and $F^d_{H(q)}(\mu, \nu) = H^d_{H(q)}(\mu, \nu)$. (See [**18**] for further details.)

For later use, we denote the product of these

$$W_{Q(q)}(\mu^{(1)}, \dots, \mu^{(k)}; \nu^{(1)}, \dots, \nu^{(\tilde{k})}) := W_{E(q)}(\mu^{(1)}, \dots, \mu^{(k)}) W_{H(q)}(\nu^{(1)}, \dots, \nu^{(\tilde{k})}). \tag{5.70}$$

6. Multispecies weighted Hurwitz numbers

6.1. Hybrid signed Hurwitz numbers at fixed genus: hybrid monotonic paths [17, 18, 26]

This case is just a hybrid product of the cases of subsection 5.2 and subsection 5.4. We choose as generating function

$$Q(w,z) := \frac{1+w}{1-z} \tag{6.1}$$

taking power series in both parameters (w, z). The associated central element is

$$Q(w,z,\mathcal{J}) = E_n(w,\mathcal{J}) H_n(z,\mathcal{J}) = \prod_{a=1}^{n} \frac{1+w\mathcal{J}_a}{1-z\mathcal{J}_a}, \tag{6.2}$$

and therefore

$$r_j^{Q(w,z)} = \frac{1+jw}{1-jz}, \tag{6.3}$$

$$r_\lambda^{Q(w,z)} = \prod_{(i,j)\in\lambda} \frac{1+(j-i)w}{1-(j-i)z} = (-w/z)^{|\lambda|}\frac{(1/w)_\lambda}{(-1/z)_\lambda}, \tag{6.4}$$

$$T_j^{Q(w,z)} = \sum_{i=1}^{j} \ln\frac{1+iw}{1-iz}, \quad T_{-j}^{Q(w,z)}(w,z) = \sum_{b=1}^{j-1} \ln\frac{1+iz}{1-iw}, \quad j>0, \tag{6.5}$$

The generating τ-function is thus [**17**, **26**]

$$\begin{aligned} \tau^{Q(w,z)}(\mathbf{t},\mathbf{s}) &= \sum_\lambda (-w/z)^{|\lambda|}\frac{(1/w)_\lambda}{(-1/z)_\lambda} s_\lambda(\mathbf{t}) s_\mu(\mathbf{s}) \\ &= \sum_{c=0}^{\infty}\sum_{d=0}^{\infty} w^c z^d \sum_{\mu,\nu,\ |\mu|=|\nu|} H_d^c(\mu,\nu) p_\mu(\mathbf{t}) p_\nu(\mathbf{s}), \end{aligned} \tag{6.6}$$

where

$$H_d^c(\mu,\nu) = \sum_{k=0}^{\infty} \sum_{\substack{\mu^{(1)} \\ \ell^*(\mu_1)=c}} (-1)^{k+d} \sideset{}{'}\sum_{\substack{\nu^{(1)},\dots,\nu^{(k)} \\ \sum_{i=1}^k \ell^*(\nu^{(i)})=d}} H(\mu^{(1)},\nu^{(1)},\dots,\nu^{(k)},\mu,\nu) \tag{6.7}$$

is the number of $n = |\mu| = |\nu| = |\mu^{(1)}| = |\nu^{(1)}| = \cdots = |\nu^{(k)}|$ sheeted branched covers with branch points of ramification type (μ,ν) at $(0,\infty)$, one further branch point, of "first class", with colength $\ell^*(\mu^{(1)}) = c$ and k further branch points, $(\nu^{1)},\dots,\nu^{(k)})$ of "second class" with total colength equal to d,

$$\sum_{i=1}^{k} \ell^*(\nu^{(i)}) = d \tag{6.8}$$

counted with sign $(-1)^{k+d}$ determined by the parity of k. Note that the sum over k is finite because of the constraint (6.8). As usual, the Riemann-Hurwitz formula

$$2-2g = \ell(\mu) + \ell(\nu) - d - \ell^*(\mu^{(1)}) \tag{6.9}$$

determines the genus g of the covering surface.

The meaning of the combinatorial Hurwitz number $F_d^c(\mu,\nu) = H_d^c(\mu,\nu)$ in this case is clear from combining the cases considered in subsection 5.2 and subsection 5.4; it is the number of $c+d$ step paths in the Cayley graph of S_n starting at an element $h \in \mathrm{cyc}(\mu)$ in the conjugacy class of type $\mathrm{cyc}(\mu)$ and ending in $\mathrm{cyc}(\nu)$ such that the first c steps are strictly monotonic and the next d steps are weakly monotonic.

This case also has a matrix integral representation [**24**], analogous to the HCIZ integral when the flow parameters $\mathbf{t}$ and $\mathbf{s}$ are again restricted to equal the trace invariants of a pair A, B of normal matrices as in (5.45), and the expansion parameters are equated to

$$w = \frac{1}{N-\alpha}, \quad z = -\frac{1}{N} \tag{6.10}$$

for some parameter α.

$$\tau^{Q(1/(N-\alpha),-1/N)}(\mathbf{t},\mathbf{s}) = \int_{U\in U(N)} \det(\mathbf{I}-\zeta UAU^{\dagger}B)^{\alpha-N} d\mu(U) \tag{6.11}$$

$$= \left(\prod_{k=0}^{N-1} \frac{k!}{(1-\alpha)_k}\right) \frac{\det\left(1-\zeta a_i b_j\right)^{\alpha-1}_{1\le i,j\le N}}{\Delta(\mathbf{a})\Delta(\mathbf{b})} \tag{6.12}$$

where

$$\zeta := \frac{N}{N-\alpha}. \tag{6.13}$$

The identification (5.46) again gives rise to a cutoff in the expansion (5.40), restricting the sum to partitions λ of length $\ell(\lambda) \le N$:

$$\tau^{Q(1/(N-\alpha),-1/N)}(\mathbf{t},\mathbf{s}) = \sum_{\lambda,\ \ell(\lambda)\le N} \left(1-\frac{\alpha}{N}\right)^{|\lambda|} \frac{(N-\alpha)_\lambda}{(N)_\lambda} s_\lambda(\mathbf{t}) s_\mu(\mathbf{s}). \tag{6.14}$$

6.2. Signed multispecies Hurwitz numbers: hybrid multimonotonic paths [26]

Now consider the multiparametric generalization of the previous example. We introduce $l+m$ expansion parameters

$$\mathbf{w} := (w_1, \dots, w_l), \quad \mathbf{z} = (z_1, \dots, z_m). \tag{6.15}$$

The weight generating functions G is chosen to be a multiple product of those for the previous cases:

$$Q^{(l,m)}(\mathbf{w},\mathbf{z}) := \prod_{\alpha=1}^{l} E(w_\alpha) \prod_{\beta=1}^{m} H(z_\beta) = \frac{\prod_{\alpha=1}^{l}(1+w_\alpha)}{\prod_{\beta=1}^{m}(1-z_\beta)} \tag{6.16}$$

The corresponding element of the center $\mathbf{Z}(\mathbf{C}[S_n])$ is

$$Q_n^{(l,m)}(\mathbf{w},\mathbf{z},\mathcal{J}) = \prod_{a=1}^{n} Q^{(l,m)}(\mathbf{w}\mathcal{J}_a, \mathbf{z}\mathcal{J}_a), \tag{6.17}$$

and therefore the eigenvalues of $Q_n^{(l,m)}(\mathbf{w},\mathbf{z},\mathcal{J})$ are

$$r_\lambda^{Q^{(l,m)}(\mathbf{w},\mathbf{z})} = \prod_{(i,j)\in\lambda} \frac{\prod_{\alpha=1}^{l}(1+(j-i)w_\alpha)}{\prod_{\beta=1}^{m}(1-(j-i)z_\beta)} = \frac{\prod_{\alpha=1}^{l}(w_\alpha)^{|\lambda|}(1/w_\alpha)_\lambda}{\prod_{\beta=1}^{m}(-z_\beta)^{|\lambda|}(-1/z_\beta)_\lambda}, \tag{6.18}$$

while the diagonal exponential fermionic coefficients are

$$\begin{aligned} T_j^{Q^{(l,m)}(\mathbf{w},\mathbf{z})} &= \sum_{i=1}^{j} \ln \frac{\prod_{\alpha=1}^{l}(1+iw_\alpha)}{\prod_{\beta=1}^{m}(1-iz_\beta)}, \\ T_{-j}^{Q^{(l,m)}(\mathbf{w},\mathbf{z})} &= -\sum_{i=0}^{j-1} \ln \frac{\prod_{\alpha=1}^{l}(1-iw_\alpha)}{\prod_{\beta=1}^{m}(1+iz_\beta)}, \quad j>0. \end{aligned} \tag{6.19}$$

The generating τ-function is thus [26]

$$\begin{aligned} \tau^{Q^{(l,m)}(\mathbf{w},\mathbf{z})}(\mathbf{t},\mathbf{s}) &= \sum_{\lambda} \frac{\prod_{\alpha=1}^{l}(w_\alpha)^{|\lambda|}(1/w_\alpha)_\lambda}{\prod_{\beta=1}^{m}(-z_\beta)^{|\lambda|}(-1/z_\beta)_\lambda} s_\lambda(\mathbf{t}) s_\lambda(\mathbf{s}) \\ &= \sum_{\mathbf{d}\in\mathbf{N}^l} \sum_{\tilde{\mathbf{d}}\in\mathbf{N}^m} \mathbf{w}^{\mathbf{d}} \mathbf{z}^{\tilde{\mathbf{d}}} \sum_{\mu,\nu,\ |\mu|=|\nu|} H^{(\mathbf{d},\tilde{\mathbf{d}})}_{Q^{(l,m)}}(\mu,\nu) p_\mu(\mathbf{t}) p_\nu(\mathbf{s}), \end{aligned} \tag{6.20}$$

where multi-index notation has been used:

$$\mathbf{w}^{\mathbf{d}} := \prod_{\alpha=1}^{l} w_{\alpha}^{d_\alpha}, \quad \mathbf{z}^{\tilde{\mathbf{d}}} := \prod_{\beta=1}^{m} z_{\beta}^{\tilde{d}_\beta} \tag{6.21}$$

with

$$\mathbf{d} := (d_1, \dots, d_l), \quad \tilde{\mathbf{d}} := (\tilde{d}_1, \dots, \tilde{d}_m), \quad d_\alpha, \tilde{d}_\beta \in \mathbf{N}. \tag{6.22}$$

Here

$$H^{(\mathbf{d},\tilde{\mathbf{d}})}_{Q^{(l,m)}}(\mu,\nu) = (-1)^D \sum_{\{k_\beta\}_{\beta=1}^m} \sum_{\substack{\{\mu^{(\alpha)}\} \\ \ell^*(\mu^{(\alpha)})=d_\alpha}} \sum_{\{\nu^{(\beta,i_\beta)}\}} \sum'_{\sum_{i_\beta=1}^{k_\beta} \ell^*(\nu^{(\beta,i_\beta)})=\tilde{d}_\beta} (-1)^C H(\{\mu^{(\alpha)}\}, \{\nu^{(\beta,i_\beta)}\}_{i_\beta=1}^{k_\beta}, \mu, \nu), \tag{6.23}$$

is the signed total number of branched coverings, weighted by the inverses of their automorphism groups, with branch points at $(0,\infty)$ having ramification profiles (μ,ν), and further branch points divided into two types: l "plain" branch points $\{\mu^{(\alpha)}\}_{\alpha=1,\dots,l}$ with colengths

$$\ell^*(\mu^\alpha) = d_\alpha \tag{6.24}$$

and

$$C = \sum_{\beta=1}^{m} k_\beta, \tag{6.25}$$

"coloured" branch points with colours labeled by $\beta = 1,\dots,m$ and ramification profiles $\{\nu^{(\beta,i_\beta)}\}_{\beta=1,\dots,m;\ i_\beta=1,\dots k_\beta}$, of total ramification type colengths

$$\sum_{i_\beta=1}^{k_\beta} \ell^*(\nu^{(\beta,i_\beta)}) = \tilde{d}_\beta \tag{6.26}$$

in each colour group, and

$$D = \sum_{\beta=1}^{m} \tilde{d}_\beta \tag{6.27}$$

is the sum of these colengths over all colours. The genus g of the covering surface is determined by the Riemann-Hurwitz formula:

$$2 - 2g = \ell(\mu) + \ell(\nu) - \sum_{\alpha=1}^{l} d_\alpha - D. \tag{6.28}$$

The combinatorial significance of the weighted Hurwitz number $F^{(\mathbf{d},\tilde{\mathbf{d}})}_{Q^{(l,m)}}(\mu,\nu)$ in this case is given (see [**26**]) by:

THEOREM 6.1. *The coefficients $H^{(\mathbf{d},\tilde{\mathbf{d}})}_{Q^{(l,m)}}(\mu,\nu) - F^{(\mathbf{d},\tilde{\mathbf{d}})}_{Q^{(l,m)}}(\mu,\nu)$ in the expansion (6.20) are equal to the number of paths in the Cayley graph of S_n generated by transpositions $(a\,b)$, $a<b$, starting at an element in the conjugacy class with cycle*

type given by the partition μ and ending in the conjugacy class with cycle type given by partition ν, such that the paths consist of a sequence of

$$k := \sum_{\alpha=1}^{l} d_\alpha + \sum_{\beta=1}^{m} \tilde{d}_\beta \tag{6.29}$$

transpositions $(a_1 b_1)\cdots(a_k b_k)$, divided into $l+m$ subsequences, the first l of which consist of $\{d_1, \dots, d_l\}$ transpositions that are strictly monotonically increasing (i.e. $b_i < b_{i+1}$ for each neighbouring pair of transpositions within the subsequence), followed by $\{\tilde{d}_1, \dots, \tilde{d}_m\}$ subsequences within each of which the transpositions are weakly monotonically increasing (i.e. $b_i \leq b_{i+1}$ for each neighbouring pair).

6.3. General weighted multispecies Hurwitz numbers [22, 23]

We may extend the multispecies signed Hurwitz numbers considered in the preceding section to general multispecies weighting [**22, 23**] by replacing the factors $E(w_\alpha)$ and $H(z_\beta)$ in the above by arbitrary weight generating functions of type $G^\alpha(w_\alpha)$ and dual type $\tilde{G}^\beta(z_\beta)$.

The partitions are divided into two classes: those corresponding to the weight factors of type $G(w)$, labelled $\{\mu^{(\alpha,u_\alpha)}\}$, and those corresponding to dual type $\tilde{G}(z)$, labelled $\{\nu^{(\beta,v_\beta}\}$, These are further subdivided into l "colours", or "species" for the first class, denoted by the label $\alpha = 1, \dots, l$ and m in the second, denoted by $\beta = 1, \dots, m$. Any given configuration $\{\{\mu^{(\alpha,u_\alpha)}\}_{1\leq u_\alpha \leq k_\alpha}, \{\nu^{(\beta,v_\beta}\}_{1\leq v_\beta\leq \tilde{k}_\beta}\}$ has k_α elements of colour α in the first class and $\tilde{k}_\beta$ elements of colour β in the second class, for a total of

$$k = \sum_{\alpha=1}^{l} k_\alpha + \sum_{\beta=1}^{m} \tilde{k}_\beta \tag{6.30}$$

partitions.

Denoting the $l+m$ expansion parameters again as

$$\mathbf{w} = (w_1, \dots, w_l), \quad \mathbf{z} = (z_1, \dots, z_m), \tag{6.31}$$

the multispecies weight generating function is formed from the product

$$G^{(l,m)}(\mathbf{w}, \mathbf{z}) := \prod_{\alpha=1}^{l} G^\alpha(w_\alpha) \prod_{\beta=1}^{m} \tilde{G}^\beta(z_\beta), \tag{6.32}$$

where each factor has an infinite product representation that is of one of the two types

$$G^\alpha(w) = \prod_{i=1}^{\infty}(1 + c_i^\alpha w), \ \alpha = 1, \dots, l \tag{6.33}$$

$$\tilde{G}^\beta(w) = \prod_{i=1}^{\infty}(1 - \tilde{c}_i^\beta w)^{-1}, \ \beta = 1, \dots, m \tag{6.34}$$

for $l+m$ infinite sequences of parameters

$$\mathbf{c}^\alpha = (c_1^\alpha, c_2^\alpha, \dots), \quad \alpha = 1, \dots, l \tag{6.35}$$

$$\tilde{\mathbf{c}}^\alpha = (\tilde{c}_1^\alpha, \tilde{c}_2^\alpha, \dots), \quad \beta = 1, \dots m. \tag{6.36}$$

The corresponding central element, denoted

$$G_n^{(l,m)}(\mathbf{w},\mathbf{z},\mathcal{J}) := \prod_{a=1}^{n}\left(\prod_{\alpha=1}^{l} G^{\alpha}(w_\alpha \mathcal{J}_a)\right)\left(\prod_{\beta=1}^{m} \tilde{G}^{\beta}(z_\beta \mathcal{J}_a)\right) \tag{6.37}$$

has eigenvalues

$$r_\lambda^{G^{(l,m)}(\mathbf{w},\mathbf{z})} = \prod_{\alpha=1}^{l} r_\lambda^{G^\alpha}(w_\alpha)\prod_{\beta=1}^{m} r_\lambda^{\tilde{G}^\beta}(z_\beta) \tag{6.38}$$

in the $\{F_\lambda\}$ basis where, as before,

$$r_\lambda^{G^\alpha}(w_\alpha) := \prod_{(ij)\in\lambda} G(w_\alpha(j-i)), \quad r_\lambda^{\tilde{G}^\beta}(z_\beta) := \prod_{(ij)\in\lambda} \tilde{G}(z_\beta(j-i)). \tag{6.39}$$

The diagonal exponential fermionic coefficients are

$$\begin{aligned} T_j^{G^{(l,m)}(\mathbf{w},\mathbf{z})} &= \sum_{i=1}^{j}\left(\prod_{\alpha=1}^{l} G^\alpha(iw_\alpha)\prod_{\beta=1}^{m}\tilde{G}^\beta(iz_\beta)\right), \\ T_{-j}^{G^{(l,m)}(\mathbf{w},\mathbf{z})} &= -\sum_{i=0}^{j-1}\ln\left(\prod_{\alpha=1}^{l} G^\alpha(-iw_\alpha)\prod_{\beta=1}^{m}\tilde{G}^\beta(-iz_\beta)\right), \quad j>0. \end{aligned} \tag{6.40}$$

The generating hypergeometric τ-function is [**22**]

$$\begin{aligned} \tau^{G^{(l,m)}(\mathbf{w},\mathbf{z})}(\mathbf{t},\mathbf{s}) &= \sum_{\lambda}\prod_{\alpha=1}^{l} r_\lambda^{G^\alpha(w_\alpha)}\prod_{\beta=1}^{m} r_\lambda^{\tilde{G}^\beta((z_\beta)} s_\lambda(\mathbf{t})s_\lambda(\mathbf{s}) \\ &= \sum_{\mathbf{d}\in\mathbf{N}^l}\sum_{\tilde{\mathbf{d}}\in\mathbf{N}^m}\mathbf{w}^{\mathbf{d}}\mathbf{z}^{\tilde{\mathbf{d}}}\sum_{\mu,\nu,\ |\mu|=|\nu|} H_{G(l,m)}^{(\mathbf{d},\tilde{\mathbf{d}})}(\mu,\nu)p_\mu(\mathbf{t})p_\nu(\mathbf{s}), \end{aligned} \tag{6.41}$$

where

$$\begin{aligned} H_{G^{(l,m)}}^{(\mathbf{d},\tilde{\mathbf{d}})}(\mu,\nu) := \sum_{k_1,\dots,k_l}\sum_{\tilde{k}_1,\dots,\tilde{k}_m} &\sum_{\substack{\{\mu^{(\alpha,u_\alpha)}\} \\ |\mu^{(\alpha,u_\alpha)}|=n \\ \sum_{u_\alpha=1}^{k_\alpha}\ell^*(\mu^{(\alpha,u_\alpha)})=d_\alpha}} \ \sum_{\substack{\{\nu^{(\beta,v_\beta)}\} \\ |\nu^{(\beta,u_\beta)}|=n \\ \sum_{v_\beta=1}^{\tilde{k}_\beta}\ell^*(\nu^{(\beta,v_\beta)})=\tilde{d}_\beta}}{}' \\ &\times W_{G^{(l,m)}}(\{\mu^{(\alpha,u_\alpha)}\},\{\nu^{(\beta,v_\beta)}\})H(\{\mu^{(\alpha,u_\alpha)}\},\{\nu^{(\beta,v_\beta)}\},\mu,\nu) \end{aligned} \tag{6.42}$$

is the geometrical multispecies Hurwitz number giving the weighted enumeration of n-sheeted branched coverings with $l+m$ branch points of type $\{\{\mu^{(\alpha,u_\alpha)}\}_{1\le u_\alpha\le k_\alpha}, \{\nu^{(\beta,v_\beta)}\}_{1\le v_\beta\le\tilde{k}_\beta}\}$ and (μ,ν) at $(0,\infty)$, with weighting factor equal to the product of those for single species

$$W_{G^{(l,m)}}(\{\mu^{(\alpha,u_\alpha)}\},\{\nu^{(\beta,v_\beta)}\}) = \prod_{\alpha=1}^{l} m_{\lambda^{(\alpha)}}(\mathbf{c}^{(\alpha)})\prod_{\beta=1}^{m} m_{\tilde{\lambda}^{(\beta)}}(\tilde{\mathbf{c}}^{\beta}). \tag{6.43}$$

Here the partitions $\{\lambda^{(\alpha)}\}_{\alpha=1,\dots,l}$, and $\{\tilde{\lambda}^{(\beta)}_{\beta=1,\dots,m}\}$ have lengths

$$\ell(\lambda^{(\alpha)}) = k_\alpha, \quad \ell(\tilde{\lambda}^{(\beta)}) = \tilde{k}_\beta, \tag{6.44}$$

weights

$$|\lambda^{(\alpha)})| = d_\alpha, \quad |\tilde{\lambda}^{(\beta)}| = \tilde{d}_\beta, \tag{6.45}$$

and parts equal to the colengths $\ell^*(\mu^{(\alpha,u_\alpha)})$ and $\ell^*(\mu^{(\beta,v_\beta)})$, for $\lambda^{(\alpha)}$ and $\tilde{\lambda}^{(\beta)}$ respectively.

The combinatorial multispecies Hurwitz number $F^{(\mathbf{d},\tilde{\mathbf{d}})}_{G^{(l,m)}}(\mu,\nu)$ is determined as follows [22]. Let D_n denote the number of partitions of weight n. For each generating function $G^\alpha(w_\alpha)$ or $\tilde{G}^\beta(z_\beta)$, let $\mathbf{F}^{d_\alpha}_{G^\alpha}$ and $\mathbf{F}^{\tilde{d}_\beta}_{\tilde{G}^\beta}$ denote the $D_n \times D_n$ matrices whose elements are $F^{d_\alpha}_{G^\alpha}(\mu,\nu)$ and $F^{\tilde{d}_\beta}_{\tilde{G}^\beta}(\mu,\nu)$, respectively, as defined in (4.43), (4.44). From the fact that the central elements $\{G^\alpha(w_\alpha,\mathcal{J}),\tilde{G}^\beta(z_\beta,\mathcal{J})\}$ all commute, it follows that the matrices $\{\mathbf{F}^{d_\alpha}_{G^\alpha},\mathbf{F}^{\tilde{d}_\beta}_{\tilde{G}^\beta}\}$ also all commute. Denoting the product of these in any order,

$$\mathbf{F}^{(\mathbf{d},\tilde{\mathbf{d}})}_{G^{(l,m)}} := \prod_{\alpha=1}^{l} \mathbf{F}^{d_\alpha}_{G^\alpha} \prod_{\beta=1}^{m} \mathbf{F}^{\tilde{d}_\beta}_{\tilde{G}^\beta}, \tag{6.46}$$

the (μ,ν) matrix element $F^{(\mathbf{d},\tilde{\mathbf{d}})}_{G^{(l,m)}}(\mu,\nu)$ is the combinatorial multispecies weighted Hurwitz number, and is equal to the geometrically defined one

$$F^{(\mathbf{d},\tilde{\mathbf{d}})}_{G^{(l,m)}}(\mu,\nu) = H^{(\mathbf{d},\tilde{\mathbf{d}})}_{G^{(l,m)}}(\mu,\nu). \tag{6.47}$$

The combinatorial meaning of $F^{(\mathbf{d},\tilde{\mathbf{d}})}_{G^{(l,m)}}(\mu,\nu)$ is as follows, Let

$$d := \sum_{\alpha=1}^{l} d_\alpha + \sum_{\beta=1}^{m} \tilde{d}_\beta. \tag{6.48}$$

Then $F^{(\mathbf{d},\tilde{\mathbf{d}})}_{G^{(l,m)}}(\mu,\nu)$, may be interpreted as the weighted sum over all sequences of d step paths in the Cayley graph from an element $h \in \mathrm{cyc}(\mu)$ in the conjugacy class of cycle type μ to one $(a_d b_d)\cdots(a_1 b_1)h$ of cycle type ν, in which the transpositions appearing are subdivided into subsets consisting of $(d_1,\dots,d_l,\tilde{d}_1,\dots,\tilde{d}_m)$ transpositions in all $\frac{d!}{(\prod_{\alpha=1}^{l} d_\alpha!)(\prod_{\beta=1}^{l} \tilde{d}_\beta!)}$ possible ways, and to each of these, if the signatures are $(\lambda^{(1)},\dots,\lambda^{(l)},\tilde{\lambda}^{(1)},\dots,\tilde{\lambda}^{(m)})$, a weight is given that is equal to the product

$$\prod_{\alpha=1}^{l} e_{\lambda^{(\alpha)}}(\mathbf{c}^\alpha) \prod_{\beta=1}^{m} h_{\tilde{\lambda}^{(\beta)}}(\tilde{\mathbf{c}}^\beta). \tag{6.49}$$

THEOREM 6.2.

$$F^{(\mathbf{d},\tilde{\mathbf{d}})}_{G^{(l,m)}}(\mu,\nu) = H^{(\mathbf{d},\tilde{\mathbf{d}})}_{G^{(l,m)}}(\mu,\nu). \tag{6.50}$$

For proofs of these results, see [22].

7. Quantum weighted Hurwitz numbers and Macdonald polynomials

Only a summary of the results will be given here; for details see [23].

7.1. Generating functions for Macdonald polynomials

Following [**36**], for two infinite sets of indeterminates $\mathbf{x} := (x_1, x_2, \dots)$, $\mathbf{y} := (y_1, y_2, \dots)$, we define the generating function

$$\Pi(\mathbf{x}, \mathbf{y}, q, t) := \prod_{a=1}^{\infty} \prod_{b=1}^{\infty} \frac{(t x_a y_b; q)_\infty}{(x_a y_b; q)_\infty}, \tag{7.1}$$

where

$$(t; q)_\infty := \prod_{i=0}^{\infty} (1 - t q^i). \tag{7.2}$$

is the infinite q-Pochhammer symbol. $\Pi(q, t, \mathbf{x}, \mathbf{y})$ can be expanded in a number of ways in terms of products of dual bases for the algebra Λ of symmetric functions

$$\Pi(\mathbf{x}, \mathbf{y}, q, t) = \sum_\lambda P_\lambda(\mathbf{x}, q, t) P_\lambda(\mathbf{y}, q, t) \tag{7.3}$$

$$= \sum_\lambda z_\mu^{-1}(q, t) p_\lambda(\mathbf{x}) p_\lambda(\mathbf{y}) \tag{7.4}$$

$$= \sum_\lambda g_\lambda(\mathbf{x}, q, t) m_\lambda(\mathbf{y}) \tag{7.5}$$

$$= \sum_\lambda g_\lambda(\mathbf{y}, q, t) m_\lambda(\mathbf{x}) \tag{7.6}$$

where

$$z_\mu(q, t) := z_\mu n_\mu(q, t), \quad n_\mu := \prod_{i=1}^{\ell(\mu)} \frac{1 - q_i^\mu}{1 - t_i^\mu}. \tag{7.7}$$

Here $\{P_\lambda(q, t, \mathbf{x})\}$ are the Macdonald symmetric functions, which are orthogonal

$$(P_\lambda(q, t), P_\mu(q, t))_{(q,t)} = 0, \quad \lambda \neq \mu \tag{7.8}$$

with respect to the inner product $(\ ,\ )_{(q,t)}$ in which the power sum symmetric functions $\{p_\lambda\}$ satisfy

$$(p_\lambda, p_\mu)_{(q,t)} = z_\mu^{-1}(q, t) \delta_{\lambda\mu}, \tag{7.9}$$

$\{m_\lambda\}$ are the basis of monomial sum symmetric functions and

$$g_\lambda(\mathbf{x}, q, t) := \prod_{i=1}^{\ell(\lambda)} g_{\lambda_i}(\mathbf{x}, q, t), \quad g_j(\mathbf{x}, q, t) := (P_j, P_j)^{-1} P_j(q, t, \mathbf{x}), \tag{7.10}$$

is the (q, t) analog of the interpolating function between the elementary $e_\lambda(\mathbf{x})$ and complete $h_\lambda(\mathbf{x})$ symmetric function bases.

7.2. Quantum families of central elements and weight generating functions

We now consider an extended infinite parametric family of generating functions $M(q, t, \mathbf{c}, z)$, depending on the infinite set of "classical" parameters $\mathbf{c}$ appearing in the infinite product representations as in (3.14), (3.18) as well as the further pair of "quantum deformation" parameters (q, t) appearing in the Macdonald polynomials [**36**].

For a dual pair of "classical" generating function $G(z)$, $\tilde{G}(z)$, with infinite product representations (3.14), (3.18) we introduce a (q,t) deformed parametric family of weight generating functions $M(q,t,\mathbf{c},z)$ as follows

$$M(q,t,\mathbf{c},z) := \prod_{k=0}^{\infty} G(tq^k z)\tilde{G}(q^k z) = \prod_{k=0}^{\infty}\prod_{i=1}^{\infty} \frac{1-tzq^k c_i}{1-zq^k c_i}. \tag{7.11}$$

The associated central element $M_n(q,t,\mathbf{c},z,\mathcal{J}) \in \mathbf{Z}(\mathbf{C}[S_n])$ is defined as

$$M_n(q,t,\mathbf{c},z\mathcal{J}) := \prod_{a=1}^{n} M(q,t,\mathbf{c},z\mathcal{J}_a) = \Pi_n(\mathbf{c},z\mathcal{J},q,t). \tag{7.12}$$

The eigenvectors are the orthogonal idempotents:

$$M_n(q,t,\mathbf{c},z\mathcal{J})F_\lambda = r_\lambda^{M(q,t,\mathbf{c},z)} F_\lambda \tag{7.13}$$

and the eigenvalues $r_\lambda^{M(q,t,\mathbf{c},z)}$ have the usual content product form

$$r_\lambda^{M(q,t,\mathbf{c},z)} = \prod_{(i,j)\in\lambda} M(q,t,\mathbf{c},z(j-i)) = \prod_{i=1}^{\infty} \frac{(ztc_i;q)_\lambda}{(zc_i;q)_\lambda} \tag{7.14}$$

where, for a partition $\lambda = (\lambda_\geq \cdots \geq \lambda_{\ell(\lambda)} > 0)$, the q-Pochhammer symbol $(t;q)_\lambda$ is defined as

$$(t;q)_\lambda := \prod_{i=1}^{\ell(\lambda)} (tq^{-i+1};q)_{\lambda_i}, \quad (t;q)_l := \prod_{k=0}^{l-1}(1-tq^k). \tag{7.15}$$

7.3. Macdonald family of quantum weighted Hurwitz numbers

Using the content product formula (7.14), we define the associated 2D Toda τ-function for $N=0$ as

$$\tau^{M(q,t,\mathbf{c},z)}(\mathbf{t},\mathbf{s}) := \sum_\lambda r_\lambda^{M(q,t,\mathbf{c},z)} s_\lambda(\mathbf{t}) s_\lambda(\mathbf{s}). \tag{7.16}$$

Substitution of the parameters $\mathbf{c}$ and the Jucys-Murphy elements $(\mathcal{J}_1,\dots,\mathcal{J}_n)$ for the indeterminates $\mathbf{x}$ and $\mathbf{y}$ in (7.5), (7.6) gives

$$M_n(q,t,\mathbf{c},z\mathcal{J}) = \sum_\lambda g_\lambda(q,t,\mathbf{c}) m_\lambda(\mathcal{J}) \tag{7.17}$$

$$= \sum_\lambda g_\lambda(q,t,\mathcal{J}) m_\lambda(\mathbf{c}). \tag{7.18}$$

Applying $M_n(q,t,\mathbf{c},z\mathcal{J})$ to C_μ and using (7.17) and Theorem 4.1, gives

$$M_n(q,t,\mathbf{c},z\mathcal{J})C_\mu = \sum_{d=0}^{\infty} z^d \sum_{\nu, |\nu|=|\mu|} F^d_{M(q,t,\mathbf{c})}(\mu,\nu) z_\nu C_\nu \tag{7.19}$$

where we define, as before, the combinatorial quantum weighted Hurwitz numbers $F^d_{M(q,t,\mathbf{c})}(\mu,\nu)$ associated to the weight generating function $M(q,t,\mathbf{c},z)$ as

$$F^d_{M(q,t,\mathbf{c})}(\mu,\nu) := \frac{1}{|n|!} \sum_{\lambda,\ |\lambda|=d} g_\lambda(\mathbf{c},q,t) m^\lambda_{\mu\nu}. \tag{7.20}$$

It follows as before that when the τ-function $\tau^{M(q,t,\mathbf{c},z)}(\mathbf{t},\mathbf{s})$ is expanded in the basis of products of power sum symmetric functions, the coefficients are the quantum weighted Hurwitz combinatorial numbers $F^d_{M(q,t,\mathbf{c})}(\mu,\nu)$:

$$\tau^{M(q,t,\mathbf{c},z)}(\mathbf{t},\mathbf{s}) = \sum_{d=0}^{\infty} \sum_{\substack{\mu,\nu \\ |\mu|=|\nu|}} z^d F^d_{M(q,t,\mathbf{c})}(\mu,\nu) p_\mu(\mathbf{t}) p_\nu(\mathbf{s}). \tag{7.21}$$

The corresponding geometrically defined quantum weighted Hurwitz numbers are somewhat more intricate. Let $\{\{\mu^{i,u_i}\}_{u_i=1,\dots,k_i}, \{\nu^{i,v_i}\}_{v_i=1,\dots,\tilde{k}_i}, \mu, \nu\}_{i=1,\dots,l}$ denote the ramification profiles of an n-sheeted covering of the Riemann sphere, with two specified branch points of ramification profile types (μ,ν), at $(0,\infty)$, and the rest divided into two classes I and II, denoted $\{\mu^{(i,u_i)}\}_{u_i=1,\dots,k_i}$ and $\{\nu^{(i,v_i)}\}_{v_i=1,\dots,\tilde{k}_i}$, respectively. These are further subdivided into l species, or "colours", labelled by $i = 1, \dots l$, the elements within each colour group distinguished by the labels $(u_i = 1, \dots, k_i)$ and $(v_i = 1, \dots, \tilde{k}_i)$. To such a grouping, we assign a partition λ of length

$$\ell(\lambda) =: l \tag{7.22}$$

and weight

$$d := |\lambda| = \sum_{i=1}^{l} \left(\sum_{u_i=1}^{k_i} \ell^*(\mu^{(i,u_i)}) + \sum_{v_i=1}^{\tilde{k}_i} \ell^*(\nu^{(i,v_i)}) \right) = \sum_{i=1}^{l} d_i, \tag{7.23}$$

whose parts $(\lambda_1 \geq \cdots \geq \lambda_l > 0)$ are equal to the total colengths

$$d_i := \sum_{u_i=1}^{k_i} \ell^*(\mu^{(i,u_i)}) + \sum_{v_i=1}^{\tilde{k}_i} \ell^*(\nu^{(i,v_i)}) \tag{7.24}$$

in weakly decreasing order. By the Riemann-Hurwitz formula, the genus g of the covering curve is given by

$$2 - 2g = \ell(\mu) + \ell(\nu) - d. \tag{7.25}$$

We now assign a weight $W_{Q(q)}(\{\mu^{(i,u_i)}, \nu^{(i,v_i)}\}, \mathbf{c})$ to each such covering as in (5.70), consisting of the product of all the weights $W_{E(q)}(\{\mu^{(i,u_i)}\}_{u_i=1,\dots,k_i})$, $W_{H(q)}(\{\nu^{(i,v_i)}\}_{v_i=1,\dots,\tilde{k}_i})$ for the subsets of different colour and class with the weight $m_\lambda(\mathbf{c})$ given by the monomial symmetric functions evaluated at the parameters $\mathbf{c}$

$$W_{Q(q)}(\{\mu^{(i,u_i)}, \nu^{(i,v_i)}\}, \mathbf{c}) := W_{Q(q)}(\{\mu^{(i,u_i)}, \nu^{(i,v_i)}\}) m_\lambda(\mathbf{c}), \tag{7.26}$$

where

$$W_{Q(q)}(\{\mu^{(i,u_i)}, \nu^{(i,v_i)}\}) := \prod_{i=1}^{l} W_{E(q)}(\{\mu^{(i,u_i)}\}_{u_i=1,\dots,k_i}) W_{H(q)}(\{\nu^{(i,v_i)}\}_{i=1,\dots,\tilde{k}_i}). \tag{7.27}$$

Using these weights, for every pair (d,e) of non-negative integers and (μ,ν) of partitions of n, we define the geometrical quantum weighted Hurwitz numbers

$H^{(d,e)}_{(\mathbf{c},q)}(\mu,\nu)$ as the sum

$$(7.28)\quad H^{(d,e)}_{(\mathbf{c},q)}(\mu,\nu) := z_\nu \sum_{l=0}^{d} \sum_{\substack{\{\mu^{(i,u_i)},\nu^{(i,v_i)}\},\ k_i\geq 1,\ \tilde{k}_i\geq 1 \\ \sum_{i=1}^{l}\sum_{u_i=1}^{k_i}\ell^*(\mu^{(i,u_i)})=e, \\ \sum_{i=1}^{l}\left(\sum_{u_i=1}^{k_i}\ell^*(\mu^{(i,u_i)})+\sum_{v_i=1}^{\tilde{k}_i}\ell^*(\nu^{(i,v_i)})\right)=d}}' W_{Q(q)}(\{\mu^{(i,u_i)},\nu^{(i,v_i)}\},\mathbf{c}) \; H(\{\mu^{(i,u_i)}\}_{\substack{u_i=1,\dots,k_i\\ i=1,\dots,l}}, \{\nu^{(i,v_i)}\}_{\substack{v_i=1,\dots,\tilde{k}_i\\ i=1,\dots,l}}, \mu,\nu).$$

We then have the following theorem, which is proved in [**23**]:

THEOREM 7.1. *The combinatorial Hurwitz numbers $F^d_{M(q,t,\mathbf{c})}(\mu,\nu)$ are degree d polynomials in t, whose coefficients are the geometrical quantum weighted Hurwitz numbers $H^{(d,e)}_{(\mathbf{c},q)}(\mu,\nu)$*

$$(7.29)\qquad F^d_{M(q,t,\mathbf{c})}(\mu,\nu) = \sum_{e=0}^{d} H^{(d,e)}_{(\mathbf{c},q)}(\mu,\nu)t^e.$$

Hence $\tau^{M(q,t,\mathbf{c},z)}(\mathbf{t},\mathbf{s})$, when expanded in the basis of products of power sum symmetric functions and power series in z and t is the generating function for the $H^{(d,e)}_{(\mathbf{c},q)}(\mu,\nu)$'s:

$$(7.30)\qquad \tau^{M(q,t,\mathbf{c},z)}(\mathbf{t},\mathbf{s}) = \sum_{d=0}^{\infty}\sum_{e=0}^{d} z^d t^e H^{(d,e)}_{(\mathbf{c},q)}(\mu,\nu)p_\mu(\mathbf{t})p_\nu(\mathbf{s}).$$

7.4. Examples

We now give several examples of special classes of weighted Hurwitz numbers that arise through restrictions or limits involving the parameters (q,t,z). The details for all these examples may be found in [**23**]; we limit ourselves to specifying the restrictions and limits involved, giving only the weight generating functions, τ-functions and quantum weighted Hurwirz formulae for each case.

7.4.1. *Elementary quantum weighted Hurwitz numbers.* In this case, we take the limits $z\to 0$, $t\to\infty$, but keep $-tz$ fixed at a finite value, that is renamed z. The resulting weight generating function is

$$(7.31)\qquad E(q,\mathbf{c},z) := \prod_{k=0}^{\infty}\prod_{i=1}^{\infty}(1+zq^k c_i) = \prod_{i=1}^{\infty}(-zc_i;q)_\infty. =: \sum_{j=0}^{\infty} e_j(q,\mathbf{c})z^j$$

Here $e_j(q,\mathbf{c})$ is the quantum deformation of the elementary symmetric function $e_j(\mathbf{c})$. The corresponding central element $E_n(q,\mathbf{c},z,\mathcal{J})\in\mathbf{Z}(\mathbf{C}[S_n])$ is:

$$(7.32)\qquad E_n(q,\mathbf{c},z\mathcal{J}) := \prod_{a=1}^{n} E(q,\mathbf{c},z\mathcal{J}_a) = \sum_\lambda z^{|\lambda|}e_\lambda(q,\mathbf{c})m_\lambda(\mathcal{J}) = \sum_\lambda z^{|\lambda|}m_\lambda(\mathcal{J})e_\lambda(q,\mathbf{c}),$$

where

$$(7.33)\qquad e_\lambda(q,\mathbf{c}) := \prod_{i=1}^{\ell(\lambda)} e_{\lambda_i}(\mathbf{c}).$$

Applying $E_n(q, \mathbf{c}, z\mathcal{J})$ to the orthogonal idempotents $\{F_\lambda\}$ to obtain the content product coefficients and to the cycle sums $\{C_\mu\}$ to obtain the Hurwitz numbers, the resulting hypergeometric $2D$ Toda τ-function is

$$\tau^{E(q,\mathbf{c},z)}(\mathbf{t}, \mathbf{s}) = \sum_\lambda r_\lambda^{E(q,\mathbf{c},z)} s_\lambda(\mathbf{t}) s_\lambda(\mathbf{s}) \tag{7.34}$$

$$= \sum_{d=0}^{\infty} z^d \sum_\lambda F^d_{E(q,\mathbf{c})}(\mu, \nu) p_\mu(\mathbf{t}) p_\nu(\mathbf{s}), \tag{7.35}$$

where

$$r_\lambda^{E(q,\mathbf{c},z)} := \prod_{(ij)\in\lambda} \prod_{k=0}^{\infty} (-zc_k; q)_\infty \tag{7.36}$$

is the content product coefficient and

$$F^d_{E(q,\mathbf{c})}(\mu, \nu) := \sum_{|\lambda|=d} e_\lambda(q, \mathbf{c}) m^\lambda_{\mu\nu} \tag{7.37}$$

is the weighted number of d-step paths in the Cayley graph of S_n generated by transpositions, starting at the conjugacy class $\mathrm{cyc}(\mu)$ and ending at $\mathrm{cyc}(\nu)$, with weight $e_\lambda(q, \mathbf{c})$ for a path of signature λ.

Now consider again n-fold branched coverings of $\mathbf{CP}^1$ with a fixed pair of branch points at $(0, \infty)$ with ramification profiles (μ, ν) and a further $\sum_{i=1}^l k_i$ branch points $\{\mu^{(i,u_i)}\}_{u_i=1,\dots,k_i}$ of l different species (or "colours"), labelled by $i = 1, \dots, l$, with non trivial ramifications profiles. The weight $W_{E^l(q)}(\{\mu^{(i,u_i)}\}_{\substack{u_i=1,\dots,k_i \\ i=1,\dots,l}}, \mathbf{c})$ for such a covering is defined by

$$W_{E^l(q)}(\{\mu^{(i,u_i)}\}_{\substack{u_i=1,\dots,k_i \\ i=1,\dots,l}}, \mathbf{c}) := W_{E^l(q)}(\{\mu^{(i,u_i)}\}_{\substack{u_i=1,\dots,k_i \\ i=1,\dots,l}})\, m_\lambda(\mathbf{c}) \tag{7.38}$$

where

$$W_{E^l(q)}(\{\mu^{(i,u_i)}\}_{\substack{u_i=1,\dots,k_i \\ i=1,\dots,l}}) := \prod_{i=1}^{l} W_{E(q)}(\{\mu^{(i,u_i)}\}_{u_i=1,\dots,k_i}). \tag{7.39}$$

It follows from the general result that

$$F^d_{E(q,\mathbf{c})}(\mu, \nu) = H^d_{E(q,\mathbf{c})}(\mu, \nu), \tag{7.40}$$

where

$$H^d_{E(\mathbf{c},q)}(\mu, \nu) := \tag{7.41}$$
$$z_\nu \sum_{l=0}^{d} \sideset{}{'}\sum_{\substack{\{\mu^{(i,u_i)}\},\ k_i\geq 1, \\ \sum_{i=1}^l \sum_{u_i=1}^{k_i} \ell^*(\mu^{(i,u_i)})=d}} W_{E^l(q)}(\{\mu^{(i,u_i)}\}_{\substack{u_i=1,\dots,k_i \\ i=1,\dots,l}}, \mathbf{c})\ H(\{\mu^{(i,u_i)}\}_{\substack{u_i=1,\dots,k_i \\ i=1,\dots,l}}, \mu, \nu)$$

is the geometrically defined quantum weighted Hurwitz number for this case.

7.4.2. *Complete quantum weighted Hurwitz numbers.* This is the dual of the preceding case, obtained by setting $t=0$. The weight generating function becomes

$$H(q,\mathbf{c},z) := \prod_{k=0}^{\infty}\prod_{i=1}^{\infty}(1-zq^k c_i)^{-1} = \prod_{i=1}^{\infty}(zc_i;q)_\infty^{-1} =: \sum_{j=0}^{\infty} h_j(q,\mathbf{c})z^j, \tag{7.42}$$

where $h_j(q,\mathbf{c})$ is the quantum deformation of the complete symmetric function $h_j(\mathbf{c})$. The corresponding central element $H_n(q,\mathbf{c},z,\mathcal{J}) \in \mathbf{Z}(\mathbf{C}[S_n])$ is:
(7.43)

$$H_n(q,\mathbf{c},z\mathcal{J}) := \prod_{a=1}^{n} H(q,\mathbf{c},\mathcal{J}_a) = \sum_{\lambda} z^{|\lambda|}h_\lambda(q,\mathbf{c})m_\lambda(\mathcal{J}) = \sum_{\lambda} z^{|\lambda|}m_\lambda(\mathbf{c})h_\lambda(q,\mathcal{J}),$$

where

$$h_\lambda(q,\mathbf{c}) := \prod_{i=1}^{\ell(\lambda)} h_{\lambda_i}(q,\mathbf{c}). \tag{7.44}$$

Specializing the general case to $t=0$, the resulting hypergeometric $2D$ Toda τ-function is

$$\tau^{H(q,\mathbf{c},z)}(\mathbf{t},\mathbf{s}) = \sum_{\lambda} r_\lambda^{H(q,\mathbf{c},z)} s_\lambda(\mathbf{t})s_\lambda(\mathbf{s}) \tag{7.45}$$

$$= \sum_{d=0}^{\infty} z^d \sum_{\lambda} F^d_{H(q,\mathbf{c})}(\mu,\nu)p_\mu(\mathbf{t})p_\nu(\mathbf{s}), \tag{7.46}$$

where

$$r_\lambda^{H(q,\mathbf{c},z)} := \prod_{(ij)\in\lambda}\prod_{k=0}^{\infty}(z(j-i)c_k;q)_\infty^{-1} \tag{7.47}$$

and

$$F^d_{H(q,\mathbf{c})}(\mu,\nu) := \sum_{|\lambda|=d} h_\lambda(q,\mathbf{c})m^\lambda_{\mu\nu} \tag{7.48}$$

is the weighted number of paths in the Cayley graph of S_n generated by transpositions, starting at the conjugacy class $\mathrm{cyc}(\mu)$ and ending at $\mathrm{cyc}(\nu)$, with weight $h_\lambda(q,\mathbf{c})$ for a path of signature λ.

Consider again the n-fold branched coverings of $\mathbf{CP}^1$, with a fixed pair of branch points at $(0,\infty)$ with ramification profiles (μ,ν) and a further $\sum_{i=1}^{l}\tilde{k}_i$ branch points $\{\nu^{(i,v_i)}\}_{v_i=1,\dots,\tilde{k}_i}$ of l different species (or "colours"), labelled by $i=1,\dots,l$, with nontrivial ramifications profiles. The weight $W_{H^l(q)}(\{\nu^{(i,v_i)}\}_{\substack{v_i=1,\dots,\tilde{k}_i\\ i=1,\dots,l}},\mathbf{c})$ for such a covering is defined by

$$W_{H^l(q)}(\{\nu^{(i,v_i)}\}_{\substack{v_i=1,\dots,\tilde{k}_i\\ i=1,\dots,l}},\mathbf{c}) := W_{H^l(q)}(\{\nu^{(i,v_i)}\}_{\substack{v_i=1,\dots,\tilde{k}_i\\ i=1,\dots,l}})\, m_\lambda(\mathbf{c}) \tag{7.49}$$

$$W_{H^l(q)}(\{\nu^{(i,v_i)}\}_{\substack{v_i=1,\dots,k_i\\ i=1,\dots,l}}) := \prod_{i=1}^{l} W_{H(q)}(\{\nu^{(i,v_i)}\}_{v_i=1,\dots,k_i}). \tag{7.50}$$

We again have

$$F^d_{H(q,\mathbf{c})}(\mu,\nu) = H^d_{H(q,\mathbf{c})}(\mu,\nu), \tag{7.51}$$

where

$$H^d_{H(\mathbf{c},q)}(\mu,\nu) := z_\nu \sum_{l=0}^{d} \sum'_{\substack{\{\nu^{(i,v_i)}\},\ \tilde{k}_i\geq 1,\\ \sum_{i=1}^{l}\sum_{v_i=1}^{k_i}\ell^*(\nu^{(i,v_i)})=d}} W_{H^l(q)}(\{\nu^{(i,v_i)}\}_{\substack{v_i=1,\dots,\tilde{k}_i\\ i=1,\dots,l}}, \mathbf{c})\ H(\{\nu^{(i,v_i)}\}_{\substack{v_i=1,\dots,\tilde{k}_i\\ i=1,\dots,l}}, \mu,\nu). \tag{7.52}$$

7.4.3. *Hall-Littlewood function weighted Hurwitz numbers.* The generating function for Hall-Littlewood polynomials $P_\lambda(\mathbf{x},t)$, is obtained by setting $q=0$ in eq. (7.11). The orthogonality relations [**36**] become

$$(P_\lambda, P_\mu)_t = \delta_{\lambda\mu}(b_\lambda(t))^{-1}, \quad b_\lambda(t) := \prod_{i\geq 1}\prod_{k=1}^{m_i(\lambda)}(1-t^k) \tag{7.53}$$

with respect to the scalar product $(\ ,\ )_t$ defined by

$$(p_\lambda, p_\mu)_t = \delta_{\lambda\mu} z_\lambda n_\lambda(t), \quad n_\lambda := \prod_{i=1}^{\ell(\lambda)} \frac{1}{1-t^{\lambda_i}}. \tag{7.54}$$

Defining, as in [**36**],

$$q_\lambda(\mathbf{x},t) := b_\lambda(t)\prod_{i=1}^{\ell(\lambda)} P_j(\mathbf{x},t), \tag{7.55}$$

we obtain the following expansion

$$L(t,\mathbf{x},\mathbf{y}) = \prod_{i,j}^{\infty}\frac{1-tx_iy_j}{1-x_iy_j} = \sum_{\lambda}^{\infty} q_\lambda(\mathbf{x},t)m_\lambda(\mathbf{y}) = \sum_{\lambda}^{\infty} q_\lambda(\mathbf{y},t)m_\lambda(\mathbf{x}). \tag{7.56}$$

The corresponding central element of $\mathbf{C}[S_n]$ is then

$$L(t,\mathbf{c},z\mathcal{J}) := \prod_{i=1}^{\infty}\prod_{a=1}^{n}\frac{1-tc_iz\mathcal{J}_a}{1-c_iz\mathcal{J}_a} = \sum_{\lambda}^{\infty} z^{|\lambda|}q_\lambda(\mathbf{c},t)m_\lambda(\mathcal{J}) = \sum_{\lambda}^{\infty} z^{|\lambda|}q_\lambda(\mathcal{J},t)m_\lambda(\mathbf{c}). \tag{7.57}$$

The hypergeometric $2D$ Toda τ-function then reduces to

$$\tau^{L(t,\mathbf{c},z)}(\mathbf{t},\mathbf{s}) := \sum_\lambda r_\lambda^{L(t,\mathbf{c},z)} s_\lambda(\mathbf{t})s_\lambda(\mathbf{s}) \tag{7.58}$$

$$= \sum_{d=0}^{\infty} z^d \sum_\lambda F^d_{L(t,\mathbf{c})}(\mu,\nu)p_\mu(\mathbf{t})p_\nu(\mathbf{s}) \tag{7.59}$$

(7.60)

where the content product coefficient $r_\lambda^{L(t,\mathbf{c},z)}$ is

$$r_\lambda^{L(t,\mathbf{c},z)} := \prod_{(ij)\in\lambda}\prod_{k=1}^{\infty}\frac{1-tz(j-i)c_k}{1-z(j-i)c_k} = \prod_{k=1}^{\infty}(-t)^{|\lambda|}\frac{(-1/(tzc_k))_\lambda}{(-1/(zc_k))_\lambda}, \tag{7.61}$$

and

$$F^d_{L(t,\mathbf{c})}(\mu,\nu) := \sum_{|\lambda|=d} q_\lambda(\mathbf{c},t)m^\lambda_{\mu\nu} \tag{7.62}$$

is the weighted number of paths in the Cayley graph with the weight $q_\lambda(\mathbf{c}, t)$ for a path of signature λ.

As in the general case, we also have

$$F^d_{L(t,\mathbf{c})}(\mu, \nu) = \sum_{e=0}^{d} H^{(d,e)}_{\mathbf{c}}(\mu, \nu) t^e \tag{7.63}$$

where, denoting the total number of branch points,

$$K := \sum_{i=1}^{l} (k_i + \tilde{k}_i) \tag{7.64}$$

the weighted generalization of the multispecies hybrid signed Hurwitz numbers studied in [**25**] is

$$H^{(d,e)}_{L(\mathbf{c})}(\mu, \nu) := \tag{7.65}$$

$$z_\nu \sum_{l=0}^{d} \sum'_{\substack{\{\mu^{(i,u_i)}, \nu^{(i,v_i)}\},\ k_i \geq 1,\ \tilde{k}_i \geq 1 \\ \sum_{i=1}^{l} \sum_{u_i=1}^{k_i} \ell^*(\mu^{(i,u_i)}) = e, \\ \sum_{i=1}^{l} \left(\sum_{u_i=1}^{k_i} \ell^*(\mu^{(i,u_i)}) + \sum_{v_i=1}^{\tilde{k}_i} \ell^*(\nu^{(i,v_i)})\right) = d}} (-1)^{K+d-e} H(\{\mu^{(i,u_i)}\}_{\substack{u_i=1,\dots,k_i \\ i=1,\dots,l}}, \{\nu^{(i,v_i)}\}_{\substack{v_i=1,\dots,\tilde{k}_i \\ i=1,\dots,l}}, \mu, \nu).$$

Its interpretation in terms of weighted enumerations of multispecies Hurwitz numbers of two classes is the same as in the general Macdonald case, with the general quantum weighting factor reducing to a sign times the standard classical one $m_\lambda(\mathbf{c})$.

7.4.4. *Jack function weighted Hurwitz numbers.* The Jack polynomials $P^{(\alpha)}_\lambda$ are obtained by setting $t = q^\alpha$ and taking the limit $q \to 1$. These satisfy the orthogonality relations [**36**]

$$\langle P^\alpha_\lambda, P^\alpha_\mu \rangle_\alpha = \delta_{\lambda\mu} z_\lambda b_\lambda(t), \quad b_\lambda(t) := \prod_{i=1}^{\ell(\lambda)} \prod_{j=1}^{\lambda_i} (1 - t^j) \tag{7.66}$$

where the scalar product $\langle\ ,\ \rangle_\alpha$ is defined by [**36**]

$$\langle p_\lambda, p_\mu \rangle_\alpha = \delta_{\lambda\mu} z_\lambda \alpha^{\ell(\lambda)}. \tag{7.67}$$

The corresponding family of weight generating functions becomes

$$J(\alpha, \mathbf{c}, z) := \prod_{k=1}^{\infty} (1 - z c_i)^{-1/\alpha} \tag{7.68}$$

and the central elements

(7.69)

$$J(\alpha, z\mathcal{J}) := \prod_{i=1}^{\infty} \prod_{a=1}^{n} (1 - z c_i \mathcal{J}_a)^{-1/\alpha} = \sum_\lambda z^{|\lambda|} g^\alpha_\lambda(\mathcal{J}) m_\lambda(\mathbf{c}) = \sum_\lambda z^{|\lambda|} g^\alpha_\lambda(\mathbf{c}) m_\lambda(\mathcal{J}),$$

where the symmetric functions $g^\alpha_\lambda(\mathbf{x})$ are the analogs of the $e_\lambda(\mathbf{x})$ or $h_\lambda(\mathbf{x})$ bases formed from products of elementary or complete symmetric functions [**36**]

$$g^\alpha_\lambda(\mathbf{x}) = \alpha^{\ell(\lambda)} \prod_{i=1}^{\ell(\lambda} P^{(\alpha)}_{(\lambda_i)}(\mathbf{x}). \tag{7.70}$$

The associated hypergeometric $2D$ Toda τ-function is

$$\tau^{J(\alpha,\mathbf{c},z)}(\mathbf{t},\mathbf{s}) = \sum_{\lambda} r_\lambda^{J(\alpha,\mathbf{c},z)} s_\lambda(\mathbf{t}) s_\lambda(\mathbf{s}) \tag{7.71}$$

where the content product coefficients are

$$r_\lambda^{J(\alpha,\mathbf{c},z)} := \prod_{(ij)\in\lambda} \prod_{k=0}^{\infty} (1 - z(j-i)c_k)^{-1/\alpha} = \prod_{k=1}^{\infty} (1 - zc_k)^{\frac{|\lambda|}{\alpha}} (-1/zc_k)_\lambda^{-1/\alpha}. \tag{7.72}$$

Expanding over products of power sum symmetric functions gives

$$\tau^{J(\alpha,\mathbf{c},z)}(\mathbf{t},\mathbf{s}) = \sum_{d=0}^{\infty} \sum_{\substack{\mu,\nu \\ |\mu|=|\nu|=n}} z^d F_{J(\alpha,\mathbf{c})}^d(\mu,\nu) p_\mu(\mathbf{t}) p_\nu(\mathbf{s}) \tag{7.73}$$

where

$$F_{J(\alpha,\mathbf{c})}^d(\mu,\nu) = \sum_{\lambda} g_\lambda^\alpha(\mathbf{c}) m_{\mu\nu}^\lambda \tag{7.74}$$

is the combinatorial Hurwitz number giving the weighted number of d-step paths of signature λ in the Cayley graph of S_n, starting in the conjugacy class $\mathrm{cyc}(\mu)$ and ending in $\mathrm{cyc}(\nu)$, with weight $g_\lambda^{\alpha)}(\mathbf{c})$.

We again have equality with the weighted geometrical Hurwitz number

$$F_{J(\alpha,\mathbf{c})}^d(\mu,\nu) = H_{J(\alpha,\mathbf{c})}^d(\mu,\nu), \tag{7.75}$$

where

$$H_{J(\alpha,\mathbf{c})}^d(\mu,\nu) := \sum_{k=0}^{\infty} \binom{-\frac{1}{\alpha}}{k} \sum_{\substack{\mu^{(1)},\dots,\mu^{((k)} \\ |\mu^{(i)}|=n \\ \sum_{i=1}^k \ell^*(\mu^{(i)})=d}} m_\lambda(\mathbf{c}) H(\mu^{(1)},\dots,\mu^{(k)},\mu,\nu) \tag{7.76}$$

with the sum is over partitions λ of length k, and weight d whose parts are $\{\ell^*(\mu^{(1)}),\dots,\ell^*(\mu^{(k)})\}$.

Acknowledgements. The author would like to thank G. Borot, L. Chekhov, M. Guay-Paquet, M. Kazarian, S. Lando and A.Yu. Orlov for helpful discussions.

References

[1] A. Alexandrov, A. Mironov, A. Morozov and S. Natanzon, "On KP-integrable Hurwitz functions", *J. High Energy Physics* 080 (2014).

[2] J. Ambjørn and L. Chekhov, "The matrix model for dessins d'enfants", *Ann. Inst. Henri Poincaré, Comb. Phys. Interact.* **1**, 337-361 (2014).

[3] J. Ambjørn and L. Chekhov, "A matrix model for hypergeometric Hurwitz numbers", *Theor. Math. Phys.* **181**, 1486-1498 (2014).

[4] G. Borot, B. Eynard, M. Mulase and B. Safnuk, "A matrix model for Hurwitz numbers and topological recursion", *J. Geom. Phys.* **61**, 522–540 (2011).

[5] G. Borot and B Eynard, "All-order asymptotics of hyperbolic knot invariants from non-perturbative topological recursion of A-polynomials", *Quantum Topology* **6**,. 39-138 (2015).

[6] V. Bouchard and M. Marinó, "Hurwitz numbers, matrix models and enumerative geometry" *Proc. Symp. Pure Math.* **78**, 263-283 (2008).

[7] P. Diaconis and C. Greene, "Applications of Murphy's elements", *Stanford Technical Report* 335 (1989).

[8] B. Eynard and N. Orantin, "Invariants of algebraic curves and topological expansion", *Commun. Number Theor. Phys.* **1**, 347-452 (2007),

[9] B. Eynard and N. Orantin, "Topological recursion in enumerative geometry and random matrices", *J. Phys. A* **42** 293001 (2009).

[10] B. Eynard, and N. Orantin, "Invariants of spectral curves and intersection theory of moduli spaces of complex curves", *Commun. Num. Theor. Phys.* **8** 541-588 (2014)

[11] G. Frobenius, "Über die Charaktere der symmetrischen Gruppe", *Sitzber. Pruess. Akad. Berlin*, 516-534 (1900).

[12] I. P. Goulden, M. Guay-Paquet and J. Novak, "Monotone Hurwitz numbers and the HCIZ Integral", *Ann. Math. Blaise Pascal* **21** 71-99 (2014).

[13] I. P. Goulden, M. Guay-Paquet and J. Novak, "Toda Equations and Piecewise Polynomiality for Mixed Double Hurwitz numbers", arXiv:1307.2137.

[14] I. P. Goulden and D. M. Jackson, "Transitive factorisation into transpositions and holomorphic mappings on the sphere", *Proc. Amer. Math. Soc.*, **125**, 51-60. no. 1 (1997).

[15] I.P. Goulden and D.M. Jackson, "The KP hierarchy, branched covers and triangulations", *Adv. Math* **219** 932-951 (2008).

[16] K.I. Gross and D.S. Richards, "Special functions of matrix arguments. I: Algebraic induction, zonal polynomials, and hypergeometric functions", *Trans. Amer Math Soc.*, **301**, 781-811 (1987).

[17] M. Guay-Paquet and J. Harnad, "2D Toda τ-functions as combinatorial generating functions", *Lett. Math. Phys.* **105** 827-852 (2015).

[18] M. Guay-Paquet and J. Harnad, "Generating functions for weighted Hurwitz numbers", arXiv:1408.6766.

[19] G. Frobenius, "Über die Charaktere der symmetrischen Gruppe", *Sitzber. Akad. Wiss., Berlin*, 516-534 (1900). Gesammelte Abhandlung III, 148-166.

[20] G. Frobenius, "Über die Charakterische Einheiten der symmetrischen Gruppe", *Sitzber. Akad. Wiss., Berlin*, 328-358 (1903). Gesammelte Abhandlung III, 244-274.

[21] Harish-Chandra, "Differential operators on a semisimple Lie algebra", *Amer. J. Math.*, **79**, 87-120 (1957).

[22] J. Harnad, "Multispecies weighted Hurwitz numbers", *SIGMA* **11** 097, (2015), arXiv:1504.07512.

[23] J. Harnad, "Quantum Hurwitz numbers and Macdonald polynomials", arXiv:1504.03311. *J. Math. Phys.* (2016, in press)

[24] J. Harnad and A. Yu. Orlov, "Fermionic construction of partition functions for two-matrix models and perturbative Schur function expansions", *J. Phys. A* **39** 8783–8809 (2006).

[25] J. Harnad and A. Yu. Orlov, "Convolution symmetries of integrable hierarchies, matrix models and τ-functions", in: *Integrable Systems, Random Matrices and Random Processes* (eds. P. Deift and P. Forrester), MSRI publications vol. **65**, 247-275 (2014).

[26] J. Harnad and A. Yu. Orlov, "Hypergeometric τ-functions, Hurwitz numbers and enumeration of paths", *Commun. Math. Phys.* **338**, 267-284 (2015).

[27] A. Hurwitz, "Über Riemann'sche Flässche mit gegebnise Verzweigungspunkten", *Math. Ann.* **39**, 1-61 (1891); Matematische Werke I, 321-384.

[28] A. Hurwitz, "Über die Anzahl der Riemann'sche Flässche mit gegebnise Verzweigungspunkten", *Math. Ann.* **55**, 53-66 (1902); Matematische Werke I, 42-505.

[29] C. Itzykson and J.-B. Zuber, "The planar approximation. II" *J. Math. Phys.* **21**, 411–21 (1980).

[30] A. A. Jucys, "Symmetric polynomials and the center of the symmetric group ring", *Rep. Math. Phys.* **5**(1) 107–112 (1974).

[31] M. Kazarian, "KP hierarchy for Hodge integrals", *Adv. Math.* **221**, 1-21 (2009).

[32] M. E. Kazarian and S.K. Lando, "An algebra-geometric proof of Witten's conjecture"', J*J. Amer. Math. Soc.* **20**, 1079-1089 (2007).

[33] M. Kazarian and P. Zograf, "Virasoro constraints and topological recursion for Grothendieck's dessin counting", *Lett. Math. Phys.* **105** 1057-1084 (2015).

[34] M. Kontsevich, "Intersection theory on the moduli space of curves and the matrix Airy function", *Comm. Math. Phys.* **147**, 1-23 (1992).

[35] S. K. Lando and A.K. Zvonkin *Graphs on Surfaces and their Applications*, Encyclopaedia of Mathematical Sciences, Volume **141**, with appendix by D. Zagier, Springer, N.Y. (2004).

[36] I. G. Macdonald, *Symmetric Functions and Hall Polynomials*, Clarendon Press, Oxford, (1995).

[37] A. Mironov and A. Morozov, "Virasoro constraints for Kontsevich-Hurwitz partition function", *JHEP*02 (2009) 024.

[38] A. D. Mironov, A. Yu. Morozov, and S. M. Natanzon, "Complete set of cut-and-join operators in the Hurwitz-Kontsevich theory" *Theor. Math, Phys.* **166**, 1-22 (2011).

[39] G. E. Murphy, "A new construction of Young's seminormal representation of the symmetric groups", *Journal of Algebra* **69**, 287–297 (1981).

[40] S. M. Natanzon and A. Yu. Orlov, "Hurwitz numbers and BKP hierarchy", arXiv:1407.8323.

[41] S. M. Natanzon and A. Yu. Orlov, "BKP and projective Hurwitz numbers", arXiv:1501.01283.

[42] A. Okounkov, "Toda equations for Hurwitz numbers", *Math. Res. Lett.* **7**, 447–453 (2000).

[43] A. Yu. Orlov and D. M. Scherbin, "Hypergeometric solutions of soliton equations", *Theoretical and Mathematical Physics* **128**, 906-926 (2001).

[44] R. Pandharipande, "The Toda Equations and the Gromov-Witten Theory of the Riemann Sphere", *Lett. Math. Phys.* **53**, 59-74 (2000).

[45] M. Sato, "Soliton equations as dynamical systems on infinite dimensional Grassmann manifolds", RIMS, Kyoto Univ. *Kokyuroku* **439**, 30-46 (1981).

[46] I. Schur "Neue Begründung' der Theorie der Gruppencharaktere", *Sitzber. Akad. Wiss., Berlin*, 406-432 (1905).

[47] M. Sato and Y. Sato, "Soliton equations as dynamical systems on infinite dimensional Grassmann manifold", in: Nonlinear PDE in Applied Science, Proc. U. S.-Japan Seminar, Tokyo 1982, Kinokuniya, Tokyo, 1983, pp. 259-271.

[48] G. Segal and G. Wilson, "Loop groups and equations of KdV type ", *Publications Mathématiques de l'IHÉS* **6**, 5-65 (1985).

[49] K. Takasaki, "Initial value problem for the Toda lattice hierarchy", in: *Group Representation and Systems of Differential Equations*, 139–163, *Adv. Stud. in Pure Math.* **4**, (1984).

[50] T. Takebe, "Representation theoretical meaning of the initial value problem for the Toda lattice hierarchy I", *Lett. Math. Phys.* **21** 77–84, (1991).

[51] K. Ueno and K. Takasaki, "Toda Lattice Hierarchy", in: *Group Representation and Systems of Differential Equations*, *Adv. Stud. in Pure Math.* **4**, , 1–95 (1984).

[52] P. Zograf, "Enumeration of Grothendieck's dessins and KP hierarchy", *Int. Math. Res. Notices* **2015**, No. 24, 13533-13544 (2015).

Centre de recherches mathématiques, Université de Montréal, C. P. 6128, succ. centre ville, Montréal, QC, Canada H3C 3J7

Department of Mathematics and Statistics, Concordia University, 1455 de Maisonneuve Blvd. W. Montréal, QC, Canada H3G 1M8

Proceedings of Symposia in Pure Mathematics
Volume **93**, 2016
http://dx.doi.org/10.1090/pspum/093/01570

Calabi–Yau threefolds with infinite fundamental group

Atsushi Kanazawa

ABSTRACT. The present article surveys our recent results (Hashimoto and Kanazawa, "Calabi-Yau threefolds of type K (I) classification"; Hashimoto and Kanazawa, "Calabi-Yau threefolds of type K (II) mirror symmetry") about Calabi–Yau threefolds with infinite fundamental group. They are relatively simple yet rich enough to display the essential complexities of Calabi–Yau geometries, and thus they provide good testing-grounds for general theories and conjectures.

1. Introduction

Throughout the article, a Calabi–Yau threefold is a smooth complex projective threefold X with trivial canonical bundle $\Omega^3_X \cong \mathcal{O}_X$ and $H^1(X, \mathcal{O}_X) = 0$. A fundamental gap in the classification of algebraic threefolds is the lack of understanding of Calabi–Yau threefolds, classification of which seems very challenging at this point. Therefore in this article we concentrate on a special class of Calabi–Yau threefolds, namely those with infinite fundamental group. Most Calabi–Yau threefolds we know have finite fundamental groups: for example, complete intersection Calabi–Yau threefolds in toric varieties or homogeneous spaces, and (resolutions of singularities of) finite quotients thereof. Calabi–Yau threefolds with infinite fundamental group were only partially explored before the pioneering work of Oguiso and Sakurai [**13**]. A Calabi–Yau threefold with infinite fundamental group always admits an étale Galois covering either by an abelian threefold (type A) or by the product of a K3 surface and an elliptic curve (type K). In [**13**], Calabi–Yau threefolds of type A were essentially classified[1], while the full classification of type K was unsettled. The main purpose of the present article is to give an overview of the full classification obtained in [**8**]. We also discuss their interesting properties with special emphasis on mirror symmetry [**9**].

As we will see, Calabi–Yau threefolds with infinite fundamental group are relatively simple yet rich enough to display the essential complexities of Calabi–Yau geometries, and we expect that they will provide good testing-grounds for general theories and conjectures. Indeed, the simplest example of type K, known as the Enriques Calabi–Yau threefold, has been one of the most tractable compact Calabi–Yau threefolds both in string theory and mathematics (for example [**1**,**6**,**10**,**11**,**14**]).

2010 *Mathematics Subject Classification.* Primary 14J28, 14J32, 14J33.
Key words and phrases. Calabi–Yau manifold, fundamental group, mirror symmetry.

[1]We refine their result further in [**8**].

We believe that many of what is known for the Enriques Calabi–Yau threefold have natural generalizations to Calabi–Yau threefolds of type K.

Acknowledgement. The author warmly thanks the organizers of the *Calabi–Yau manifolds and their moduli conference 2014* for their hospitality. Special thanks go to Kenji Hashimoto who kindly allowed him to talk about their joint work as if it were his own. The author is supported by the Center of Mathematical Sciences and Applications at Harvard University.

2. Classification

In this section we will provide the full classification of Calabi–Yau threefolds with infinite fundamental group. Let X be a Calabi–Yau threefold with infinite fundamental group. Then the Bogomolov decomposition theorem implies that X admits an étale Galois covering either by an abelian threefold or by the product of a K3 surface and an elliptic curve. We call X of type A in the former case and of type K in the latter case. Among many candidates of such coverings, we can always find a unique smallest one, up to isomorphism as a covering [**3**]. We call the smallest covering the minimal splitting covering of X.

2.1. Calabi–Yau threefolds of type K. We begin with Calabi–Yau threefolds of type K. In this case the full classification is given in terms of the Galois group of the minimal splitting covering as follows.

THEOREM 2.1 ([**8**]). *There exist exactly eight Calabi–Yau threefolds of type K, up to deformation equivalence. The equivalence class is uniquely determined by the Galois group G of the minimal splitting covering. Moreover, the Galois group is isomorphic to one of the following combinations of cyclic and dihedral groups*

$$C_2,\ C_2 \times C_2,\ C_2 \times C_2 \times C_2,\ D_6,\ D_8,\ D_{10},\ D_{12},\ \textit{or}\ C_2 \times D_8.$$

Moreover, an explicit presentation of the deformation class of each Calabi–Yau threefold of type K is obtained in [**8**]. In the following, we shall briefly summarize the construction of the deformation classes. Let X be a Calabi–Yau threefold of type K and $\pi : S \times E \to X$ its minimal splitting covering with Galois group G. There exists a canonical isomorphism $\mathrm{Aut}(S \times E) \cong \mathrm{Aut}(S) \times \mathrm{Aut}(E)$, which induces a faithful G-action on each S and E:

$$\begin{array}{ccccc} \mathrm{Aut}(S) & \xleftarrow{p_1} & \mathrm{Aut}(S \times E) & \xrightarrow{p_2} & \mathrm{Aut}(E) \\ \cup & & \cup & & \cup \\ p_1(G) & \xleftarrow[p_1|_G]{\cong} & G & \xrightarrow[p_2|_G]{\cong} & p_2(G), \end{array}$$

where p_i denotes i-th projection with respect to the isomorphism $\mathrm{Aut}(S \times E) \cong \mathrm{Aut}(S) \times \mathrm{Aut}(E)$.

PROPOSITION 2.2 ([**8, 13**]). *Let $H := \mathrm{Ker}(G \to \mathrm{GL}(H^{2,0}(S)))$ and take any $\iota \in G \setminus H$. Then the following hold:*

(1) $\mathrm{ord}(\iota) = 2$ *and $G = H \rtimes \langle \iota \rangle$, where the semi-direct product structure is given by $\iota h \iota = h^{-1}$ for all $h \in H$;*
(2) *g acts on S as an Enriques involution if $g \in G \setminus H$;*

(3) *ι acts on E as -1_E and H as translations of the form $\langle t_a\rangle \times \langle t_b\rangle \cong C_n \times C_m$ under an appropriate origin of E. Here t_a and t_b are translations of order n and m respectively for some $(n,m) \in \{(1,k)(1 \leq k \leq 6),\ (2,2),\ (2,4)\}$.*

Conversely, such a G-action on the product $S \times E$ yields a Calabi–Yau threefold $X := (S \times E)/G$ of type K. Proposition 2.2 provides us with a complete understanding of the G-action on E (see [**8**, Section 3] for details), and therefore a classification of Calabi–Yau threefolds of type K essentially reduces to that of K3 surfaces equipped with actions described above, which we shall call *Calabi–Yau*. In what follows, G is always one of the finite groups described in Proposition 2.2. A basic example of a K3 surface which the reader could bear in mind is the following.

PROPOSITION 2.3 (Horikawa model [**2**, Section V, 23]). *Consider the double covering $\pi : S \to \mathbb{P}^1 \times \mathbb{P}^1$ branching along a bidegree $(4,4)$-divisor B. Then S is a K3 surface if it is smooth. We denote by θ the covering involution on S. Assume that B is invariant under the involution λ of $\mathbb{P}^1 \times \mathbb{P}^1$ given by $(x,y) \mapsto (-x,-y)$, where $x$and y are the inhomogeneous coordinates of $\mathbb{P}^1 \times \mathbb{P}^1$. The involution λ lifts to a symplectic involution of S. Then $\theta \circ \lambda$ is an involution of S without fixed points unless B passed through one of four fixed points of λ on $\mathbb{P}^1 \times \mathbb{P}^1$. The quotient surface $T = S/\langle \theta \circ \lambda\rangle$ is therefore an Enriques surface.*

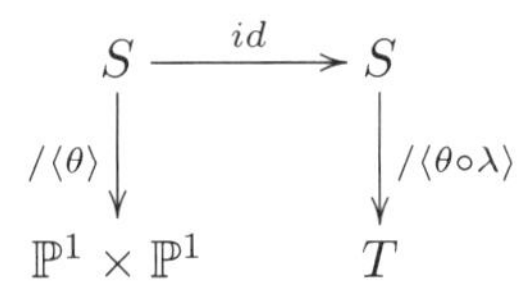

The classical theory of Enriques surfaces says that any generic K3 surface with an Enriques involution is realized as a Horikawa model ([**2**, Propositions 18.1, 18.2]).

EXAMPLE 2.4 (Enriques Calabi–Yau threefold). *Let S be a K3 surface with an Enriques involution ι and E an elliptic curve with negation -1_E. The free quotient $X := (S \times E)/\langle(\iota, -1_E)\rangle$ is the simplest Calabi–Yau threefold of type K, known as the Enriques Calabi–Yau threefold.*

In order to obtain other Calabi–Yau threefolds of type K, we consider special classes of the Horikawa model as follows. Let $\rho_1, \rho_2 : G \to \mathrm{PGL}(2,\mathbb{C})$ be 2-dimensional complex projective representations of $G := H \rtimes C_2$, which we do not specify at this point. Let λ be the generator of the second factor C_2. We then get a G-action $\rho_1 \times \rho_2$ on $\mathbb{P}^1 \times \mathbb{P}^1$. Suppose that there exists a G-stable smooth curve B of bidegree $(4,4)$. We then obtain a Horikawa K3 surface S as the double covering $\pi : S \to \mathbb{P}^1 \times \mathbb{P}^1$ branching along B and the G-action on $\mathbb{P}^1 \times \mathbb{P}^1$ lifts to S as a symplectic G-action. We further assume that the curve B does not pass through any of fixed points of $g \in G$, $g \neq 1$. With the same notation as in Proposition 2.3, it can be checked that the symplectic G-action and the covering transformation θ commute.

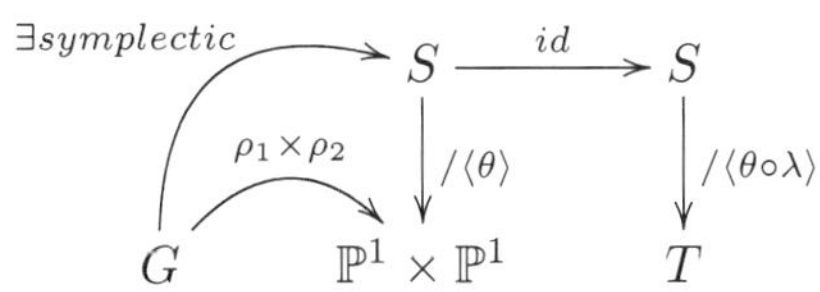

By twisting λ by θ, we obtain a new G-action on S, i.e.

$$\mathrm{Aut}(S) \supset G \times \langle\theta\rangle \supset H \rtimes \langle \theta \circ \lambda\rangle \cong G$$

The new G-action on S turns out to be a Calabi–Yau action. A fundamental result of [**8**, Section 3.3] confirms that a generic K3 surface equipped with a Calabi–Yau action is realized (not necessarily uniquely) in this way. To put it another way, there always exist 2-dimensional complex projective representations $\rho_1, \rho_2 : G \to \mathrm{PGL}(2, \mathbb{C})$ which satisfy all the assumptions mentioned above. Here are two examples.

EXAMPLE 2.5. *Suppose that* $G \cong D_{12} := \langle a, b | a^6 = b^2 = baba = 1\rangle$. *For* $i = 1, 2$, *we define* $\rho_i : D_{12} \to \mathrm{PGL}(2, \mathbb{C})$ *by*

$$a \mapsto \begin{bmatrix} \zeta_{12}^i & 0 \\ 0 & \zeta_{12}^{12-i} \end{bmatrix}, \quad b \mapsto \begin{bmatrix} 0 & 1 \\ 1 & 0 \end{bmatrix},$$

where ζ_k *denotes a primitive* k*-th root of unity. A basis of* D_{12}*-invariant polynomials of bidegree* $(4, 4)$ *are given by* $x^4z^4 + y^4w^4, x^4zw^3 + y^4z^3w, x^2y^2z^2w^2$. *A generic linear combination of these cuts out a desired smooth curve of bidegree* $(4, 4)$.

EXAMPLE 2.6. *Suppose that* $G \cong D_8 \times C_2 = \langle a, b, c | a^4 = b^2 = baba = 1, ac = ca, bc = cb\rangle$ *For* $i = 1, 2$, *we define* $\rho_i : D_8 \times C_2 \to \mathrm{PGL}(2, \mathbb{C})$ *by*

$$a \mapsto \begin{bmatrix} \zeta_8 & 0 \\ 0 & \zeta_8^7 \end{bmatrix}, \quad b \mapsto \begin{bmatrix} 0 & 1 \\ 1 & 0 \end{bmatrix}, \quad c \mapsto \begin{bmatrix} \sqrt{-1}^{\,i-1} & 0 \\ 0 & \sqrt{-1}^{\,1-i} \end{bmatrix}.$$

A basis of $D_8 \times C_2$*-invariant polynomials of bidegree* $(4, 4)$ *are given by* $x^4z^4 + y^4w^4, x^4w^4 + y^4z^4, x^2y^2z^2w^2$. *A generic linear combination of these cuts out a desired smooth curve of bidegree* $(4, 4)$.

2.2. Calabi–Yau threefolds of type A. We now turn to Calabi–Yau threefolds of type A. Let $A := \mathbb{C}^d/\Lambda$ be a d-dimensional complex torus. There is a natural semi-direct decomposition $\mathrm{Aut}(A) = A \rtimes \mathrm{Aut}_{\mathrm{Lie}}(A)$, where the first factor is the translation group of A and $\mathrm{Aut}_{\mathrm{Lie}}(A)$ consists of elements that fix the origin of A. We call the second factor of $g \in \mathrm{Aut}(A)$ the Lie part of g and denote it by g_0.

THEOREM 2.7 ([**8**]). *Let* X *be a Calabi–Yau threefold of type A. Then* X *is isomorphic to the étale quotient* A/G *of an abelian threefold* A *by an action of a finite group* G, *where* A *and* G *are given by the following.*

(1) $A = A'/T$, *where* A' *is the direct product of three elliptic curves* E_1, E_2 *and* E_3:

$$A' := E_1 \times E_2 \times E_3, \quad E_i := \mathbb{C}/(\mathbb{Z} \oplus \mathbb{Z}\tau_i), \quad \tau_i \in \mathbb{H}$$

and T *is one of the subgroups of* A' *in the following table, which consists of* 2*-torsion points of* A'.

T_1	T_2	T_3	T_4
0	$\langle(0, 1/2, 1/2)_{A'}\rangle$	$\langle(1/2, 1/2, 0)_{A'}, (1/2, 0, 1/2)_{A'}\rangle$	$\langle(1/2, 1/2, 1/2)_{A'}\rangle$

Here $(z_1, z_2, z_3)_{A'}$ *denotes the image of* $(z_1, z_2, z_3) \in \mathbb{C}^3$ *in* A'.

(2) $G \cong C_2 \times C_2$ *or* D_8.

(a) *If* $G = \langle a\rangle \times \langle b\rangle \cong C_2 \times C_2$, *then* G *is generated by*

$$a\colon (z_1, z_2, z_3)_A \mapsto (z_1 + \tau_1/2, -z_2, -z_3)_A,$$
$$b\colon (z_1, z_2, z_3)_A \mapsto (-z_1, z_2 + \tau_2/2, -z_3 + \tau_3/2)_A.$$

(b) *If* $G = \langle a, b \mid a^4 = b^2 = abab = 1\rangle \cong D_8$, *then* $\tau_2 = \tau_3 =: \tau$, $T = T_2$ *or* T_3, *and* G *is generated by*

$$a\colon (z_1, z_2, z_3)_A \mapsto (z_1 + \tau_1/4, -z_3, z_2)_A,$$
$$b\colon (z_1, z_2, z_3)_A \mapsto (-z_1, z_2 + \tau/2, -z_3 + (1+\tau)/2)_A.$$

Moreover, each case really occurs.

In contrast to Calabi–Yau threefolds of type K, Calabi–Yau threefolds of type A are not classified by the Galois groups of the minimal splitting coverings. They are classified by the minimal totally splitting coverings, where abelian threefolds A which cover X split into the product of three elliptic curves.

2.3. Summary. Theorem 2.1 and Theorem 2.7 finally complete the full classification of Calabi–Yau threefolds with infinite fundamental group with a very explicit description of each deformation class.

THEOREM 2.8 ([**8**]). *There exist precisely fourteen Calabi–Yau threefolds with infinite fundamental group, up to deformation equivalence. To be more precise, six of them are of type A and eight of them are of type K.*

3. Mirror Symmetry

Calabi–Yau threefolds of type K are, by construction, topologically self-mirror threefolds. However, mirror symmetry should involve more than the mere exchange of Hodge numbers. In this section, we focus on Calabi–Yau threefolds of type K, whose mirror symmetry bears a resemblance to mirror symmetry of Borcea–Voisin threefolds [**4**,**15**]. Mirror symmetry of Borcea–Voisin threefolds relies on the strange duality of K3 surfaces with anti-symplectic involution discovered by Nikulin [**12**].

Let $M_G := (H^2(S,\mathbb{Z})^H)^\iota$ and $N_G := (H^2(S,\mathbb{Z})^H)_\iota$. We begin with the case $G \cong C_2$. A generic K3 surface S with an Enriques involution is a self-mirror K3 surface in the sense of Dolgachev [**5**], i.e.

$$U \oplus NS(S) \cong T(S) \tag{3.1}$$

as lattices, where U stands for the hyperbolic lattice ($NS(S)$ and $T(S)$ are given by M_G and N_G respectively in this case). Thus it is no wonder that the corresponding Enriques Calabi–Yau threefold (Example 2.4) is an example of a self-mirror threefold. For $G \not\cong C_2$, it turns out that the corresponding K3 surface cannot be self-mirror symmetric as $\mathrm{rank}(T(S)) < 12$. In general M_G and N_G do not manifest symmetry over integers $\mathbb{Z}$, but the duality

$$U \oplus M_G \cong N_G. \tag{3.2}$$

still holds over rational numbers $\mathbb{Q}$ (or some extension of $\mathbb{Z}$) [**9**].

G	M_G	N_G
C_2	$U(2) \oplus E_8(-2)$	$U \oplus U(2) \oplus E_8(-2)$
$C_2 \times C_2$	$U(2) \oplus D_4(-2)$	$U(2)^{\oplus 2} \oplus D_4(-2)$
$C_2 \times C_2 \times C_2$	$U(2) \oplus \langle -4\rangle^{\oplus 2}$	$U(2)^{\oplus 2} \oplus \langle -4\rangle^{\oplus 2}$
D_6	$U(2) \oplus A_2(-2)$	$U(3) \oplus U(6) \oplus A_2(-2)$
D_8	$U(2) \oplus \langle -4\rangle$	$U(4)^{\oplus 2} \oplus \langle -4\rangle$
D_{10}	$U(2)$	$U(5) \oplus U(10)$
D_{12}	$U(2)$	$U(6)^{\oplus 2}$
$C_2 \times D_8$	$U(2)$	$U(4) \oplus \langle 4\rangle \oplus \langle -4\rangle$

It is worth noting that M_G is not equal to $NS(S)$ but to the G-invariant part $NS(S)^G$, while $T(S) = N_G$ always holds. The duality (3.2) can be thought of as an H-equivariant version of the duality (3.1) since we have $M_G = (H^2(S,\mathbb{Z})^H)^\iota$ and $N_G = (H^2(S,\mathbb{Z})^H)_\iota$. Based on this new duality, we showed in [**9**] that Calabi–Yau threefolds of type K are self-mirror symmetric while the corresponding K3 surfaces are not in general. We also obtained several results parallel to what is known for Borcea–Voisin threefolds: Voisin's work on Yukawa couplings [**15**], and Gross and Wilson's work on special Lagrangian fibrations [**7**].

THEOREM 3.1 ([**9**]). *Let X be a Calabi–Yau threefold of type K. The asymptotic behavior of the A-Yukawa coupling Y_A^X around the large volume limit coincides with that of the B-Yukawa coupling Y_B^X around a large complex structure limit.*

THEOREM 3.2 ([**9**]). *Any Calabi–Yau threefold X of type K admits a special Lagrangian T^3-fibration $\pi : X \to B$, where B is topologically either S^3 or an S^1-bundle over $\mathbb{RP}^2$.*

Another important result obtained in [**9**] is the computation of the Brauer groups $\mathrm{Br}(X) := \mathrm{Tor}(H^2(X, \mathcal{O}_X^\times))$, which are topological invariants for the Calabi–Yau threefolds.

THEOREM 3.3 ([**9**]). *Let X be a Calabi–Yau threefold of type K with Galois group G. Then we have $\mathrm{Br}(X) \cong \mathbb{Z}_2^{\oplus m}$, where m is given by the following.*

G	C_2	$C_2 \times C_2$	$C_2 \times C_2 \times C_2$	D_6	D_8	D_{10}	D_{12}	$C_2 \times D_8$
m	1	2	3	1	2	1	2	3

The Brauer groups are believed to play an important role in mirror symmetry but very little is known at this point. We hope that our work provides new examples of interesting mirror symmetry with non-trivial Brauer groups.

There are a number of natural questions that arise from [**8, 9**]. Higher genus mirror symmetry of Calabi–Yau threefolds of type K would be very interesting to investigate. The simplest example (Enriques Calabi–Yau threefold) has previously been worked out both in physics and mathematics [**10, 11**]. In fact, the rich fibration structure makes the study of the special Kähler geometry of the complex moduli space particularly promising.

References

[1] P. S. Aspinwall, *An $N = 2$ dual pair and a phase transition*, Nuclear Phys. B **460** (1996), no. 1, 57–76, DOI 10.1016/0550-3213(95)00611-7. MR1374750 (97f:81113)

[2] W. P. Barth, K. Hulek, C. A. M. Peters, and A. Van de Ven, *Compact complex surfaces*, 2nd ed., Ergebnisse der Mathematik und ihrer Grenzgebiete. 3. Folge. A Series of Modern Surveys in Mathematics [Results in Mathematics and Related Areas. 3rd Series. A Series of Modern Surveys in Mathematics], vol. 4, Springer-Verlag, Berlin, 2004. MR2030225 (2004m:14070)

[3] A. Beauville, *Some remarks on Kähler manifolds with $c_1 = 0$*, Classification of algebraic and analytic manifolds (Katata, 1982), Progr. Math., vol. 39, Birkhäuser Boston, Boston, MA, 1983, pp. 1–26, DOI 10.1007/BF02592068. MR728605 (86c:32031)

[4] C. Borcea, *K3 surfaces with involution and mirror pairs of Calabi-Yau manifolds*, Mirror symmetry, II, AMS/IP Stud. Adv. Math., vol. 1, Amer. Math. Soc., Providence, RI, 1997, pp. 717–743. MR1416355 (97i:14023)

[5] I. V. Dolgachev, *Mirror symmetry for lattice polarized K3 surfaces*, J. Math. Sci. **81** (1996), no. 3, 2599–2630, DOI 10.1007/BF02362332. Algebraic geometry, 4. MR1420220 (97i:14024)

[6] S. Ferrara, J. A. Harvey, A. Strominger, and C. Vafa, *Second-quantized mirror symmetry*, Phys. Lett. B **361** (1995), no. 1-4, 59–65, DOI 10.1016/0370-2693(95)01074-Z. MR1362446 (97b:81097)

[7] M. Gross and P. M. H. Wilson, *Mirror symmetry via 3-tori for a class of Calabi-Yau threefolds*, Math. Ann. **309** (1997), no. 3, 505–531, DOI 10.1007/s002080050125. MR1474205 (99k:14066)

[8] K. Hashimoto and A. Kanazawa, Calabi–Yau threefolds of type K (I) Classification, arXiv:1409.7601

[9] K. Hashimoto and A. Kanazawa, Calabi–Yau threefolds of type K (II) Mirror symmetry, preprint.

[10] A. Klemm and M. Mariño, *Counting BPS states on the Enriques Calabi-Yau*, Comm. Math. Phys. **280** (2008), no. 1, 27–76, DOI 10.1007/s00220-007-0407-z. MR2391190 (2009d:81331)

[11] D. Maulik and R. Pandharipande, *New calculations in Gromov-Witten theory*, Pure Appl. Math. Q. **4** (2008), no. 2, Special Issue: In honor of Fedor Bogomolov., 469–500, DOI 10.4310/PAMQ.2008.v4.n2.a7. MR2400883 (2009d:14073)

[12] V. V. Nikulin, *Discrete reflection groups in Lobachevsky spaces and algebraic surfaces*, Proceedings of the International Congress of Mathematicians, Vol. 1, 2 (Berkeley, Calif., 1986), Amer. Math. Soc., Providence, RI, 1987, pp. 654–671. MR934268 (89d:11032)

[13] K. Oguiso and J. Sakurai, *Calabi-Yau threefolds of quotient type*, Asian J. Math. **5** (2001), no. 1, 43–77. MR1868164 (2002i:14043)

[14] R. P. Thomas, *Examples of bundles on Calabi-Yau 3-folds for string theory compactifications*, Adv. Theor. Math. Phys. **4** (2000), no. 1, 231–247. MR1807601 (2001m:14060)

[15] C. Voisin, *Miroirs et involutions sur les surfaces* $K3$ (French), Astérisque **218** (1993), 273–323. Journées de Géométrie Algébrique d'Orsay (Orsay, 1992). MR1265318 (95j:14051)

Department of Mathematics, Harvard University, 1 Oxford Street, Cambridge, Massachusetts 02138

E-mail address: kanazawa@cmsa.fas.harvard.edu

Proceedings of Symposia in Pure Mathematics
Volume **93**, 2016
http://dx.doi.org/10.1090/pspum/093/01571

Logarithmic invariants of links

Jun Murakami

Abstract. We generalize the logarithmic knot invariant for links with several components. We also investigate its relation to the hyperbolic volume of the corresponding cone manifold for the figure-eight knot, the 5_2 knot, the Whitehead link and the Borromean rings.

Introduction

In this report, the logarithmic invariant of knots is generalized to links, and its relation to the hyperbolic volume is also investigated. This relation between the colored Jones invariant and the hyperbolic volume is known as the volume conjecture.

The Jones polynomial is the invariant of knots discovered by V. Jones [**4**], which is defined by the skein relation

$$q^{-1}\, V_{K_+}(q) - q\, V_{K_-}(q) = (q^{1/2} - q^{-1/2})\, V_{K_0}(q),$$

K_+ : , K_- : , K_0 : 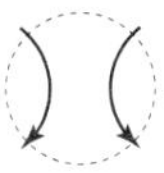.

where K_+, K_-, K_0 are the knots which are equal except in the disc where they are as above. This invariant is related to a solvable lattice model and quantum R matrix of the quantum group $\mathcal{U}_q(sl_2)$. By using such relations, the Jones polynomial is generalized to various invariants, which are now called quantum invariants. The colored Jones invariant is one of these invariants corresponding to a finite dimensional irreducible representation of sl_2.

In the mid 1990's, R. Kashaev constructed a new quantum invariant of knots from the quantum dilogarithm, and he observed a relation between his invariant and the hyperbolic volume of the complement of the knot [**5**]. Let N be a positive integer greater than 1, K be a hyperbolic knot and $\langle K \rangle_N$ be Kashaev's invariant. The relation he found is

$$\lim_{N\to\infty} \frac{2\,\pi\, \log |\langle K \rangle_N|}{N} = \mathrm{Vol}_{\mathrm{hyp}}(S^3 \setminus K).$$

Here $S^3 \setminus K$ has a canonical hyperbolic structure and $\mathrm{Vol}_{\mathrm{hyp}}(S^3 \setminus K)$ is the volume with respect to this structure. He observed this relation for the simplest three

2010 *Mathematics Subject Classification*. Primary 57M27; Secondary 51M10, 58J28.
Key words and phrases. Knots, 3-manidfolds, Chern-Simons invariants.
This work was partially supported by JSPS KAKENHI Grant Numbers 90157751, 25610022.

knot		hyperbolic volume	CS invariant
4_1		$2.0298\ldots$	0
5_2		$2.8281\ldots$	$-0.1532\ldots$
6_1		$3.1639\ldots$	$0.1559\ldots$

FIGURE 1. The hyperbolic volumes and the Chern-Simons invariants of the three simplest hyperbolic knots

hyperbolic knots in Figure 1. Kashaev's invariant $\langle K \rangle_N$ turned out to be equal to the colored Jones invariant corresponding to the N dimensional irreducible representation at N-th root of unity, and Kashaev's conjecture is generalized in [**8**] as follows.

CONJECTURE 1. (Complexified volume conjecture.)

$$J_N(K)_{q=\xi_N} \underset{N\to\infty}{\sim} \exp\left(\frac{N}{2\,\pi}\left(\mathrm{Vol}_{\mathrm{hyp}}(S^3\setminus K)+\sqrt{-1}\ \mathrm{CS}(S^3\setminus K)\right)\right),$$

where $J_d(K)$ be the colored Jones invariant corresponding to the d dimensional irreducible representation of sl_2 and $\xi_N = \exp(\pi\sqrt{-1}/N)$. Here J_d is normalized as $J_d(\bigcirc) = 1$ for a trivial knot $\bigcirc$.

The above conjecture implies that $J_N(K)_{q=\xi_N}$ grows exponentially with respect to N, while $J_{\lfloor \alpha N\rfloor}(K)$ does not have such a property for $0 < \alpha < 1$. However, by taking a certain derivative of $J_{\lfloor \alpha N\rfloor}(K)$, we get an invariant which grows exponentially and relates to the hyperbolic volume of certain cone manifold along K. This is the logarithmic invariant we would like to explain.

We first review the logarithmic invariant of knots in Section 1, and then extend it to links in Section 2.

Acknowledgement. I would like to thank Y. Kabaya for a useful discussion. Mathematica and SnapPy were used for numerical computations.

1. Logarithmic invariants of knots

The logarithmic invariant of knots is introduced in [**10**], and its relation to the colored Jones invariant is given in [**9**]. For a positive integer k, let $V_k(K)$ be the colored Jones invariant without normalization satisfying $V_k(\bigcirc) = d_k$, where d_k is the quantum dimension

$$d_k = (-1)^{k-1}\,\frac{q^k - q^{-k}}{q - q^{-1}}.$$

Note that $V_k(K) = d_k\, J_k(K)$. Sometimes k is called the color of the invariant.

DEFINITION 1.1. Let N be an integer greater than 1 and k be an integer with $0 \leq k \leq N$. The logarithmic invariant $\gamma_{N,k}(K)$ is defined by

$$\gamma_{N,0}(K) = \left.\frac{V_{2N}(K)}{[2N]}\right|_{q=\xi_N},$$

$$\gamma_{N,N}(K) = \left.\frac{V_N(K)}{[N]}\right|_{q=\xi_N},$$

$$\gamma_{N,k}(K) = \frac{\xi_N}{2N}\frac{d}{dq}\{1\}\left(V_k(K) + V_{2N-k}(K)\right)\bigg|_{q=\xi_N}, \qquad (1 \leq k \leq N-1)$$

where $\{n\} = q^n - q^{-n}$, $[n] = \frac{\{n\}}{\{1\}}$ and $\xi_N = \exp(\pi\sqrt{-1}/N)$.

The colored Jones invariant $V_k(K)$ is expressed by Habiro [**2**] as follows.

$$V_m(K) = \sum_{j=0}^{\infty} a_j(K)\,\frac{\{m+j\}\{m+j-1\}\cdots\{m-j\}}{\{1\}}. \tag{1.1}$$

Note that $a_j(K)$ does not depend on k, and the sum is finite since the terms for $j \geq m$ vanish. From (1.1), we have

$$V_k(K)|_{q=\xi_N} + V_{2N-k}(K)|_{q=\xi_N} = 0$$

for $1 \leq k \leq N$. Moreover, we have the following.

THEOREM 1.2. *The logarithmic invariant $\gamma_{N,k}(K)$ is given by a derivative of the colored Jones invariant with respect to the color k as follows*

$$\gamma_{N,0}(K) = \frac{N\{1\}}{2\pi\sqrt{-1}}\frac{d}{dm}V_m(K)\bigg|_{\substack{m=2N\\ q=\xi_N}},$$

$$\gamma_{N,k}(K) = \frac{N\{1\}}{\pi\sqrt{-1}}\frac{d}{dm}V_m(K)\bigg|_{\substack{m=k\\ q=\xi_N}},$$

$$\gamma_{N,N}(K) = \frac{N\{1\}}{2\pi\sqrt{-1}}\frac{d}{dm}V_m(K)\bigg|_{\substack{m=N\\ q=\xi_N}}.$$

Here we consider (1.1) as a formal infinite sum with respect to the parameter m, differentiate each terms and substitute $m = 0$, k or N, and $q = \xi_N$. Then all but a finite number of terms vanish.

We propose the following conjecture for the logarithmic invariant.

CONJECTURE 2. Let K be a hyperbolic knot in S^3. For $0 < \theta < \pi$, let K_θ be the cone manifold along K with the cone angle θ, and $k_N = \lfloor \frac{2\pi-\theta}{2\pi}N \rfloor$, the largest integer less than or equal to $\frac{2\pi-\theta}{2\pi}N$. Then the following holds

$$\gamma_{N,k_N}(K) \underset{N\to\infty}{\sim} \exp\left(\frac{N}{2\pi}\left(\mathrm{Vol}_{\mathrm{hyp}}(K_\theta) + \sqrt{-1}\,\mathrm{CS}(K_\theta)\right)\right).$$

Here $\mathrm{CS}(K_\theta)$ is the classical Chern Simons invariant in [**1**]. If the cone angle θ is equal to 0, then K_0 is equal to the complement of K, and $\mathrm{CS}(K)$ is equal to the topological Chern-Simons invariant in [**8**] given by $CS(K_\theta) = -2\pi^2\mathrm{cs}(K_\theta)$.

REMARK 1.3. Since $\gamma_{N,N}(K) = J_N(K)|_{q=\xi_N}$, this conjecture is a generalization of Conjecture 1. An analogous generalization is proposed by H. Murakami in [**7**] for the colored Jones invariant, which is called the parametrized volume conjecture.

REMARK 1.4. For a hyperbolic three manifold M, $\mathrm{Vol}_{\mathrm{hyp}}(M) + \sqrt{-1}\,\mathrm{CS}(M)$ is called the complex volume of M. For the complement $S^3 \setminus K$ of a hyperbolic knot K, its complete hyperbolic structure is uniquely defined. However, if we consider an incomplete hyperbolic structure, the hyperbolic structure has a natural deformation space which is parametrized by one complex parameter ([**11**], [**13**]), and K_θ is contained in this deformation space. The complex volume can be considered as an analytic function of the deformation parameter, and the Chern-Simons invariant is naturally extended to the cone manifold K_θ.

REMARK 1.5. The logarithmic invariants of knots can be generalized to invariants of a framed knot K. Let f be the framing of K and K_0 be the 0 framed knot which is isotopic to K as a non-framed knot, then the framed version $\gamma^{(f)}_{N,k}(K)$ is defined by

$$\gamma^{(f)}_{N,k}(K) = q^{f(k^2-1)/2}\,\gamma_{N,k}(K_0).$$

Then Conjecture 1 is reformulated for a framed knot as

$$\gamma^{(f)}_{N,k_N}(K) \underset{N\to\infty}{\sim} \exp\left(\frac{N}{2\,\pi}\left(\mathrm{Vol}_{\mathrm{hyp}}(K_\theta) + \sqrt{-1}\,\mathrm{CS}^{(f)}(K_\theta)\right)\right).$$

The invariants $\mathrm{CS}(K_\theta)$ and $\mathrm{CS}^{(f)}(K_\theta)$ are the Chern-Simons invariants coming from different Chern-Simons actions. If $\theta = \frac{2\pi}{n}$ for a positive integer n, then K_θ is an orbifold. In this case,

$$\mathrm{CS}(K_{\frac{2\pi}{n}}) \equiv \mathrm{CS}^{(f)}(K_{\frac{2\pi}{n}}) \pmod{\frac{1}{2n}}$$

for any f, and the Chern-Simons invariant of $K_{\frac{2\pi}{n}}$ is well-defined modulo $\frac{1}{2n}$ [**3**].

In the rest of this section, we show some examples of Conjecture 2. The first example is the figure eight knot 4_1 in Figure 1. For 4_1, the coefficients $a_j(4_1)$ ($j = 0, 1, 2, \dots$) in (1.1) are all equal to 1, and the colored Jones invariants $V_k(4_1)|_{q=\xi_N}$ ($k = 1, 2, \dots$) are all real numbers. Moreover, the logarithmic invariant $\gamma_{N,k}(4_1)$ ($k = 1, 2, \dots$) are also real numbers. The values $\frac{2\pi}{N}\log|V_k(4_1)|$, $\frac{2\pi}{N}\log|\gamma_{N,k}(4_1)|$ and the volume of the cone manifolds along 4_1 with cone angle $2\pi\frac{N-k}{N}$ are plotted in Figure 2 for $N = 200$ and $100 \le k < 200$. The volume is computed by a formula in [**6**].

The second example is the knot 5_2 in Figure 1. For 5_2, $a_j(5_2)$ in (1.1) is given as follows

$$a_j(5_2) = (-1)^j\,q^{3j^2+5j}\sum_{l=0}^{j} q^{l^2-2l-3jl}\,\frac{\{j\}\,\{j-1\}\,\dots\,\{j-l+1\}}{\{l\}\,\{l-1\}\,\dots\,\{1\}}.$$

The moduli of the logarithmic invariants and the hyperbolic volume of the corresponding cone manifolds are plotted in Figure 3. The volume is computed by SnapPy, which is a computer software to investigate the hyperbolic structure of knot complements.

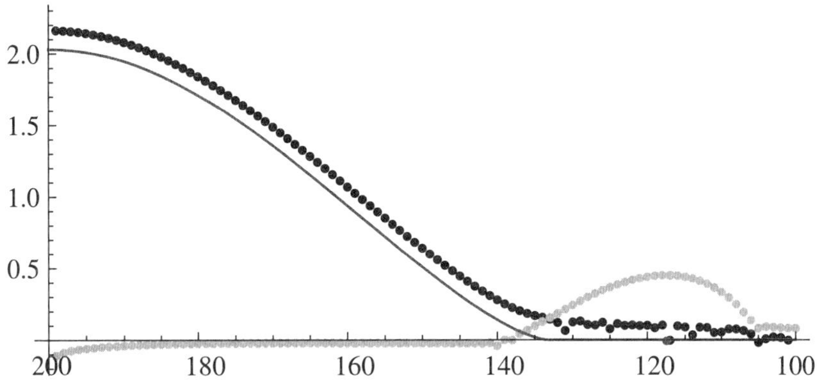

$\bullet$: $\frac{2\pi}{N}\log|\gamma_{N,k}(4_1)|$, $\bullet$: $\frac{2\pi}{N}\log|\, V_k(4_1)|_{q=\xi_N}$,

$-$: $\mathrm{Vol}_{\mathrm{hyp}}\left((4_1)_{2\pi\frac{N-k}{N}}\right)$.

FIGURE 2. The colored Jones invariants and the logarithmic invariants for $N = 200$, and the hyperbolic volumes of the corresponding cone manifolds

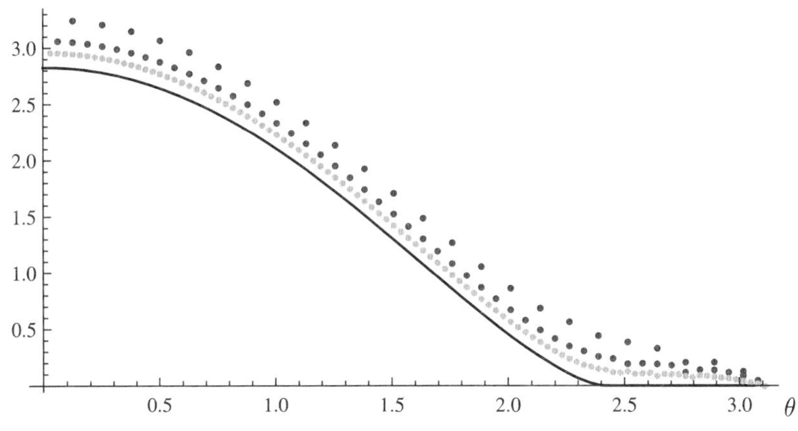

$\bullet$: $\frac{2\pi}{N}\,|\gamma_{N,\lfloor N(1-\frac{\theta}{2\pi})\rfloor}|$ for $N = 50$, $\bullet$: $\frac{2\pi}{N}\,|\gamma_{N,\lfloor N(1-\frac{\theta}{2\pi})\rfloor}|$ for $N = 100$,

$\bullet$: $\frac{2\pi}{N}\,|\gamma_{N,\lfloor N(1-\frac{\theta}{2\pi})\rfloor}|$ for $N = 200$, $-$: Hyperbolic volume of $(5_2)_\theta$

FIGURE 3. Moduli of the logarithmic invariants of the knot 5_2 for $N = 50, 100, 200$ and the hyperbolic volumes of the corresponding cone manifolds

The arguments of the ratios $\frac{\gamma_{N+1,k+1}(5_2)}{\gamma_{N,k}(5_2)}$ are plotted in Figure 4. Since $\frac{k}{N}$ and $\frac{k+1}{N+1}$ are different, we interpolate arg $\frac{\gamma_{N+1,k}(5_2)}{\gamma_{N,k}(5_2)}$ and arg $\frac{\gamma_{N+1,k+1}(5_2)}{\gamma_{N,k}(5_2)}$ linearly to get the value corresponding to $\frac{k}{N}$. Note that the Chern-Simons invariant is defined only for a hyperbolic manifold, and the conemanifod $(5_2)_\theta$ for a large θ is not

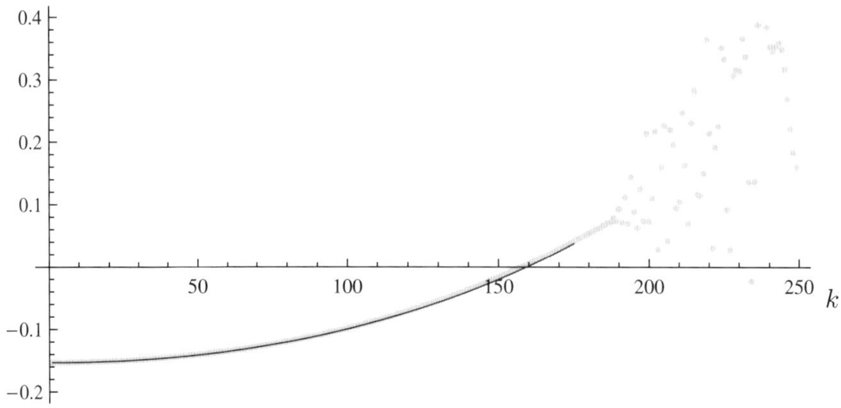

$\bullet$: $\frac{1}{\pi}\left(\left(\frac{k}{N}-\frac{k}{N+1}\right)\arg\frac{\gamma_{N+1,k+1}(5_2)}{\gamma_{N,k}(5_2)}+\left(\frac{k+1}{N+1}-\frac{k}{N}\right)\arg\frac{\gamma_{N+1,k}(5_2)}{\gamma_{N,k}(5_2)}\right)$ mod 1 at $N=500$

$-$: The Cherm-Simons invariants of $(5_2)_\theta$ $\left(\theta=2\pi(1-\frac{k}{N})\right)$.

FIGURE 4. Arguments of the logarithmic invariants of the knot 5_2 for $N=500$ and the Chern-Simons invariant of the corresponding cone manifolds

hyperbolic. The black line indicating the Chern-Simons invariants overlaps the dots corresponding to the logarithmic invariants.

2. Logarithmic invariants of links

Now we generalize the logarithmic invariant to links having several components.

DEFINITION 2.1. Let L be an l component link with components L_1, L_2, $\cdots$, L_l. Let N be an integer greater than 1 and k_1, k_2, $\cdots$, k_l be l integers with $0\le k_j\le N$. Then the logarithmic invariant $\gamma_{N,k_1,k_2,\dots,k_l}(L)$ is defined by

$$\gamma_{N,k_1,k_2,\dots,k_l}(L)=\\ \frac{d}{dq}\{1\}\sum_{\varepsilon_1,\cdots,\varepsilon_l=\pm1}\left(\prod_{j=1}^{l}(-1)^{\varepsilon_j}\,2^{-2\delta_{k_j,N}-4\delta_{k_j,0}}\right)V_{f_N(k_1,\varepsilon_1),\cdots,f_N(k_l,\varepsilon_j)}(L),$$

where

$$f_N(k_j,\varepsilon_n)=\begin{cases}2\,N, & (k_j=0)\\ N-\varepsilon_j(N-k_j), & (1\le k_j\le N-1)\\ N, & (k_j=N)\end{cases}$$

and $\delta_{i,j}=1$ $(i=j)$, $0(i\ne j)$.

For this invariant, we propose the following conjecture.

CONJECTURE 3. Let L be a hyperbolic link in S^3. For $0<\theta_j<\pi$ $(j=1,\cdots,l)$, let $L_{\theta_1,\cdots,\theta_l}$ be the cone manifold along the components of L with the cone angle θ_j around L_j, and $k_{N,j}=\lfloor\frac{2\pi-\theta_j}{2\pi}N\rfloor$, the largest integer less or equal to $\frac{2\pi-\theta_j}{2\pi}N$.

Then the following holds

$$\gamma_{N,k_{N,1},\cdots,k_{N,l}}(L) \underset{N\to\infty}{\sim} e^{\frac{N}{2\pi}\left(\mathrm{Vol}_{\mathrm{hyp}}(L_{\theta_1,\cdots,\theta_l})+\sqrt{-1}\,\mathrm{CS}(L_{\theta_1,\cdots,\theta_l})\right)}.$$

We investigate the above relation for some examples. Let L_{WH} be the Whitehead link given in Figure 5. The colored Jones invariant of L_{WH} is given in [**2**] as follows

$$V_{k_1,k_2}(L_{\mathrm{WH}}) = \sum_{j=0}^{\min(k_1,k_2)} q^{-j(j+3)} \frac{\{k_1+j\}!\,\{k_2+j\}!\,\{j\}!}{\{k_1-j-1\}!\,\{k_2-j-1\}!\,\{2j+1\}!\,\{1\}},$$

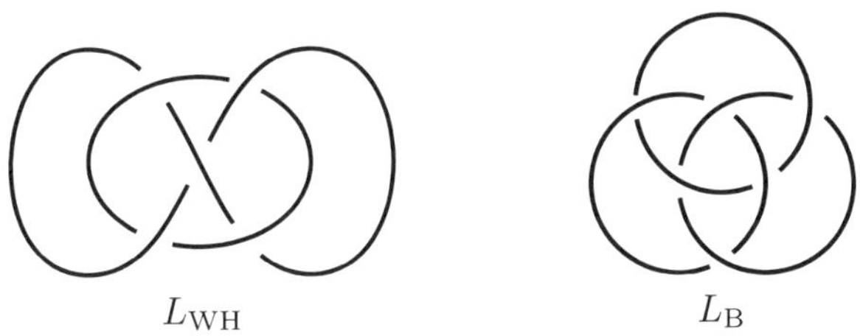

FIGURE 5. The Whitehead link L_{WH} and the Borromean rings L_{B}

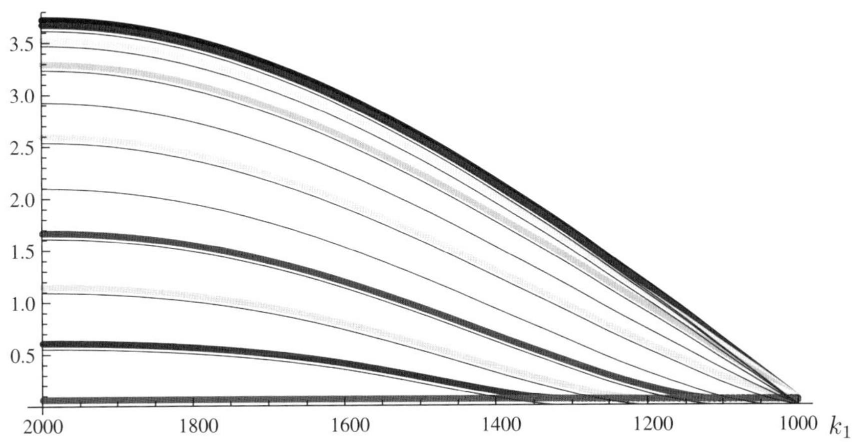

$|\gamma_{2000,k_1,k_2}(L_{\mathrm{WH}})|$:

- : $k_2 = 2000$ ($\theta_2 = 0$), • : $k_2 = 1900$ ($\theta_2 = \frac{\pi}{10}$), • : $k_2 = 1800$ ($\theta_2 = \frac{\pi}{5}$),
- : $k_2 = 1700$ ($\theta_2 = \frac{3\pi}{10}$), : $k_2 = 1600$ ($\theta_2 = \frac{2\pi}{5}$), • : $k_2 = 1500$ ($\theta_2 = \frac{\pi}{2}$),
- : $k_2 = 1400$ ($\theta_2 = \frac{3\pi}{5}$), • : $k_2 = 1300$ ($\theta_2 = \frac{7\pi}{10}$), • : $k_2 = 1200$ ($\theta_2 = \frac{4\pi}{5}$),
- : $k_2 = 1100$ ($\theta_2 = \frac{9\pi}{10}$), • : $k_2 = 1000$ ($\theta_2 = \pi$),

– : volume of $(L_{\mathrm{WH}})_{\theta_1,\theta_2}$

FIGURE 6. Values of $\frac{2\pi}{N}\log|\gamma_{N,k_1,k_2}(L_{\mathrm{WH}})|$ at $N = 2000$ and the volumes of the cone manifolds $(L_{\mathrm{WH}})_{\theta_1,\theta_2}$

where $\{j\}! = \{j\}\{j-1\}\cdots\{2\}\{1\}$. Hence $\gamma_{N,k_1,k_2}(L_{\mathrm{WH}})$ is given by

$$\gamma_{N,k_1,k_2}(L_{\mathrm{WH}}) = \frac{d}{dq}\sum_{\varepsilon_1,\varepsilon_2=\pm1}(-1)^{\varepsilon_1+\varepsilon_2}\,2^{-2\delta_{k_1,N}-2\delta_{k_2,N}-4\delta_{k_1,0}-4\delta_{k_2,0}}\times$$
$$\sum_{j=0}^{\min(k_1,k_2)} q^{-j(j+3)}\frac{\{f_N(k_1,\varepsilon_1)+j\}!\,\{f_N(k_2,\varepsilon_2)+j\}!\,\{j\}!}{\{f_N(k_1,\varepsilon_1)-j-1\}!\,\{f_N(k_2,\varepsilon_2)-j-1\}!\,\{2j+1\}!}.$$

The values of $|\gamma_{N,k_1,k_2}(L_{\mathrm{WH}})|$ are plotted in Figure 6. The volumes of the corresponding cone manifolds are computed by a formula in [**6**].

Let L_{B} the Borromean rings in Figure 5. The colored Jones invariant of L_{B} is given in [**2**] as follows

$$V_{k_1,k_2,k_3}(L_{\mathrm{B}}) =$$
$$\sum_{j=0}^{\min(k_1,k_2,k_3)}\frac{\{k_1+j\}!\,\{k_2+j\}!\,\{k_3+j\}!\,(\{j\}!)^2}{\{k_1-j-1\}!\,\{k_2-j-1\}!\,\{k_3-j-1\}!\,(\{2j+1\}!)^2\,\{1\}}.$$

Therefore, we have

$$\gamma_{N,k_1,k_2,k_3}(L_{\mathrm{B}}) =$$
$$\frac{d}{dq}\sum_{\varepsilon_1,\varepsilon_2,\varepsilon_3=\pm1}(-1)^{\varepsilon_1+\varepsilon_2+\varepsilon_3}\,2^{-2(\delta_{k_1,N}+\delta_{k_2,N}+\delta_{k_3,N})-4(\delta_{k_1,0}+\delta_{k_2,0}+\delta_{k_3,0})}\times$$
$$\sum_{j=0}^{\min(k_1,k_2,k_3)}\frac{\{f_N(k_1,\varepsilon_1)+j\}!\,\{f_N(k_2,\varepsilon_2)+j\}!\,\{f_N(k_3,\varepsilon_3)+j\}!\,(\{j\}!)^2}{\{f_N(k_1,\varepsilon_1)-j-1\}!\,\{f_N(k_2,\varepsilon_2)-j-1\}!\,\{f_N(k_3,\varepsilon_3)-j-1\}!\,(\{2j+1\}!)^2}.$$

The values of $|\gamma_{N,k_1,k_2,k_3}(L_{\mathrm{B}})|$ and the volumes of the corresponding cone manifolds $(L_{\mathrm{B}})_{\theta_1,\theta_2,\theta_3}$ are plotted as in Figure 7. The volumes are computed by a formula in [**6**].

REMARK 2.2. For $L = L_{\mathrm{WH}}$ or L_{B}, we can check from the actual formulas of $\gamma_{N,k_1,\cdots,k_l}(L)$ that the following formula holds

$$\gamma_{N,k_1,\cdots,k_l}(L) = \text{const.}\,\frac{d}{dm_j}V_{m_1,\cdots,m_l}(L)\bigg|_{m_1=k_1,\cdots,m_l=k_l}$$

for any $m_j \in \{m_1,\cdots,m_l\}$. This is a generalization of Theorem 1.

References

[1] S. Gukov and H. Murakami, SL(2, $\mathbb{C}$) *Chern-Simons theory and the asymptotic behavior of the colored Jones polynomial*, Lett. Math. Phys. **86** (2008), no. 2-3, 79–98, DOI 10.1007/s11005-008-0282-3. MR2465747 (2010m:58037)

[2] K. Habiro, *On the quantum* sl_2 *invariants of knots and integral homology spheres*, Invariants of knots and 3-manifolds (Kyoto, 2001), Geom. Topol. Monogr., vol. 4, Geom. Topol. Publ., Coventry, 2002, pp. 55–68 (electronic), DOI 10.2140/gtm.2002.4.55. MR2002603 (2004g:57023)

[3] H. M. Hilden, M. T. Lozano, and J. M. Montesinos-Amilibia, *On volumes and Chern-Simons invariants of geometric* 3*-manifolds*, J. Math. Sci. Univ. Tokyo **3** (1996), no. 3, 723–744. MR1432115 (98h:57056)

[4] V. F. R. Jones, *A polynomial invariant for knots via von Neumann algebras*, Bull. Amer. Math. Soc. (N.S.) **12** (1985), no. 1, 103–111, DOI 10.1090/S0273-0979-1985-15304-2. MR766964 (86e:57006)

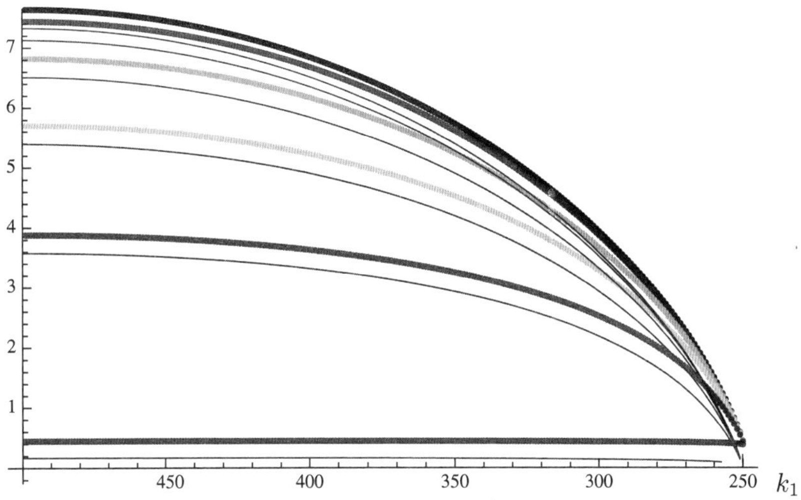

$k_3 = 500$ $(\theta_3 = 0)$

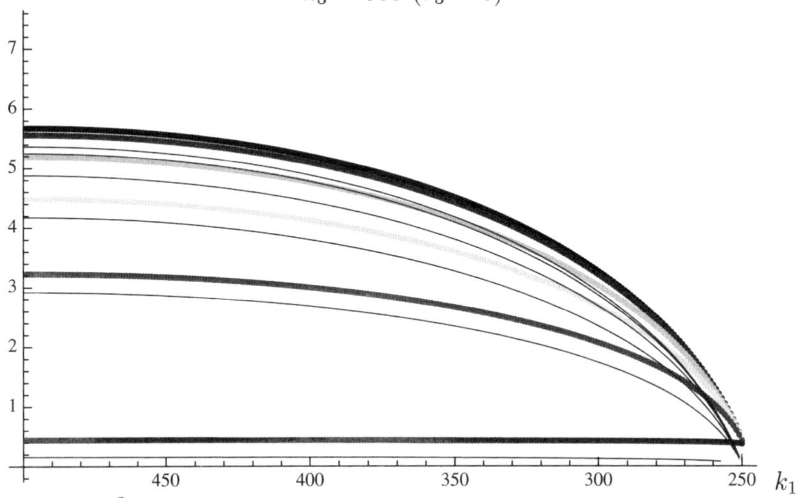

$k_3 = 350$ $(\theta_3 = \frac{3\pi}{5})$

$|\gamma_{500,k_1,k_2,k_3}(L_{\mathrm{B}})|$:

• : $k_2 = 500$ $(\theta_2 = 0)$ • : $k_2 = 450$ $(\theta_2 = \frac{\pi}{10})$ • : $k_2 = 400$ $(\theta_2 = \frac{\pi}{5})$

• : $k_2 = 350$ $(\theta_2 = \frac{3\pi}{10})$ • : $k_2 = 300$ $(\theta_2 = \frac{2\pi}{5})$ • : $k_2 = 250$ $(\theta_2 = \frac{\pi}{2})$

– : volume of $(L_{\mathrm{B}})_{\theta_1,\theta_2,\theta_3}$

FIGURE 7. Values of $|\gamma_{N,k_1,k_2,k_3}(L_{\mathrm{B}})|$ at $N = 500$ and the volumes of the cone manifolds $(L_{\mathrm{B}})_{\theta_1,\theta_2,\theta_3}$

[5] R. M. Kashaev, *The hyperbolic volume of knots from the quantum dilogarithm*, Lett. Math. Phys. **39** (1997), no. 3, 269–275, DOI 10.1023/A:1007364912784. MR1434238 (98b:57012)

[6] A. D. Mednykh, *On hyperbolic and spherical volumes for knot and link cone-manifolds*, Kleinian groups and hyperbolic 3-manifolds (Warwick, 2001), London Math. Soc. Lecture Note Ser., vol. 299, Cambridge Univ. Press, Cambridge, 2003, pp. 145–163, DOI 10.1017/CBO9780511542817.008. MR2044549 (2005a:57016)

[7] H. Murakami, *A version of the volume conjecture*, Adv. Math. **211** (2007), no. 2, 678–683, DOI 10.1016/j.aim.2006.09.005. MR2323541 (2009b:57027)

[8] H. Murakami, J. Murakami, M. Okamoto, T. Takata, and Y. Yokota, *Kashaev's conjecture and the Chern-Simons invariants of knots and links*, Experiment. Math. **11** (2002), no. 3, 427–435. MR1959752 (2004a:57016)

[9] J. Murakami, *From colored Jones invariants to logarithmic invariants.* preprint, arXiv:1406.1287.

[10] J. Murakami and K. Nagatomo, *Logarithmic knot invariants arising from restricted quantum groups*, Internat. J. Math. **19** (2008), no. 10, 1203–1213, DOI 10.1142/S0129167X08005060. MR2466562 (2009h:57022)

[11] W. D. Neumann and D. Zagier, *Volumes of hyperbolic three-manifolds*, Topology **24** (1985), no. 3, 307–332, DOI 10.1016/0040-9383(85)90004-7. MR815482 (87j:57008)

[12] Y. Yokota, *On the potential functions for the hyperbolic structures of a knot complement*, Invariants of knots and 3-manifolds (Kyoto, 2001), Geom. Topol. Monogr., vol. 4, Geom. Topol. Publ., Coventry, 2002, pp. 303–311 (electronic), DOI 10.2140/gtm.2002.4.303. MR2002618 (2004g:57037)

[13] T. Yoshida, *The η-invariant of hyperbolic 3-manifolds*, Invent. Math. **81** (1985), no. 3, 473–514, DOI 10.1007/BF01388583. MR807069 (87f:58153)

Department of Mathematics, Faculty of Science and Engineering, Waseda University, 3-4-1 Ohkubo, Shinjuku-ku, Tokyo 169-8555, Japan

E-mail address: murakami@waseda.jp

Proceedings of Symposia in Pure Mathematics
Volume **93**, 2016
http://dx.doi.org/10.1090/pspum/093/01573

Positivity of Hochster theta over $\mathbb{C}$

Mohammad Reza Rahmati

ABSTRACT. M. Hochster defines an invariant namely $\Theta(M, N)$ associated to two finitely generated modules over a hypersurface ring $R = P/f$, where $P = k\{x_0, ..., x_n\}$ or $k[X_0, ..., x_n]$, for k a field and f is a germ of holomorphic function or a polynomial, having isolated singularity at 0. This invariant can be lifted to the Grothendieck group $G_0(R)_{\mathbb{Q}}$ and is compatible with the Chern character and cycle class map, according to the works of W. Moore, G. Piepmeyer, S. Spiroff, M. Walker. They prove that it is semi-definite when f is a homogeneous polynomial, using Hodge theory on projective varieties. It is a conjecture that the same holds for general isolated singularity f. We give a proof of this conjecture using Hodge theory of isolated hypersurface singularities when $k = \mathbb{C}$. We apply this result to give a positivity criteria for intersection multiplicty of proper intersections in the variety of f.

Introduction

One of the important ways to generalize known facts about smooth algebraic varieties in algebraic geometry, is to try to extend them over singular varieties. Most of the valid methods in the smooth category are hard to be worked out in the singular category. An example of this is the definition of intersection multiplicity as the Tor-formula in algebraic geometry by an Euler characteristic. The Riemann-Roch formula provides a definition of this characteristic class by the Chern character map on the K-group. The useful observation here is the intersection theory is in fact a theory on K-groups. In the singular set up both of the definitions of K-groups and the Chern character become very complicated at the first glance. Although trying to generalize the homology theory over singular varieties provides some type of solution for this case, trying to extend the natural concepts from the smooth category to the singular one remains still complicated and difficult.

In this short note we provide a very basic example of this situation and try to approach to the solution using asymptotic Hodge theory. In this way we employ some standard methods in the extensions of polarized mixed Hodge structures to answer questions about singularities in algebraic geometry. Thus the reader may divide the text into three parts, the first is a singular set up in algebraic geometry, the second is on variation of mixed Hodge structures, and the third is an application of the results mentioned in part 2.

2010 *Mathematics Subject Classification*. Primary 14Lxx.

Key words and phrases. Matrix factorization, Riemann-Hodge bilinear relations, Residue pairing, Cyclic homology.

1. Modules over Hypersurface rings

A hypersurface ring is a ring of the form $R := P/(f)$, where P is an arbitrary ring and f a non-zero divisor. Localizing we may assume P is a local ring of dimension $n+1$. As according to the title we assume $P = \mathbb{C}\{x_0, ..., x_n\}$ and f a holomorphic germ, or $P = \mathbb{C}[x_0, ..., x_n]$ and then f would be a polynomial. Then we are mainly interested in studying finitely generated modules over these rings. Consider $f : \mathbb{C}^{n+1} \to \mathbb{C}$ in this form, and choose a representative for the Milnor fibration as $f : X \to T$, where T is the disc around 0.

Throughout the remainder of this text we assume $0 \in \mathbb{C}^{n+1}$ is the only singularity of f.

A matrix factorization of f in P is a pair of matrices A and B such that $AB = BA = f.\ id$. It is equivalent to the data of a pair of finitely generated free P-modules

$$d_0 : X^0 \leftrightarrows X^1 : d_1, \qquad d_0 d_1 = d_1 d_0 = f.\ id.$$

It is a basic fact, discovered by D. Eisenbud, that the R-modules have a minimal resolution that is eventually 2-periodic. Specifically, in a free resolution of such a module M, we see that after n-steps we have an exact sequence of the following form,

$$0 \to M' \to F_{n-1} \to F_{n-2} \to ... \to F_0 \to M \to 0 \tag{1}$$

where the F_i are free R-modules of finite rank and $depth_R(M') = n$. If $M' = 0$ then M has a free resolution of finite length., If $M' \neq 0$, then M' is a maximal Cohen-Macaulay module, that is $depth_R(M') = n$. So "up to free modules" any R-module can be replaced by a maximal Cohen-Macaulay module. If M is a maximal Cohen-Macaulay R-module that is minimally generated by p elements, its resolution as P-module has the form

$$\begin{array}{ccccccccc} 0 & \to & P^p & \overset{A}{\to} & P^p & \to & M & \to & 0 \\ & & & B & & & & & \\ & & \downarrow & \swarrow & \downarrow & & \downarrow 0 & & \\ 0 & \to & P^p & \overset{A}{\to} & P^p & \to & M & \to & 0 \end{array}$$

where A is some $p \times p$ matrix with $det(A) = f^q$. The fact that multiplication by f acts as 0 on M produces a matrix B such that $A.B = B.A = f.I$, where I is the identity matrix. In other words, we find a matrix factorization (A, B) of f determined uniquely up to base change in the free module P^p, by M. The matrix factorization not only determines M but also a resolution of M as R-module.

$$.... \to R^p \to R^p \to R^p \to M \to 0.$$

So a minimal resolution of M looks in general as follows

$$... \to G \to F \to G \to F_{n-1} \to ... \to F_0 \to M \to 0.$$

As a consequence all the homological invariants like $Tor_k^R(M, N), Ext_R^k(M, n)$ are 2-periodic, [**BVS**], [**EP**].

The category of matrix factorizations of f over R, namely $MF(R, f)$; is defined to be the differential $\mathbb{Z}/2$-graded category, whose objects are pairs (X, d), where $X = X^0 \oplus X^1$ is a free $\mathbb{Z}/2$-graded R-module of finite rank equipped with an R-linear map d of odd degree satisfying $d^2 = f.\ id_X$. Here the degree is calculated in

$\mathbb{Z}/2$. With regard to the first definition

$$d = \begin{pmatrix} 0 & d_0 \\ d_1 & 0 \end{pmatrix}, \qquad d^2 = f.\ id.$$

The morphisms $MF(X, X')$ are given by $\mathbb{Z}/2$-graded R-module maps from X to X' (equivalently between the components X^0 and X^1) provided that the differential is given by

$$d(f) = d_{X'} \circ f - (-1)^{|f|} f \circ d_X. \tag{2}$$

Here d_X or d'_X may be considered as the matrix given above or to be separately d_0 and d_1, and also it is evident that $d(f)^2 = 0$. By choosing bases for X^0 and X^1 we reach to the former definition, [**EP**].

2. Hochster theta Function

M. Hochster in his study of direct summand conjecture defined the following invariant for the hypersurface ring $R = P/(f)$, namely Θ-invariant.

DEFINITION 2.1 (Hochster theta pairing). *The theta pairing of two R-modules M and N over a hypersurface ring $R/(f)$ is*

$$\Theta(M, N) := l(Tor^R_{2k}(M, N)) - l(Tor^R_{2k+1}(M, N)), \qquad k >> 0$$

The definition makes sense as soon as the lengths appearing are finite. This certainly happens if R has an isolated singular point.

EXAMPLE 2.2 ([**BVS**]). *Take $f = xy - z^2, M = \mathbb{C}[[xyz]]/(x, y)$. A matrix factorization (A, B) associated to M is given by*

$$A = \begin{pmatrix} y & -z \\ -z & x \end{pmatrix}, \qquad B = \begin{pmatrix} x & z \\ z & y \end{pmatrix}.$$

The $Tor^R_k(M, M)$ is the homology of the complex

$$..... \to \mathbb{C}[[y]]^2 \to \mathbb{C}[[y]]^2 \to \mathbb{C}[[y]] \to 0$$

where

$$\alpha = \begin{pmatrix} y & 0 \\ 0 & 0 \end{pmatrix}, \qquad \beta = \begin{pmatrix} 0 & 0 \\ 0 & y \end{pmatrix}.$$

So we find that $\Theta(M, M) = 0$.

The Hochster theta pairing is additive on short exact sequences in each argument, and thus determines a $\mathbb{Z}$-valued pairing on $G(R)$, the Grothendieck group of finitely generated R-modules. One loses no information by tensoring with $\mathbb{Q}$ and so often theta is interpreted as a symmetric bilinear form on the rational vector space $G(R)_{\mathbb{Q}}$. It is basic that theta would vanish if either M or N be Artinian or have finite projective dimension [**MPSW**], [**BVS**]. The Θ-invariant has different interpretations as intersection multiplicity in the singular category.

THEOREM 2.3 ([**BVS**]). *When $M = \mathcal{O}_Y = R/I, N = \mathcal{O}_Z = R/J$, where $Y, Z \subseteq X_0$ are the sub-varieties defined by the ideals I, J respectively, then*

$$\Theta(\mathcal{O}_Y, \mathcal{O}_Z) = i(0; Y, Z)$$

in case that $Y \cap Z = 0$. Here $i(0; \cdot, \cdot)$ is the ordinary intersection multiplicity in $\mathbb{C}^{n+1}$.

By additivity over short exact sequences and the fact that any module admits a finite filtration with sub-quotients of the form R/I, knowing $\Theta(\mathcal{O}_Y, \mathcal{O}_Z)$ determines $\Theta(M, N)$ for all modules M, N.

THEOREM 2.4 ([**BVS**]). *Assume $f \in \mathbb{C}[x_1, ..., x_{2m+2}]$ is a homogeneous polynomial of degree d, and $X_0 = f^{-1}(0) \in \mathbb{C}^{2m+2}$ and $T = V(f) \in \mathbb{P}^{2m+1}$ the associated projective cone of degree d. Let Y and Z be also codimension m cycles in T. If Y, Z intersect transversely, then*

$$\Theta(\mathcal{O}_Y, \mathcal{O}_Z) = -\tfrac{1}{d}[[Y]] \cdot [[Z]]$$

where $[[Y]] := d[Y] - deg(Y) \cdot h^m$ is the primitive class of $[Y]$, with $h \in H^1(T)$ the hyperplane class.

The primitive class of a cycle Y is the projection of its fundamental class $[Y] \in H^m(T)$ into the orthogonal complement to h^m with respect to the intersection pairing into $H^{2m}(T) = \mathbb{C}$. As $h^m \cdot h^m = d = deg(T)$ and $[Y] \cdot h^m = deg(Y)$ the description of the primitive class follows. Substituting the claim can be reformulated

$$\Theta(\mathcal{O}_Y, \mathcal{O}_Z) = -\tfrac{1}{d}[[Y]] \cdot [[Z]] = -d[Y] \cdot [Z] + deg(Y) deg(Z)$$

where $[Y].[Z]$ denotes the intersection form on the cohomology of the projective space, [**BVS**].

When f in consideration is a homogeneous polynomial of degree d, such that $X := Proj(R)$ is a smooth k-variety, the theta pairing is induced, via Chern character map, from the pairing on the primitive part of de Rham cohomology

$$\frac{H^{(n-1)/2}(X, \mathbb{C})}{\mathbb{C}.\gamma^{(n-1)/2}} \times \frac{H^{(n-1)/2}(X, \mathbb{C})}{\mathbb{C}.\gamma^{(n-1)/2}} \to \mathbb{C}$$

given by

$$(a, b) \to (\textstyle\int_X a \cup \gamma^{(n-1)/2})(\int_X a \cup \gamma^{(n-1)/2}) - d(\int_X a \cup b)$$

where γ is the class of a hyperplane section and theta would vanish for rings of this type having even dimensions. When $n = 1$ by γ^0 we mean $1 \in H^0(X, \mathbb{C})$, [**MPSW**].

THEOREM 2.5 ([**MPSW**]). *For R and X as above, if n is odd there is a commutative diagram*

$$\text{(3)} \qquad \begin{array}{ccc} G(R)_{\mathbb{Q}}^{\otimes 2} & \longleftarrow & (\frac{K(X)_{\mathbb{Q}}}{\alpha})^{\otimes 2} \\ \Theta \downarrow & & \downarrow (ch^{n-1/2})^{\otimes 2} \\ \mathbb{C} & \xleftarrow[\theta]{} & (\frac{H^{(n-1)/2}(X, \mathbb{C})}{\mathbb{C}.\gamma^{(n-1)/2}})^{\otimes 2} \end{array}$$

where $\alpha = \mathcal{O}_H(1)$ with $H \subset X$ a hyperplane section.

THEOREM 2.6 ([**MPSW**]). *For R and X as above and n odd the restriction of the pairing $(-1)^{(n+1)/2}\Theta$ to*

$$im(ch^{\frac{n-1}{2}}) : K(X)_{\mathbb{Q}}/\alpha \to \frac{H^{(n-1)/2}(X, \mathbb{C})}{\mathbb{C}.\gamma^{\frac{n-1}{2}}}$$

is positive definite. i.e. $(-1)^{(n+1)/2}\Theta(v, v) \geq 0$ with equality holding if and only if $v = 0$. In this way Θ is semi-definite on $G(R)$.

PROOF. [**MPSW**]. Define

$$W = H^{n-1}(X(\mathbb{C}), \mathbb{Q}) \cap H^{\frac{n-1}{2},\frac{n-1}{2}}(X(\mathbb{C})).$$

It is classical that the image of $ch^{(n-1)/2}$ is contained in W. Define $e : W/\mathbb{Q}.\gamma^{(n-1)/2} \hookrightarrow H^{n-1}(X, \mathbb{Q})$ by

$$e(a) = a - \frac{\int_X a \cup \gamma^{(n-1)/2}}{d}.\gamma^{(n-1)/2} \in W.$$

We know that $\Theta(a, b) = -d.I^{coh}(e(a), e(b))$ Now the theorem follows from the polarization properties of cup product on cohomology of projective varieties. □

3. Hodge theory and Residue form

Assume $f : \mathbb{C}^{n+1} \to \mathbb{C}$ is a germ of an isolated singularity. We choose a representative $f : X \to T$ over a small disc T according to the milnor fibration theorem. It is possible to embed the Milnor fibration $f : X \to T$ into a compactified (projective) fibration $f_Y : Y \to T$ such that the fiber Y_t sits in $\mathbb{P}^{n+1}$ for $t \neq 0$. The projective fibration f_Y has a unique singularity at $0 \in Y_0$ over $t = 0$. Then, there exists a short exact sequence

$$0 \to H^n(Y_0, \mathbb{Q}) \to H^n(Y_t, \mathbb{Q}) \xrightarrow{i^*} H^n(X_t, \mathbb{Q}) \to 0, \qquad t \neq 0. \tag{4}$$

We have $H^n(Y_0, \mathbb{Q}) = \ker(M_Y - id)$, by the invariant cycle theorem, where M_Y is the monodromy of f_Y. W. Schmid (resp. J. Steenbrink) has defined a canonical MHS on $H^n(Y_t, \mathbb{Q})$ (resp. on $H^n(Y_t, \mathbb{Q})$) namely limit MHS, which make the above sequence an exact sequence of MHS's. In the following we shall assume $H^n(Y_t, \mathbb{Q})$ (resp. on $H^n(Y_t, \mathbb{Q})$) are equipped with their limit MHS. The form $S_Y := (-1)^{n(n-1)/2} I_Y^{coh} : H^n(Y_t, \mathbb{Q}) \times H^n(Y_t, \mathbb{Q}) \to \mathbb{Q}$ is the polarization form of pure Hodge structure on $H^n(Y_t, \mathbb{C})$, $t \in T'$ (the punctured disc).

Let $H^n(X_t, \mathbb{Q}) = H^n(X_t, \mathbb{Q})_{\neq 1} \oplus H^n(X_t, \mathbb{Q})_1$ (similarly for Y) be the decomposition into generalized eigenspaces of monodromy M. Suppose $M = M_s.M_u$ is the Jordan decomposition and $N = \log(M_u)$. In the short exact sequence, the map i^* is an isomorphism on $H^n(Y_t, \mathbb{Q})_{\neq 1} \to H^n(X_t, \mathbb{Q})_{\neq 1}$ giving $S = (-1)^{n(n-1)/2} I^{coh} = (-1)^{n(n-1)/2} I_Y^{coh} = S_Y$ on $H^n(X_t, \mathbb{Q})_{\neq 1}$.

The above short exact sequence restricts to the following one,

$$0 \to \ker\{N_Y : H^n(Y_t, \mathbb{Q})_1 \to H^n(Y_t, \mathbb{Q})_1\} \to H^n(Y_t, \mathbb{Q})_1 \to H^n(X_t, \mathbb{Q})_1 \to 0 \tag{5}$$

So $a, b \in H^n(X_t, \mathbb{Q})_1$ have pre-images $a_Y, b_Y \in H^n(Y_t, \mathbb{Q})_1$ and

$$S(a, b) = S_Y(a_Y, (-N_Y)b_Y) \tag{6}$$

is independent of the lifts of a_Y, b_Y, by the fact that N_Y is an infinitesimal isometry for S_Y. The equation (10) defines the desired polarization on $H^n(X_t, \mathbb{Q})_1$. The polarization form S is M-invariant, non-degenerate, $(-1)^n$-symmetric on $H^n(X_t, \mathbb{Q})_{\neq 1}$ and $(-1)^{n+1}$-symmetric on $H^n(X_t, \mathbb{Q})_1$, [**H1**]. Suppose,

$$H^n(X_\infty, \mathbb{C}) = \bigoplus_{p,q,\lambda} I_\lambda^{p,q}$$

is the Deligne-Hodge bigrading, and generalized eigenspaces of vanishing cohomology, and also $\lambda = \exp(-2\pi i\alpha)$ with $\alpha \in (-1, 0]$. Consider the isomorphism obtained by composing the three maps,

$$\Phi_\lambda^{p,q} : I_\lambda^{p,q} \xrightarrow{\hat{\Phi}_\lambda} Gr_V^{\alpha+n-p} H'' \xrightarrow{pr} Gr_V^\bullet H''/\partial_t^{-1} H'' \xrightarrow{\cong} \Omega_f \tag{7}$$

where

$$\hat{\Phi}_\lambda^{p,q} := \partial_t^{p-n} \circ \psi_\alpha | I_\lambda^{p,q}$$

$$\Phi = \bigoplus_{p,q,\lambda} \Phi_\lambda^{p,q}, \qquad \Phi_\lambda^{p,q} = pr \circ \hat{\Phi}_\lambda^{p,q}$$

where $I^{p,q}$ stands for the bigrading, ∂_t is the Gauss-Manin connection and ψ_α is the nearby map cf. [**11**]. The map Φ is obviously an isomorphism because both of the ψ_α and ∂_t^{-1} are isomorphisms, cf. [**H1**], [**V**].

DEFINITION 3.1 (MHS on Ω_f). *We define a mixed Hodge structure on Ω_f using the isomorphism Φ. This means that all the data of the Steenbrink MHS on $H^n(X_\infty, \mathbb{C})$ such as the $\mathbb{Q}$ or $\mathbb{R}$-structure, the weight filtration $W_\bullet \Omega_{f,\mathbb{Q}}$ and the Hodge filtration $F^\bullet \Omega_{f,\mathbb{C}}$ is defined via the isomorphism Φ. Specifically; in this way we obtain a conjugation map*

$$\bar{\ } : \Omega_{f,\mathbb{Q}} \otimes \mathbb{C} \to \Omega_{f,\mathbb{Q}} \otimes \mathbb{C}, \qquad \Omega_{f,\mathbb{Q}} := \Phi^{-1} H^n(X_\infty, \mathbb{Q}) \tag{8}$$

defined from the conjugation on $H^n(X_\infty, \mathbb{C})$ via this isomorphism.

Recall that the limit (Steenbrink) mixed Hodge structure, is defined by

$$F^p H^n(X_\infty, \mathbb{C})_\lambda = \psi_\alpha^{-1}(Gr_V^\alpha \partial_t^{n-p} H'')$$

This justifies the power of ∂_t^{-1} applied in the definition of Φ.

THEOREM 3.2 ([**11**]). *Assume $f : (\mathbb{C}^{n+1}, 0) \to (\mathbb{C}, 0)$, is a holomorphic germ with isolated singularity at 0. Then, the isomorphism Φ makes the following diagram commutative up to a complex constant;*

$$\begin{array}{ccc} \widehat{Res}_{f,0} : \Omega_f \times \Omega_f & \longrightarrow & \mathbb{C} \\ \Big\downarrow (\Phi^{-1}, \Phi^{-1}) & & \Big\downarrow \times * \\ S : H^n(X_\infty) \times H^n(X_\infty) & \longrightarrow & \mathbb{C} \end{array} \qquad * \neq 0 \tag{9}$$

where,

$$\widehat{Res}_{f,0} = res_{f,0}\ (\bullet, \tilde{C}\ \bullet)$$

and $\tilde{C}$ is defined relative to the Deligne decomposition of Ω_f, via the isomorphism Φ. If $J^{p,q} = \Phi^{-1} I^{p,q}$ is the corresponding subspace of Ω_f, then

$$\Omega_f = \bigoplus_{p,q} J^{p,q} \qquad \tilde{C}|_{J^{p,q}} = (-1)^p \tag{10}$$

In other words;

$$S(\Phi^{-1}(\omega), \Phi^{-1}(\eta)) = * \times\ res_{f,0}(\omega, \tilde{C}.\eta), \qquad 0 \neq * \in \mathbb{C} \tag{11}$$

The proof of the Theorem 3.2 is a generalization of an argument in [**CIR**], in the quasi-homogeneous case.

COROLLARY 3.3 ([**11**, Riemann-Hodge bilinear relations for Grothendieck pairing on Ω_f]). *Assume $f : \mathbb{C}^{n+1} \to \mathbb{C}$ is a holomorphic germ with isolated singularity. Suppose $\mathfrak{f}$ is the corresponding map to N on $H^n(X_\infty)$, via the isomorphism Φ. Define*

$$P_l = PGr_l^W := \ker(\mathfrak{f}^{l+1} : Gr_l^W \Omega_f \to Gr_{-l-2}^W \Omega_f)$$

Going to W-graded pieces;

$$\widehat{Res}_l : PGr_l^W \Omega_f \otimes_{\mathbb{C}} PGr_l^W \Omega_f \to \mathbb{C} \tag{12}$$

is non-degenerate and according to Lefschetz decomposition

$$Gr_l^W \Omega_f = \bigoplus_r \mathfrak{f}^r P_{l-2r}$$

we will obtain a set of non-degenerate bilinear forms,

$$\widehat{Res}_l \circ (id \otimes \mathfrak{f}^l) : PGr_l^W \Omega_f \otimes_{\mathbb{C}} PGr_l^W \Omega_f \to \mathbb{C}, \tag{13}$$

$$\widehat{Res}_l = res_{f,0}\ (id \otimes \tilde{C}.\ \mathfrak{f}^l)$$

where $\tilde{C}$ is as in 3.2, such that the corresponding hermitian form associated to these bilinear forms is positive definite. In other words,

- $\widehat{Res}_l(x, y) = 0, \qquad x \in P_r,\ y \in P_s,\ r \neq s$
- *If $x \neq 0$ in P_l,*

$$Const \times res_{f,0}\ (C_l x, \tilde{C}.\ \mathfrak{f}^l.\bar{x}) > 0, \qquad Const \in \mathbb{C}$$

  where C_l is the corresponding Weil operator, and the conjugation is as in (7).

Note that the map

$$A_f = \frac{\mathcal{O}_X}{\partial f} \to \Omega_f, \qquad f \mapsto f dx_0 ... dx_n$$

is an isomorphism. Thus, the above corollary would state similarly for A_f.

4. Chern character (Denis trace map) for isolated hypersurface singularities

The Hochschild chain complex of $MF(R, f)$ is quasi-isomorphic to the Koszul complex of the regular sequence $\partial_0 f, ..., \partial_n f$. In particular the Hochschild homology (and also the Hochschild cohomology) of 2-periodic dg-category $MF(R, f)$ is isomorphic to the module of relative differentials or the Jacobi ring of f, [**D**].

THEOREM 4.1 ([**D**, T. Dykerhoff], [**PV**]). *The canonical bilinear form on the Hochschild homology of category of matrix factorizations $\mathcal{C} = MF(P, f)$ of f, after the identification*

$$HH_* MF(P, f) \cong A_f \otimes dx[n] \tag{14}$$

coincides with

$$\langle g \otimes dx, h \otimes dx \rangle = (-1)^{n(n-1)/2} res_{f,0}(g, h) \tag{15}$$

The work of Polishchuck and Vaintrob ([**PV**] Theorem 4.1.4) establishes a relation as

$$\Theta(M, N) = \pm res_{f,0}(ch(M), ch(N))$$

where the sign is dependent to the complex dimension of the fibers of the function f ([**CW**] page 2).
The Chern character or Denis trace map is a ring homomorphism

$$ch : K_0'(X) \to HH_0(X) \cong \Omega_f \tag{16}$$

where K_0' is free abelian group on the isomorphism classes of finitely generated modules modulo relations obtained from short exact sequences. The construction of the Chern character map or Chern classes is functorial w.r.t flat pull back. In the special case of $i : X \hookrightarrow Y$ the compactification, the following diagram commutes,

$$\begin{array}{ccccc} K_0'(Y_0) & \xrightarrow{ch_Y} & HH_0(Y_0) \cong \Omega_f^Y & \xrightarrow{\Phi_Y^{-1}} & H^n(Y_\infty) \\ {\scriptstyle i^*}\downarrow & & \downarrow{\scriptstyle i^*} & & \downarrow{\scriptstyle i^*} \\ K_0'(X_0) & \xrightarrow[ch_X]{} & HH_0(X_0) \cong \Omega_f^X & \xrightarrow[\Phi_X^{-1}]{} & H^n(X_\infty). \end{array} \tag{17}$$

Given a matrix factorization (A, B) for a maximal Cohen-Macaulay M, one can find de Rham representatives for the Chern classes. Consider $\mathbb{C}[[x_0, ..., x_n]]$ as a $\mathbb{C}[[t]]$-module with t acting as multiplication by f. Denote by Ω^p the module of germs of p-forms on $\mathbb{C}^{n+1}$, and let $\Omega_f^p = \Omega^p/(df \wedge \Omega^{p-1})$. One puts $\omega(M) = dA \wedge dB$. The components of the Chern character

$$ch_M := tr(\exp(\omega(M))) = \sum_i \frac{1}{i!}\omega^i(M) \tag{18}$$

are well-defined classes

$$\omega^i(M) = tr((dA \wedge dB)^i) \in \Omega_f^{2i}/(df \wedge \Omega^{2i-1}) \tag{19}$$

There are also odd degree classes

$$\eta^i(M) := tr(AdB(dA \wedge dB)^i) \in \Omega_f^{2i+1}/\Omega_f^{2i}.$$

The group $\Omega_f^{2i+1}/d\Omega_f^{2i}$ can be identified with the cyclic homology $HC_i(P/\mathbb{C}\{t\})$. They fit into the following short exact sequence such that $d\eta^{i-1} = \omega^i(M)$.

$$0 \to \Omega_f^{2i-1}/\Omega_f^{2i-2} \to \Omega^{2i}/(df \wedge \Omega^{2i-1}) \to \Omega^{2i}/\Omega^{2i-1} \to 0.$$

If the number of variables $n+1$ is even, then a top degree form sits in the Brieskorn module

$$\mathcal{H}_f^{(0)} = \Omega^n/(df \wedge d\Omega^{n-1})$$

a free $\mathbb{C}[[t]]$-module of rank μ. The higher residue pairing

$$K : \mathcal{H}_f^{(0)} \times \mathcal{H}_f^{(0)} \to \mathbb{C}[t, t^{-1}]$$

of K. Saito can be seen as the de Rham realization of the Seifert form of the singularity, [**BVS**].

5. Positivity of theta pairing-Main Result

The following theorem was conjectured in [**MPSW**].

THEOREM 5.1. *Let S be an isolated hypersurface singularity of dimension n. If n is odd, then $(-1)^{(n+1)/2}\Theta$ is positive semi-definite on $G(R)_{\mathbb{Q}}$, i.e $(-1)^{(n+1)/2}\Theta$ $(M, M) \geq 0$.*

PROOF. By additivity of Θ on each variable, we may replace M, N by maximal Cohen-Macaulay modules. According to 4.1 and 3.2 the determination of the sign of Θ amounts to understanding how the image of Chern classes look like in the MHS of Ω_f. By Theorem 3.2 it amounts to the same things for the image in $H^n(X_\infty)$ under the isomorphism Φ. The following diagram is commutative by the functorial properties of Chern character.

$$\begin{array}{ccc} K_0'(Y_0) & \xrightarrow{\Phi_Y^{-1} o ch_Y} & H^n(Y_\infty) \\ {\scriptstyle i^*}\downarrow & & \downarrow{\scriptstyle i^*} \\ K_0'(X_0) & \xrightarrow[\Phi_X^{-1} o ch_X]{} & H^n(X_\infty). \end{array} \tag{20}$$

We are assuming that i^* is surjective. The Chern class we are concerned with, is a Hodge cycle and the commutativity of the above diagram allows us to replace the pre-image of the Chern character for X, with similar cycle upstairs. The polarization form S_X was defined via that of S_Y. Thus, if

$$H^n(Y_\infty) = \oplus_{p+q=n} H^{p,q}$$

is the Hodge decomposition, the only non trivial contribution in the cup product will be for the $H^{n/2,n/2}$, and the polarization form is evidently definite on this subspace (Hodge cycles). Because the map N_Y is of type $(-1,-1)$ for the Hodge structure of $H^n(Y_\infty)$ and the polarization $S_Y(H^{n/2,n/2}, H^{n/2-1,n/2-1}) = 0$ for obvious reasons, the corresponding Chern class should lie in $H^n_{\neq 1}$. In this way one only needs to prove the positivity statement for Hochster Θ when the Chern character is in $H_{Y,\neq 1}$, and this is the content of Theorem 2.8. □

Theorem 6.1 recover a very special case of Serre multiplicity conjecture, for intersections at the isolated singularity (proved in general by O. Gabber).

COROLLARY 5.2. *The Serre intersection multiplicity $\chi(M, N)$ is always non-negative for proper intersections at 0, where M and N are finitely generated $R = P/(f)$-modules with $l(M \otimes N) < \infty$.*

References

[AGV] V. I. Arnold, S. M. Gusein-Zade, and A. N. Varchenko, *Singularities of differentiable maps. Volume 2*, Modern Birkhäuser Classics, Birkhäuser/Springer, New York, 2012. Monodromy and asymptotics of integrals; Translated from the Russian by Hugh Porteous and revised by the authors and James Montaldi; Reprint of the 1988 translation. MR2919697

[BVS] R.-O. Buchweitz and D. Van Straten, *An index theorem for modules on a hypersurface singularity* (English, with English and Russian summaries), Mosc. Math. J. **12** (2012), no. 2, 237–259, 459. MR2978754

[CW] O. Celikbas and M. E. Walker, *Hochster's theta pairing and algebraic equivalence*, Math. Ann. **353** (2012), no. 2, 359–372, DOI 10.1007/s00208-011-0688-6. MR2915540

[CIR] A. Chiodo, H. Iritani, and Y. Ruan, *Landau-Ginzburg/Calabi-Yau correspondence, global mirror symmetry and Orlov equivalence*, Publ. Math. Inst. Hautes Études Sci. **119** (2014), 127–216, DOI 10.1007/s10240-013-0056-z. MR3210178

[D] T. Dyckerhoff, *Compact generators in categories of matrix factorizations*, Duke Math. J. **159** (2011), no. 2, 223–274, DOI 10.1215/00127094-1415869. MR2824483 (2012h:18014)

[EP] D. Eisenbud, I. Peeva, Matrix factorizations for complete intersections, and minimal free resolutions, arxiv:math 2013

[G3] P. Griffiths and J. Harris, *Principles of algebraic geometry*, Wiley Classics Library, John Wiley & Sons, Inc., New York, 1994. Reprint of the 1978 original. MR1288523 (95d:14001)

[H1] C. Hertling, *Classifying spaces for polarized mixed Hodge structures and for Brieskorn lattices*, Compositio Math. **116** (1999), no. 1, 1–37, DOI 10.1023/A:1000638508890. MR1669448 (2000d:32048)

[MPSW] W. Moore, G. Piepmeyer, S. Spiroff, M. Walker, Hochster theta pairing and Riemann-Hodge bilinear relations arxiv 0910.1289v2

[PV] A. Polishchuk and A. Vaintrob, *Chern characters and Hirzebruch-Riemann-Roch formula for matrix factorizations*, Duke Math. J. **161** (2012), no. 10, 1863–1926, DOI 10.1215/00127094-1645540. MR2954619

[SC2] J. Scherk and J. H. M. Steenbrink, *On the mixed Hodge structure on the cohomology of the Milnor fibre*, Math. Ann. **271** (1985), no. 4, 641–665, DOI 10.1007/BF01456138. MR790119 (87b:32014)

[SCH] W. Schmid, *Variation of Hodge structure: the singularities of the period mapping*, Invent. Math. **22** (1973), 211–319. MR0382272 (52 #3157)

[11] [R] M. Rahmati, On the mixed Hodge structure associated to isolated hypersurface singularities, arxiv.org, Feb. 2014.

[V] A. Varchenko, On the local residue and the intersection form on the vanishing cohomology, Math. USSR Izv. Volume 26, Number 1, 1986

Centro de Investigacion en Matematicas, A.C.

E-mail address: mrahmati@cimat.mx

Proceedings of Symposia in Pure Mathematics
Volume **93**, 2016
http://dx.doi.org/10.1090/pspum/093/01589

Cohomological Donaldson–Thomas theory

Balázs Szendrői

ABSTRACT. This review gives an introduction to cohomological Donaldson–Thomas theory: the study of a cohomology theory on moduli spaces of sheaves on Calabi–Yau threefolds, and of complexes in 3-Calabi–Yau categories, categorifying their numerical DT invariant. Local and global aspects of the theory are both covered, including representations of quivers with potential. We will discuss the construction of the DT sheaf, a nontrivial topological coefficient system on such a moduli space, along with some cohomology computations. The Cohomological Hall Algebra, an algebra structure on cohomological DT spaces, will also be introduced. The review closes with some recent appearances, and extensions, of the cohomological DT story in the theory of knot invariants, of cluster algebras, and elsewhere.

CONTENTS

Introduction

Cohomological Donaldson–Thomas theory is the cohomological study of moduli spaces of sheaves on three-dimensional Calabi–Yau varieties. In its most developed form, it is a theory tied to a specific class of geometries in a specific number of dimensions. It is hoped that the richness of the theory to be presented will alleviate concerns over the specialized starting point.

2010 *Mathematics Subject Classification.* Primary 14N35, 14J32, 14C30.

Key words and phrases. Calabi-Yau threefolds, Donaldson-Thomas theory, vanishing cycle sheaf.

The aim of this review is to give a tutorial introduction to this subject, at a level accessible to graduate students wanting to get a glimpse and looking for pointers to the literature on specific topics. The discussion will start with an overview of *numerical* DT theory, effectively the Euler characteristic shadow of the full theory, which was also discussed recently in the review [**88**], in particular its Sections $3\frac{1}{2}$ and $4\frac{1}{2}$. While I will not assume knowledge of [**88**], those hoping to learn about different facets of the subject should peruse both this review and [**88**].

We will mostly work in the language of schemes, all schemes assumed to be of finite type over the base $\mathbb{C}$ of complex numbers. Algebraic stacks will make the occasional appearance, but most of our stacks will be global quotient stacks of the form $[M/G]$ with G an affine algebraic group acting on a scheme M without a geometric (scheme) quotient. It has recently become clear that the most natural context for the subject is that of derived symplectic geometry [**89**]; because of (lack of) expertise, and the intention of keeping technicalities to a necessary minimum, I will not give a full treatment of derived geometry, restricting instead to the occasional side remark. The same applies to the main source of motivation for the historical development of this area of research: supersymmetric physics.

Acknowledgements. This review is a write-up of my lectures at the String Math Summer School held at the University of British Columbia in June 2014. I would like to thank its organizer Jim Bryan for the invitation to speak there. Thanks are also due to Kai Behrend, Tom Bridgeland, Jim Bryan, Ben Davison, Ian Grojnowski, Dominic Joyce, Alastair King, Maxim Kontsevich, Davesh Maulik, Andrew Morrison, Sergei Mozgovoy, Rahul Pandharipande, Yan Soibelman and Richard Thomas for many valuable conversations on DT theory over the years. Support from the Leverhulme Trust (Royal Society Leverhulme Trust Senior Research Fellowship) and EPSRC (Programme Grant EP/I033343/1) during the preparation of this review is also gratefully acknowledged.

1. An overview of DT theory

1.1. Calabi–Yau threefolds. We work over the field of complex numbers. A *Calabi–Yau threefold* is a quasiprojective three-dimensional non-singular algebraic variety Y over $\mathbb{C}$, whose canonical (dualizing) line bundle ω_Y is trivial: $\omega_Y \cong \mathcal{O}_Y$. Sometimes simple connectedness of Y or the weaker condition $H^1(\mathcal{O}_Y) = 0$ will be assumed also, but we will not need this in general. As examples, it will be useful to keep the following list in mind, with varying amounts of projectivity.

(1) We can require Y to be projective; there are hundreds of millions of families of examples, starting with the quintic threefold $Y = Y_5 \subset \mathbb{P}^4$. This will be the hardest case for the general theory to handle, so we will often study simpler geometries first.
(2) Let S be an algebraic surface, then $Y = \mathrm{Tot}(\omega_S)$, the total space of the canonical line bundle of S, is a Calabi–Yau threefold. A special case is $S = \mathbb{P}^2$, with $Y = \mathrm{Tot}(\mathcal{O}_{\mathbb{P}^2}(-3))$.
(3) Let C be a projective curve, and $\mathcal{L}_1, \mathcal{L}_2$ line bundles on C. Then the total space $Y = \mathrm{Tot}(\mathcal{L}_1 \oplus \mathcal{L}_2)$ is a Calabi–Yau threefold if and only if $\mathcal{L}_1 \otimes \mathcal{L}_2 \cong \omega_C$. A much-studied special case is the resolved conifold geometry, when $C = \mathbb{P}^1$ and $\mathcal{L}_i = \mathcal{O}_{\mathbb{P}^1}(-1)$. We will denote this space by $Y = \mathcal{O}_{\mathbb{P}^1}(-1, -1)$.

(4) The simplest Calabi–Yau threefold of all, by our definition, is $Y = \mathbb{C}^3$.

(5) A more general construction, giving some of the examples from (2)-(4) and others, is the following. Let $\Gamma < \mathrm{SL}(3, \mathbb{C})$ be a finite group, and $\bar{Y} = \mathbb{C}^3/\Gamma$ the quotient, a singular variety with Gorenstein singularities and $\omega_{\bar{Y}} \cong \mathcal{O}_{\bar{Y}}$. Then there exist, usually several, quasi-projective Calabi–Yau resolutions $Y_i \to \bar{Y}$. A further generalization is to consider resolutions of other varieties $\bar{Y}$ with Gorenstein singularities and trivial canonical bundle.

1.2. Sheaves on Calabi–Yau threefolds and DT invariants. Let us for the time being restrict to projective Calabi–Yau threefolds Y satisfying the extra condition $H^1(\mathcal{O}_Y) = 0$. One set of traditional, topological invariants attached to such a geometry Y are the topological Betti numbers $b_i(Y) = \dim H^i(Y, \mathbb{Q})$ and the Hodge numbers $h^{p,q}(Y) = \dim H^q(Y, \Omega^p_Y)$, the dimension of the coherent cohomology of Y with coefficients in the sheaf of holomorphic p-forms. Under the assumption $H^1(\mathcal{O}_Y) = 0$, these two sets of numbers determine each other, so only depend on the underlying differentiable manifold. The initial aim of Donaldson–Thomas theory [**36, 99**] was to define more general numerical invariants of a Calabi–Yau threefold Y, which are sensitive to its complex structure, while still being invariant in the sense that they are unchanged under small deformations of the complex structure. These invariants are attached to various moduli spaces of sheaves $\mathcal{M} = \mathcal{M}(Y)$ on Y.

To be slightly more precise, let us fix a cohomology class $\alpha \in H^{\mathrm{even}}(Y, \mathbb{Q})$. Then we can consider the moduli *stack* $\mathcal{M}(Y, \alpha)$ of coherent sheaves on Y with Chern character α. This is a non-Hausdorff Artin stack, usually of infinite type, and as such does not admit sensible numerical invariants. Imposing a *stability condition* σ may, in an ideal situation, give a proper, finite type and Hausdorff open substack $\mathcal{M}^\sigma(Y, \alpha) \subset \mathcal{M}(Y, \alpha)$ parametrizing σ-stable sheaves, with a *coarse moduli scheme* which by abuse of notation I will also denote $\mathcal{M}^\sigma(Y, \alpha)$. (Let me be vague for now on what sort of stability conditions one might impose.) This moduli scheme of σ-stable sheaves is a better candidate for defining numerical invariants, but there is still a serious issue: in all but a handful of cases, $\mathcal{M}^\sigma(Y, \alpha)$ has very bad singularities. Its tangent space at a point $[E] \in \mathcal{M}^\sigma(Y, \alpha)$ corresponding to a coherent sheaf E on Y is $T_{[E]}\mathcal{M}^\sigma(Y, \alpha) \cong \mathrm{Ext}^1(E, E)$, but there is a complicated higher-order map, the Kuranishi map, from this tangent space to the *obstruction space* $\mathrm{Ext}^2(E, E)$, whose zero set gives a local model for $\mathcal{M}^\sigma(Y, \alpha)$ near its point $[E]$.

The special feature for Calabi–Yau threefolds is that the tangent and obstruction spaces are naturally dual using Serre duality and $\omega_Y \cong \mathcal{O}_Y$:

$$\mathrm{Ext}^2(E, E) \cong \mathrm{Ext}^1(E, E \otimes \omega_Y)^\vee \cong \mathrm{Ext}^1(E, E)^\vee.$$

This observation was the starting point DT theory, formalized in the result [**99**] that $\mathcal{M}^\sigma(Y, \alpha)$ carries a *perfect obstruction theory* [**5**] of dimension 0, in fact a symmetric obstruction theory [**6**]. A dimension 0 perfect obstruction theory gives rise to a virtual fundamental class [**88**, Appendix]

$$[\mathcal{M}^\sigma(Y, \alpha)]^{\mathrm{vir}} \in H_0(\mathcal{M}^\sigma(Y, \alpha), \mathbb{Z}),$$

whose degree

$$\deg[(\mathcal{M}^\sigma(Y, \alpha))]^{\mathrm{vir}} \in \mathbb{Z}$$

is the sought-for numerical invariant, the *DT invariant* of Y in the class α and stability σ. It is important to understand that this is a *global* construction of the invariant, requiring $\mathcal{M}^\sigma(Y,\alpha)$ to be proper.

It was proved by Thomas [**99**] that this quantity is indeed an invariant of the underlying threefold in the sense explained above: if we fix a smooth family $\{Y_t\}$ of projective Calabi–Yau threefolds, a cohomology class α on the underlying smooth manifold, and a stability condition σ, the resulting number is constant in the family.

The right stability condition to choose depends on the specifics of the situation under study. One option involves choosing some standard stability condition for sheaves such as Gieseker stability with respect to an ample line bundle. The numerical DT invariants change under change of stability condition following precise *wall-crossing relations*, as has been established as the outcome of extensive research whose highlights include [**55**, **57**, **66**].

EXAMPLE 1.1. The equality of the dimensions of the tangent and obstruction spaces of the moduli scheme $\mathcal{M}^\sigma(Y,\alpha)$ at all its points says that its *virtual dimension* is 0. This suggests that in sufficiently general, transversal situations, $\mathcal{M}^\sigma(Y,\alpha)$ might consist of a finite number of points, perhaps even reduced points. In the latter case, and only in that case, the DT invariant is simply the number of points. While in algebraic geometry such transversality is extremely rare, this indeed being the point of introducing the virtual fundamental class, there are some interesting cases, discussed in [**99**, Thm. 3.55 and Sect. 4], where generically the moduli scheme is a finite set of reduced points. See Example 8.1 below for another example. On the other hand, as soon as the DT invariant is negative, such as in the explicit examples in Section 2.4 and Example 5.7, this cannot happen.

EXAMPLE 1.2. For the case of sheaves of rank 1, one can be more explicit about the stability condition involved. Given a torsion-free sheaf E of rank 1 on Y, there is an injective map of sheaves $E \to \det(E)$ to the determinant of E, which must be an invertible sheaf. Up to tensoring by this invertible sheaf, we can assume that the determinant is trivial, and then we get an embedding $E \to \mathcal{O}_Y$ with cokernel the structure sheaf $Z \subset Y$ of a subscheme of Y, necessarily supported in codimension two. The parameter space of such embedded subschemes is the *Hilbert scheme* of codimension-two subschemes of Y; under the assumption $H^1(\mathcal{O}_Y) = 0$, the moduli scheme of sheaves is isomorphic to the Hilbert scheme and is in particular separated and of finite type (for fixed topological invariants). See [**88**, Section $3\frac{1}{2}$] for further discussion of this important special case, which plays a major role in the historical development of the subject, as well as being responsible for important connections to other counting problems.

REMARK 1.3. If the moduli problem of sheaves of class α involves σ-semistable objects, then the situation changes radically. In this case, $\mathcal{M}^\sigma(Y,\alpha)$ is only a coarse moduli space, and its tangent space (as a scheme) can no longer be identified simply as an Ext-group. In this case, DT invariants have to be defined by a complicated roundabout procedure [**57**], and they take values in $\mathbb{Q}$. We refer for more details about the issues, and their solution, to [**57**].

1.3. Local interpretation: the Behrend function. The above definition of the DT invariant was global: it needed a proper moduli space $\mathcal{M} = \mathcal{M}^\sigma(Y,\alpha)$, so that the degree of its virtual fundamental class is defined. In the case of invariants attached to symmetric obstruction theories, Behrend [**3**] gave a radically different, local interpretation. What this work showed was that given any scheme X, associated to X is a canonical *constructible* function $\nu_X \colon X \to \mathbb{Z}$, defined in terms of local data, at a nonsingular point of X its value being $(-1)^{\dim X}$. We can then define the *virtual Euler characteristic* of X to be the integer

$$\chi_{\rm vir}(X) = \int_X \nu_X d\chi \in \mathbb{Z}.$$

The latter integral is the integral of the constructible function ν_X against the (usual, topological) Euler characteristic, defined by the formula

$$\int_X \nu_X d\chi = \sum_{n\in\mathbb{Z}} n \cdot \chi\left(\nu_X^{-1}(n)\right).$$

We then have the following important result.

THEOREM 1.4. [**3**] *Let $\mathcal{M}$ be a proper scheme equipped with a symmetric perfect obstruction theory, such as a proper moduli space of stable sheaves on a projective Calabi–Yau threefold, so that $\mathcal{M}$ has a virtual fundamental class $[\mathcal{M}]^{\rm vir}$. Then*

$$\deg[\mathcal{M}]^{\rm vir} = \chi_{\rm vir}(\mathcal{M}).$$

The importance of this result is twofold. First of all, even if $\mathcal{M}$ is proper, it allows for stratification ("cut and paste") arguments since the right hand side (but not the left) behaves well under stratifications. In particular, the right hand side is defined for any scheme, whether it is proper or not. This is the second advantage: the right hand side of this formula defines DT invariants of non-compact Calabi–Yau threefolds as well. For such threefolds, we can still often define moduli spaces $\mathcal{M} = \mathcal{M}^\sigma(Y, \alpha)$, so long as we suitably restrict the set of objects under study. For example, we could ask for moduli of sheaves which themselves have compact support inside Y. We then obtain finite type moduli schemes, but they are frequently non-compact. An extended example will be discussed in Section 2.

1.4. Invariants from other categories with 3-Calabi–Yau properties. The discussion above used moduli schemes, and defined numerical invariants, starting from the category $\mathcal{C} = \mathrm{Coh}(Y)$ of coherent sheaves on a Calabi–Yau threefold Y. We relied on three facts. The first, geometric fact was that objects $E \in \mathcal{C}$ have a well-defined moduli theory. The second, homological fact was Serre duality with trivial dualizing sheaf, leading to the duality of the tangent space $\mathrm{Ext}^1(E, E)$ and the obstruction space $\mathrm{Ext}^2(E, E)$. A final important requirement was that the category $\mathcal{C}$ admits suitable notions of stability. One can hope to repeat the discussion for other categories $\mathcal{C}$ with these properties. We leave the most general definition to other sources [**66**]; we restrict to examples.

(1) For a Calabi–Yau threefold Y, we can consider $\mathcal{C} = \mathcal{D}^b(\mathrm{Coh}\, Y)$, the *derived category* of coherent sheaves on Y. Serre duality works in exactly the same way as for sheaves. The most general notion of stability in this context is (Douglas-)Bridgeland stability [**15**].

(2) Let $\mathcal{A}$ be a finitely generated associative, not necessarily commutative algebra over $\mathbb{C}$. Let $\mathcal{C} = \mathcal{A}\text{-}\mathrm{mod}$ be the category of finite dimensional right $\mathcal{A}$-modules. Then the homological condition is satisfied if $\mathcal{A}$ is a *3-Calabi–Yau algebra* [**44**]. This means in particular that

   (a) for $M, N \in \mathcal{C}$, $\mathrm{Ext}^i_{\mathcal{A}}(M, N)$ is finite dimensional, and vanishes unless $0 \le i \le 3$;

(b) for $M, N \in \mathcal{C}$, there are perfect bifunctorial pairings

$$\mathrm{Ext}^i_{\mathcal{A}}(M,N) \times \mathrm{Ext}^{3-i}_{\mathcal{A}}(N,M) \to \mathbb{C}.$$

As a subcase, one can consider 3-Calabi–Yau algebras arising from *quivers with potential* (see Section 3); the objects in $\mathcal{A}$ then readily admit a moduli theory, and (King) stability conditions [**63**].

DT invariants in these situations have been defined in [**66**], respectively [**67**, **76**, **97**].

EXAMPLE 1.5. An important special case of (1) is the stable pairs theory of Pandharipande and Thomas [**87**]. This is a variant of Example 1.2, where we parametrize objects of $\mathcal{D}^b(\mathrm{Coh}\, Y)$ which are generically ideal sheaves of embedded codimension-two subschemes $Z \subset X$ (curves), but with a stability condition different from considering torsion-free sheaves. A detailed discussion of this story can be found in [**88**, Sect. $4\frac{1}{2}$].

EXAMPLE 1.6. Interesting examples of (2) come from derived equivalences of the form

$$\mathcal{D}^b_c(\mathrm{Coh}\, Y) \cong \mathcal{D}^b(\mathcal{A} - \mathrm{mod}),$$

for Y a (noncompact) Calabi–Yau threefold, $\mathcal{D}^b_c(\mathrm{Coh}\, Y)$ the derived category of coherent sheaves with compact support on Y and $\mathcal{A}$ a finitely-generated commutative, usually non-commutative algebra. Such examples arise in the context of the McKay correspondence [**16**], and more generally for resolutions of three-dimensional Gorenstein singularities [**100**].

1.5. Generating series of DT invariants and BPS invariants. More than on individual invariants, the emphasis of the subject has been on the study of generating series of the DT invariants. One reason for this is the origin of these investigations in string theory, where it is natural to form generating series (or partition functions) of individual invariants, and study these as functions of the variables. In mathematical terms, alongside fixing the Calabi–Yau geometry Y and a stability parameter σ, we also fix an affine sublattice $\Lambda \subset H^{\mathrm{even}}(Y, \mathbb{Q})$ of rank n. We then consider an expression of the form

$$Z_{Y,\Lambda}(\mathbf{t}) = \sum_{\alpha \in \Lambda} \chi_{\mathrm{vir}}\left(\mathcal{M}^\sigma(Y,\alpha)\right) \mathbf{t}^\alpha \in \mathbb{Z}[[\mathbf{t}]].$$

Here $\mathbf{t} = \{t_1, \ldots, t_n\}$ is a set of auxiliary variables, and $\mathbf{t}^\alpha$ is multi-index notation.

It was conjectured by [**57**, **66**], based on earlier calculations and physical considerations [**45**], that the generating series of DT invariants $Z_{Y,\Lambda}(\mathbf{t})$ take the form of certain infinite products with integral exponents $\Omega^\sigma(\alpha) \in \mathbb{Z}$. These invariants, sometimes called *BPS invariants*, are the fundamental structural quantities governing DT generating series and their wall-crossing. The name comes from their interpretation as the number of BPS states (states saturating a bound on their mass) in certain supersymmetric theories. The integrality of $\Omega^\sigma(\alpha)$ remains conjectural in general. See [**87**, Section $2\frac{1}{2}$] for the parallel story in Gromov–Witten theory. See also Section 7.2 below.

1.6. Categorifying DT invariants. Given a moduli space of stable sheaves (or objects) $\mathcal{M}$ on a Calabi–Yau threefold (or CY3 category), the theory explained so far associates an integer invariant $\chi_{\mathrm{vir}}(\mathcal{M})$ to $\mathcal{M}$, which, as explained above, can be expressed as a weighted Euler characteristic. The next question one can ask is whether this construction can be *categorified*. The first level of categorification would associate to $\mathcal{M}$ some finite-dimensional *graded vector space* $\mathcal{H}^*(\mathcal{M})$ so that

$$\chi_{\mathrm{vir}}(\mathcal{M}) = \sum_j (-1)^j \dim \mathcal{H}^j(\mathcal{M}). \tag{1.1}$$

If $\mathcal{M}$ is nonsingular, with the Behrend function identically $(-1)^{\dim M}$, then we could just take for $\mathcal{H}^*(\mathcal{M})$ the (shifted) classical cohomology $H^*(\mathcal{M}, \mathbb{Q}_{\mathcal{M}}[\dim \mathcal{M}])$.

In the following sections, I will explain the construction of the graded vector space $\mathcal{H}^*(\mathcal{M})$ in general, starting from the moduli interpretation of $\mathcal{M}$. This vector space will depend on more than just the scheme structure of $\mathcal{M}$; in modern language, we need to retain the derived moduli scheme structure (and a version of a symplectic form), not just the classical scheme structure. We will use a classical truncation below, which will give this structure entirely in non-derived terms.

A next level of categorification would involve associating a *category* $\mathcal{C}(\mathcal{M})$ to the moduli space $\mathcal{M}$, some linearization of $\mathcal{C}(\mathcal{M})$ being $\mathcal{H}^*(\mathcal{M})$. I will make some comments about this below (see in particular Remark 5.6); in general, it is not currently known how to build such a category.

REMARK 1.7. Closely related to categorified DT invariants are *refined DT invariants*. Here the invariant attached to a moduli space $\mathcal{M}$ still takes values in a ring (and not a category), but it retains more information about $\mathcal{M}$ than just a number. Refined invariants in DT theory were conjectured to exist in the theoretical physics literature first [**51**, **53**, **54**]. These works exploited the idea that in theories built from local Calabi–Yau threefolds, there are certain extra symmetries that allow for a refinement (or quantization) of the numerical invariant $\chi_{\mathrm{vir}}(\mathcal{M}) \in \mathbb{Z}$ to an invariant in $\mathbb{Z}[q, q^{-1}]$. In the above language, a reasonable guess might be to associate to $\mathcal{M}$ the expression $\sum_j q^j \dim \mathcal{H}^j(\mathcal{M})$. As we will explain below (see Section 6.1), this turns out to be the wrong thing to do in general, for reasons of Hodge theory, but the correct answer is closely related.

DT invariants can be defined in the most general ring where the cut-and-paste relation satisfied by an Euler characteristic-type invariant holds, which is some version of the *Grothendieck ring of varieties* $K(\mathrm{Var}_{\mathbb{C}})$. Invariants taking values in such rings are referred to as *motivic DT invariants*, introduced in [**66**]. We will return to this point later.

2. A salient example

I now introduce one salient example, in some sense the simplest infinite family of moduli spaces of sheaves on a Calabi–Yau threefold. Many of the features of the general theory appear here in transparent form. The Calabi–Yau threefold in question is simply affine three-space $Y = \mathbb{C}^3$. As we saw in the previous section, to be able to talk about moduli spaces, we need to assume some compactness of the objects we study; so in this example, we will be parametrizing collections of points of affine three-space. In other words, we will work in the context of Example 1.2, parametrizing rank-one torsion-free sheaves on $Y = \mathbb{C}^3$ with trivial determinant and compactly supported quotient.

2.1. The Hilbert scheme of points. There are different ways of parametrizing collections of (unlabelled) points of a variety Y. The simplest way is to consider the symmetric product

$$S(Y) = \bigsqcup_{n=1}^{\infty} S^n(Y)$$

where S^n is the n-th symmetric product

$$S^n(Y) = Y^n/\Sigma_n,$$

the quotient by the symmetric group Σ_n of the n-fold Cartesian product of Y. A point in $S(Y)$ corresponds to a collection of points of Y with multiplicities. However, $S(Y)$ is not a moduli space: its points do not parametrize objects on Y which vary in a flat deformation family. From this point of view, better behaved is the *Hilbert scheme (of points)* of Y

$$\mathrm{Hilb}(Y) = \bigsqcup_{n=1}^{\infty} \mathrm{Hilb}^n(Y).$$

Here as a set, $\mathrm{Hilb}^n(Y)$, the n-th Hilbert scheme (of points) of Y, is given by

$$\mathrm{Hilb}^n(Y) = \{\mathcal{I} \lhd \mathcal{O}_Y\colon\ \dim H^0(\mathcal{O}_Y/\mathcal{I}) = n\}.$$

$\mathrm{Hilb}(Y)$ parametrizes flat families of finite-codimension ideals of the structure sheaf $\mathcal{O}_Y$ with fixed codimension; by a theorem of Grothendieck, its connected components $\mathrm{Hilb}^n(Y)$ are (represented by) a separated scheme, which are quasiprojective if Y is. It is well known that $\mathrm{Hilb}(Y)$ is nonsingular if $\dim Y \leq 2$; the first case when singularities appear on the Hilbert scheme is when $\dim Y = 3$. Starting in this dimension, the singularities are really severe for large number of points n; in particular the $\mathrm{Hilb}^n(Y)$ are reducible, and not known to be reduced.

2.2. The Hilbert scheme of points of $\mathbb{C}^d$. Let us work out in detail what the definition of $\mathrm{Hilb}^n(Y)$ says for the specific case of affine space $Y = \mathbb{C}^d$. I am using set-theoretic notation, but it is easy to check that all the steps can be performed in families. First, since $\mathbb{C}^d$ is affine, we can forget about sheaves and think about rings instead.

$$\begin{aligned}
\mathrm{Hilb}^n(\mathbb{C}^d) &= \{\mathcal{I} \lhd \mathcal{O}_{\mathbb{C}^d}\colon\ \dim H^0(\mathcal{O}_{\mathbb{C}^d}/\mathcal{I}) = n\} \\
&= \{I \lhd \mathbb{C}[\mathcal{O}_{\mathbb{C}^d}]\colon\ \dim \mathbb{C}[\mathcal{O}_{\mathbb{C}^d}]/I = n\} \\
&= \{I \lhd \mathbb{C}[x_1, \dots, x_d]\colon\ \dim \mathbb{C}[x_1, \dots, x_d]/I = n\}.
\end{aligned}$$

Here $\mathbb{C}[\mathcal{O}_{\mathbb{C}^d}]$ is the ring of functions on affine d-space, and in the last line a set of affine coordinates were chosen. We can turn things round now: think of the n-dimensional vector space $V = \mathbb{C}[x_1, \dots, x_d]/I$ as the primary object, which comes with a $\mathbb{C}[x_1, \dots, x_d]$-module structure as extra data. This can be encoded in operators $X_1, \dots, X_d \in \mathrm{GL}(V)$, which are required to commute. Also, $\mathbb{C}[x_1, \dots, x_d]/I$ is not an arbitrary module, but it is *cyclic*: it is generated as a $\mathbb{C}[x_1, \dots, x_d]$-module by the image $v \in V$ of $1 \in \mathbb{C}[x_1, \dots, x_d]$. This vector v is often called a *cyclic* or *framing vector.*

Fixing an n-dimensional vector space V, we can consider

$$M_n = \{(X_1, \dots, X_d, v)\colon [X_i, X_j] = 0, \langle X_1, \dots, X_d\rangle v = V\} \subset \mathrm{GL}(V)^d \times V,$$

where $[\ ,\ ]$ denotes the standard commutator on matrices. Since the choice of (X_i, v) is only determined up to automorphisms of V, we obtain a set-theoretic description

$$\mathrm{Hilb}^n(\mathbb{C}^d) = M_n/\,\mathrm{GL}(V).$$

This concrete description turns out to be very useful; let us analyze it. Separate the data into two parts: there is the tuple $(X_1, \ldots, X_d, v)$ together with the cyclicity condition; and the relations $[X_i, X_j] = 0$. The first part gives an auxiliary space

$$N_n = \{(X_1, \ldots, X_d, v) \colon \langle X_1, \ldots, X_d\rangle v = V\} \subset \mathrm{GL}(V)^d \times V,$$

the parameter space of cyclic modules for the free *non-commutative* $\mathbb{C}$-algebra generated by non-commuting variables $X_1, \ldots, X_d$, and its quotient

$$\mathrm{NCHilb}_d^n = N_n / \mathrm{GL}(V).$$

PROPOSITION 2.1. *The noncommutative Hilbert scheme* NCHilb_d^n *is a smooth quasiprojective variety. The (commutative) Hilbert scheme* $\mathrm{Hilb}^n(\mathbb{C}^d)$ *is the subscheme of* NCHilb_d^n *cut out by the relations* $[X_i, X_j] = 0$.

PROOF. Quasiprojectivity of NCHilb_d^n follows from the fact that $N_n/\mathrm{GL}(V)$ is a GIT quotient of the space of d-tuples of matrices under the action of $\mathrm{GL}(V)$; cyclicity corresponds to a particularly pleasant choice of stability condition [**77**, Sect. 1]. It is easy to see that cyclic triples are all stable with no non-trivial automorphisms, hence NCHilb_d^n is also nonsingular. Finally, the commutation relations cut out the Hilbert scheme as a subscheme inside NCHilb_d^n, in other words the scheme of commuting framed endomorphisms represents the functor of families of finite codimension ideals in $\mathbb{C}[x_1, \ldots, x_d]$; the proof is given in the proof of [**81**, Thm. 1.9]. □

The space NCHilb_d^n is often called the *non-commutative Hilbert scheme*, the moduli space of codimension-n ideals in the non-commutative ring $\mathbb{C}\langle X_1, \ldots, X_d\rangle$. For $d = 2$, Proposition 2.1 gives the well known description of the Hilbert scheme of the affine plane as the framed space of pairs of commuting matrices [**81**, Thm. 1.9].

2.3. The Hilbert scheme of points of $\mathbb{C}^3$. Let us now restrict to the case $d = 3$, corresponding to the local Calabi–Yau threefold $Y = \mathbb{C}^3$. Consider the function $\widetilde{W}_n \colon N_n \to \mathbb{C}$ given by

$$(X_1, X_2, X_3, v) \mapsto \operatorname{Tr} X_1[X_2, X_3].$$

From properties of Tr it follows that this map is more symmetric than it looks, being cyclically symmetric in the indices. Now let us think about what is the degeneracy locus $\mathrm{Crit}(\widetilde{W}_n)$ of this function, the locus on N_n given by the vanishing of the one-form $\mathrm{d}\widetilde{W}_n$. Forgetting the vector v which does not enter into the function $\widetilde{W}_n$, local coordinates on N_n are just the matrix entries of the X_i. In matrix entries, we have

$$\widetilde{W}_n(X_1, X_2, X_3) = \sum_{i,j,k} (X_1)_{ij}(X_2)_{jk}(X_3)_{ki} - \sum_{i,j,k} (X_1)_{ij}(X_3)_{jk}(X_2)_{ki}.$$

Differentiating with respect to a matrix entry $(X_1)_{ij}$, the corresponding component of the one-form $\mathrm{d}\widetilde{W}_n$ is precisely the (j, i) entry of the commutator $[X_2, X_3]$. By cyclicity, this also holds for all the other differentiations. The vanishing of the one-form therefore precisely recovers the commutator relations, and so the subset (subscheme) $\mathrm{Crit}(\widetilde{W}_n)$ of N_n is exactly M_n. Finally, $\tilde{W}_n$ is $\mathrm{GL}(V)$-invariant, so it descends to a function

$$W_n \colon \mathrm{NCHilb}_3^n \to \mathbb{C}$$

with degeneracy locus given by the quotient of the degeneracy locus on N_n, which is just $\mathrm{Hilb}^n(\mathbb{C}^3)$. We can summarize the discussion in the following result.

PROPOSITION 2.2. *The smooth noncommutative Hilbert scheme* NCHilb_3^n *carries the function* $W_n\colon \mathrm{NCHilb}_3^n \to \mathbb{C}$. *The commutative Hilbert scheme* $\mathrm{Hilb}^n(\mathbb{C}^3)$ *is the scheme-theoretic critical locus* $\mathrm{Crit}(W_n)$ *of the function* W_n *on* NCHilb_3^n, *the subscheme given by the equations* $\{\mathrm{d}W_n = 0\}$.

REMARK 2.3. Instead of considering framed triples, we can also look at the space of triples of commuting matrices

$$M_n^0 = \{(X_1, X_2, X_3)\colon [X_i, X_j] = 0\} \subset \mathrm{GL}(V)^3,$$

and its quotient $M_n^0/\mathrm{GL}(V)$. Since the action is no longer very pleasant, it is best to think of this as the stack quotient $[M_n^0/\mathrm{GL}(V)]$. An analogous argument to the above gives that this is isomorphic to the stack $\mathrm{Tors}^n(\mathbb{C}^3)$ of torsion sheaves of length n on $\mathbb{C}^3$, yet another way to parametrize n points on $\mathbb{C}^3$. This again is a degeneracy locus, now in the stacky sense.

The significance of these results is that, as I will discuss below, being a (scheme-theoretic) critical locus is exactly the right local structure for a moduli space of sheaves on a Calabi–Yau threefold; this in turn will lead us to cohomological DT theory.

2.4. Numerical DT invariants. While the method of computation is not strictly relevant to the direction we are going to take, for completeness I mention the well known result for DT invariants. Let

$$Z_{\mathbb{C}^3}(t) = 1 + \sum_{n\geq 1} \chi_{\mathrm{vir}}\left(\mathrm{Hilb}^n(\mathbb{C}^3)\right) t^n.$$

THEOREM 2.4. [**6**, **73**] *We have*

$$Z_{\mathbb{C}^3}(t) = \prod_{m\geq 1} \left(1 - (-t)^n\right)^{-n}.$$

One proof proceeds by exploiting the torus action on $\mathbb{C}^3$, which extends to each $\mathrm{Hilb}^n(\mathbb{C}^3)$ and has isolated fixed points parametrized by three-dimensional partitions of n. It is a non-trivial result of [**6**] that each fixed point contributes $(-1)^n$, leading to the above formula via combinatorics. The result can also be deduced from Theorem 6.1 below.

3. Quivers with potential and their representations

Motivated by the previous example, I will now introduce a larger class of examples of DT-like moduli spaces, coming from algebra rather than geometry. We need to start with some non-commutative algebra.

3.1. Quivers with potential. A *quiver* $Q = (V(Q), E(Q), t, h)$ consists of a finite set $V(Q)$ of *vertices*, and a finite set $E(Q)$ of *arrows* between them, with each arrow $a \in E(Q)$ having a tail $t(a) \in V(Q)$ and a head $h(a) \in V(Q)$. For now, no further conditions are assumed, so multiple arrows and arrows with head and tail at the same vertex are both allowed. We call arrows a_1, a_2 composable if $h(a_1) = t(a_2)$, and we write their composition as $a_2 a_1$. A *path* is a composition $a_n \dots a_1$ of composable arrows; there is then an obvious notion of composable paths also. A *loop* based at a vertex v is a path such that $v = h(a_n) = t(a_1)$. We allow an empty loop s_v at each vertex $v \in V(Q)$.

The path algebra $\mathbb{C}Q$ of Q is defined to be the $\mathbb{C}$-vector space generated by all paths, with the product defined on generators to be the composite of two paths if they are composable, otherwise zero. It has a semisimple subalgebra generated by the empty loops s_v at the vertices, which are idempotents summing to an identity element $1 \in \mathbb{C}Q$.

The path algebra itself, from our point of view, is too non-commutative; we wish to impose some relations. The specific relations we have in mind arise as follows. Let $\mathbb{C}Q_{\text{cycl}}$ be the vector space spanned by loops, with two loops considered equal if they can be rotated into each other by cyclic permutation. Let $W \in \mathbb{C}Q_{\text{cycl}}$ be an element of this space, a finite linear combination of loops up to cyclic permutation; we call W a *potential*.

Given an arrow $a \in E(Q)$, we define formal partial differentiation

$$\partial_a \colon \mathbb{C}Q_{\text{cycl}} \to \mathbb{C}Q$$

to be the linear map given on simple loops without repeating edges by

$$\partial_a(a_n \dots a_1) = \begin{cases} a_{i-1} \dots a_2 a_1 a_n \dots a_{i+1} & \text{if } a = a_i, \\ 0 & \text{otherwise.} \end{cases}$$

We extend ∂_a to $\mathbb{C}Q_{\text{cycl}}$ using $\mathbb{C}$-linearity and the Leibnitz rule. We then define the *Jacobi algebra* corresponding to the pair (Q, W) to be the quotient

$$J(Q, W) = \mathbb{C}Q / \langle\langle \partial_a W | \; a \in E(Q) \rangle\rangle.$$

3.2. Examples.

EXAMPLE 3.1. Let Q have one vertex and three loop edges a_1, a_2, a_3. Then $\mathbb{C}Q \cong \mathbb{C}\langle X_1, X_2, X_3 \rangle$ is the free non-commutative polynomial algebra on three variables. Let $W = a_1 a_2 a_3 - a_2 a_1 a_3$. Then it is easy to check that, in cyclic notation,

$$\partial_{a_i} W = [a_{i+1}, a_{i+2}],$$

and so $J(Q, W) \cong \mathbb{C}[x_1, x_2, x_3]$, the free *commutative* polynomial algebra on three variables. This example is clearly related to the discussion in Section 2; further connections will be revealed later.

EXAMPLE 3.2. For a substantial generalization of this example, consider the following situation. Let $\Gamma < \mathrm{SL}(3, \mathbb{C})$ be a finite abelian group of order r acting on $\mathbb{C}^3$ by linear maps of determinant 1. Without loss of generality, we can assume that Γ acts by diagonal matrices. Following McKay, attach a quiver to $\Gamma < \mathrm{SL}(3, \mathbb{C})$ as follows.

(1) Let the vertex set $V(Q)$ of Q be the set of irreducible representations $\{\rho_0, \dots, \rho_{r-1}\}$ of Γ.
(2) The diagonal embedding $\Gamma < \mathrm{SL}(3, \mathbb{C})$ corresponds to the existence of indices d_1, d_2, d_3 such that the action of Γ on $\mathbb{C}^3$ is via
$$\mathbb{C}^3 \simeq V = \rho_{d_1} \oplus \rho_{d_2} \oplus \rho_{d_3}.$$
(3) Set the numbers of arrows in Q from ρ_i from ρ_j to be $\dim \mathrm{Hom}_\Gamma(\rho_i, \rho_j \otimes V)$. In other words, the number of arrows from ρ_i from ρ_j counts the number of copies of ρ_i in $\rho_j \otimes V$.

Note that we have a decomposition of the set of arrows

$$E(Q) = E(Q)_1 \sqcup E(Q)_2 \sqcup E(Q)_3$$

according to which summand of V a particular arrow came from. The resulting oriented graph Q is oriented three-regular, with each vertex being the head, and the tail, of exactly one arrow from each $E(Q)_i$.

To define a potential on the quiver Q, we take the potential of the previous example, and make it equivariant. In other words, let $\{a_{j3}a_{j2}a_{j1}\} \in \mathbb{C}Q_{\rm cycl}$ be the set of all loops in Q up to cyclic permutation such that $a_{ji} \in E(Q)_i$; let $\{b_{k3}b_{k1}b_{k2}\} \in \mathbb{C}Q_{\rm cycl}$ be the set of all loops in Q up to cyclic permutation with $b_{ki} \in E(Q)_i$. Now set

$$W = \sum_j a_{j3}a_{j2}a_{j1} - \sum_k b_{k3}b_{k1}b_{k2} \in \mathbb{C}Q_{\rm cycl}.$$

PROPOSITION 3.3. *The Jacobi algebra $J(Q,W)$ defined by $\Gamma < \mathrm{SL}(3,\mathbb{C})$ is isomorphic to the* twisted coordinate ring $\mathbb{C}[V]\star\mathbb{C}[\Gamma]$, *where elements of* $\mathbb{C}[\Gamma]$ *commute past polynomials in* $\mathbb{C}[V]$ *using the* Γ*-action on* V. *This is a non-commutative algebra which is finite as a module over its centre* $Z(J(Q,W)) \cong \mathbb{C}[V]^\Gamma$, *the invariant ring.*

REMARK 3.4. If Γ is a nonabelian subgroup of $\mathrm{SL}(3,\mathbb{C})$, the definition of the McKay quiver remains the same, and much of the discussion generalizes; the obtained Jacobi algebra will still be Morita equivalent, though not necessarily isomorphic, to the twisted coordinate ring [**44**, Theorem 4.4.6].

EXAMPLE 3.5. Let Q have two vertices $V(Q) = \{1,2\}$ and four arrows, with a_1, a_2 from 1 to 2, and b_1, b_2 from 2 to 1. Following [**65**], set $W = a_1b_1a_2b_2 - a_1b_2a_2b_1$.

PROPOSITION 3.6. *The Jacobi algebra $J(Q,W)$ is a non-commutative algebra which is finite as a module over its centre*

$$Z(J(Q,W)) \cong \mathbb{C}[x_{11},x_{12},x_{21},x_{22}]/\langle x_{11}x_{22} - x_{12}x_{21}\rangle,$$

with $x_{ij} = a_ib_j + b_ja_i$.

The following common features of the examples above stand out.

(1) The non-commutative algebra $J(Q,W)$ is module-finite over its centre $Z(J(Q,W))$.
(2) The centre $Z(J(Q,W))$ is isomorphic to the coordinate ring of an *affine three-dimensional Gorenstein singularity*, in other words a local singular Calabi–Yau threefold. In the first case, this threefold is $\mathbb{C}^3$; in the second case, it is the quotient $\mathbb{C}^3/\Gamma$; in the last case, the threefold ordinary double point singularity $\{x_{11}x_{22} - x_{12}x_{21}\} \subset \mathbb{C}^4$.
(3) The algebra $J(Q,W)$ is a 3-Calabi–Yau algebra [**44**], in particular satisfying the conditions (a)-(b) of Section 1.4 (2).

If properties (1)-(3) hold, then $J(Q,W)$ is called a *non-commutative crepant resolution* [**100**] of the singular threefold $\bar{Y} = \operatorname{Spec} Z(J(Q,W))$. Properties (1)-(3) hold more generally in case (Q,W) is constructed from a toric Gorenstein threefold singularity $\bar{Y}$, as proved in [**17**, Thm. 8.6]. Given a general (Q,W), some of the properties can fail. For further details of aspects of the general theory, see for example [**9, 23, 76**].

REMARK 3.7. While quiver algebras have been studied extensively in mathematics, the precise construction explained in this section originated in supersymmetric physics, in studies aiming to understand string theory on D-branes in Calabi–Yau threefolds in the language of gauge theory. While the literature is too vast for us to survey here, important papers include [**8, 37, 38, 43, 49**]. The latter papers develop, in the toric case, a dual combinatorial construction called the dimer model [**17**]. The combinatorics of dimer models plays a role in computing numerical DT invariants on the corresponding toric models [**76, 97**].

3.3. Representations of quivers with potential. Let us continue to work with a general (Q, W), whether or not properties (1)-(3) of Section 3.2 hold. Given a *dimension vector* $\mathbf{d} = (d_i) \in \mathbb{N}^{V(Q)}$, a *representation of the quiver* Q is a collection of vector spaces $\{U_i, i \in V(Q)\}$ attached to the vertices with $\dim U_i = d_i$, and linear maps $\{\phi_a\colon U_{t(a)} \to U_{h(a)}, a \in E(Q)\}$ attached to the arrows. There is an obvious notion of morphisms, and thus isomorphisms, of representations of Q. Fixing $\mathbf{d} \in \mathbb{N}^{V(Q)}$ and the set $\{U_i\}$, the space of all possible representations is

$$N_{\mathbf{d}} = \prod_{a \in E(Q)} \operatorname{Hom}(U_{t(a)}, U_{h(a)}),$$

which is acted on by the group

$$G_{\mathbf{d}} = \prod_{i \in V(Q)} \operatorname{GL}(U_i).$$

The quotient stack $\mathcal{R}_{\mathbf{d}}(W) = [N_{\mathbf{d}}/G_{\mathbf{d}}]$ parametrizes isomorphism classes of representations of Q of dimension vector $\mathbf{d}$. We can think of such representations alternatively as (finite-dimensional) representations of the algebra $\mathbb{C}Q$, and it is easy to see that all finite-dimensional representations arise in this way.

It is not difficult to put the potential W into the picture. Because W is a linear combination of loops up to cyclic rotation, its trace $\operatorname{Tr} W$ can be unambiguously evaluated on any representation. The partial derivatives $\partial_a W$ consist of a linear combination of paths, and hence they can also be evaluated on any representation of Q. Just as before, requiring these combinations to vanish gives the degeneracy locus of the function $\operatorname{Tr} W$. On the other hand, such representations are precisely the representations of the Jacobi algebra $J(Q, W)$. So set

$$M_{\mathbf{d}} = \{(\phi_e) \in N_{\mathbf{d}}\colon \partial_a(W)(\phi_e) = 0 \text{ for all } a \in E(Q)\} \subset N_{\mathbf{d}}.$$

The quotient stack $\mathcal{R}_{\mathbf{d}}(W, Q) = [M_{\mathbf{d}}/G_{\mathbf{d}}]$ then parametrizes isomorphism classes of representations of $J(Q, W)$ of dimension vector $\mathbf{d}$.

Finally, if one prefers spaces to stacks, one way to go is to fix a framing vertex $j \in V(Q)$ and consider representations with a cyclic vector $v \in U_j$, meaning that the images of v under all possible combinations of ϕ_a span $\oplus_i U_i$. Thus set

$$N_{\mathbf{d}}^j = \{((\phi_a), v)\colon v \text{ is a cyclic vector for } (\phi_a)\} \subset \prod_{a \in E(Q)} \operatorname{Hom}(U_{t(a)}, U_{h(a)}) \times U_j$$

to be the space of linear maps with choice of framing vector, and let $M_{\mathbf{d}}^j$ be the subspace of representations satisfying the relations $\partial_a(W)(\phi_e) = 0$ for all $a \in E(Q)$.

PROPOSITION 3.8. (1) *The stack* $\mathcal{R}_{\mathbf{d}}(Q) = [N_{\mathbf{d}}/G_{\mathbf{d}}]$ *is a non-singular stack, equipped with a regular function* $W_{\mathbf{d}} = \operatorname{Tr} W\colon \mathcal{R}_{\mathbf{d}}(W) \to \mathbb{C}$ *given* *by evaluating the trace of W in a given representation. Its stack-theoretic critical locus is exactly $\mathcal{R}_{\mathbf{d}}(W,Q)$.*

(2) *Given a framing vertex $j \in V(Q)$, the quotient $\mathcal{R}^j_{\mathbf{d}}(Q) = N^j_{\mathbf{d}}/G_{\mathbf{d}}$ is a non-singular quasiprojective variety, equipped with a regular function $W^j_{\mathbf{d}} = \operatorname{Tr} W\colon \mathcal{R}^j_{\mathbf{d}}(Q) \to \mathbb{C}$. The scheme-theoretic critical locus of $W^j_{\mathbf{d}}$ is the subscheme $\mathcal{R}^j_{\mathbf{d}}(W,Q) = M^j_{\mathbf{d}}/G_{\mathbf{d}}$ of $\mathcal{R}^j_{\mathbf{d}}(Q)$.*

REMARK 3.9. Choosing a framing vertex and considering cyclic representations is a special case of a more general construction, that of choosing a stability parameter and considering stable representations. For quivers, the right notion of stability is King stability [**63**]. Fixing a dimension vector $\mathbf{d}$, there is a space of (King) stability parameters

$$H_{\mathbf{d}} = \{\theta \in \mathbb{Z}^{V_Q} : \theta \cdot \mathbf{d} = 0\} \subset \mathbb{Z}^{V(Q)}.$$

Choosing a generic $\theta \in H_{\mathbf{d}}$ we get the moduli spaces $\mathcal{R}^\theta_{\mathbf{d}}(Q) \supset \mathcal{R}^\theta_{\mathbf{d}}(Q,W)$ of θ-stable representations of $\mathbb{C}Q$, respectively $J(Q,W)$; both are quasi-projective GIT quotients, the former non-singular. Since we will not need the details of this construction, we will not recall them and refer instead to [**63**]. The framing case is when the θ-weight of the vertex j is chosen to be large positive, and all the other weights sufficiently negative.

We thus obtain a large supply of moduli spaces with the property that they arise as critical loci of regular functions on smooth schemes or stacks. In the examples of the previous section, these spaces are sometimes familiar objects.

(1) For the three-loop quiver of Example 3.1 and dimension vector $\mathbf{d} = (n)$, as the framed moduli space we recover the Hilbert scheme of n points on $\mathbb{C}^3$ from Section 2. The stack of unframed representations is the stack of torsion sheaves.
(2) For the McKay quiver of Example 3.2, from a finite abelian subgroup $\Gamma \in \mathrm{SL}(3,\mathbb{C})$, choose the framing vertex to be at the trivial representation ρ_0. Then for dimension vector $\mathbf{d} = (1,\dots,1)$, the space $\mathcal{R}^{\rho_0}_{\mathbf{d}}(Q,W)$ is Nakamura's Γ-Hilbert scheme [**82**]. More general dimension vectors lead to equivariant (or stacky) Hilbert schemes of the quotient stack $[\mathbb{C}^3/\Gamma]$. Keeping $\mathbf{d} = (1,\dots,1)$ but changing the stability condition leads to various moduli spaces of Γ-constellations [**22**].
(3) Finally in the conifold case of Example 3.5, for certain specific stability parameters θ the spaces $\mathcal{R}^\theta_{\mathbf{d}}(W,Q)$ can be identified [**79**] with geometric DT and PT (pairs) moduli spaces [**73, 87**] of the resolved conifold geometry $Y = \mathcal{O}_{\mathbb{P}^1}(-1,-1)$.

4. Local and global structure of the moduli space

Let us turn to the question in what generality the structure found in Section 2 and generalized algebraically in Section 3 can be expected to hold in geometric cases. Fix a projective Calabi–Yau threefold Y, and let $\mathcal{M}$ be a moduli space of simple coherent sheaves on Y.

4.1. Local structure. It is now known that, locally, the space $\mathcal{M}$ is a critical locus. Here is the precise result.

THEOREM 4.1. *Let $\mathcal{M}$ be a moduli space of simple coherent sheaves on the projective Calabi–Yau threefold Y. Let $p \in \mathcal{M}$ be a closed point. Then there exists a quadruple (R, U, f, i) with the following properties:*

- *$p \in R$ is a Zariski open neighbourhood of p in $\mathcal{M}$;*
- *$f \colon U \to \mathbb{C}$ is a regular function on a smooth scheme U;*
- *$i \colon R \to U$ is an embedding, such that $i(R) = \mathrm{Crit}(f)$ as subschemes of U.*

This is a hard theorem. In a gauge-theoretic, infinite-dimensional setup, it has been known and exploited for some time that the moduli space of holomorphic bundles on a Calabi–Yau threefold can be described by the critical points of the Chern–Simons functional on the space of connections [**36**, **101**]. While it is possible to cut this result down to a finite-dimensional model [**57**], the result one obtains is necessarily weaker (non-algebraic). A finite-dimensional algebraic result which was weaker in a different way has been known for some time, due to Behrend [**3**]: the moduli space is the zero-locus of a so-called *almost closed one-form*. The full statement above follows from combining two difficult results. The first one states that the moduli scheme $\mathcal{M}$ has the structure of a *smooth derived scheme with a (-1)-shifted symplectic structure* [**89**]; I refer to the original source for the definition of all these terms. The second result is that a smooth (-1)-shifted symplectic derived scheme is locally of the stated form [**11**, **12**].

Theorem 4.1 continues to hold if instead of simple coherent sheaves, we consider simple complexes, objects of the category $\mathcal{D}^b(\mathrm{Coh}\, Y)$. There is also an extension to stacks, for which we refer to [**2**]. Two variants of this result that would be useful, which however do not appear to be in the literature, is the existence of a (-1)-shifted symplectic structure on $\mathcal{M}$ in case

- Y is only quasiprojective, but we restrict the objects parametrized so that $\mathcal{M}$ is still of finite type;
- $\mathcal{M}$ is a moduli of objects in a 3-Calabi–Yau category $\mathcal{C}$, perhaps with some extra properties.

See [**18**, Thm.7.3.2] for a partial result in the first case.

It appears to be a difficult problem to find explicit local models (R, U, f, i) for specific geometric situations.

EXAMPLE 4.2. A set of examples well studied in the literature [**1**, **59**] where explicit potentials can be written down is the case of (ideal sheaves of) rational curves $C \cong \mathbb{P}^1$ in (local) Calabi–Yau threefolds Y. Here is one example from [**1**, Sect. 5.3]. Start with the Calabi–Yau threefold

$$Y = \mathbb{C}^3_{x,y_1,y_2} \cup \mathbb{C}^3_{w,z_1,z_2}$$

glued from affine charts using the gluing map

$$\left\{ \begin{array}{ccc} w & = & x^{-1} \\ z_1 & = & x^3 y_1 + y_2^2 \\ z_2 & = & x^{-1} y_2 \end{array} \right\}.$$

This threefold contains the curve $C = \mathbb{A}^1_x \cup \mathbb{A}^1_w \cong \mathbb{P}^1$, with usual gluing map $w = x^{-1}$, given in the affine charts by $y_i = 0$, respectively $z_i = 0$. The normal bundle is $\mathcal{N}_C = \mathcal{O}_{\mathbb{P}^1}(-3, 1)$, as seen from the first terms on the right hand side. A general section of $\mathcal{N}_C$, giving a deformation of the curve, is given in the first affine chart by the equations $y_1 = 0$, $y_2 = a + bx$, where a, b are coordinates on the two-dimensional first order tangent space. The deformation space R of the curve

C is given by $R = \mathrm{Crit}(f)$ for $f(a,b) = ab^2$ on $U = \mathbb{C}^2_{a,b}$. It is in fact easy to write down the whole one-dimensional family of curves corresponding to the reduced part $\mathbb{A}^1_a$ of R; just consider the map

$$\begin{array}{rcl} \phi\colon \mathbb{A}^1_a \times \mathbb{P}^1 & \to & Y = \mathbb{C}^3_{x,y_1,y_2} \cup \mathbb{C}^3_{w,z_1,z_2} \\ (a,x) & \mapsto & (x,0,a) \\ (a,w) & \mapsto & (w,a^2,aw). \end{array} \tag{4.1}$$

For $a = 0$, we have our original $(-3,1)$-curve $C_0 = C$, whereas for $a \neq 0$, we have a $(-2,0)$-curve C_a moving locally in a smooth one-parameter family.

If one changes the gluing data of Y to

$$\left\{ \begin{array}{rcl} w & = & x^{-1} \\ z_1 & = & x^3y_1 + y_2^2 + x^2y_2^{2n+1} \\ z_2 & = & x^{-1}y_2 \end{array} \right\},$$

then one gets the famous Laufer example [**68**] of a non-deforming, contractible $(1,-3)$-curve $C \subset Y$, whose deformation space R is the non-reduced, isolated critical locus $\mathrm{Crit}(f)$ of the function f on $U = \mathbb{C}^2_{a,b}$ given by $f(a,b) = ab^2 + \frac{a^{2n+2}}{2n+2}$.

4.2. Global structure. We call a quadruple (R,U,f,i) as in Theorem 4.1 a *critical chart* at $p \in \mathcal{M}$. We thus know that the moduli space $\mathcal{M}$ is covered by critical charts. The question is what is the right "minimal" global structure which allows us to choose a compatible set of critical charts.

This problem was solved by Joyce in [**56**], where he introduced the notion of a *d-critical locus*, a pair (X,s), where X is a scheme, and s is the section of a certain sheaf of vector spaces on X. A d-critical locus is, locally in the Zariski topology, covered by critical charts (R_j,U_j,f_j,i_j). The section s essentially remembers f_j up to second order on its critical locus $i_j(R_j)$, and serves to glue the various charts consistently. I refer to [**56**] for further details. A d-critical locus has a well-defined *d-canonical line bundle* $K_{X,s}$, living on the reduced scheme X^{red}, with the property that for every critical chart (R_j,U_j,f_j,i_j), there is a natural isomorphism

$$K_{X,s}|_{R_{j,\mathrm{red}}} \cong i_j^*(K_{U_j}^{\otimes 2})|_{R_{j,\mathrm{red}}}.$$

These isomorphisms are local, and are consequences of the fact that X is cut out of in each local U_j by a section of Ω_{U_j}. An *orientation* of the d-critical locus (X,s) is a global choice of square root line bundle $K^{1/2}_{X,s}$ of the d-canonical line bundle $K_{X,s}$. We then have

THEOREM 4.3. *Let $\mathcal{M}$ be a moduli space of simple coherent sheaves on a projective Calabi–Yau threefold Y. Then $\mathcal{M}$ obtains a d-critical locus structure from the moduli interpretation, which can be oriented.*

The existence of the d-critical structure follows from [**12**], whereas the existence of square roots of the canonical bundle is proved in [**52, 83**].

EXAMPLE 4.4. Joyce [**56**, Ex.2.39] gives an example of a d-critical locus which cannot arise as a moduli space of simple sheaves or objects on a projective Calabi–Yau threefold, since it does not admit an orientation. Let X be the non-reduced scheme with $X_{\mathrm{red}} = \mathbb{P}^1$ containing a single embedded point $p \in \mathbb{P}^1$. Let $f\colon \mathbb{C}^2 \to \mathbb{C}$ be given by $f(x,y) = x^2y$. This defines a critical chart $(\mathrm{Crit}(f), \mathbb{C}^2, f, i)$ on X near p. Away from p, we can take the trivial critical chart $(\mathbb{A}^1, \mathbb{A}^1, 0, j)$, with

$j\colon \mathbb{A}^1 \cong \mathbb{P}^1 \setminus \{p\}$. We obtain a d-critical locus structure (X, s) on X with $K_{X,s}|_{X_{\mathrm{red}}} \cong \mathcal{O}_{\mathbb{P}^1}(-5)$, hence non-orientable.

Note that the critical chart $(\mathrm{Crit}(f), \mathbb{C}^2, f, i)$ itself was shown to be a moduli of curves and thus sheaves on a Calabi–Yau threefold Y in Example 4.2. In this example, it must be the case that in the compact moduli space $\mathcal{M}$ of the ideal sheaf of the curve, there is also singular behaviour over the compactifying point. It can in fact be argued directly from the geometry that this must be the case: the family (4.1) of curves cannot be completed with a smooth $(0, -2)$-curve over $\infty \in \mathcal{M}_{\mathrm{red}} \cong \mathbb{P}^1$.

An extension of Theorem 4.3 to stacks can be found in [**2**, **56**].

5. The DT sheaf and its cohomology

Suppose that $\mathcal{M}$ is an oriented d-critical locus, for example coming from a moduli problem on a Calabi–Yau threefold. As explained in Section 1.6, we are looking for a cohomology theory $\mathcal{H}^*(\mathcal{M})$ on $\mathcal{M}$, whose Euler characteristic is the DT invariant of $\mathcal{M}$ computed from the Behrend function as in (1.1).

5.1. Local construction. Suppose first that we are in the ideal situation of Sections 2-3, when the moduli scheme $\mathcal{M}$ is covered by a single critical chart $(R = \mathcal{M}, U, f, i)$. In this case, a candidate for a coefficient system on $\mathcal{M}$, and thus the relevant cohomology theory, can be read off from Behrend's paper [**3**]. Namely, he proves that in this case the value of the Behrend function $\nu_{\mathcal{M}}(p)$ at a point $p \in \mathcal{M}$ agrees with the (signed) Euler characteristic of the reduced Milnor fibre of the function $f\colon U \to \mathbb{C}$ at $p \in \mathrm{Crit}(f) = \mathcal{M}$. This in turn is known to agree with the Euler characteristic, with supports at p, of the *sheaf of vanishing cycles* of f, see Section A.2. We can take this perverse sheaf

$$\Phi = \phi_f \mathbb{Q}_U[\dim U]$$

as our coefficient system, the *DT sheaf*, on $\mathcal{M}$, and consider its (hyper)cohomology $H^*(\mathcal{M}, \Phi)$ as our *cohomological Donaldson–Thomas invariant*. For now, this is a graded vector space; it also carries the action of the monodromy endomorphism T. There is also a duality isomorphism $\Phi \cong \mathbb{D}_{\mathcal{M}}\Phi$ of Φ with its Verdier dual.

REMARK 5.1. A pair (U, f) of a smooth variety and a regular function gives rise to a plethora of cohomology theories, both algebraic and topological; see for example [**67**, Sections 4 and 7]. The cohomology of the sheaf of vanishing cycles, called *critical cohomology* by Kontsevich and Soibelman, is but one possibility. Its main advantage over all other definitions is that is is defined "on shell", in terms of data that can be described as living on the critical locus $\mathrm{Crit}(f)$ only. All other variants seem to require global data of the pair (U, f). Of course the study of pairs (U, f) has received a lot of recent attention in the form of the study of Landau–Ginzburg models; the literature is too vast for us to survey it here.

EXAMPLE 5.2. For a harmless example, assume that $\mathcal{M}$ is nonsingular (in particular reduced). We can simply take the critical chart $(R = \mathcal{M}, U = \mathcal{M}, f = 0, i = \mathrm{id}_{\mathcal{M}})$. Then $\Phi = \mathbb{Q}_{\mathcal{M}}[\dim \mathcal{M}]$ is just the constant perverse sheaf, and the cohomological DT invariant is just the (shifted) classical cohomology of $\mathcal{M}$.

EXAMPLE 5.3. Let $U = \mathbb{C}^2$, $f(x, y) = xy$. Then $R = \mathrm{Crit}(f) = p \subset \mathbb{C}^2$ is the reduced origin, and it can be checked easily that $\Phi = \mathbb{Q}_p$. This example generalizes to split quadratic forms, which always give the trivial coefficient system on the (reduced) critical locus. This then becomes the same example as the previous one, but in a different embedding. We will get rid of the dependence on the embedding in the next section.

EXAMPLE 5.4. Here is a small example with singular critical locus. Let $U = \mathbb{C}^3$ and $f(x,y,z) = xyz$. Then $R = \mathrm{Crit}(f) = \mathbb{C}_x \cup \mathbb{C}_y \cup \mathbb{C}_z$ is the union of the three coordinate axes meeting at a point $p \in R$, the origin in $\mathbb{C}^3$. The vanishing cycle sheaf $\Phi = \phi_f \mathbb{Q}_{\mathbb{C}^3}[3]$, an object in the category $\mathrm{Perv}(R)$, must have simple factors which are intersection cohomology sheaves of strata in R (see Section A.1). An easy computation shows that it has a composition series with simple factors $\mathbb{Q}_p$ (the trivial sheaf at the singular point), the intersection cohomology sheaf IC_R of the equidimensional R, and once again $\mathbb{Q}_p$.

EXAMPLE 5.5. For a second non-trivial singular example, historically one of the first to be computed in this context, let $U = \mathbb{C}^{15}$, thought of as the space of 6×6 skew-symmetric matrices. Let $f \colon U \to \mathbb{C}$ be given by the Pfaffian of the corresponding matrix. Then $R = \mathrm{Crit}(f)$ is reduced and irreducible, with a single singular point $p \in R$. It is easy to say more: the Pfaffian is a cubic function, and its partial derivatives with respect to the matrix entries are the Plücker relations defining, after projectivisation, the subvariety $\mathrm{Gr}(2,6) \subset \mathbb{P}^{14} = \mathbb{P}U$. Thus R is the affine cone over the Grassmannian, with its singular point $p \in R$ the cone point.

The space R, by virtue of its construction, carries an oriented d-critical locus structure. As proved in [**34**], this structure arises from a genuine DT problem: up to a constant factor $\mathbb{C}^3$, it describes the singular locus of the Hilbert scheme $\mathrm{Hilb}^4(\mathbb{C}^3)$, the first time the Hilbert scheme is singular. The singular points of $\mathrm{Hilb}^4(\mathbb{C}^3)$ are exactly at the colength-4 ideals $\mathbf{m}_q^2$, the squares of maximal ideals of points $q \in \mathbb{C}^3$. This locus is a copy of $\mathbb{C}^3$, and the local structure is exactly given by the cone over the Grassmannian [**58**].

The DT sheaf Φ can also be analyzed in very concrete terms. It turns out to be non-semisimple, once again with a composition series with simple factors $\mathbb{Q}_p$ (the trivial sheaf at the singular point), the intersection cohomology sheaf IC_R, and $\mathbb{Q}_p$.

Let us finally discuss the simplest stacky situation, which will be important later. Let $\mathcal{M} = [M/G]$ be a global quotient stack, with G reductive, arising as a stacky critical locus $\mathcal{M} = \mathrm{Crit}(f) \subset [N/G]$ where $f \colon N \to \mathbb{C}$ is a regular, G-invariant function. In this case, the vanishing cycle sheaf $\Phi = \phi_f \mathbb{Q}_N[\dim N]$ becomes an equivariant object [**46**] on M, and we can take its equivariant cohomology $H^*_G(M,\Phi)$ [**67**] as our definition of the cohomological DT invariant of $\mathcal{M}$. It is naturally a module over $H^*_G(\mathrm{pt},\mathbb{Q})$.

REMARK 5.6. As explained in Section 1.6, the next level of categorification would involve building a category $\mathcal{C}(\mathcal{M})$ out of the moduli space $\mathcal{M}$. When $\mathcal{M}$ is covered by a single critical chart $(R = \mathcal{M}, U, f, i)$, the relevant category is likely to be some version of the *category of matrix factorizations* [**70, 84**] $\mathrm{MF}(U,f)$ of f on U. The periodic cyclic homology of this (dg) category is known to be [**41**] isomorphic to the cohomology of the vanishing cycle sheaf of f.

5.2. Global issues. Suppose that X is a scheme equipped with a d-critical locus structure, covered by a collection of critical charts $\{(R_j, U_j, f_j, i_j) : j \in J\}$. Locally, we can take the perverse sheaves $\Phi_j = \phi_{f_j} \mathbb{Q}_{U_j}[\dim U_j]$; the question is whether the d-critical structure carries enough information to glue these local Φ_j to a global sheaf Φ on X. We are certainly helped by the fact that perverse sheaves are known to glue uniquely, once compatible glueing isomorphisms are specified. As it turns out [**14**, Thm. 6.9], this almost suffices: the d-critical structure allows one to specify glueing isomorphisms *up to a sign*; an oriented d-critical structure gives a unique glueing Φ on X.

Given a moduli space $\mathcal{M}$ of simple sheaves on a projective Calabi–Yau threefold Y, by Theorem 4.3 it carries a d-critical structure and an orientation (a square root of its canonical bundle). Choosing an orientation, we can glue the local vanishing cycle sheaves to a global *DT sheaf* Φ on $\mathcal{M}$; it is important to remember that Φ depends on this further choice. Given all that though, we finally have a definition of the *cohomological DT invariant of* $\mathcal{M}$ [**14**, **62**] as the (hyper)cohomology $H^*(\mathcal{M}, \Phi)$. This is a graded vector space, equipped with the action of an endomorphism T glued from the local monodromy endomorphisms. The local duality isomorphisms also glue to an isomorphism $\Phi \cong \mathbb{D}_{\mathcal{M}}\Phi$ of Φ with its Verdier dual.

The case when $\mathcal{M}$ is a stack is covered again in [**2**]. The results contained in [**2**] allow one to define cohomological DT invariants for (finite type) moduli Artin stacks of sheaves (or complexes without negative Ext groups) on Calabi–Yau threefolds. Note that in the presence of semistables, such a stack is not isomorphic to the moduli *scheme* of semistables, so the connection of this construction to results on moduli spaces containing semistables, obtained by wall crossing [**57**], is not immediately clear.

EXAMPLE 5.7. Let Y be a projective Calabi–Yau threefold containing a divisor E which is itself a $\mathbb{P}^1$-bundle $\pi : E \to C$ over a nonsingular projective curve C of genus g. Assume also that X admits a contraction to a singular projective threefold $\bar{Y}$ which contracts E along π and is an isomorphism outside E. Let $\beta \in H_2(Y, \mathbb{Z})$ be the class of the fibre of π. Then for the lowest possible value of the Euler characteristic n, the moduli space of ideal sheaves $\mathcal{M}(Y, \beta, n)$ is isomorphic to C itself, each point of C parametrizing the ideal sheaf of the corresponding fibre of π.

This is a non-singular (reduced) moduli space, and one d-critical locus structure on it is given by the choice $(C, C, 0, \mathrm{id}_C)$, with d-canonical divisor $K_C^{\otimes 2}$. This admits the obvious square root K_C, giving the trivial coefficient system $\mathbb{Q}_C[1]$ on C, the cohomological DT invariant $H^*(C, \mathbb{Q}_C[1])$, and the numerical DT invariant $2g - 2$. Note that for $g = 0$, this numerical DT invariant is negative.

There are other choices of orientation: for any 2-torsion line bundle $L \in \mathrm{Pic}^0(C)$, we get another orientation $K_C \otimes L$. A nontrivial L corresponds to an unramified double cover $q : D \to C$. There is correspondingly a decomposition $q_*\mathbb{Q}_D = \mathbb{Q}_C \oplus \mathcal{L}$, with $\mathcal{L}$ a nontrivial rank-one local system on the curve C. With this choice of orientation data, the cohomological DT invariant becomes the cohomology $H^*(C, \mathcal{L}[1])$, a kind of Prym cohomology.

5.3. Hodge structure on cohomological DT and its purity. As explained in the Appendix, given an algebraic variety or scheme X, there is a category whose objects carry richer information than perverse sheaves on X: the category of mixed Hodge modules. In the case of a single critical chart (R, U, f, i), we can consider the mixed Hodge module

$$\phi_f \mathbb{Q}_U(\dim U/2)[\dim U] \in \mathrm{MMHM}(R)_{\frac{1}{2}}$$

to be our Hodge-theoretic DT sheaf; see Section A.3 for the definition of the category $\mathrm{MMHM}(R)_{\frac{1}{2}}$ of *monodromic* mixed Hodge modules on R and its half-twist functor, which is needed in the definition above to get duality right. Its cohomology carries a (polarized) mixed Hodge structure with monodromy action. Given a d-critical locus structure on a scheme $\mathcal{M}$, the glueing procedure described above, leading to the DT sheaf Φ, can be performed on the level of mixed Hodge modules with monodromy [**14**]. Consequently the cohomological DT invariant acquires a mixed Hodge structure too; the cohomology groups $H^i(\mathcal{M}, \Phi)$ and dually $H^i_c(\mathcal{M}, \Phi)$ become objects in the category $\mathrm{MMHS}_{\frac{1}{2}}$ of monodromic mixed Hodge structures.

EXAMPLE 5.8. Let $f\colon \mathbb{C}^n \to \mathbb{C}$ be a polynomial, with an isolated critical point at $p \in \mathbb{C}^n$ which we will assume without loss of generality to be at the origin. By shrinking $\mathbb{C}^n$ to an open set U, we can assume that for $f \in U \to \mathbb{C}$, we have $\mathrm{Crit}(f)_{\mathrm{red}} = \{p\}$ the origin. In this case, the cohomology $H^*(\phi_f \mathbb{Q}_U[\dim U])$ is essentially the cohomology of the Milnor fibre of f at p. This cohomology carries a natural Hodge structure by work of Steenbrink [**94**]. In general, this Hodge structure is mixed and not pure: it carries a non-trivial weight filtration. By another result of Steenbrink [**95**] however, the mixed Hodge structure is pure if f is *quasi-homogeneous*, that is if there is a set of integral weights on the coordinates of $\mathbb{C}^n$ which makes f homogeneous of positive weight.

A natural question in the theory is what would in general guarantee the purity of the cohomological DT invariant. Recall that in the classical case of constant coefficients, the cohomology $H^k(X, \mathbb{Q})$ of a variety X carries a mixed Hodge structure which is pure of weight k if X is smooth and projective. As mentioned above, DT-style moduli spaces $\mathcal{M}$ are smooth in the appropriate (derived) context, and so it would have perhaps been reasonable to expect that as soon as they are projective also, the cohomological DT invariant carries a pure Hodge structure. As Example 5.8 shows, this already fails when the reduced scheme underlying $\mathcal{M}$ is a single point. However, the final result quoted in the example also shows that one can hope for purity in the presence of a $\mathbb{C}^*$-action. It is also known in the literature on Landau–Ginzburg models that the presence of suitable $\mathbb{C}^*$-actions leads to stronger results than are possible in the general case. In this context, the following is the best result known.

PROPOSITION 5.9. [**29**] *Let $f\colon U \to \mathbb{C}$ be a regular function on a smooth variety U. Assume that U carries a $\mathbb{C}^*$-action, so that f is equivariant with respect to the weight d action of $\mathbb{C}^*$ on $\mathbb{C}$, with $d > 0$. Assume also that the critical locus $R = \mathrm{Crit}(f)$ of f is proper. Then the mixed Hodge structure on $H^*(R, \phi_f \mathbb{Q}_U)$ is pure.*

Note that, as before, purity has to be interpreted appropriately in the monodromic sense. This is not a difficult result given Saito's technology [**91, 92**]; it is also far from optimal. It seems reasonable to expect that if $\mathcal{M}$ is a projective scheme with a $\mathbb{C}^*$-action and a $\mathbb{C}^*$-equivariant oriented d-critical locus structure, then the cohomological DT invariant attached to $\mathcal{M}$ carries a pure Hodge structure.

EXAMPLE 5.10. To give a non-trivial example of Proposition 5.9, consider a compactification of Example 5.4: let $X = \mathbb{P}^1 \cup_p \mathbb{P}^1 \cup_p \mathbb{P}^1$ be the union of three projective lines meeting at a point $p \in R$. This scheme can be given a d-critical locus structure by using R from Example 5.4 as one of the charts, and three trivial charts in the neighbourhoods of compactifying points. In turns out however [**40**, Sect. 6] that X can even be covered by a single critical chart (R', U', f', i'), using an auxiliary quiver construction. For this chart, all conditions of Proposition 5.9 are satisfied. Indeed, $H^*(X, \Phi)$ can be computed to be isomorphic to the cohomology of two disjoint copies of $\mathbb{P}^1$, and thus indeed pure.

EXAMPLE 5.11. Another interesting example arises in the stable pairs theory of the local resolved conifold geometry $Y = \mathcal{O}_{\mathbb{P}^1}(-1,-1)$, discussed first in [**87**, Section 4.1]. For the lowest degree curve class, all moduli spaces are nonsingular. For the case of the curve class having multiplicity two, for the lowest value of the Euler characteristic the moduli space is still nonsingular. The next moduli space $\mathcal{M}$ is more interesting. The reduced variety underlying $\mathcal{M}$ is isomorphic to $\mathbb{P}^3$. As discussed in [**87**, Section 4.1] however, this cannot be the full answer: the numerical pairs invariant equals 4 and not -4, which would be the answer in case the moduli space were just a smooth $\mathbb{P}^3$. The moduli space $\mathcal{M}$ is in fact a non-reduced scheme, the thickening of $\mathbb{P}^3$ along the embedded quadric $Q \subset \mathbb{P}^3$, with Zariski tangent spaces of dimension 4 along the quadric.

The corresponding DT sheaf was computed, under an assumption about the local form of the potential, in [**98**, Ex.4.5]. Indeed, the simplest potential that gives the quoted non-reduced behaviour is of the form $f = x^2 y$ in some local variables x, y on a smooth space U. Assuming that this is indeed the local form of the potential at all points of the embedded quadric in $\mathbb{P}^3$, it can be shown that the DT sheaf Φ on $\mathcal{M}$ is the intersection cohomology sheaf $\mathrm{IC}(\mathcal{L})$ for a rank-one local system $\mathcal{L}$ on $\mathbb{P}^3 \setminus Q$ with nontrivial $\mathbb{Z}/2$ monodromy.

On the other hand, it follows from [**79**] that this space $\mathcal{M}$ is also a global degeneracy locus, and for the corresponding critical chart all conditions of Proposition 5.9 are satisfied. Indeed, $H^*(\mathcal{M}, \Phi)$ is isomorphic to the cohomology of the quadric Q, and thus pure.

5.4. Summary of properties of the cohomological DT invariant. In summary, let $\mathcal{M}$ be a scheme equipped with an oriented d-critical locus structure; let Φ be the corresponding DT (perverse) sheaf on $\mathcal{M}$. Then the cohomological DT invariant $H^*(\mathcal{M}, \Phi)$ attached to $\mathcal{M}$ is a graded vector space with the following properties and carrying the following structures.

(1) The Euler characteristic of the cohomological DT invariant
$$\sum_i (-1)^i \dim H^i(\mathcal{M}, \Phi) = \int_{\mathcal{M}} \nu_{\mathcal{M}} d\chi$$
agrees with the numerical DT invariant computed from the Behrend function $\nu_{\mathcal{M}}$ of $\mathcal{M}$.
(2) $H^*(\mathcal{M}, \Phi)$ carries the action of the monodromy endomorphism T.
(3) The self-duality of Φ under Verdier duality implies that there is a duality isomorphism between $H^*(\mathcal{M}, \Phi)$ and $H^*_c(\mathcal{M}, \Phi)$. In particular, if $\mathcal{M}$ is proper, then $H^*(\mathcal{M}, \Phi)$ is self-dual (satisfies Poincaré duality).
(4) $H^*(\mathcal{M}, \Phi)$ carries a monodromic mixed Hodge structure.
(5) If $\mathcal{M}$ is proper, and covered by a single $\mathbb{C}^*$-equivariant critical chart with positive $\mathbb{C}^*$-weight on the base, then the monodromic mixed Hodge structure on $H^*(\mathcal{M}, \Phi)$ is pure.

This in particular applies when $\mathcal{M}$ is a projective moduli scheme of stable sheaves on a smooth projective Calabi–Yau threefold Y. The cohomological, as opposed to the numerical, DT invariant is generally *not* invariant under deformations of the underlying Calabi–Yau threefold Y; see Section 8.1 below for further discussion.

6. Some computational results

6.1. The cohomological DT invariant in the K-group. It appears difficult to compute cohomological DT invariants directly. Most existing computations use an indirect route, computing on an appropriate locally closed stratification of the moduli space. Such results will give weaker answers than the computation of the full cohomology of the DT sheaf, since cohomology is not additive on stratifications. To be able to use the long exact sequence on cohomology, we also switch to compactly supported cohomology (which, by Poincaré duality, carries the same information). So given $(\mathcal{M}, \Phi)$, we set

$$[H_c^*(\mathcal{M}, \Phi)] = \sum_i (-1)^i [H_c^i(\mathcal{M}, \Phi)] \in K(\mathrm{MMHS}_{\frac{1}{2}})$$

to be the formal alternating sum of cohomologies inside the K-group of monodromic mixed Hodge structures. For this invariant, given a decomposition $\mathcal{M} = U \cup Z$ into an open subscheme and its complement, the long exact sequence in cohomology indeed gives

$$[H_c^*(\mathcal{M}, \Phi)] = [H_c^*(U, \Phi|_U)] + [H_c^*(Z, \Phi|_Z)] \in K(\mathrm{MMHS}_{\frac{1}{2}}). \tag{6.1}$$

Because of cancellations in the K-group, an element $\alpha \in K(\mathrm{MMHS}_{\frac{1}{2}})$ does not allow one to reconstruct the original complex A^* of monodromic mixed Hodge structures giving rise to it, or even the dimensions of the various pieces of cohomology. One piece of information that survives is the *weight polynomial*, defined in Section A.4, living in $\mathbb{Z}[q^{\pm\frac{1}{2}}]$. The corresponding one-variable deformation or quantization of the numerical DT invariant is often referred to as a *refined DT invariant* [**35, 54**].

On the other hand, the situation changes completely if we know that the element $\alpha = [H_c^*(\mathcal{M}, \Phi)]$ comes from $H_c^*(\mathcal{M}, \Phi)$ with *pure* Hodge structure. In that case, there can be no cancellation between different terms in (6.1), and so the knowledge of the K-theory element enables one to fully determine the dimensions of the cohomology groups $H_c^*(\mathcal{M}, \Phi)$, and hence $H^*(\mathcal{M}, \Phi)$. In this case, the weight polynomial becomes simply the Poincaré polynomial.

In the next two subsections, we give the answers in the K-group to some cohomological DT problems from earlier, and discuss what is known about the full cohomology.

6.2. The cohomological DT invariants of $\mathbb{C}^3$. Consider

$$Z_{\mathbb{C}^3}(t) = 1 + \sum_{n \geq 1} [H_c^*(\mathrm{Hilb}^n(\mathbb{C}^3), \Phi_n)] t^n \in K(\mathrm{MMHS}_{\frac{1}{2}})[[t]],$$

with Φ_n the DT sheaf on $\mathrm{Hilb}^n(\mathbb{C}^3)$, obtained from the critical locus interpretation of Proposition 2.2. This series turns out to admit an elegant infinite product form in terms of the cohomology $\mathbb{L}$ of the affine line (see Section A.4).

THEOREM 6.1. [**4**] *We have*

$$Z_{\mathbb{C}^3}(t) = \prod_{m=1}^{\infty} \prod_{k=0}^{m-1} \left(1 - \mathbb{L}^{k+2-\frac{m}{2}} t^m\right)^{-1} \in K(\mathrm{MMHS}_{\frac{1}{2}})[[t]].$$

This is the precise analogue of Göttsche's formula, which computes the cohomology of the Hilbert scheme of points of a smooth quasiprojective surface, for the space $Y = \mathbb{C}^3$. It also allows one to confirm a computation of refined DT invariants in the physics literature [**54**]. Setting $\mathbb{L}^{\frac{1}{2}} = -1$, we recover Theorem 2.4. It was very recently proved in [**25**] that the critical cohomology carries a pure Hodge structure, so this formula encodes all the cohomological DT invariants in this example.

EXAMPLE 6.2. For the case of four points, an expansion gives

$$[H_c^*(\mathrm{Hilb}^4(\mathbb{C}^3), \Phi_4)] = \mathbb{L}^6 + \mathbb{L}^5 + 3\mathbb{L}^3 + 3\mathbb{L}^3 + 3\mathbb{L}^2 + \mathbb{L} + 1 \in K(\mathrm{MMHS}_{\frac{1}{2}}).$$

This result was proved earlier in [**34**].

6.3. Other examples. Other local cases where values of cohomological DT invariants are known, at least in the K-group, include [**27, 74, 75**]; the last reference gives a computation which involves nontrivial monodromy on the DT invariant. I spell out just one more example. Let $Y = \mathcal{O}_{\mathbb{P}^1}(-1,-1)$ be the resolved conifold geometry, containing the zero-section $Y \supset C \cong \mathbb{P}^1$. Let $\mathcal{N}_{d[C],n}$ be the moduli space of stable pairs [**87**], supported on the curve class $d[C]$ and with Euler characteristic n, carrying the DT sheaf $\Phi_{d,n}$. Let

$$Z_Y(t,T) = \sum_{n,d} \left[H^*\left(\mathcal{N}_{d[C],n}, \Phi_{d,n}\right)\right] T^d t^n.$$

THEOREM 6.3. [**74**] *We have*

$$Z_Y(t,T) = \prod_{m\geq 1} \prod_{j=0}^{m-1} \left(1 - \mathbb{L}^{-\frac{m}{2}+\frac{1}{2}+j} t^m T\right).$$

In this case, it is known from the quiver description of these moduli spaces [**79**] that Proposition 5.9 applies, and the underlying cohomology is pure; therefore this result gives a full computation of the cohomology. The cohomological DT invariant of Example 5.11 can be read off from expanding this series.

7. The Kontsevich–Soibelman cohomological Hall algebra

One of the reasons to be interested in a cohomological extension of Donaldson–Thomas theory is that since this invariant takes values in a category rather than a ring, one can define maps between different cohomology groups. Indeed, starting from a quiver with potential, Kontsevich–Soibelman [**67**] showed how one can define an associative algebra structure on cohomological DT. This construction can be viewed as a precise mathematical realization of the Harvey–Moore algebra of BPS states [**50**].

7.1. Quivers without potential. Let us first cover the case when we start with a quiver Q with potential $W = 0$. As in Section 3.3, given a dimension vector $\mathbf{d} = (d_i) \in \mathbb{N}^{V(Q)}$, we fix vector spaces $\{U_i\}$ of the right dimension. The space of all representations

$$N_{\mathbf{d}} = \prod_{a\in E(Q)} \mathrm{Hom}(U_{t(a)}, U_{h(a)}),$$

is acted on by the group

$$G_{\mathbf{d}} = \prod_{i \in V(Q)} \mathrm{GL}(U_i).$$

We get the quotient stack $\mathcal{R}_{\mathbf{d}}(W) = [N_{\mathbf{d}}/G_{\mathbf{d}}]$ and its cohomology $H^*_{G_{\mathbf{d}}}(N_{\mathbf{d}}, \mathbb{Q})$. Let

$$\mathcal{H}_Q = \bigoplus_{\mathbf{d} \in \mathbb{N}^{V(Q)}} H^*_{G_{\mathbf{d}}}(N_{\mathbf{d}}, \mathbb{Q}).$$

THEOREM 7.1. [**67**] *The vector space $\mathcal{H}_{\mathbb{Q}}$ admits a product operation, under which it becomes an associative unital $\mathbb{N}^{V(Q)}$-graded algebra.*

We refer to the original paper for the definition of the product, which essentially arises from the standard correspondence involving subrepresentations and quotient representations, familiar from Hall algebra theory. Several examples are computed in [**67**]. For example, it is clear that when Q has only one vertex, $\mathcal{H}_Q$ is isomorphic as a vector space to $\oplus_{n>0} H^*(\mathrm{BGL}(n, \mathbb{C})))$. The algebra structure depends on the number of loops; for the case of no loop, respectively one loop, the algebra $\mathcal{H}_Q$ is isomorphic [**67**, Sect. 2.5] to the exterior, respectively symmetric algebra on infinitely many generators.

A quiver Q is called *symmetric* if for any $i, j \in V(Q)$ it has the same number of oriented edges from vertex i to vertex j as from vertex j to vertex i. In the special case when Q is symmetric, it is possible to use the cohomological grading, together with a slight twist of the multiplication, to endow $\mathcal{H}_Q$ with a $\mathbb{N}^{V(Q)} \times \mathbb{Z}$-graded supercommutative structure. It was conjectured in [**67**], and proved in [**39**], that in fact this algebra is *free* supercommutative.

REMARK 7.2. It is important to note that the COHA itself only depends on a quiver (possibly with potential, see next section), not on any notion of stability. Choosing a stability function gives a decomposition (involving a spectral sequence) of the COHA into an ordered product of more elementary algebras, similar to a decomposition for universal enveloping algebras which comes from the Poincaré–Birkhoff–Witt theorem. For details, see [**67**, Section 5].

7.2. Inserting the potential. Let us equip Q with a nonzero potential W. As in Proposition 3.8, we get the function $W_{\mathbf{d}} \colon N_{\mathbf{d}} \to \mathbb{C}$, its critical locus $M_{\mathbf{d}} = \mathrm{Crit}(W_{\mathbf{d}})$, and the stack $[M_{\mathbf{d}}/G_{\mathbf{d}}]$ of representations of the Jacobi algebra $J(Q, W)$. We set

$$\mathcal{H}_{Q,W} = \bigoplus_{\mathbf{d} \in \mathbb{N}^{V(Q)}} H^*_{G_{\mathbf{d}}}(M_{\mathbf{d}}, \phi_{W_{\mathbf{d}}} \mathbb{Q}_{N_{\mathbf{d}}}[\dim N_{\mathbf{d}} - \dim G_{\mathbf{d}}]).$$

Once again, this becomes [**24**, **67**] a graded associative algebra.

The structure of this algebra was studied further recently in [**28**], giving in particular an algebraic meaning to the BPS invariants $\Omega^\sigma(\alpha)$ of Section 1.5 in these quiver cases. There are essentially no cases where the structure of this algebra is fully known in concrete terms, though the dimensional reduction technique of [**4**], adapted to the COHA in [**24**, **67**], should allow interesting computations.

7.3. Representations of the COHA. One of the most exciting potential directions is to find representations of the COHA associated to a quiver (Q, W), and related algebras, on the cohomological DT invariants attached to *stable* representations of (Q, W) under appropriate stability conditions [**96**, **98**].

EXAMPLE 7.3. Let (Q, W) be the three-loop quiver with potential from Example 3.1. Then we have the cohomological Hall algebra $\mathcal{H}_{Q,W}$, which should have an action on the vector space

$$\mathcal{A} = \bigoplus_{n\geq 0} H^*(\mathrm{Hilb}^n(\mathbb{C}^3), \Phi_n).$$

Here Φ_n is the DT sheaf on $\mathrm{Hilb}^n(\mathbb{C}^3)$ obtained from the critical locus interpretation of Proposition 2.2. More precisely, there should be *two* actions, corresponding to adding and subtracting a representation, giving "creation" and "annihilation" operators. The resulting action of a double algebra $\widetilde{\mathcal{H}}_{Q,W}$ would be the three-dimensional analogue of the Grojnowski–Nakajima action [**48, 80**] of an infinite-dimensional Heisenberg algebra on $\bigoplus_{n\geq 0} H^*(\mathrm{Hilb}^n(\mathbb{C}^2), \mathbb{Q})$. Theorem 6.1 could in turn be interpreted as a (refined) character formula for this action.

For the much simpler example of the A_1 quiver, the double COHA was explicitly computed recently in [**102**].

7.4. The global case. Constructing algebra structures on globally defined cohomological DT spaces, involving moduli spaces covered by several critical charts, appears to be a very interesting challenge. Essentially nothing is known in this direction.

8. Further directions

Without giving all, or indeed sometimes any, details, and without aiming for completeness, I discuss some further research directions related to cohomological Donaldson–Thomas theory.

8.1. Deformation (in)variance. As explained in Section 1, one of the properties of numerical DT invariants is that they are constant in a smooth deformation family of projective Calabi–Yau threefolds. One can ask more generally about deformation properties of other flavours of DT invariants. For the case of the Hilbert scheme of points of a general threefold, the result of [**4**] shows that the motivic invariants only depend, additionally, on the motive of the underlying threefold Y, as one would minimally expect. Thus, for example, the weight polynomial refinement of DT remains constant in a smooth deformation family, since it depends only on the weight polynomial of Y, which is constant in a family. More recently, [**60**] observed (conjectural) constancy of the weight polynomial realization of motivic stable pairs theory on thickened K3 surfaces.

Refined DT invariants do not always need to be deformation invariant however.

EXAMPLE 8.1. Let us return to the example studied in Example 5.7, with Y a projective Calabi–Yau threefold containing a divisor E which is itself a $\mathbb{P}^1$-bundle $\pi : E \to C$ over a curve C of genus g. It is well known that for $g > 0$, the generic deformation Y_t of $Y = Y_0$ contains $2g - 2$ isolated rational curves in the homology class of the fibre of the fibration π. Thus the moduli space deforms from $\mathcal{M}_0 = C$ to $\mathcal{M}_t$ being a disjoint union of $2g - 2$ reduced points. Clearly the numerical DT invariant $2g - 2 = (-1)^{\dim C}\chi(C)$ is unchanged, but the weight polynomial, using *trivial* orientation data, changes from the (shifted) Poincaré polynomial of C to the constant polynomial $2g - 2$ and thus jumps. Interestingly, it can be readily checked that with any choice of *nontrivial* (Prym) orientation data, the weight polynomial (though not the Hodge polynomial) remains *unchanged*.

In a different direction, a jump in refined DT invariants can also be observed in a local situation, under homogeneous changes of potentials on certain simple quivers [**19**].

8.2. Knot invariants. Start with a singular curve C with planar singularities. One can then form the generating series of (ordinary, topological) Euler characteristics of the Hilbert schemes of points $\mathrm{Hilb}^n(C)$ of C. This series, and various refinements, were conjectured by Oblomkov–(Rasmussen–)Shende [**85, 86**] to be given by formulae derived from the HOMFLY(PT) invariants of the links around the different singular points. These connections were later re-phrased and generalized [**31, 32**] via embedding C in a local Calabi–Yau threefold, and studying certain DT invariants of this threefold. Indeed, Maulik [**71**] recently proved the original conjectures of [**85**] using wall crossing in DT theory (and other tools). Cohomological DT invariants in this story may be related to categorifications of knot invariants such as Khovanov homology.

8.3. Quantum cluster positivity. Starting with a quiver Q and a fixed vertex $j \in V(Q)$, Fomin–Zelevinsky [**42**] introduced a combinatorial operation called *mutation*, leading to another quiver $\mu_j(Q)$. This is an operation that can then be repeated on other vertices, usually leading to an infinite number of other quivers mutation equivalent to Q. The mutation operation also associates certain Laurent polynomials to vertices of all these quivers, related by the so-called cluster transformation rule. It was a conjecture of Fomin–Zelevinsky, recently proved in [**69**], that all these Laurent polynomials have positive coefficients.

The cluster mutation story has two extensions relevant for us. In one direction, [**30**] proved that mutation extends under certain conditions to an operation on quivers with potential. In [**61**] it was shown that this gives rise to very natural derived equivalences between 3-Calabi–Yau (dg) categories. Kontsevich–Soibelman [**66**] and Nagao [**78**] then showed that wall crossing of DT invariants of these categories reproduces a lot of the structure of the cluster polynomials.

In another direction, Berenstein–Zelevinsky [**7**] introduced a quantization of the cluster story, introducing a version of the cluster transformation rule in a mildly non-commutative, quantized Laurent polynomial ring. This allows one to formulate a quantum cluster positivity conjecture, saying that even in the quantized setting, all coefficients in the Laurent polynomials are positive.

The two lines of thought were brought together in work of Efimov [**40**], who showed that certain very specific motivic DT invariants exactly reproduce the cluster transformations of *quantized* Laurent polynomials. As noticed in that paper, a purity result on *cohomological* DT invariants would give positivity for quantum cluster polynomials. By Proposition 5.9 above, purity holds in certain $\mathbb{C}^*$-equivalent situations, leading to the result [**29**] that quantum cluster positivity is true for quivers which can be equipped with a suitably non-degenerate *quasi-homogeneous* potential. The proof of the quantum cluster positivity conjecture in complete generality, relying still on cohomological Donaldson-Thomas theory, was very recently announced in [**26**].

8.4. Localization. Since the seminal paper [**47**], many computations involving virtual fundamental classes were performed in the presence of a high degree of (torus) symmetry, using localization. In the case of symmetric obstruction theories, localization is even more effective [**6**]; as mentioned, Theorem 2.4 was first proved using torus localization. Away from the numerical case however, localization has so far proven much less effective. Many refined DT results, starting with [**12**], have used a dimensional reduction trick, which again relies on a certain kind of torus action; but this does not reduce the computation to torus-fixed points directly. Recently, there has been progress on this problem: Maulik [**72**] introduced a framework which can, at least in theory, compute refined DT invariants via localization in substantial examples, such as the case investigated earlier using a hands-on definition in [**20**]. It would be interesting to generalize this work to cohomological DT invariants, where, as for intersection cohomology [**64**], hyperbolic localization [**13**] is likely to play an important role.

8.5. K-theoretic DT invariants. A radically new point of view on refinements of DT theory was presented recently by Nekrasov and Okounkov in [**83**]. This work defines curve and sheaf counting on Calabi–Yau threefolds with values in K-theory, taking the K-theory class of the *virtual structure sheaf* on the relevant moduli stack of sheaves. This construction is a holomorphic variant of the constructible DT-sheaf discussed in Section 5. Moreover, [**83**] also discusses a conjectural, not fully constructed theory of curve- (more precisely brane-)counting on a Calabi–Yau fivefold Z, which connects to K-theoretic DT invariants on Calabi–Yau threefolds arising as torus-fixed loci inside the fivefold Z. Maulik's work [**72**], in turn, connects the K-theoretic and motivic DT invariants under some (currently strong) hypotheses. The relationships between these various flavours of DT theory, as well as connections to geometric engineering, definitely deserve further study.

8.6. DT theory in higher dimensions. Following on from one theme of the previous subsection, one may well wonder whether a sensible theory of sheaf counting exists on Calabi–Yau d-folds of arbitrary dimension greater than 3. One hint that such a theory might exist is that the product form of the generating series of motivic DT invariants of Hilbert schemes of points of a threefold Y, presented in [**4**], admits a natural generalization for $\dim Y = d$ arbitrary, and gives the correct result both for $\dim Y < 3$ and for small numbers of points. We thus have the answer for Hilbert schemes in general, we only need to formulate the right question! More importantly perhaps, the derived geometry results of [**11, 12, 89**] are all dimension-independent: the general results are that moduli stacks of sheaves on a smooth d-dimensional projective Calabi–Yau variety carry a $(d-2)$-shifted symplectic structure, and there are normal forms for such structures generalizing the critical locus description for $d = 3$. These results were used as the starting point to explore four-dimensional DT theory in [**10**]. A related approach relying on gauge theory was presented in [**21**]. It would be interesting to compute invariants of substantial examples in either of these frameworks.

Appendix A. Perverse sheaves, vanishing cycles, and Hodge modules

In this Appendix, I recall aspects of the categories of constructible and perverse sheaves and Hodge modules on complex algebraic varieties, and the vanishing cycle construction, mainly to fix notation. Further details can be found in [**33, 90**]; I will not give original references in this section.

A.1. Constructible and perverse sheaves. Let X be a complex algebraic variety or scheme. The category $\mathrm{Sh}(X)$ of sheaves of $\mathbb{Q}$-vector spaces in the classical (complex) topology has a subcategory $\mathrm{Const}(X)$ consisting of *constructible* objects, sheaves which are local systems when restricted to different strata of an *algebraic* stratification of X. Let $\mathcal{D}_c^b(X)$ denote the derived category of bounded complexes of sheaves on X whose cohomology sheaves are in $\mathrm{Const}(X)$.

If $f\colon X \to Y$ is a morphism of schemes, then we have pushforward functors $f_*, f_!\colon \mathcal{D}_c^b(X) \to \mathcal{D}_c^b(Y)$, the direct image and direct image with compact supports functors. There are also pullback functors $f^*, f^!\colon \mathcal{D}_c^b(Y) \to \mathcal{D}_c^b(X)$, pullback and extraordinary pullback. We also have a tensor product $\overset{L}{\otimes}$, and a (Verdier) duality functor $\mathbb{D}_X\colon \mathcal{D}_c^b(X) \to \mathcal{D}_c^b(X)^{\mathrm{op}}$. These satisfy all the rules of the six operation formalism; in particular, Verdier duality intertwines between pushforward and compactly supported pushforward.

For any scheme X with structure morphism $a\colon X \to \mathrm{pt}$, and the constant vector space $\mathbb{Q} \in \mathcal{D}_c^b(\mathrm{pt})$, let $\mathbb{Q}_X = a^*\mathbb{Q}$. If X is smooth, then we have an isomorphism $\mathbb{D}_X\mathbb{Q}_X \cong \mathbb{Q}_X[2\dim X]$. On the other hand, the cohomologies of $a_*\mathbb{Q}_X$, respectively $a_!\mathbb{Q}_X$ are isomorphic to the cohomology $H^*(X)$ and compactly supported cohomology $H_c^*(X)$ of X respectively. Thus, for example, we get the classical fact that for smooth X, cohomology and compactly supported cohomology are naturally dual.

The derived category $\mathcal{D}_c^b(X)$ has an interesting and nontrivial heart $\mathrm{Perv}(X)$, called the *category of perverse sheaves on* X even though its objects are not sheaves but certain complexes defined by support conditions. This category is preserved by Verdier duality, pushforward under closed inclusions, (shifted) pullback by smooth maps, and tensor product. If X is smooth, then $\mathbb{Q}_X[\dim X] \in \mathrm{Perv}(X)$. Also $\mathrm{Perv}(X)$ is artinian and noetherian, with every object having a finite filtration with quotients being *intersection cohomology sheaves* $\mathrm{IC}_Y(\mathcal{L})$ where $Y \subset X$ is a smooth, locally closed, irreducible subvariety, and $\mathcal{L}$ is a local system on Y.

Despite not being sheaves but complexes, perverse sheaves on complex schemes form a *stack*: locally defined perverse sheaves can be glued to a global object, given compatible isomorphisms.

A.2. The sheaf of vanishing cycles. Given a regular function $f\colon U \to \mathbb{C}$ with zero-locus $U_0 = f^{-1}(0)$, there is a functor, the *vanishing cycle functor* $\phi_f\colon \mathcal{D}_c^b(U) \to \mathcal{D}_c^b(U_0)$, mapping $\mathrm{Perv}(U)$ to $\mathrm{Perv}(U_0)$. For any $\mathcal{F} \in \mathcal{D}_c^b(U)$ the image $\phi_f\mathcal{F}$ carries an extra piece of structure, the monodromy endomorphism T. In the case of the constant perverse sheaf $\mathbb{Q}_U[\dim U]$ on a smooth variety U, the only case needed here, the vanishing cycle perverse sheaf can be viewed as an object in $\mathrm{Perv}(X)$ where $X = \mathrm{Crit}(f)$ is the critical locus of f. Since Verdier duality commutes with taking vanishing cycles, the vanishing cycle sheaf of the constant sheaf is equipped with an isomorphism

$$\phi_f\mathbb{Q}_U[\dim U] \cong \mathbb{D}_X\phi_f\mathbb{Q}_U[\dim U]$$

with its Verdier dual.

A.3. Hodge modules. Given a smooth complex variety U, there is an enriched category, the category of *(polarized) mixed Hodge modules* [**91**, **92**] $\mathrm{MHM}(U)$ with a faithful functor $\mathrm{rat}\colon \mathrm{MHM}(U) \to \mathrm{Perv}(U)$. With a very substantial amount of work, the six-functor formalism extends to this category. For singular schemes X, the category of mixed Hodge modules $\mathrm{MHM}(X)$ needs to be defined via embedding X into a smooth variety.

For $U = \mathrm{pt}$, the category of mixed Hodge modules on U is the category MHS of mixed Hodge structures. Thus the advantage of this generalization is that given $\mathcal{F} \in$

MHM(X), its cohomology and cohomology with compact support, which coincide with its pushforward (with compact support) to the point, acquire mixed Hodge structures.

A further complication is that while for $f\colon U \to \mathbb{C}$, the vanishing cycle functor ϕ_f extends to mixed Hodge modules, the image objects under this functor acquire the action of the monodromy endomorphism. Thus the image lies in a new category of *monodromic* mixed Hodge modules MMHM(U_0). One can either treat this category directly as a category of objects with endomorphism [**93**], or one can treat it geometrically using an auxiliary disc [**67**]. Here I will informally adopt the former approach, so objects of MMHM(U) are simply pairs of objects $(\mathcal{F}, T)$ of MHM(U_0) and an endomorphism T. However, one needs to be careful about the definition of various functors. For example, the monodromic Verdier duality functor $\mathbb{D}_X^M\colon \mathrm{MMHM}(U_0) \to \mathrm{MMHM}(U_0)^{\mathrm{op}}$ involves a certain shift. See [**14**, Sect. 2.10] for a slightly more detailed but still concise treatment.

The vanishing cycle cohomology of the pair $(U = \mathbb{C}, f = x^2)$ has tensor square isomorphic to the cohomology $\mathbb{Q}(-1)[-2]$ of the affine line. This object itself can therefore be considered as a half-twist object $\mathbb{Q}(-1/2)[-1] \in \mathrm{MMHS}$. One can formally extend the category MMHS with a tensor inverse $\mathbb{Q}(1/2)[1]$ to this object; we denote this new category by $\mathrm{MMHS}_{\frac{1}{2}}$. By pullback, we get extended categories of monodromic mixed Hodge modules $\mathrm{MMHM}(X)_{\frac{1}{2}}$ on general complex schemes X.

A.4. The K-group of mixed Hodge structures. The K-group $K(\mathrm{MMHS}_{\frac{1}{2}})$ is a ring, though one has to be careful with the definition of multiplication in the presence of monodromy. It contains a distinguished element $\mathbb{L} \in K(\mathrm{MMHS}_{\frac{1}{2}})$, the cohomology of the affine line $\mathbb{C}$. This element admits a square root $\mathbb{L}^{\frac{1}{2}}$, the vanishing cycle cohomology of the pair $(U = \mathbb{C}, f = x^2)$, which has a formal inverse $\mathbb{L}^{-\frac{1}{2}}$ in $K(\mathrm{MMHS}_{\frac{1}{2}})$.

There is a ring homomorphism $\chi\colon K(\mathrm{MMHS}_{\frac{1}{2}}) \to \mathbb{Z}$ given by computing the Euler characteristic of a complex of mixed Hodge structures. This maps $\mathbb{L}$ to 1 and, to be consistent with conventions used elsewhere, $\mathbb{L}^{\frac{1}{2}}$ to -1. There are other invariants however which survive. The Poincaré polynomial (with coefficients the dimension of cohomology in different degrees) is not one of them. However, the *weight polynomial* or Poincaré–Serre polynomial gives a homomorphism $\mathrm{wt}\colon K(\mathrm{MMHS}_{\frac{1}{2}}) \to \mathbb{Z}[q^{\pm\frac{1}{2}}]$. This is defined as follows: for an ordinary complex of mixed Hodge structures A^*, define

$$\mathrm{wt}(A^*;q) = \sum_{j\in\mathbb{Z}}\sum_{i\in\mathbb{Z}}(-1)^i \dim\left(Gr_W^j H^i(A^*)\right) \in \mathbb{Z}[q^{\pm\frac{1}{2}}],$$

with $Gr_W^j H^i(A^*)$ the weight-j part of $\mathrm{H}^i(A^*)$; let also $\mathrm{wt}(\mathbb{L}^{-\frac{1}{2}}) = q^{-\frac{1}{2}}$. For a monodromic complex, define

$$\mathrm{wt}(A^*;q) = \mathrm{wt}(A_1^*;q) + q\cdot \mathrm{wt}(A_{\neq 1}^*;q) \in \mathbb{Z}[q^{\pm\frac{1}{2}}];$$

here, A_1^* is the part of the complex where the monodromy acts by eigenvalue 1, and $A_{\neq 1}^*$ is a complement. There is also a Hodge version of this polynomial, taking into account the Hodge filtration also, which I will not spell out in detail. The Euler characteristic can of course be recovered from the weight polynomial by setting $q^{\frac{1}{2}} = -1$.

References

[1] P. S. Aspinwall and S. Katz, *Computation of superpotentials for D-branes*, Comm. Math. Phys. **264** (2006), no. 1, 227–253, DOI 10.1007/s00220-006-1527-6. MR2212222 (2007b:81178)

[2] O. Ben-Bassat, C. Brav, V. Bussi, and D. Joyce, *A 'Darboux theorem' for shifted symplectic structures on derived Artin stacks, with applications*, Geom. Topol. **19** (2015), no. 3, 1287–1359, DOI 10.2140/gt.2015.19.1287. MR3352237

[3] K. Behrend, *Donaldson-Thomas type invariants via microlocal geometry*, Ann. of Math. (2) **170** (2009), no. 3, 1307–1338, DOI 10.4007/annals.2009.170.1307. MR2600874 (2011d:14098)

[4] K. Behrend, J. Bryan, and B. Szendrői, *Motivic degree zero Donaldson-Thomas invariants*, Invent. Math. **192** (2013), no. 1, 111–160, DOI 10.1007/s00222-012-0408-1. MR3032328

[5] K. Behrend and B. Fantechi, *The intrinsic normal cone*, Invent. Math. **128** (1997), no. 1, 45–88, DOI 10.1007/s002220050136. MR1437495 (98e:14022)

[6] K. Behrend and B. Fantechi, *Symmetric obstruction theories and Hilbert schemes of points on threefolds*, Algebra Number Theory **2** (2008), no. 3, 313–345, DOI 10.2140/ant.2008.2.313. MR2407118 (2009c:14006)

[7] A. Berenstein and A. Zelevinsky, *Quantum cluster algebras*, Adv. Math. **195** (2005), no. 2, 405–455, DOI 10.1016/j.aim.2004.08.003. MR2146350 (2006a:20092)

[8] D. Berenstein and M. Douglas, Seiberg duality for quiver gauge theories, arXiv:hep-th/0207027.

[9] R. Bocklandt, *Graded Calabi Yau algebras of dimension 3*, J. Pure Appl. Algebra **212** (2008), no. 1, 14–32, DOI 10.1016/j.jpaa.2007.03.009. MR2355031 (2008h:16013)

[10] D. Borisov and D. Joyce, Virtual fundamental classes for moduli spaces of sheaves of Calabi–Yau four-folds, manuscript.

[11] J. Bouaziz and I. Grojnowski, A d-shifted Darboux theorem, arXiv:1309.2197.

[12] C. Brav, V. Bussi and D. Joyce, A 'Darboux theorem' for derived schemes with shifted symplectic structure, arXiv:1305.6302.

[13] T. Braden, *Hyperbolic localization of intersection cohomology*, Transform. Groups **8** (2003), no. 3, 209–216, DOI 10.1007/s00031-003-0606-4. MR1996415 (2004f:14037)

[14] C. Brav, V. Bussi, D. Dupont, D. Joyce, and B. Szendrői, *Symmetries and stabilization for sheaves of vanishing cycles*, J. Singul. **11** (2015), 85–151. With an appendix by Jörg Schürmann. MR3353002

[15] T. Bridgeland, *Stability conditions on triangulated categories*, Ann. of Math. (2) **166** (2007), no. 2, 317–345, DOI 10.4007/annals.2007.166.317. MR2373143 (2009c:14026)

[16] T. Bridgeland, A. King, and M. Reid, *The McKay correspondence as an equivalence of derived categories*, J. Amer. Math. Soc. **14** (2001), no. 3, 535–554 (electronic), DOI 10.1090/S0894-0347-01-00368-X. MR1824990 (2002f:14023)

[17] N. Broomhead, *Dimer models and Calabi-Yau algebras*, Mem. Amer. Math. Soc. **215** (2012), no. 1011, viii+86, DOI 10.1090/S0065-9266-2011-00617-9. MR2908565

[18] V. Bussi, *Derived symplectic structures in generalized Donaldson–Thomas theory and categorification*, ProQuest LLC, Ann Arbor, MI, 2014. Thesis (D.Phil.)–University of Oxford (United Kingdom). MR3389261

[19] A. Cazzaniga, A. Morrison, B. Pym and B. Szendrői, work in progress.

[20] J. Choi, S. Katz, and A. Klemm, *The refined BPS index from stable pair invariants*, Comm. Math. Phys. **328** (2014), no. 3, 903–954, DOI 10.1007/s00220-014-1978-0. MR3201216

[21] Y. Cao and N. C. Leung, Donaldson-Thomas theory for Calabi–Yau 4-folds, arXiv:1407.7659.

[22] A. Craw and A. Ishii, *Flops of G-Hilb and equivalences of derived categories by variation of GIT quotient*, Duke Math. J. **124** (2004), no. 2, 259–307, DOI 10.1215/S0012-7094-04-12422-4. MR2078369 (2005g:14030)

[23] B. Davison, *Consistency conditions for brane tilings*, J. Algebra **338** (2011), 1–23, DOI 10.1016/j.jalgebra.2011.05.005. MR2805177 (2012e:14110)

[24] B. Davison, The critical CoHA of a quiver with potential, arXiv:1311.7172v4.

[25] B. Davison, The integrality conjecture and the cohomology of preprojective stacks, arXiv:1602.02110.

[26] B. Davison, Positivity for quantum cluster algebras, arXiv:1601.07918.

[27] B Davison and S. Meinhardt, The motivic Donaldson-Thomas invariants of (−2)-curves, arXiv:1208.2462.

[28] B Davison and S. Meinhardt, Donaldson-Thomas theory for categories of homological dimension one with potential, arXiv:1512.08898.

[29] B. Davison, D. Maulik, J. Schuermann and B. Szendrői, Purity for graded potentials and cluster positivity, arXiv:1307.3379, to appear Comp. Math.

[30] H. Derksen, J. Weyman, and A. Zelevinsky, *Quivers with potentials and their representations. I. Mutations*, Selecta Math. (N.S.) **14** (2008), no. 1, 59–119, DOI 10.1007/s00029-008-0057-9. MR2480710 (2010b:16021)

[31] D.-E. Diaconescu, Z. Hua, and Y. Soibelman, *HOMFLY polynomials, stable pairs and motivic Donaldson-Thomas invariants*, Commun. Number Theory Phys. **6** (2012), no. 3, 517–600, DOI 10.4310/CNTP.2012.v6.n3.a1. MR3021320

[32] D.-E. Diaconescu, V. Shende, and C. Vafa, *Large N duality, Lagrangian cycles, and algebraic knots*, Comm. Math. Phys. **319** (2013), no. 3, 813–863, DOI 10.1007/s00220-012-1563-3. MR3040377

[33] A. Dimca, *Sheaves in topology*, Universitext, Springer-Verlag, Berlin, 2004. MR2050072 (2005j:55002)

[34] A. Dimca and B. Szendrői, *The Milnor fibre of the Pfaffian and the Hilbert scheme of four points on* $\mathbb{C}^3$, Math. Res. Lett. **16** (2009), no. 6, 1037–1055, DOI 10.4310/MRL.2009.v16.n6.a12. MR2576692 (2011a:32047)

[35] T. Dimofte and S. Gukov, *Refined, motivic, and quantum*, Lett. Math. Phys. **91** (2010), no. 1, 1–27, DOI 10.1007/s11005-009-0357-9. MR2577296 (2011m:14094)

[36] S. K. Donaldson and R. P. Thomas, *Gauge theory in higher dimensions*, The geometric universe (Oxford, 1996), Oxford Univ. Press, Oxford, 1998, pp. 31–47. MR1634503 (2000a:57085)

[37] M. R. Douglas, S. Govindarajan, T. Jayaraman, and A. Tomasiello, *D-branes on Calabi-Yau manifolds and superpotentials*, Comm. Math. Phys. **248** (2004), no. 1, 85–118, DOI 10.1007/s00220-004-1091-x. MR2104607 (2006c:81155)

[38] M. Douglas and G. Moore, D-branes, quivers and ALE Instantons, arXiv:hep-th/9603167.

[39] A. I. Efimov, *Cohomological Hall algebra of a symmetric quiver*, Compos. Math. **148** (2012), no. 4, 1133–1146, DOI 10.1112/S0010437X12000152. MR2956038

[40] A. Efimov, Quantum cluster variables via vanishing cycles, arXiv:1112.3601v2.

[41] A. Efimov, Cyclic homology of categories of matrix factorizations, arXiv:1212.2859.

[42] S. Fomin and A. Zelevinsky, *Cluster algebras. I. Foundations*, J. Amer. Math. Soc. **15** (2002), no. 2, 497–529 (electronic), DOI 10.1090/S0894-0347-01-00385-X. MR1887642 (2003f:16050)

[43] S. Franco, A. Hanany, D. Vegh, B. Wecht, and K. D. Kennaway, *Brane dimers and quiver gauge theories*, J. High Energy Phys. **1** (2006), 096, 48 pp. (electronic), DOI 10.1088/1126-6708/2006/01/096. MR2201227 (2007b:81191)

[44] V. Ginzburg, Calabi–Yau algebras, arXiv:math/0612139.

[45] R. Gopakumar and C. Vafa, M-Theory and Topological Strings I, II, arXiv:hep-th/9809187, arXiv:hep-th/9812127.

[46] M. Goresky, R. Kottwitz, and R. MacPherson, *Equivariant cohomology, Koszul duality, and the localization theorem*, Invent. Math. **131** (1998), no. 1, 25–83, DOI 10.1007/s002220050197. MR1489894 (99c:55009)

[47] T. Graber and R. Pandharipande, *Localization of virtual classes*, Invent. Math. **135** (1999), no. 2, 487–518, DOI 10.1007/s002220050293. MR1666787 (2000h:14005)

[48] I. Grojnowski, *Instantons and affine algebras. I. The Hilbert scheme and vertex operators*, Math. Res. Lett. **3** (1996), no. 2, 275–291, DOI 10.4310/MRL.1996.v3.n2.a12. MR1386846 (97f:14041)

[49] A. Hanany and D. Vegh, *Quivers, tilings, branes and rhombi*, J. High Energy Phys. **10** (2007), 029, 35, DOI 10.1088/1126-6708/2007/10/029. MR2357949 (2008m:81153)

[50] J. A. Harvey and G. Moore, *On the algebras of BPS states*, Comm. Math. Phys. **197** (1998), no. 3, 489–519, DOI 10.1007/s002200050461. MR1652775 (2001a:81229)

[51] T. Hollowood, A. Iqbal, and C. Vafa, *Matrix models, geometric engineering and elliptic genera*, J. High Energy Phys. **3** (2008), 069, 81, DOI 10.1088/1126-6708/2008/03/069. MR2391051 (2009m:81235)

[52] Z. Hua, Orientation data on moduli space of sheaves on Calabi–Yau threefold, arXiv:1212.3790.

[53] A. Iqbal and A.-K. Kashani-Poor, SU(N) *geometries and topological string amplitudes*, Adv. Theor. Math. Phys. **10** (2006), no. 1, 1–32. MR2222220 (2007d:81208)

[54] A. Iqbal, C. Kozçaz, and C. Vafa, *The refined topological vertex*, J. High Energy Phys. **10** (2009), 069, 58, DOI 10.1088/1126-6708/2009/10/069. MR2607441 (2011d:81266)

[55] D. Joyce, *Configurations in abelian categories. IV. Invariants and changing stability conditions*, Adv. Math. **217** (2008), no. 1, 125–204, DOI 10.1016/j.aim.2007.06.011. MR2357325 (2009d:18015b)

[56] D. Joyce, *A classical model for derived critical loci*, J. Differential Geom. **101** (2015), no. 2, 289–367. MR3399099

[57] D. Joyce and Y. Song, *A theory of generalized Donaldson-Thomas invariants*, Mem. Amer. Math. Soc. **217** (2012), no. 1020, iv+199, DOI 10.1090/S0065-9266-2011-00630-1. MR2951762

[58] S. Katz, *The desingularization of* $\mathrm{Hilb}^4\mathbf{P}^3$ *and its Betti numbers*, Zero-dimensional schemes (Ravello, 1992), de Gruyter, Berlin, 1994, pp. 231–242. MR1292488 (95f:14004)

[59] S. Katz, *Versal deformations and superpotentials for rational curves in smooth threefolds*, Symposium in Honor of C. H. Clemens (Salt Lake City, UT, 2000), Contemp. Math., vol. 312, Amer. Math. Soc., Providence, RI, 2002, pp. 129–136, DOI 10.1090/conm/312/04989. MR1941578 (2003i:14053)

[60] S. Katz, A. Klemm and R. Pandharipande, On the motivic stable pairs invariants of K3 surfaces, arXiv:1407.3181.

[61] B. Keller and D. Yang, *Derived equivalences from mutations of quivers with potential*, Adv. Math. **226** (2011), no. 3, 2118–2168, DOI 10.1016/j.aim.2010.09.019. MR2739775 (2012d:16043)

[62] Y.-H. Kiem and J. Li, Categorification of Donaldson-Thomas invariants via perverse sheaves, arXiv:1212.6444.

[63] A. D. King, *Moduli of representations of finite-dimensional algebras*, Quart. J. Math. Oxford Ser. (2) **45** (1994), no. 180, 515–530, DOI 10.1093/qmath/45.4.515. MR1315461 (96a:16009)

[64] F. Kirwan, *Intersection homology and torus actions*, J. Amer. Math. Soc. **1** (1988), no. 2, 385–400, DOI 10.2307/1990921. MR928263 (89c:14072)

[65] I. R. Klebanov and E. Witten, *Superconformal field theory on threebranes at a Calabi-Yau singularity*, Nuclear Phys. B **536** (1999), no. 1-2, 199–218, DOI 10.1016/S0550-3213(98)00654-3. MR1666725 (99k:81253)

[66] M. Kontsevich and Y. Soibelman, Stability structures, motivic Donaldson–Thomas invariants and cluster transformations, 2008, arXiv:0811.2435.

[67] M. Kontsevich and Y. Soibelman, *Cohomological Hall algebra, exponential Hodge structures and motivic Donaldson-Thomas invariants*, Commun. Number Theory Phys. **5** (2011), no. 2, 231–352, DOI 10.4310/CNTP.2011.v5.n2.a1. MR2851153 (2012k:14079)

[68] H. B. Laufer, *On* $\mathbf{C}P^1$ *as an exceptional set*, Recent developments in several complex variables (Proc. Conf., Princeton Univ., Princeton, N. J., 1979), Ann. of Math. Stud., vol. 100, Princeton Univ. Press, Princeton, N.J., 1981, pp. 261–275. MR627762 (82m:32012)

[69] K. Lee and R. Schiffler, *Positivity for cluster algebras*, Ann. of Math. (2) **182** (2015), no. 1, 73–125, DOI 10.4007/annals.2015.182.1.2. MR3374957

[70] K. H. Lin and D. Pomerleano, *Global matrix factorizations*, Math. Res. Lett. **20** (2013), no. 1, 91–106, DOI 10.4310/MRL.2013.v20.n1.a9. MR3126725

[71] D. Maulik, Stable pairs and the HOMFLY polynomial, arXiv:1210.6323.

[72] D. Maulik, Motivic residues and Donaldson–Thomas theory, manuscript.

[73] D. Maulik, N. Nekrasov, A. Okounkov, and R. Pandharipande, *Gromov-Witten theory and Donaldson-Thomas theory. I*, Compos. Math. **142** (2006), no. 5, 1263–1285, DOI 10.1112/S0010437X06002302. MR2264664 (2007i:14061)

[74] A. Morrison, S. Mozgovoy, K. Nagao, and B. Szendrői, *Motivic Donaldson-Thomas invariants of the conifold and the refined topological vertex*, Adv. Math. **230** (2012), no. 4-6, 2065–2093, DOI 10.1016/j.aim.2012.03.030. MR2927365

[75] A. Morrison and K. Nagao, *Motivic Donaldson-Thomas invariants of small crepant resolutions*, Algebra Number Theory **9** (2015), no. 4, 767–813, DOI 10.2140/ant.2015.9.767. MR3352820

[76] S. Mozgovoy and M. Reineke, *On the noncommutative Donaldson-Thomas invariants arising from brane tilings*, Adv. Math. **223** (2010), no. 5, 1521–1544, DOI 10.1016/j.aim.2009.10.001. MR2592501 (2011e:16026)

[77] D. Mumford and K. Suominen, *Introduction to the theory of moduli*, Algebraic geometry, Oslo 1970 (Proc. Fifth Nordic Summer-School in Math.), Wolters-Noordhoff, Groningen, 1972, pp. 171–222. MR0437531 (55 #10455)

[78] K. Nagao, *Donaldson-Thomas theory and cluster algebras*, Duke Math. J. **162** (2013), no. 7, 1313–1367, DOI 10.1215/00127094-2142753. MR3079250

[79] K. Nagao and H. Nakajima, *Counting invariant of perverse coherent sheaves and its wall-crossing*, Int. Math. Res. Not. IMRN **17** (2011), 3885–3938, DOI 10.1093/imrn/rnq195. MR2836398 (2012h:14141)

[80] H. Nakajima, *Heisenberg algebra and Hilbert schemes of points on projective surfaces*, Ann. of Math. (2) **145** (1997), no. 2, 379–388, DOI 10.2307/2951818. MR1441880 (98h:14006)

[81] H. Nakajima, *Lectures on Hilbert schemes of points on surfaces*, University Lecture Series, vol. 18, American Mathematical Society, Providence, RI, 1999. MR1711344 (2001b:14007)

[82] I. Nakamura, *Hilbert schemes of abelian group orbits*, J. Algebraic Geom. **10** (2001), no. 4, 757–779. MR1838978 (2002d:14006)

[83] N. Nekrasov and A. Okounkov, Membranes and sheaves, arXiv:1404.2323.

[84] D. O. Orlov, *Triangulated categories of singularities and D-branes in Landau-Ginzburg models* (Russian, with Russian summary), Tr. Mat. Inst. Steklova **246** (2004), no. Algebr. Geom. Metody, Svyazi i Prilozh., 240–262; English transl., Proc. Steklov Inst. Math. **3 (246)** (2004), 227–248. MR2101296 (2006i:81173)

[85] A. Oblomkov and V. Shende, *The Hilbert scheme of a plane curve singularity and the HOMFLY polynomial of its link*, Duke Math. J. **161** (2012), no. 7, 1277–1303, DOI 10.1215/00127094-1593281. MR2922375

[86] A. Oblomkov, J. Rasmussen and V. Shende, The Hilbert scheme of a plane curve singularity and the HOMFLY homology of its link, arXiv:1201.2115.

[87] R. Pandharipande and R. P. Thomas, *Curve counting via stable pairs in the derived category*, Invent. Math. **178** (2009), no. 2, 407–447, DOI 10.1007/s00222-009-0203-9. MR2545686 (2010h:14089)

[88] R. Pandharipande and R. P. Thomas, 13/2 ways of counting curves, in: Moduli spaces (eds. L. Brambila-Paz, P. Newstead, O. Garcia-Prada, R. P. Thomas), CUP, 2014.

[89] T. Pantev, B. Toën, M. Vaquié, and G. Vezzosi, *Shifted symplectic structures*, Publ. Math. Inst. Hautes Études Sci. **117** (2013), 271–328, DOI 10.1007/s10240-013-0054-1. MR3090262

[90] C. A. M. Peters and J. H. M. Steenbrink, *Mixed Hodge structures*, Ergebnisse der Mathematik und ihrer Grenzgebiete. 3. Folge. A Series of Modern Surveys in Mathematics [Results in Mathematics and Related Areas. 3rd Series. A Series of Modern Surveys in Mathematics], vol. 52, Springer-Verlag, Berlin, 2008. MR2393625 (2009c:14018)

[91] M. Saito, *Modules de Hodge polarisables* (French), Publ. Res. Inst. Math. Sci. **24** (1988), no. 6, 849–995 (1989), DOI 10.2977/prims/1195173930. MR1000123 (90k:32038)

[92] M. Saito, *Mixed Hodge modules*, Publ. Res. Inst. Math. Sci. **26** (1990), no. 2, 221–333, DOI 10.2977/prims/1195171082. MR1047415 (91m:14014)

[93] M. Saito, Thom–Sebastiani theorem for mixed Hodge modules, preprint 1990/2011.

[94] J. H. M. Steenbrink, *Mixed Hodge structure on the vanishing cohomology*, Real and complex singularities (Proc. Ninth Nordic Summer School/NAVF Sympos. Math., Oslo, 1976), Sijthoff and Noordhoff, Alphen aan den Rijn, 1977, pp. 525–563. MR0485870 (58 #5670)

[95] J. Steenbrink, *Intersection form for quasi-homogeneous singularities*, Compositio Math. **34** (1977), no. 2, 211–223. MR0453735 (56 #11995)

[96] Y. Soibelman, Remarks on cohomological Hall algebras and their representations, arXiv:1404.1606.

[97] B. Szendrői, *Non-commutative Donaldson-Thomas invariants and the conifold*, Geom. Topol. **12** (2008), no. 2, 1171–1202, DOI 10.2140/gt.2008.12.1171. MR2403807 (2009e:14100)

[98] B. Szendrői, *Nekrasov's partition function and refined Donaldson-Thomas theory: the rank one case*, SIGMA Symmetry Integrability Geom. Methods Appl. **8** (2012), Paper 088, 16, DOI 10.3842/SIGMA.2012.088. MR3007271

[99] R. P. Thomas, *A holomorphic Casson invariant for Calabi-Yau 3-folds, and bundles on $K3$ fibrations*, J. Differential Geom. **54** (2000), no. 2, 367–438. MR1818182 (2002b:14049)

[100] M. van den Bergh, *Non-commutative crepant resolutions*, The legacy of Niels Henrik Abel, Springer, Berlin, 2004, pp. 749–770. MR2077594 (2005e:14002)

[101] E. Witten, *Chern-Simons gauge theory as a string theory*, The Floer memorial volume, Progr. Math., vol. 133, Birkhäuser, Basel, 1995, pp. 637–678. MR1362846 (97j:57052)

[102] X. Xiao, The double of representations of Cohomological Hall Algebra for A_1-quiver, arXiv:1407.7593.

MATHEMATICAL INSTITUTE, UNIVERSITY OF OXFORD

E-mail address: `szendroi@maths.ox.ac.uk`

Selected Published Titles in This Series

93 **Vincent Bouchard, Charles Doran, Stefan Méndez-Diez, and Callum Quigley, Editors,** String-Math 2014, 2016

90 **Ron Donagi, Sheldon Katz, Albrecht Klemm, and David R. Morrison, Editors,** String-Math 2012, 2015

89 **D. Dolgopyat, Y. Pesin, M. Pollicott, and L. Stoyanov, Editors,** Hyperbolic Dynamics, Fluctuations and Large Deviations, 2015

88 **Ron Donagi, Michael R. Douglas, Ljudmila Kamenova, and Martin Rocek, Editors,** String-Math 2013, 2014

87 **Helge Holden, Barry Simon, and Gerald Teschl, Editors,** Spectral Analysis, Differential Equations and Mathematical Physics: A Festschrift in Honor of Fritz Gesztesy's 60th Birthday, 2013

86 **Kailash C. Misra, Daniel K. Nakano, and Brian J. Parshall, Editors,** Recent Developments in Lie Algebras, Groups and Representation Theory, 2012

85 **Jonathan Block, Jacques Distler, Ron Donagi, and Eric Sharpe, Editors,** String-Math 2011, 2012

84 **Alex H. Barnett, Carolyn S. Gordon, Peter A. Perry, and Alejandro Uribe, Editors,** Spectral Geometry, 2012

83 **Hisham Sati and Urs Schreiber, Editors,** Mathematical Foundations of Quantum Field Theory and Perturbative String Theory, 2011

82 **Michael Usher, Editor,** Low-dimensional and Symplectic Topology, 2011

81 **Robert S. Doran, Greg Friedman, and Jonathan Rosenberg, Editors,** Superstrings, Geometry, Topology, and C^*-algebras, 2010

80 **D. Abramovich, A. Bertram, L. Katzarkov, R. Pandharipande, and M. Thaddeus, Editors,** Algebraic Geometry, 2009

79 **Dorina Mitrea and Marius Mitrea, Editors,** Perspectives in Partial Differential Equations, Harmonic Analysis and Applications, 2008

78 **Ron Y. Donagi and Katrin Wendland, Editors,** From Hodge Theory to Integrability and TQFT, 2008

77 **Pavel Exner, Jonathan P. Keating, Peter Kuchment, Toshikazu Sunada, and Alexander Teplyaev, Editors,** Analysis on Graphs and Its Applications, 2008

76 **Fritz Gesztesy, Percy Deift, Cherie Galvez, Peter Perry, and Wilhelm Schlag, Editors,** Spectral Theory and Mathematical Physics: A Festschrift in Honor of Barry Simon's 60th Birthday, 2007

75 **Solomon Friedberg, Daniel Bump, Dorian Goldfeld, and Jeffrey Hoffstein, Editors,** Multiple Dirichlet Series, Automorphic Forms, and Analytic Number Theory, 2006

74 **Benson Farb, Editor,** Problems on Mapping Class Groups and Related Topics, 2006

73 **Mikhail Lyubich and Leon Takhtajan, Editors,** Graphs and Patterns in Mathematics and Theoretical Physics, 2005

72 **Michel L. Lapidus and Machiel van Frankenhuijsen, Editors,** Fractal Geometry and Applications: A Jubilee of Benoît Mandelbrot, 2004

71 **Gordana Matić and Clint McCrory, Editors,** Topology and Geometry of Manifolds, 2003

70 **Michael D. Fried and Yasutaka Ihara, Editors,** Arithmetic Fundamental Groups and Noncommutative Algebra, 2002

69 **Anatole Katok, Rafael de la Llave, Yakov Pesin, and Howard Weiss, Editors,** Smooth Ergodic Theory and Its Applications, 2001

68 **Robert S. Doran and V. S. Varadarajan, Editors,** The Mathematical Legacy of Harish-Chandra, 2000

ISBN: 978-1-4704-1992-9

9 781470 419929

PSPUM/93